Hopf Algebras
in Noncommutative
Geometry and
Physics

PURE AND APPLIED MATHEMATICS

A Program of Monographs, Textbooks, and Lecture Notes

EXECUTIVE EDITORS

Earl J. Taft
Rutgers University
New Brunswick, New Jersey

Zuhair Nashed
University of Central Florida
Orlando, Florida

EDITORIAL BOARD

M. S. Baouendi
University of California,
San Diego

Jane Cronin
Rutgers University

Jack K. Hale
Georgia Institute
of Technology

S. Kobayashi
University of California,
Berkeley

Marvin Marcus
University of California,
Santa Barbara

W. S. Masse
Yale University

Anil Nerode
Cornell University

Donald Passman
University of Wisconsin,
Madison

Fred S. Roberts
Rutgers University

David L. Russell
Virginia Polytechnic Institute
and State University

Walter Schempp
Universität Siegen

Mark Teply
University of Wisconsin,
Milwaukee

MONOGRAPHS AND TEXTBOOKS IN
PURE AND APPLIED MATHEMATICS

1. *K. Yano*, Integral Formulas in Riemannian Geometry (1970)
2. *S. Kobayashi*, Hyperbolic Manifolds and Holomorphic Mappings (1970)
3. *V. S. Vladimirov*, Equations of Mathematical Physics (A. Jeffrey, ed.; A. Littlewood, trans.) (1970)
4. *B. N. Pshenichnyi*, Necessary Conditions for an Extremum (L. Neustadt, translation ed.; K. Makowski, trans.) (1971)
5. *L. Narici et al.*, Functional Analysis and Valuation Theory (1971)
6. *S. S. Passman*, Infinite Group Rings (1971)
7. *L. Dornhoff*, Group Representation Theory. Part A: Ordinary Representation Theory. Part B: Modular Representation Theory (1971, 1972)
8. *W. Boothby and G. L. Weiss, eds.*, Symmetric Spaces (1972)
9. *Y. Matsushima*, Differentiable Manifolds (E. T. Kobayashi, trans.) (1972)
10. *L. E. Ward, Jr.*, Topology (1972)
11. *A. Babakhanian*, Cohomological Methods in Group Theory (1972)
12. *R. Gilmer*, Multiplicative Ideal Theory (1972)
13. *J. Yeh*, Stochastic Processes and the Wiener Integral (1973)
14. *J. Barros-Neto*, Introduction to the Theory of Distributions (1973)
15. *R. Larsen*, Functional Analysis (1973)
16. *K. Yano and S. Ishihara*, Tangent and Cotangent Bundles (1973)
17. *C. Procesi*, Rings with Polynomial Identities (1973)
18. *R. Hermann*, Geometry, Physics, and Systems (1973)
19. *N. R. Wallach*, Harmonic Analysis on Homogeneous Spaces (1973)
20. *J. Dieudonné*, Introduction to the Theory of Formal Groups (1973)
21. *I. Vaisman*, Cohomology and Differential Forms (1973)
22. *B.-Y. Chen*, Geometry of Submanifolds (1973)
23. *M. Marcus*, Finite Dimensional Multilinear Algebra (in two parts) (1973, 1975)
24. *R. Larsen*, Banach Algebras (1973)
25. *R. O. Kujala and A. L. Vitter, eds.*, Value Distribution Theory: Part A; Part B: Deficit and Bezout Estimates by Wilhelm Stoll (1973)
26. *K. B. Stolarsky*, Algebraic Numbers and Diophantine Approximation (1974)
27. *A. R. Magid*, The Separable Galois Theory of Commutative Rings (1974)
28. *B. R. McDonald*, Finite Rings with Identity (1974)
29. *J. Satake*, Linear Algebra (S. Koh et al., trans.) (1975)
30. *J. S. Golan*, Localization of Noncommutative Rings (1975)
31. *G. Klambauer*, Mathematical Analysis (1975)
32. *M. K. Agoston*, Algebraic Topology (1976)
33. *K. R. Goodearl*, Ring Theory (1976)
34. *L. E. Mansfield*, Linear Algebra with Geometric Applications (1976)
35. *N. J. Pullman*, Matrix Theory and Its Applications (1976)
36. *B. R. McDonald*, Geometric Algebra Over Local Rings (1976)
37. *C. W. Groetsch*, Generalized Inverses of Linear Operators (1977)
38. *J. E. Kuczkowski and J. L. Gersting*, Abstract Algebra (1977)
39. *C. O. Christenson and W. L. Voxman*, Aspects of Topology (1977)
40. *M. Nagata*, Field Theory (1977)
41. *R. L. Long*, Algebraic Number Theory (1977)
42. *W. F. Pfeffer*, Integrals and Measures (1977)
43. *R. L. Wheeden and A. Zygmund*, Measure and Integral (1977)
44. *J. H. Curtiss*, Introduction to Functions of a Complex Variable (1978)

45. *K. Hrbacek and T. Jech*, Introduction to Set Theory (1978)
46. *W. S. Massey*, Homology and Cohomology Theory (1978)
47. *M. Marcus*, Introduction to Modern Algebra (1978)
48. *E. C. Young*, Vector and Tensor Analysis (1978)
49. *S. B. Nadler, Jr.*, Hyperspaces of Sets (1978)
50. *S. K. Segal*, Topics in Group Rings (1978)
51. *A. C. M. van Rooij*, Non-Archimedean Functional Analysis (1978)
52. *L. Corwin and R. Szczarba*, Calculus in Vector Spaces (1979)
53. *C. Sadosky*, Interpolation of Operators and Singular Integrals (1979)
54. *J. Cronin*, Differential Equations (1980)
55. *C. W. Groetsch*, Elements of Applicable Functional Analysis (1980)
56. *I. Vaisman*, Foundations of Three-Dimensional Euclidean Geometry (1980)
57. *H. I. Freedan*, Deterministic Mathematical Models in Population Ecology (1980)
58. *S. B. Chae*, Lebesgue Integration (1980)
59. *C. S. Rees et al.*, Theory and Applications of Fourier Analysis (1981)
60. *L. Nachbin*, Introduction to Functional Analysis (R. M. Aron, trans.) (1981)
61. *G. Orzech and M. Orzech*, Plane Algebraic Curves (1981)
62. *R. Johnsonbaugh and W. E. Pfaffenberger*, Foundations of Mathematical Analysis (1981)
63. *W. L. Voxman and R. H. Goetschel*, Advanced Calculus (1981)
64. *L. J. Corwin and R. H. Szczarba*, Multivariable Calculus (1982)
65. *V. I. Istratescu*, Introduction to Linear Operator Theory (1981)
66. *R. D. Järvinen*, Finite and Infinite Dimensional Linear Spaces (1981)
67. *J. K. Beem and P. E. Ehrlich*, Global Lorentzian Geometry (1981)
68. *D. L. Armacost*, The Structure of Locally Compact Abelian Groups (1981)
69. *J. W. Brewer and M. K. Smith, eds.*, Emmy Noether: A Tribute (1981)
70. *K. H. Kim*, Boolean Matrix Theory and Applications (1982)
71. *T. W. Wieting*, The Mathematical Theory of Chromatic Plane Ornaments (1982)
72. *D. B.Gauld*, Differential Topology (1982)
73. *R. L. Faber*, Foundations of Euclidean and Non-Euclidean Geometry (1983)
74. *M. Carmeli*, Statistical Theory and Random Matrices (1983)
75. *J. H. Carruth et al.*, The Theory of Topological Semigroups (1983)
76. *R. L. Faber*, Differential Geometry and Relativity Theory (1983)
77. *S. Barnett*, Polynomials and Linear Control Systems (1983)
78. *G. Karpilovsky*, Commutative Group Algebras (1983)
79. *F. Van Oystaeyen and A. Verschoren*, Relative Invariants of Rings (1983)
80. *I. Vaisman*, A First Course in Differential Geometry (1984)
81. *G. W. Swan*, Applications of Optimal Control Theory in Biomedicine (1984)
82. *T. Petrie and J. D. Randall*, Transformation Groups on Manifolds (1984)
83. *K. Goebel and S. Reich*, Uniform Convexity, Hyperbolic Geometry, and Nonexpansive Mappings (1984)
84. *T. Albu and C. Nastasescu*, Relative Finiteness in Module Theory (1984)
85. *K. Hrbacek and T. Jech*, Introduction to Set Theory: Second Edition (1984)
86. *F. Van Oystaeyen and A. Verschoren*, Relative Invariants of Rings (1984)
87. *B. R. McDonald*, Linear Algebra Over Commutative Rings (1984)
88. *M. Namba*, Geometry of Projective Algebraic Curves (1984)
89. *G. F. Webb*, Theory of Nonlinear Age-Dependent Population Dynamics (1985)
90. *M. R. Bremner et al.*, Tables of Dominant Weight Multiplicities for Representations of Simple Lie Algebras (1985)
91. *A. E. Fekete*, Real Linear Algebra (1985)
92. *S. B. Chae*, Holomorphy and Calculus in Normed Spaces (1985)
93. *A. J. Jerri*, Introduction to Integral Equations with Applications (1985)

94. *G. Karpilovsky*, Projective Representations of Finite Groups (1985)
95. *L. Narici and E. Beckenstein*, Topological Vector Spaces (1985)
96. *J. Weeks*, The Shape of Space (1985)
97. *P. R. Gribik and K. O. Kortanek*, Extremal Methods of Operations Research (1985)
98. *J. A. Chao and W. A. Woyczynski, eds.*, Probability Theory and Harmonic Analysis (1986)
99. *G. D. Crown et al.*, Abstract Algebra (1986)
100. *J. H. Carruth et al.*, The Theory of Topological Semigroups, Volume 2 (1986)
101. *R. S. Doran and V. A. Belfi*, Characterizations of C*-Algebras (1986)
102. *M. W. Jeter*, Mathematical Programming (1986)
103. *M. Altman*, A Unified Theory of Nonlinear Operator and Evolution Equations with Applications (1986)
104. *A. Verschoren*, Relative Invariants of Sheaves (1987)
105. *R. A. Usmani*, Applied Linear Algebra (1987)
106. *P. Blass and J. Lang*, Zariski Surfaces and Differential Equations in Characteristic $p > 0$ (1987)
107. *J. A. Reneke et al.*, Structured Hereditary Systems (1987)
108. *H. Busemann and B. B. Phadke*, Spaces with Distinguished Geodesics (1987)
109. *R. Harte*, Invertibility and Singularity for Bounded Linear Operators (1988)
110. *G. S. Ladde et al.*, Oscillation Theory of Differential Equations with Deviating Arguments (1987)
111. *L. Dudkin et al.*, Iterative Aggregation Theory (1987)
112. *T. Okubo*, Differential Geometry (1987)
113. *D. L. Stancl and M. L. Stancl*, Real Analysis with Point-Set Topology (1987)
114. *T. C. Gard*, Introduction to Stochastic Differential Equations (1988)
115. *S. S. Abhyankar*, Enumerative Combinatorics of Young Tableaux (1988)
116. *H. Strade and R. Farnsteiner*, Modular Lie Algebras and Their Representations (1988)
117. *J. A. Huckaba*, Commutative Rings with Zero Divisors (1988)
118. *W. D. Wallis*, Combinatorial Designs (1988)
119. *W. Wieslaw*, Topological Fields (1988)
120. *G. Karpilovsky*, Field Theory (1988)
121. *S. Caenepeel and F. Van Oystaeyen*, Brauer Groups and the Cohomology of Graded Rings (1989)
122. *W. Kozlowski*, Modular Function Spaces (1988)
123. *E. Lowen-Colebunders*, Function Classes of Cauchy Continuous Maps (1989)
124. *M. Pavel*, Fundamentals of Pattern Recognition (1989)
125. *V. Lakshmikantham et al.*, Stability Analysis of Nonlinear Systems (1989)
126. *R. Sivaramakrishnan*, The Classical Theory of Arithmetic Functions (1989)
127. *N. A. Watson*, Parabolic Equations on an Infinite Strip (1989)
128. *K. J. Hastings*, Introduction to the Mathematics of Operations Research (1989)
129. *B. Fine*, Algebraic Theory of the Bianchi Groups (1989)
130. *D. N. Dikranjan et al.*, Topological Groups (1989)
131. *J. C. Morgan II*, Point Set Theory (1990)
132. *P. Biler and A. Witkowski*, Problems in Mathematical Analysis (1990)
133. *H. J. Sussmann*, Nonlinear Controllability and Optimal Control (1990)
134. *J.-P. Florens et al.*, Elements of Bayesian Statistics (1990)
135. *N. Shell*, Topological Fields and Near Valuations (1990)
136. *B. F. Doolin and C. F. Martin*, Introduction to Differential Geometry for Engineers (1990)
137. *S. S. Holland, Jr.*, Applied Analysis by the Hilbert Space Method (1990)
138. *J. Okniński*, Semigroup Algebras (1990)
139. *K. Zhu*, Operator Theory in Function Spaces (1990)
140. *G. B. Price*, An Introduction to Multicomplex Spaces and Functions (1991)
141. *R. B. Darst*, Introduction to Linear Programming (1991)

142. *P. L. Sachdev*, Nonlinear Ordinary Differential Equations and Their Applications (1991)
143. *T. Husain*, Orthogonal Schauder Bases (1991)
144. *J. Foran*, Fundamentals of Real Analysis (1991)
145. *W. C. Brown*, Matrices and Vector Spaces (1991)
146. *M. M. Rao and Z. D. Ren*, Theory of Orlicz Spaces (1991)
147. *J. S. Golan and T. Head*, Modules and the Structures of Rings (1991)
148. *C. Small*, Arithmetic of Finite Fields (1991)
149. *K. Yang*, Complex Algebraic Geometry (1991)
150. *D. G. 'Hoffman et al.*, Coding Theory (1991)
151. *M. O. González*, Classical Complex Analysis (1992)
152. *M. O. González*, Complex Analysis (1992)
153. *L. W. Baggett*, Functional Analysis (1992)
154. *M. Sniedovich*, Dynamic Programming (1992)
155. *R. P. Agarwal*, Difference Equations and Inequalities (1992)
156. *C. Brezinski*, Biorthogonality and Its Applications to Numerical Analysis (1992)
157. *C. Swartz*, An Introduction to Functional Analysis (1992)
158. *S. B. Nadler, Jr.*, Continuum Theory (1992)
159. *M. A. Al-Gwaiz*, Theory of Distributions (1992)
160. *E. Perry*, Geometry: Axiomatic Developments with Problem Solving (1992)
161. *E. Castillo and M. R. Ruiz-Cobo*, Functional Equations and Modelling in Science and Engineering (1992)
162. *A. J. Jerri*, Integral and Discrete Transforms with Applications and Error Analysis (1992)
163. *A. Charlier et al.*, Tensors and the Clifford Algebra (1992)
164. *P. Biler and T. Nadzieja*, Problems and Examples in Differential Equations (1992)
165. *E. Hansen*, Global Optimization Using Interval Analysis (1992)
166. *S. Guerre-Delabrière*, Classical Sequences in Banach Spaces (1992)
167. *Y. C. Wong*, Introductory Theory of Topological Vector Spaces (1992)
168. *S. H. Kulkarni and B. V. Limaye*, Real Function Algebras (1992)
169. *W. C. Brown*, Matrices Over Commutative Rings (1993)
170. *J. Loustau and M. Dillon*, Linear Geometry with Computer Graphics (1993)
171. *W. V. Petryshyn*, Approximation-Solvability of Nonlinear Functional and Differential Equations (1993)
172. *E. C. Young*, Vector and Tensor Analysis: Second Edition (1993)
173. *T. A. Bick*, Elementary Boundary Value Problems (1993)
174. *M. Pavel*, Fundamentals of Pattern Recognition: Second Edition (1993)
175. *S. A. Albeverio et al.*, Noncommutative Distributions (1993)
176. *W. Fulks*, Complex Variables (1993)
177. *M. M. Rao*, Conditional Measures and Applications (1993)
178. *A. Janicki and A. Weron*, Simulation and Chaotic Behavior of a-Stable Stochastic Processes (1994)
179. *P. Neittaanmäki and D. Tiba*, Optimal Control of Nonlinear Parabolic Systems (1994)
180. *J. Cronin*, Differential Equations: Introduction and Qualitative Theory, Second Edition (1994)
181. *S. Heikkilä and V. Lakshmikantham*, Monotone Iterative Techniques for Discontinuous Nonlinear Differential Equations (1994)
182. *X. Mao*, Exponential Stability of Stochastic Differential Equations (1994)
183. *B. S. Thomson*, Symmetric Properties of Real Functions (1994)
184. *J. E. Rubio*, Optimization and Nonstandard Analysis (1994)
185. *J. L. Bueso et al.*, Compatibility, Stability, and Sheaves (1995)
186. *A. N. Michel and K. Wang*, Qualitative Theory of Dynamical Systems (1995)
187. *M. R. Darnel*, Theory of Lattice-Ordered Groups (1995)
188. *Z. Naniewicz and P. D. Panagiotopoulos*, Mathematical Theory of Hemivariational Inequalities and Applications (1995)

189. *L. J. Corwin and R. H. Szczarba*, Calculus in Vector Spaces: Second Edition (1995)
190. *L. H. Erbe et al.*, Oscillation Theory for Functional Differential Equations (1995)
191. *S. Agaian et al.*, Binary Polynomial Transforms and Nonlinear Digital Filters (1995)
192. *M. I. Gil'*, Norm Estimations for Operation-Valued Functions and Applications (1995)
193. *P. A. Grillet*, Semigroups: An Introduction to the Structure Theory (1995)
194. *S. Kichenassamy*, Nonlinear Wave Equations (1996)
195. *V. F. Krotov*, Global Methods in Optimal Control Theory (1996)
196. *K. I. Beidar et al.*, Rings with Generalized Identities (1996)
197. *V. I. Arnautov et al.*, Introduction to the Theory of Topological Rings and Modules (1996)
198. *G. Sierksma*, Linear and Integer Programming (1996)
199. *R. Lasser*, Introduction to Fourier Series (1996)
200. *V. Sima*, Algorithms for Linear-Quadratic Optimization (1996)
201. *D. Redmond*, Number Theory (1996)
202. *J. K. Beem et al.*, Global Lorentzian Geometry: Second Edition (1996)
203. *M. Fontana et al.*, Prüfer Domains (1997)
204. *H. Tanabe*, Functional Analytic Methods for Partial Differential Equations (1997)
205. *C. Q. Zhang*, Integer Flows and Cycle Covers of Graphs (1997)
206. *E. Spiegel and C. J. O'Donnell*, Incidence Algebras (1997)
207. *B. Jakubczyk and W. Respondek*, Geometry of Feedback and Optimal Control (1998)
208. *T. W. Haynes et al.*, Fundamentals of Domination in Graphs (1998)
209. *T. W. Haynes et al.*, eds., Domination in Graphs: Advanced Topics (1998)
210. *L. A. D'Alotto et al.*, A Unified Signal Algebra Approach to Two-Dimensional Parallel Digital Signal Processing (1998)
211. *F. Halter-Koch*, Ideal Systems (1998)
212. *N. K. Govil et al.*, eds., Approximation Theory (1998)
213. *R. Cross*, Multivalued Linear Operators (1998)
214. *A. A. Martynyuk*, Stability by Liapunov's Matrix Function Method with Applications (1998)
215. *A. Favini and A. Yagi*, Degenerate Differential Equations in Banach Spaces (1999)
216. *A. Illanes and S. Nadler*, Jr., Hyperspaces: Fundamentals and Recent Advances (1999)
217. *G. Kato and D. Struppa*, Fundamentals of Algebraic Microlocal Analysis (1999)
218. *G. X.-Z. Yuan*, KKM Theory and Applications in Nonlinear Analysis (1999)
219. *D. Motreanu and N. H. Pavel*, Tangency, Flow Invariance for Differential Equations, and Optimization Problems (1999)
220. *K. Hrbacek and T. Jech*, Introduction to Set Theory, Third Edition (1999)
221. *G. E. Kolosov*, Optimal Design of Control Systems (1999)
222. *N. L. Johnson*, Subplane Covered Nets (2000)
223. *B. Fine and G. Rosenberger*, Algebraic Generalizations of Discrete Groups (1999)
224. *M. Väth, Volterra* and Integral Equations of Vector Functions (2000)
225. *S. S. Miller and P. T. Mocanu*, Differential Subordinations (2000)
226. *R. Li et al.*, Generalized Difference Methods for Differential Equations: Numerical Analysis of Finite Volume Methods (2000)
227. *H. Li and F. Van Oystaeyen*, A Primer of Algebraic Geometry (2000)
228. *R. P. Agarwal*, Difference Equations and Inequalities: Theory, Methods, and Applications, Second Edition (2000)
229. *A. B. Kharazishvili*, Strange Functions in Real Analysis (2000)
230. *J. M. Appell et al.*, Partial Integral Operators and Integro-Differential Equations (2000)
231. *A. I. Prilepko et al.*, Methods for Solving Inverse Problems in Mathematical Physics (2000)
232. *F. Van Oystaeyen*, Algebraic Geometry for Associative Algebras (2000)
233. *D. L. Jagerman*, Difference Equations with Applications to Queues (2000)
234. *D. R. Hankerson et al.*, Coding Theory and Cryptography: The Essentials, Second Edition, Revised and Expanded (2000)
235. *S. Dascalescu et al.*, Hopf Algebras: An Introduction (2001)

236. *R. Hagen et al.*, C*-Algebras and Numerical Analysis (2001)
237. *Y. Talpaert*, Differential Geometry: With Applications to Mechanics and Physics (2001)
238. *R. H. Villarreal*, Monomial Algebras (2001)
239. *A. N. Michel et al.*, Qualitative Theory of Dynamical Systems: Second Edition (2001)
240. *A. A. Samarskii*, The Theory of Difference Schemes (2001)
241. *J. Knopfmacher and W.-B. Zhang*, Number Theory Arising from Finite Fields (2001)
242. *S. Leader*, The Kurzweil-Henstock Integral and Its Differentials (2001)
243. *M. Biliotti et al.*, Foundations of Translation Planes (2001)
244. *A. N. Kochubei*, Pseudo-Differential Equations and Stochastics over Non-Archimedean Fields (2001)
245. *G. Sierksma*, Linear and Integer Programming: Second Edition (2002)
246. *A. A. Martynyuk*, Qualitative Methods in Nonlinear Dynamics: Novel Approaches to Liapunov's Matrix Functions (2002)
247. B. G. Pachpatte, Inequalities for Finite Difference Equations (2002)
248. *A. N. Michel and D. Liu*, Qualitative Analysis and Synthesis of Recurrent Neural Networks (2002)
249. *J. R. Weeks*, The Shape of Space: Second Edition (2002)
250. *M. M. Rao and Z. D. Ren*, Applications of Orlicz Spaces (2002)
251. *V. Lakshmikantham and D. Trigiante*, Theory of Difference Equations: Numerical Methods and Applications, Second Edition (2002)
252. *T. Albu*, Cogalois Theory (2003)
253. *A. Bezdek*, Discrete Geometry (2003)
254. *M. J. Corless and A. E. Frazho*, Linear Systems and Control: An Operator Perspective (2003)
255. *I. Graham and G. Kohr*, Geometric Function Theory in One and Higher Dimensions (2003)
256. *G. V. Demidenko and S. V. Uspenskii*, Partial Differential Equations and Systems Not Solvable with Respect to the Highest-Order Derivative (2003)
257. *A. Kelarev*, Graph Algebras and Automata (2003)
258. *A. H. Siddiqi*, Applied Functional Analysis: Numerical Methods, Wavelet Methods, and Image Processing (2004)
259. *F. W. Steutel and K. van Harn*, Infinite Divisibility of Probability Distributions on the Real Line (2004)
260. *G. S. Ladde and M. Sambandham*, Stochastic versus Deterministic Systems of Differential Equations (2004)
261. *B. J. Gardner and R. Wiegandt*, Radical Theory of Rings (2004)
262. *J. Haluska*, The Mathematical Theory of Tone Systems (2004)
263. *C. Menini and F. Van Oystaeyen*, Abstract Algebra: A Comprehensive Treatment (2004)
264. *E. Hansen and G. W. Walster*, Global Optimization Using Interval Analysis: Second Edition, Revised and Expanded (2004)
265. *M. M. Rao*, Measure Theory and Integration: Second Edition, Revised and Expanded (2004)
266. *W. J. Wickless*, A First Graduate Course in Abstract Algebra (2004)
267. *R. P. Agarwal, M. Bohner, and W-T Li*, Nonoscillation and Oscillation Theory for Functional Differential Equations (2004)
268. *J. Galambos and I. Simonelli*, Products of Random Variables: Applications to Problems of Physics and to Arithmetical Functions (2004)

Additional Volumes in Preparation

Hopf Algebras in Noncommutative Geometry and Physics

edited by

Stefaan Caenepeel
Free University of Brussels
VUB, Belgium

Freddy Van Oystaeyen
University of Antwerp
UIA, Belgium

CRC Press
Taylor & Francis Group
Boca Raton London New York

CRC Press is an imprint of the
Taylor & Francis Group, an **informa** business

First published 2005 by Marcel Dekker

Published 2018 by CRC Press
Taylor & Francis Group
6000 Broken Sound Parkway NW, Suite 300
Boca Raton, FL 33487-2742

First issued in hardback 2018

© 2005 by Taylor & Francis Group, LLC
CRC Press is an imprint of Taylor & Francis Group, an Informa business

No claim to original U.S. Government works

ISBN 13: 978-1-138-45430-9 (hbk)
ISBN 13: 978-0-8247-5759-5 (pbk)

This book contains information obtained from authentic and highly regarded sources. Reasonable efforts have been made to publish reliable data and information, but the author and publisher cannot assume responsibility for the validity of all materials or the consequences of their use. The authors and publishers have attempted to trace the copyright holders of all material reproduced in this publication and apologize to copyright holders if permission to publish in this form has not been obtained. If any copyright material has not been acknowledged please write and let us know so we may rectify in any future reprint.

Except as permitted under U.S. Copyright Law, no part of this book may be reprinted, reproduced, transmitted, or utilized in any form by any electronic, mechanical, or other means, now known or hereafter invented, including photocopying, microfilming, and recording, or in any information storage or retrieval system, without written permission from the publishers.

For permission to photocopy or use material electronically from this work, please access www. copyright.com (http://www.copyright.com/) or contact the Copyright Clearance Center, Inc. (CCC), 222 Rosewood Drive, Danvers, MA 01923, 978-750-8400. CCC is a not-for-profit organization that provides licenses and registration for a variety of users. For organizations that have been granted a photocopy license by the CCC, a separate system of payment has been arranged.

Trademark Notice: Product or corporate names may be trademarks or registered trademarks, and are used only for identification and explanation without intent to infringe.

Visit the Taylor & Francis Web site at
http://www.taylorandfrancis.com

and the CRC Press Web site at
http://www.crcpress.com

Library of Congress Cataloging-in-Publication Data
A catalog record for this book is available from the Library of Congress.

Preface

This volume contains the proceedings of meeting "Hopf algebras and quantum groups", which was held from May 28 until June 1, 2002, at the *Koninklijke Vlaamse Academie van België voor Wetenschappen en Kunsten*, the Royal Academy located in the historic centre of Brussels. The conference focussed at new results on classical Hopf algebras, with emphasis on the classification theory of finite dimensional Hopf algebras, categorical aspects of Hopf algebras, connections with Mathematical Physics, and the recent developments in the theory of corings and of quasi-Hopf algebras.

The meeting was supported by: the project "Noncommutative Geometry" (NOG) of the European Science Foundation (ESF); the bilateral project "New computational, geometric and algebraic methods applied to quantum groups and differential operators" of the Flemish and Chinese governments; the bilateral project "Hopf algebras in Algebra, Topology, Geometry and Physics" of the Flemish and Romanian governments; the Flemish Science Foundation FWO-Vlaanderen; the Royal Flemish Academy of Belgium (KVAB); the Free University of Brussels (VUB); the TMR project "Algebraic Lie Representations". We wish to thank the ESF, the Flemish government, the FWO-Vlaanderen, the KVAB, VUB and TMR for enabling us to organize this meeting.

Our thanks also go to the participants, the speakers and the authors. J.-P. Tignol kindly gave us the permission to use his "Fesart style file", used previously in volume 208 of the Lecture Notes in Pure and Applied Mathematics Series.

All papers have been individually refereed, and we thank the anonymous referees.

Stefaan Caenepeel
Freddy Van Oystaeyen
December 2003

Contents

Preface
Conference Participants

Morita Contexts for Corings and Equivalences   1
J. Abuhlail

Hopf Order Module Algebra Orders   21
F. Aly and F. Van Oystaeyen

An alternative Notion of Hopf Algebroid   31
G. Böhm

Topological Hopf Algebras, Quantum Groups and Deformation Quantization   55
Ph. Bonneau and D. Sternheimer

On Coseparable and Biseparable Corings   71
T. Brzeziński, L. Kadison and R. Wisbauer

More Properties of Yetter-Drinfeld Modules over Quasi-Hopf Algebras   89
D. Bulacu, S.Caenepeel and F. Panaite

Rationality Properties for Morita Contexts associated to Corings   113
S. Caenepeel, J. Vercruysse and S.H. Wang

Morita Duality for Corings over Quasi-Frobenius Rings   137
L. El Kaoutit and J. Gómez-Torrecillas

Quantized Coinvariants at Transcendental q 155
K. R. Goodearl and T. H. Lenagan

Classification of Differentials on Quantum Doubles and Finite
Noncommutative Geometry   167
S. Majid

Noncommutative Differentials and Yang-Mills on Permutation Groups S_n 189
S. Majid

The Affineness Criterion for Doi-Koppinen Modules   215
C. Menini and G. Militaru

Algebra Properties invariant under Twisting   229
S. Montgomery

Quantum $SL(3, \mathbb{C})$'s: the missing case   245
C. Ohn

Cuntz Algebras and Dynamical Quantum Group $SU(2)$ 257
A. Paolucci

On Symbolic Computations in Braided Monoidal Categories   269
B. Pareigis

Quotients of Finite Quasi-Hopf Algebras   281
P. Schauenburg

Adjointable Monoidal Functors and Quantum Groupoids   291
K. Szlachányi

On Galois Corings   309
R. Wisbauer

Conference Participants

Abuhlail, Jawad (Birzeit); abuhlail@kfupm.edu.sa

Adriaenssens, Jan (Antwerp); jan.adriaenssens@ua.ac.be

Ardizzoni, Alessandro (Ferrara); ardiz@dm.unife.it

Backelin, Erik (Antwerp); backelin@uia.ua.ac.be

Balan, Adriana (Bucharest); steleanu@lycos.com

Beattie, Margaret (Mount Allison, Sackville); mbeattie@mta.ca

Böhm, Gabriella (Budapest); bgabr@rmki.kfki.hu

Bonneau, Philippe (Bourgogne); bonneau@u-Bourgogne.fr

Brown, Ken (Glasgow); kab@maths.gla.ac.uk

Brzezinski, Thomas (Wales at Swansea); T.Brzezinski@swansea.ac.uk

Bulacu, Daniel (Bucharest); dbulacu@al.math.unibuc.ro

Caenepeel, Stef (Brussels); scaenepe@vub.ac.be

Chin, William (De Paul, Chicago); wchin@condor.depaul.edu

Cibils, Clause (Montpellier); cibils@math.univ-montp2.fr

Cohen, Mia (Beer Sheva); mia@cs.bgu.ac.il

Cuadra, Juan (Antwerp and Almería; juan.cuadra@ua.ac.be, jcdiaz@ual.es

De Groot, Erwin (Brussels); edegroot@vub.ac.be

Didt, Daniel (München); didt@mathematik.uni-muenchen.de

Farinati, Marco (Buenos Aires); mfarinat@dm.uba.ar

Frønsdahl, Christian (UCLA); fronsdal@physics.ucla.edu

Garcia, Socorro (Tenerife); sgarcia@ull.es

Gomez Torrecillas, José (Granada); torrecil@ugr.es

Gomez, Xavier (Queen Mary, London); X.Gomez@qmul.ac.uk

Gonzalez, Ramon (Vigo); rgrodri@correo.uvigo.es

Guédénon, Thomas (Brussels); thomas.guedenon@vub.ac.be

Hajac, Pjotr (München); Piotr.Hajac@fuw.edu.pl

Heckenberger, Istvan (Leipzig): heckenbe@mathematik.uni-leipzig.de

Iovanov, Miodrag (Bucharest); myo30@lycos.com

Iyer, Uma (Bonn); uiyer@mpim-bonn.mpg.de

Janviere, Ndirahista (Antwerp); Ndirahisha.Janviere@ua.ac.be

Kadison, Lars (Göteborg) ; kadison@cisunix.unh.edu

Kolb, Stefan (Leipzig) ; kolb@itp.uni-leipzig.de

Lebruyn, Lieven (Antwerp); lieven.lebruyn@ua.ac.be

Legagneux, Jean-Louis ; Legagneux.Coudier@Wanadoo.fr

Lenagan, Tom (Edinburgh); tom@maths.ed.ac.uk

Majid, Shawn (Queen Mary, London); s.majid@qmul.ac.uk

Mendoza, Judit (Tenerife); jmendoza@ull.es

Menini, Claudia (Ferrara); men@dns.unife.it

Militaru, Gigel (Bucharest); gmilit@al.math.unibuc.ro

Montgomery, Susan (Southern California, Los Angeles); smontgom@mtha.usc.edu

Năstăsescu, Constantin (Bucharest); cnastase@al.math.unibuc.ro

Natale, Sonja (Còrdoba); natale@mate.uncor.edu

Ohn, Christian (Reims); christian.ohn@univ-reims.fr
Panaite, Florin (Bucharest); florin.panaite@imar.ro
Paolucci, Anna (Leeds) ; paolucci@amsta.leeds.ac.uk
Pareigis, Bodo (München); pareigis@lmu.de
Ruan, Zhong-Jin (Illinois at Urbana); ruan@math.uiuc.edu
Schauenburgh, Peter (München); schauen@rz.mathematik.uni-muenchen.de
Schneider, Hans-Jürgen (München); Hans-Juergen.Schneider@mathematik.uni-muenchen.de
Slingerland, Joost Amsterdam; slinger@science.uva.nl
Solotar, Andrea (Buenos Aires); asolotar@dm.uba.ar
Sommerhauser, Yorck (München); sommerh@rz.mathematik.uni-muenchen.de
Sternheimer, Daniel (Bourgogne); Daniel.Sternheimer@u-Bourgogne.fr
Szlachanyi, Kornel (Budapest); szlach@rmki.kfki.hu
Ufer, Stefan (München); ufer@rz.mathematik.uni-muenchen.de
Van Oystaeyen, Fred (Antwerp); fred.vanoystaeyen@ua.ac.be
Vercruysse, Joost (Brussels); joost.vercruysse@vub.ac.be
Vidunas, Raimundas (Antwerp); vidunas@math.kyushu-u.ac.jp
Wang, Dingguo (Brussels and Henan); diwang@vub.ac.be
Wisbauer, Robert (Düsseldorf); wisbauer@math.uni-duesseldorf.de
Wisniewski, Piotr (Torun); pikonrad@mat.uni.torun.pl
Zhang, Yinhuo (Antwerp and Fiji); Yinhuo.Zhang@vuw.ac.nz

Program

Tuesday, May 28

10.00-10.50	B. Pareigis (München) *On Symbolic Computations with Elements in Braided Monoidal Categories*
10.50-11.20	Coffee
11.20-11.55	L. Kadison (Göteborg) *Antipodes at Depth Two*
12.00-12.35	A. Solotar (Buenos Aires) *Hochschild homology oftrivial extensions*
12.35-14.20	Lunch
14.20-15.10	T. Brzeziński (Wales at Swansea) *Why corings?*
15.15-15.50	Y. Sommerhäuser (München) *Self-dual modules of semisimple Hopf algebras*
15.50-16.20	Coffee
16.20-16.55	M. Farinati (Buenos Aires) *Hochschild homology of Generalized Weyl Algebras*
17.00-17.35	S. Natale (Córdoba) *On the semi-solvability of semisimple Hopf algebras of small dimension*
17.45-19.00	Reception in the Marble Room

Wednesday, May 29

09.00-09.50	H.-J. Schneider (München) *On the classification of pointed Hopf algebras*
09.55-10.30	R. Wisbauer (Düsseldorf) *Galois corings*
10.30-11.00	Coffee
11.00-11.35	J. Gomez Torrecillas (Granada) *Semisimple corings*
11.40-12.15	S. Caenepeel (Brussels) *Cleft entwining structures and Morita contexts associated to a coring*
12.15-14.00	Lunch
14.00-14.35	P. Schauenburgh (München) *Some properties of finite quasi-Hopf algebras*
14.40-15.15	T. Lenagan (Edinburgh) *Coinvariants and co-orbits for quantum matrices*
15.20-15.55	P. Hajac (München) *Fredholm Index and locally trivial noncommutative Hopf fibration*
16.00-16.30	Coffee

Thursday, May 30

09.30-10.20	S. Montgomery (Univ. of Southern California) *Properties of algebras invariant under twisting*
10.20-10.50	Coffee
10.50-11.25	C. Menini (Ferrara) *Quantum groups of dimension 16 with the Chevalley property*
11.30-12.05	D. Bulacu (Bucharest) *Hopf modules for quasi-Hopf algebras*
12.05-14.00	Lunch break
14.00-14.50	K. Szlachányi (Budapest) *Adjointable lax monoidal functors and bialgebroids*
14.55-15.30	G. Böhm (Budapest) *Hopf algebroids with bijective antipode*
15.30-16.00	Coffee
16.00-16.35	J. Cuadra (Almería and Antwerp) *The Brauer group of the Dihedral group respect to a quasi-triangular structure*
16.40-17.15	Y. Zhang (Antwerp and Fiji) *The equivariant Brauer group of an infinite group*
17.20-17.55	D. Wang (Brussels and Qufu Univ.) *Twistings, crossed coproducts and Hopf Galois coextensions*
20.00	Congress Dinner

Friday, May 31

09.30-10.20	S. Majid (Queen Mary College, London) *Dirac operators and Riemannian geometry on Hopf algebras*
10.20-10.50	Coffee
10.50-11.25	W. Chin (De Paul Univ., Chicago) *Prime Spectra of Quantized Hyperalgebras*
11.30-12.05	C. Cibils (Montpellier) *Hochschild cohomology of Hopf algebras*
12.05-14.00	Lunch break
14.00-14.35	M. Beattie (Mount Allison University) *Hopf algebras of dimension 14*
14.40-15.15	G. Militaru (Bucharest)
15.15-15.45	Coffee

15.45-16.20	F. Panaite (Bucharest)
	Hopf bimodules as modules over a diagonal crossed product algebra
16.25-17.00	X. Gomez (Queen Mary College, London)
	Relating Woronowicz's quantum Lie algebras and Majid's braided Lie algebras
17.05-17.40	P. Wisniewski (Torun)
	The Lasker-Noether theorem for commutative and noetherian module algebras over a pointed Hopf algebra

Saturday, June 1

09.30-10.20	D. Sternheimer and Ph. Bonneau (Univ. de Bourgogne)
	Topological Hopf algebras, quantum groups and deformation quantization
10.20-10.50	Coffee
10.50-11.25	C. Frønsdahl (UCLA)
11.30-12.05	A. Paolucci (Leeds)
	Cuntz Algebras and Dynamical Quantum Group $SL(2)$
12.10-12.40	C. Ohn (Reims)
	Quantum $SL(3)$'s: the missing case

Morita Contexts for Corings and Equivalences

JAWAD ABUHLAIL[*] Mathematics Department, Birzeit University
PO Box 14, Birzeit, Palestine
e-mail: *jabuhlail@birzeit.edu, abuhlail@kfupm.edu.sa*

ABSTRACT. In this note we study Morita contexts and Galois extensions for corings. For a coring $\mathcal{C}$ over a (not necessarily commutative) ground ring A we give equivalent conditions for $\mathcal{M}^{\mathcal{C}}$ to satisfy the weak. resp. the strong structure theorem. We also characterize the so called *cleft $\mathcal{C}$-Galois extensions* over commutative rings. Our approach is similar to that of Y. Doi and A. Masuoka in their work on (cleft) H-Galois extensions (see for example [10, 11]).

1. INTRODUCTION

Let $\mathcal{C}$ be a coring over a not necessarily commutative ring A and assume A to be a right $\mathcal{C}$-comodule through $\varrho_A : A \longrightarrow A \otimes_A \mathcal{C} \simeq \mathcal{C}$, $a \mapsto \mathbf{x}a$ for some grouplike element $\mathbf{x} \in \mathcal{C}$ (see [4, Lemma 5.1]). In Section 2, we study from the viewpoint of Morita theory the relationship between A and its subring of coinvariants $B := A^{co\mathcal{C}} = \{b \in A \mid \varrho(b) = b\mathbf{x}\}$. We consider the A-ring $^*\mathcal{C} := \mathrm{Hom}_{A-}(\mathcal{C}, A)$ and its left ideal $Q := \{q \in {}^*\mathcal{C} \mid \sum c_1 q(c_2) = q(c)\mathbf{x} \text{ for all } c \in \mathcal{C}\}$ and show that B and $^*\mathcal{C}$ are connected via a Morita context using $_B A_{\cdot\mathcal{C}}$ and $_{\cdot\mathcal{C}}Q_B$ as connecting bimodules. Our Morita context is in fact a generalization of Doi's Morita context presented in [11].

In Section 3, we introduce the weak (resp. the strong) structure theorem for $\mathcal{M}^{\mathcal{C}}$. In the case where $_A\mathcal{C}$ is locally projective in the sense of Zimmermann-Huisgen [26], we characterize A being a generator (a progenerator) in the category of right $\mathcal{C}$-comodules by $\mathcal{M}^{\mathcal{C}}$ satisfying the weak (resp. the strong) structure theorem. Here the notion of Galois corings introduced by T. Brzeziński [4] plays an important role. The results and proofs are essentially module theoretic and similar to those of [21] for the catgeory $\mathcal{M}(H)_A^C$ of Doi-Koppinen modules corresponding to a right-right Doi-Koppinen structure (H, A, C) (see also [19] for the case $C = H$).

The notion of a C-Galois extension A of a ring B was introduced by T. Brzeziński and S. Majid in [3] and is related to the so called entwining structures introduced in the same paper. In the third section we give equivalent conditions for a C-Galois extension A/B to be cleft. Our results generalize results of [5] from the case of a base field to the case of a commutative ground ring. In the special case $\varrho(a) = \sum a_\psi \otimes x^\psi$, for some grouplike element $x \in C$, we get a complete generalization of [10, Theorem 1.5] (and [11, Theorem 2.5]).

2000 *Mathematics Subject Classification.* 16D90, 16S40, 16W30, 13B02.

Key words and phrases. Morita Contexts, Corings, Hopf Algebras, (Cleft) Galois Extensions, Entwining Structures, Entwined Module.

[*]current address: Dr. Jawad Abuihlail, Box # 281- KFUPM, 31261 Dhahran, Saudi Arabia.

With A we denote a not necessarily commutative ring with $1_A \neq 0_A$ and with $\mathcal{M}_A$ (resp. $_A\mathcal{M}$, $_A\mathcal{M}_A$) the category of *unital* right A-modules (resp. left A-modules, A-bimodules). For every right A-module W we denote by $\mathrm{Gen}(W_A)$ (resp. $\sigma[W_A]$) the class of W-generated (resp. W-subgenerated) right A-modules. We refer to [25] for a detailed discussion of the theory of categories of type $\sigma[W]$.

A left A-module W is called *locally projective* (in the sense of [26]), if for every diagram with exact rows in $_A\mathcal{M}$

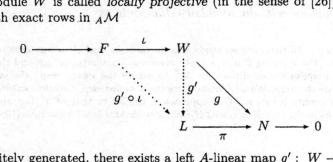

with $_AF$ finitely generated, there exists a left A-linear map $g' : W \to L$ such that $\pi \circ g' \circ \iota = g \circ \iota$. Note that every projective A-module is locally projective. By [26, Theorem 2.1] a left A-module W is locally projective, if and only if for every right A-module M the following map is injective:

$$\alpha_M^W : M \otimes_A W \longrightarrow \mathrm{Hom}_{-A}(^*W, M),\ m \otimes_A w \mapsto [f \mapsto mf(w)].$$

It's easy then to see that every locally projective A-module is flat and A-cogenerated. Let $\mathcal{C}$ be an A-coring. We consider the canonical A-bimodule $^*\mathcal{C} := \mathrm{Hom}_{A-}(\mathcal{C}, A)$ as an A-ring with the canonical A-bimodule structure, multiplication $(f \cdot g)(c) := \sum g(c_1 f(c_2))$ and unity $\varepsilon_{\mathcal{C}}$. If $_A\mathcal{C}$ is locally projective, then we have an isomorphism of categories $\mathcal{M}^{\mathcal{C}} \simeq \sigma[\mathcal{C}_{*\mathcal{C}}]$ (in particular $\mathcal{M}^{\mathcal{C}} \subseteq \mathcal{M}_{*\mathcal{C}}$ is a full subcategory) and we have a left exact functor $\mathrm{Rat}^{\mathcal{C}}(-) : \mathcal{M}_{*\mathcal{C}} \to \mathcal{M}^{\mathcal{C}}$ assigning to every right $^*\mathcal{C}$-module its maximal $\mathcal{C}$-*rational* $^*\mathcal{C}$-submodule, which turns to be a right $\mathcal{C}$-comodule. Moreover $\mathcal{M}^{\mathcal{C}} = \mathcal{M}_{*\mathcal{C}}$ iff $_A\mathcal{C}$ is f.g. and projective. For more results on the $\mathcal{C}$-rational $^*\mathcal{C}$-modules see [1].

After this paper was finished, it turned out that some results were discovered independently by S. Caenepeel, J. Vercruysse and S. Wang [8].

2. MORITA CONTEXTS

In this Section, we fix the following: $\mathcal{C}$ is an A-coring with grouplike element $\mathbf{x}$ and A is a right $\mathcal{C}$-comodule with structure map (see for example [4, Lemma 5.1])

$$\varrho_A : A \longrightarrow A \otimes_A \mathcal{C} \simeq \mathcal{C},\ a \mapsto \mathbf{x}a.$$

Then $A \in \mathcal{M}_{*\mathcal{C}}$, with $a \leftharpoonup g = \sum a_{<0>} g(a_{<1>}) = g(\mathbf{x}a)$ for all $a \in A$ and $g \in {}^*\mathcal{C}$. For $M \in \mathcal{M}_{*\mathcal{C}}$ put

$$M^{\mathbf{x}} := \{m \in M \mid mg = mg(\mathbf{x})\ \text{for all}\ g \in {}^*\mathcal{C}\}.$$

In particular $A^{\mathbf{x}} := \{a \in A \mid a \leftharpoonup g = ag(\mathbf{x})\ \text{for all}\ g \in {}^*\mathcal{C}\} \subset A$ is a subring. For $M \in \mathcal{M}^{\mathcal{C}}$, we set

$$M^{co\mathcal{C}} := \{m \in M \mid \varrho(m) = m \otimes_A \mathbf{x}\} \subseteq M^{\mathbf{x}}.$$

MORITA CONTEXTS FOR CORINGS AND EQUIVALENCES

Obviously $B := A^{co\mathcal{C}} = \{b \in A \mid bx = xb\} \subseteq A^x$ is a subring and ϱ_A is (B, A)-bilinear. For $M \in \mathcal{M}^{\mathcal{C}}$ we have $M^{co\mathcal{C}} \in \mathcal{M}_B$. Moreover we set

$$Q = \{q \in {}^*\mathcal{C} \mid \sum c_1 q(c_2) = q(c)x \text{ for all } c \in \mathcal{C}\} \subseteq ({}^*\mathcal{C})^x$$

LEMMA 2.1. *1. For every right $^*\mathcal{C}$-module M we have an isomorphism of right B-modules*

$$\omega_M : \mathrm{Hom}_{-\,{}^*\mathcal{C}}(A, M) \longrightarrow M^x, \ f \mapsto f(1_A)$$

with inverse $m \mapsto [a \mapsto ma]$.
2. Let $_A\mathcal{C}$ be locally projective. If $M \in \mathcal{M}^{\mathcal{C}}$, then $M^{co\mathcal{C}} = M^x \simeq \mathrm{Hom}_{-\,{}^*\mathcal{C}}(A, M) = \mathrm{Hom}^{\mathcal{C}}(A, M)$. Hence

$$\Psi_M : M^{co\mathcal{C}} \otimes_B A \longrightarrow M, \ m \otimes_B a \mapsto ma$$

is surjective (resp. injective, bijective), iff

$$\Psi'_M : \mathrm{Hom}^{\mathcal{C}}(A, M) \otimes_B A \longrightarrow M, \ f \otimes_B a \mapsto f(a)$$

is surjective (resp. injective, bijective).
3. We have $\mathrm{Hom}_{-\,{}^*\mathcal{C}}(A, {}^*\mathcal{C}) \simeq ({}^*\mathcal{C})^x$. If moreover $_A\mathcal{C}$ is A-cogenerated (resp. locally projective and ${}^\square\mathcal{C} := \mathrm{Rat}^{\mathcal{C}}({}^*\mathcal{C}_{\cdot\mathcal{C}})$), then $Q = ({}^*\mathcal{C})^x$ (resp. $Q = ({}^\square\mathcal{C})^{co\mathcal{C}}$).
4. For every $M \in \mathcal{M}_{\cdot\mathcal{C}}$ (resp. $M \in \mathcal{M}^{\mathcal{C}}$) and all $m \in M, q \in Q$ we have $mq \in M^x$ (resp. $mq \in M^{co\mathcal{C}}$).

Proof. 1. is obvious and 2. is trivial.

3. Considering $^*\mathcal{C}$ as a right $^*\mathcal{C}$-module via right multiplication we get $\mathrm{Hom}_{-\,{}^*\mathcal{C}}(A, {}^*\mathcal{C}) \simeq ({}^*\mathcal{C})^x$ by (1). If $q \in ({}^*\mathcal{C})^x$, then we have for all $g \in {}^*\mathcal{C}$ and $c \in \mathcal{C}$:

$$g(\sum c_1 q(c_2)) = \sum g(c_1 q(c_2)) = (q \cdot g)(c) = (qg(x))(c) = q(c)g(x) = g(q(c)x),$$

i.e. $\sum c_1 q(c_2) - q(c)x \in \mathrm{Re}(\mathcal{C}, A) := \bigcap\{\mathrm{Ke}(g) \mid g \in \mathrm{Hom}_{A-}(\mathcal{C}, A)\}$. If $_A\mathcal{C}$ is A-cogenerated, then $\mathrm{Re}(\mathcal{C}, A) = 0$, hence $Q = ({}^*\mathcal{C})^x$.
Assume that $_A\mathcal{C}$ is locally projective. Then we have for all $q \in Q, g \in {}^*\mathcal{C}$ and $c \in \mathcal{C}$:

$$(q \cdot g)(c) = \sum g(c_1 q(c_2)) = g(q(c)x) = q(c)g(x) = (qg(x))(c),$$

hence $q \in {}^\square\mathcal{C}$, with $\varrho(q) = q \otimes_A x$, i.e. $q \in ({}^\square\mathcal{C})^{co\mathcal{C}}$. On the other hand, if $q \in ({}^\square\mathcal{C})^{co\mathcal{C}}$, then for all $g \in {}^*\mathcal{C}$ we have $q \cdot g = qg(x)$, i.e. $q \in ({}^*\mathcal{C})^x = Q$.

4. Let $M \in \mathcal{M}_{\cdot\mathcal{C}}$. Then we have for all $q \in Q, g \in {}^*\mathcal{C}$ and $m \in M$:

$$(mq)g = m(q \cdot g) = m(qg(x)) = (mq)g(x),$$

i.e. $mq \in M^{\times}$. If $M \in \mathcal{M}^C$, then we have for all $m \in M$ and $q \in Q$:

$$
\begin{aligned}
\varrho_M(mq) &= \varrho_M\left(\sum m_{<0>}q(m_{<1>})\right) \\
&= \sum m_{<0><0>} \otimes_A m_{<0><1>}q(m_{<1>}) \\
&= \sum m_{<0>} \otimes_A m_{<1>1}q(m_{<1>2}) \\
&= \sum m_{<0>} \otimes_A q(m_{<1>})\mathbf{x} \\
&= \sum m_{<0>}q(m_{<1>}) \otimes_A \mathbf{x} \\
&= \sum mq \otimes_A \mathbf{x},
\end{aligned}
$$

i.e. $mq \in M^{\mathrm{coC}}$. $\qquad\square$

LEMMA 2.2.   1. *With the canonical actions, A is a $(B, {}^*C)$-bimodule.*
2. *Q is a $({}^*C, B)$-bimodule.*

Proof. 1. By assumption, $A \in \mathcal{M}^C \subseteq \mathcal{M}_{\cdot C}$. For all $b \in B$, $a \in A$ and $g \in {}^*C$ we have

$$
b(a \leftharpoonup g) = bg(\mathbf{x}a) = g(b(\mathbf{x}a)) = g(\mathbf{x}(ba)) = (ba) \leftharpoonup g.
$$

2. For all $a \in A$, $q \in Q$ and $c \in C$, we have

$$
\sum c_1(aq)(c_2) = \sum c_1 q(c_2 a) = \sum (ca)_1 q((ca)_2) = q(ca)\mathbf{x} = (aq)(c)\mathbf{x}.
$$

For all $q \in Q$, $b \in B$ and $c \in C$, we have

$$
\sum c_1(qb)(c_2) = \sum c_1 q(c_2)b = q(c)\mathbf{x}b = q(c)b\mathbf{x} = (qb)(c)\mathbf{x}.
$$

On the other hand we have, for all $q \in Q$, $g \in {}^*C$ and $c \in C$:

$$
\begin{aligned}
\sum c_1(g \cdot q)(c_2) &= \sum c_1 q(c_{21}g(c_{22})) = \sum c_{11}q(c_{12}g(c_2)) \\
&= \sum c_{11}(g(c_2)q)(c_{12}) = \sum (g(c_2)q)(c_1)\mathbf{x} \\
&= \sum q(c_1 g(c_2))\mathbf{x} = (g \cdot q)(c)\mathbf{x}.
\end{aligned}
$$

Moreover, we have for all $b \in B$, $q \in Q$, $g \in {}^*C$ and $c \in C$:

$$
\begin{aligned}
((g \cdot q)b)(c) &= (g \cdot q)(c)b = \sum q(c_1 g(c_2))b \\
&= \sum (qb)(c_1 g(c_2)) = (g \cdot qb)(c).
\end{aligned}
$$

$\qquad\square$

THEOREM 2.3. *With notation as above, we have*

1. $(A^{\times}, {}^*C, A, ({}^*C)^{\times}, \widetilde{F}, \widetilde{G})$ *is a Morita context, where*

$$
\begin{aligned}
\widetilde{F} &: ({}^*C)^{\times} \otimes_{A^{\times}} A \longrightarrow {}^*C, \quad q \otimes_{A^{\times}} a \mapsto qa, \\
\widetilde{G} &: A \otimes_{\cdot C} ({}^*C)^{\times} \longrightarrow A^{\times}, \quad a \otimes_{\cdot C} q \mapsto a \leftharpoonup q.
\end{aligned}
$$

2. $(B, {}^*C, A, Q, F, G)$ *is a Morita context, where*

$$
\begin{aligned}
F &: Q \otimes_B A \longrightarrow {}^*C, \quad q \otimes_B a \mapsto qa, \\
G &: A \otimes_{\cdot C} Q \longrightarrow B, \quad a \otimes_{\cdot C} q \mapsto a \leftharpoonup q.
\end{aligned}
$$

If ${}_A C$ is locally projective, then the two Morita contexts coincide.

Proof. 1. By Lemma 2.1 we have $\text{End}(A_{\cdot C}) \simeq A^{\times}$, $(^*C)^{\times} \simeq \text{Hom}_{-\cdot C}(A, {}^*C)$, and the result follows by [16, Proposition 12.6].

2. By Lemma 2.2 A is a $(B, {}^*C)$-bimodule and Q is a $(^*C, B)$-bimodule. For all $q \in Q, g \in {}^*C$, $a \in A$ and $c \in C$ we have

$$F(g \cdot q \otimes_B a)(c) = \sum q(c_2 g(c_1))a = (g \cdot qa)(c) = (g \cdot F(q \otimes_B a))(c)$$

and

$$
\begin{aligned}
F(q \otimes_B a \leftharpoonup g)(c) &= q(c)(a \leftharpoonup g) = q(c)g(\mathbf{x}a) \\
&= g(q(c)\mathbf{x}a) = \sum g(c_1 q(c_2)a) \\
&= \sum g(c_1(qa)(c_2)) = (F(q \otimes_B a) \cdot g)(c),
\end{aligned}
$$

hence F is $^*C$-bilinear. Note that by Lemma 2.1 G is well defined and is obviously B-bilinear. Moreover we have for all $a, \widetilde{a} \in A$ and $q, \widetilde{q} \in Q$ the following associativity relations:

$$
\begin{aligned}
(F(q \otimes_B a) \cdot \widetilde{q})(c) &= \sum \widetilde{q}(c_1 q(c_2)a) = \widetilde{q}(q(c)\mathbf{x}a) \\
&= q(c)\widetilde{q}(\mathbf{x}a) = (qG(a \otimes_{\cdot C} \widetilde{q}))(c), \\
G(a \otimes_{\cdot C} q)\widetilde{a} &= q(\mathbf{x}a)\widetilde{a} = (q\widetilde{a})(\mathbf{x}a) \\
&= F(q \otimes_B \widetilde{a})(\mathbf{x}a) = a \leftharpoonup F(q \otimes_B \widetilde{a}).
\end{aligned}
$$

If ${}_A C$ is locally projective, then $A^{\times} = A^{coC}$, $(^*C)^{\times} = Q$ by Lemma 2.1 and the two contexts coincide. $\qquad\square$

2.4. [4, Definition 5.3] An A-coring C is called *Galois*, if there exists an A-coring isomorphism $\chi : A \otimes_B A \longrightarrow C$ such that $\chi(1_A \otimes_B 1_A) = \mathbf{x}$. Recall that $A \otimes_B A$ is an A-coring with the canonical A-bimodule structure, comultiplication

$$\Delta : A \otimes_B A \longrightarrow (A \otimes_B A) \otimes_A (A \otimes_B A), \ \widetilde{a} \otimes_B a \mapsto (\widetilde{a} \otimes_B 1_A) \otimes_A (1_A \otimes_B a)$$

and counit $\varepsilon_{A \otimes_B A} : A \otimes_B A \longrightarrow A, \ \widetilde{a} \otimes_B a \mapsto \widetilde{a}a$.

2.5. Consider the functors

$$(-)^{coC} : \mathcal{M}^C \longrightarrow \mathcal{M}_B \text{ and } - \otimes_B A : \mathcal{M}_B \longrightarrow \mathcal{M}^C.$$

By [4, Proposition 5.2], $(- \otimes_B A, (-)^{coC})$ is an adjoint pair of covariant functors; the adjunctions are given by

$$\Phi_N : N \longrightarrow (N \otimes_B A)^{coC}, \ n \mapsto n \otimes_B 1_A \tag{1}$$

and

$$\Psi_M : M^{coC} \otimes_B A \longrightarrow M, \ m \otimes_B a \mapsto ma. \tag{2}$$

If Ψ_M is an isomorphism for all $M \in \mathcal{M}^C$, then we say that $\mathcal{M}^C$ satisfies the *weak structure theorem*. If in addition Φ_N is an isomorphism for all $N \in \mathcal{M}_B$, then we say that $\mathcal{M}^C$ satisfies the *strong structure theorem* (in this case $(-)^{coC}$ and $- \otimes_B A$ give an equivalence of categories $\mathcal{M}^C \simeq \mathcal{M}_B$).

2.6. Let $W \in \mathcal{M}_A$ and consider the canonical right $\mathcal{C}$-comodule $W \otimes_A \mathcal{C}$. Then $W \simeq (W \otimes_A \mathcal{C})^{co\mathcal{C}}$ via $w \mapsto w \otimes_A \mathbf{x}$ with inverse $w \otimes_A c \mapsto w\varepsilon_{\mathcal{C}}(c)$ and we define

$$\beta_W := \Psi_{W \otimes_A \mathcal{C}} : W \otimes_B A \longrightarrow W \otimes_A \mathcal{C}, \quad w \otimes_B a \mapsto w \otimes_A \mathbf{x}a. \tag{3}$$

In particular we have for $W = A$ the morphism of A-corings

$$\beta := \Psi_{A \otimes_A \mathcal{C}} : A \otimes_B A \longrightarrow A \otimes_A \mathcal{C} \simeq \mathcal{C}, \quad \tilde{a} \otimes_B a \mapsto \tilde{a}\mathbf{x}a. \tag{4}$$

If β is bijective, then $\mathcal{C}$ is a Galois A-coring and we call the ring extension A/B $\mathcal{C}$-Galois.

THEOREM 2.7. *For the Morita context* $(B, {}^*\mathcal{C}, A, Q, F, G)$ *the following statements are equivalent:*

1. $G : A \otimes_{\cdot_\mathcal{C}} Q \longrightarrow B$ *is surjective (bijective and* $B = A^{\mathbf{x}}$*);*
2. *there exists* $\hat{q} \in Q$, *such that* $\hat{q}(\mathbf{x}) = 1_A$;
3. *for every right* ${}^*\mathcal{C}$*-module* M *we have a* B*-module isomorphism* $M \otimes_{\cdot_\mathcal{C}} Q \simeq M^{\mathbf{x}}$.
4. *for every right* $\mathcal{C}$*-comodule* M *we have* $M \otimes_{\cdot_\mathcal{C}} Q \simeq M^{co\mathcal{C}}$ *as* B*-modules.*

If $_A\mathcal{C}$ *is locally projective, then (1)-(4) are equivalent to:*

5. $A_{\cdot_\mathcal{C}}$ *is (f.g.) projective.*

Proof. 1. $\Rightarrow$ 2. Assume that G is surjective. Then there exist $a_1, \cdots, a_k$ and $q_1, \cdots, q_k \in Q$, such that $G(\sum_{i=1}^{k} a_i \otimes_{\cdot_\mathcal{C}} q_i) = 1_A$. Set $\hat{q} := \sum_{i=1}^{k} a_i q_i \in Q$. Then we have

$$\hat{q}(\mathbf{x}) = (\sum_{i=1}^{k} a_i q_i)(\mathbf{x}) = \sum_{i=1}^{k} q_i(\mathbf{x}a_i) = \sum_{i=1}^{k} (a_i \leftharpoonup q_i) = G(\sum_{i=1}^{k} a_i \otimes_{\cdot_\mathcal{C}} q_i) = 1_A.$$

2. $\Rightarrow$ 3. Consider the B-module morphism

$$\xi_M : M \otimes_{\cdot_\mathcal{C}} Q \longrightarrow M^{\mathbf{x}}, \quad m \otimes_{\cdot_\mathcal{C}} q \mapsto mq.$$

Let $\hat{q} \in Q$ with $\hat{q}(\mathbf{x}) = 1_A$ and define $\tilde{\xi}_M : M^{\mathbf{x}} \longrightarrow M \otimes_{\cdot_\mathcal{C}} Q$, $m \mapsto m \otimes_{\cdot_\mathcal{C}} \hat{q}$. For every $n \in M^{\mathbf{x}}$ we have

$$(\xi_M \circ \tilde{\xi}_M)(n) = \xi_M(n \otimes_{\cdot_\mathcal{C}} \hat{q}) = n \leftharpoonup \hat{q} = n\hat{q}(\mathbf{x}) = n.$$

On the other hand we have for all $m \in M$ and $q \in Q$:

$$\begin{aligned}(\tilde{\xi}_M \circ \xi_M)(m \otimes_{\cdot_\mathcal{C}} q) &= \tilde{\xi}_M(m \leftharpoonup q) = m \leftharpoonup q \otimes_{\cdot_\mathcal{C}} \hat{q} = m \otimes_{\cdot_\mathcal{C}} q \cdot \hat{q} \\ &= m \otimes_{\cdot_\mathcal{C}} q\hat{q}(\mathbf{x}) = m \otimes_{\cdot_\mathcal{C}} q,\end{aligned}$$

i.e. ξ_M is bijective with inverse $\tilde{\xi}_M$.

3. $\Rightarrow$ 4. Let $M \in \mathcal{M}^{\mathcal{C}}$. By Lemma 2.1 we have $\xi_M(M \otimes_{\cdot_\mathcal{C}} Q) \subseteq M^{co\mathcal{C}} \subseteq M^{\mathbf{x}}$. By assumption $\xi_M : A \otimes_{\cdot_\mathcal{C}} Q \longrightarrow M^{\mathbf{x}}$ is bijective. Hence $M^{\mathbf{x}} = M^{co\mathcal{C}}$ and $M \otimes_{\cdot_\mathcal{C}} Q \overset{\xi_M}{\simeq} M^{co\mathcal{C}}$.

4. $\Rightarrow$ 1. We are done since $G = \xi_A$.

MORITA CONTEXTS FOR CORINGS AND EQUIVALENCES

If $_AC$ is locally projective, then $B \simeq \mathrm{End}(A_{\bullet C})$, $Q \simeq \mathrm{Hom}_{-\bullet C}(A,{}^*C)$ and the equivalence 1. $\Leftrightarrow$ 5. follows by [16, Corollary 12.8]. $\square$

COROLLARY 2.8. *For the Morita context $(B,{}^*C, A, Q, F, G)$ assume there exists $\widehat{q} \in Q$ with $\widehat{q}(\mathbf{x}) = 1_A$ (equivalently $G : Q \otimes_B A \longrightarrow {}^*C$ is surjective). Then:*

1. *For every $N \in \mathcal{M}_B$, Φ_N is an isomorphism.*
2. *B is a left B-direct summand of A.*
3. *$_BA$ and Q_B are generators.*
4. *$A_{\bullet C}$ and $_{\bullet C}Q$ are f.g. and projective.*
5. *$F : Q \otimes_B A \longrightarrow {}^*C$ induces bimodule isomorphisms*

$$A \simeq \mathrm{Hom}_{\bullet C-}(Q,{}^*C) \ and \ Q \simeq \mathrm{Hom}_{-\bullet C}(A,{}^*C).$$

6. *The bimodule structures above induce ring isomorphisms*

$$B \simeq \mathrm{End}(A_{\bullet C}) \ and \ B \simeq \mathrm{End}(_{\bullet C}Q)^{\mathrm{op}}.$$

Proof. 1. Let $N \in \mathcal{M}_B$. Then we have by Theorem 2.7 the isomorphisms $G : A \otimes_{\bullet C} Q \longrightarrow B$ and $\xi_{N \otimes_B A} : (N \otimes_B A) \otimes_{\bullet C} Q \longrightarrow (N \otimes_B A)^{\mathrm{co}C}$. Moreover Φ_N is given by the canonical isomorphisms

$$N \simeq N \otimes_B B \simeq N \otimes_B (A \otimes_{\bullet C} Q) \simeq (N \otimes_B A) \otimes_{\bullet C} Q \simeq (N \otimes_B A)^{\mathrm{co}C}.$$

2. The map $\mathrm{tr}_A : A \longrightarrow B$, $a \mapsto a {\leftharpoondown} \widehat{q}$ is left B-linear with $\mathrm{tr}_A(b) = b$ for all $b \in B$.

3.-6. follow from standard Morita theory arguments, see e.g. [16, Proposition 12.7]). $\square$

PROPOSITION 2.9. *Consider the Morita context $(B,{}^*C, A, Q, F, G)$ and assume that $F : Q \otimes_{\bullet C} A \longrightarrow {}^*C$ is surjective. Then:*

1. *$A_{\bullet C}$ is a generator, $Q \simeq \mathrm{Hom}_{B-}(A, B)$ as bimodules and ${}^*C \simeq \mathrm{End}(Q_B)$.*
2. *$\mathcal{M}^C$ satisfies the weak structure theorem, in particular A/B is C-Galois.*

Proof. 1. follows by standard Morita Theory arguments, see e.g. [16, Proposition 12.7].

2. By assumption $\varepsilon_C = F(\sum\limits_{i=1}^{k} q_i \otimes_B a_i)$ for some $\{(q_i, a_i)\}_{i=1}^{k} \subseteq Q \times A$. In this case $\Psi_M : M^{\mathrm{co}C} \otimes_B A \longrightarrow M$ is bijective with inverse $\widetilde{\Psi}_M : M \longrightarrow M^{\mathrm{co}C} \otimes_B A$, $m \mapsto \sum\limits_{i=1}^{k} mq_i \otimes_B a_i$. Indeed, for all $m \in M$, $n \in M^{\mathrm{co}C}$ and $a \in A$, we have

$$
\begin{aligned}
(\Psi_M \circ \widetilde{\Psi}_M)(m) &= \sum_{i=1}^{k}(mq_i)a_i = \sum_{i=1}^{k}(m_{<0>}q_i(m_{<1>})a_i \\
&= \sum_{i=1}^{k} m_{<0>}(q_i a_i)(m_{<1>}) = \sum_{i=1}^{k} m_{<0>}\varepsilon_C(m_{<1>}) = m
\end{aligned}
$$

and

$$(\widetilde{\Psi}_M \circ \Psi_M)(n \otimes_B a) = \sum_{i=1}^{k} (na)q_i \otimes_B a_i = \sum_{i=1}^{k} nq_i(\mathbf{x}a) \otimes_B a_i$$

$$= \sum_{i=1}^{k} n \otimes_B q_i(\mathbf{x}a)a_i = \sum_{i=1}^{k} n \otimes_B (q_i a_i)(\mathbf{x}a)$$

$$= n \otimes_B \varepsilon_C(\mathbf{x}a) = n \otimes_B a.$$

$\square$

THEOREM 2.10. *For the Morita context* $(B, {}^*C, A, Q, F, G)$, *the following statements are equivalent:*

1. $F: Q \otimes_B A \longrightarrow {}^*C$ *is surjective (and, a fortiori, bijective);*
2. (a) Q_B *is f.g. and projective;*
   (b) $\Omega: A \longrightarrow \mathrm{Hom}_{-B}(Q, B)$, $a \mapsto [q \mapsto a {\leftharpoonup} q]$ *is a bimodule isomorphism;*
   (c) ${}_{\cdot C}Q$ *is faithful.*

If ${}_A C$ *is* A-*cogenerated, then 1. and 2. are also equivalent to:*

3. (a) ${}_B A$ *is finitely generated and projective;*
   (b) $\Lambda: {}^*C \longrightarrow \mathrm{End}({}_B A)^{\mathrm{op}}$, $g \mapsto [a \mapsto a {\leftharpoonup} g]$ *is a ring isomorphism.*
4. $A_{\cdot C}$ *is a generator.*

If ${}_A C$ *is finitely generated and projective, then 1.-4. are equivalent to:*

5. $\mathcal{M}^C$ *satisfies the weak structure theorem.*

Proof. The implications 1. $\Rightarrow$ 2., 3., 4. follow without any finiteness conditions on C by standard Morita Theory arguments (see e.g. [16, Proposition 12.7]). Note that ${}_{\cdot C}Q$ is faithful by the embedding ${}^*C \hookrightarrow \mathrm{End}(Q_B)$ (see Proposition 2.9 (1)).

2. $\Rightarrow$ 1. Let $\{(q_i, p_i)\}_{i=1}^{k} \subset Q \times \mathrm{Hom}_{-B}(Q, B)$ be a dual basis for Q_B. By (b) there exist $a_1, ..., a_k \in A$, such that $\Omega(a_i) = q_i$ for $i = 1, ..., k$. For every $q \in Q$ we have then $(\sum_{i=1}^{k} q_i a_i) \cdot q = \sum_{i=1}^{k} q_i(a_i {\leftharpoonup} q) = \sum_{i=1}^{k} q_i p_i(q) = q$, hence $\sum_{i=1}^{k} q_i a_i = \varepsilon_C$ by (c) and the ${}^*C$-bilinear morphism $F: Q \otimes_{\cdot C} A \longrightarrow {}^*C$ is surjective.

Assume ${}_A C$ to be A-cogenerated.
3. $\Rightarrow$ 1. Let $\{(a_i, p_i)\}_{i=1}^{k} \subset A \times \mathrm{Hom}_{B-}(A, B)$ be a dual basis of ${}_B A$. By (b), there exist $g_1, ..., g_k \in {}^*C$, such that $\Lambda(g_i) = p_i$ for $i = 1, ..., k$. **Claim:** $g_1, ..., g_k \in Q$. For all $f \in {}^*C$ and $i = 1, ..., k$ we have

$$\Lambda(g_i \cdot f)(a) = a {\leftharpoonup} (g_i \cdot f) = (a {\leftharpoonup} g_i) {\leftharpoonup} f$$

$$= p_i(a) {\leftharpoonup} f = f(\mathbf{x}p_i(a))$$

$$= f(p_i(a)\mathbf{x}) = p_i(a)f(\mathbf{x})$$

$$= (p_i f(\mathbf{x}))(a) = \Lambda(g_i f(\mathbf{x}))(a),$$

hence $g_i \cdot f = g_i f(\mathbf{x})$, i.e. $g_i \in ({}^*C)^{\mathbf{x}} = Q$, by Lemma 2.1 (2). Moreover, for every $a \in A$ we have: $\Lambda(\sum_{i=1}^{k} g_i a_i)(a) = \sum_{i=1}^{k} a {\leftharpoonup} g_i a_i = \sum_{i=1}^{k} p_i(a)a_i = a$, i.e. $\sum_{i=1}^{k} g_i a_i = \varepsilon_C$

and the $^*C$-bilinear morphism F is surjective.

4. $\Rightarrow$ 1. Since $Q \simeq \mathrm{Hom}_{-\,^*C}(A,\,^*C)$, we have $\mathrm{Im}(F) = \mathrm{tr}(A,\,^*C) := \sum\{\mathrm{Im}(h) : h \in \mathrm{Hom}_{-\,^*C}(A,\,^*C)\}$, hence $\mathrm{Im}(F) = \,^*C$ iff $A_{\bullet C}$ is a generator (e.g. [25, Page 154]).

Assume $_AC$ to be finitely generated and projective.
1. $\Rightarrow$ 5. follows without any finiteness conditions on C by Proposition 2.9 (2).

5. $\Rightarrow$ 1. Since $_AC$ is f.g. and projective, we have $\mathcal{M}^C \simeq \mathcal{M}_{\bullet C}$ (e.g. [4, Lemma 4.3]), hence $^*C \in \mathcal{M}^C$, $Q = (^*C)^{coC}$ and $F = \Psi_{\bullet C}$. $\qquad\square$

3. GALOIS EXTENSIONS AND EQUIVALENCES

We keep the notation introduced in Section 2. For every $M \in \mathcal{M}^C$, we have a C-colinear morphism

$$\Psi'_M : \mathrm{Hom}^C(A, M) \otimes_B A \longrightarrow M, \ f \otimes_B a \mapsto f(a).$$

In this Section, we characterize A being a generator (resp. a progenerator) in $\mathcal{M}^C$, under the assumption that $_AC$ is locally projective. We use the same approach as in [21]; in the special case of the category of Doi-Koppinen modules $\mathcal{M}(H)_A^C$, we recover the results from [21].

LEMMA 3.1. *Assume that $_AC$ is locally projective, $_BA$ is flat and A/B is C-Galois. Then*

1. *A is a subgenerator in $\mathcal{M}^C$, i.e. $\sigma[A_{\bullet C}] = \sigma[C_{\bullet C}]$;*
2. *Ψ'_M is injective, for every $M \in \mathcal{M}^C$;*
3. *Ψ'_M is an isomorphism, for every A-generated $M \in \mathcal{M}^C$.*

Proof. 1. Since A/B is C-Galois, $\beta' := \Psi'_C$ is an isomorphism, hence C is A-generated. Consequently $\sigma[A_{\bullet C}] \subseteq \sigma[C_{\bullet C}] \subseteq \sigma[A_{\bullet C}]$, i.e. $\sigma[A_{\bullet C}] = \sigma[C_{\bullet C}]$.
2. With slight modifications, the proof of [21, Lemma 3.22] applies.
3. If $M \in \mathcal{M}^C$ is A-generated, then Ψ'_M is surjective, hence bijective by (2). $\qquad\square$

The following result is a generalization of [5, Proposition 3.13] (which is itself a generalization of [14, Theorem 2.11]).

PROPOSITION 3.2. *Assume that A/B is C-Galois.*

1. *If $_BA$ is flat, then $\mathcal{M}^C$ satisfies the weak structure theorem.*
2. *Assume that there exists $\widehat{q} \in Q$, such that $\widehat{q}(\mathbf{x}) = 1_A$. If $_BA$ is flat, or for all $b \in B$ and $c \in C$ we have $\widehat{q}(cb) = \widehat{q}(c)b$, then $\mathcal{M}^C$ satisfies the strong structure theorem.*

Proof. 1. This is the first part of the proof of [4, Theorem 5.6].
2. By assumption and Corollary 2.8, Φ_N is an isomorphism for all $N \in \mathcal{M}_B$. If $_BA$ is flat, then $\mathcal{M}^C$ satisfies the weak structure theorem by (1). On the other hand, if for all $b \in B$ and $c \in C$ we have $\widehat{q}(cb) = \widehat{q}(c)b$, then an argument similar to the one used in the proof of [5, Proposition 3.13] shows that $\mathcal{M}^C$ satisfies the weak structure theorem. $\qquad\square$

10   J. ABUHLAIL

THEOREM 3.3. *Assume that $_AC$ is locally projective. Then the following statements are equivalent.*

1. $\mathcal{M}^C$ *satisfies the weak structure theorem;*
2. $_BA$ *is flat and A/B is C-Galois;*
3. $_BA$ *is flat and $\beta' := \Psi'_C$ is an isomorphism;*
4. $_BA$ *is flat and for every A-generated $M \in \mathcal{M}^C$, Ψ'_M is bijective;*
5. *for every $M \in \mathcal{M}^C = \sigma[C \cdot_C]$, the C-colinear morphism Ψ'_M is bijective;*
6. $\sigma[C \cdot_C] = \text{Gen}(A \cdot_C);$
7. $_BA$ *is flat, $\sigma[C \cdot_C] = \sigma[A \cdot_C]$ and $\text{Hom}_{- \cdot_C}(A, -) : \text{Gen}(A \cdot_C) \longrightarrow \mathcal{M}_B$ is full faithful;*
8. $\text{Hom}^C(A, -) : \mathcal{M}^C \longrightarrow \mathcal{M}_B$ *is faithful;*
9. A *is a generator in $\mathcal{M}^C$.*

Proof. 1. $\Leftrightarrow$ 5. and 2. $\Leftrightarrow$ 3. follow by Lemma 2.1. The equivalences 4. $\Leftrightarrow$ 5. $\Leftrightarrow$ 6. $\Leftrightarrow$ 7. follow by [21, Theorem 2.3]. The equivalence 8. $\Leftrightarrow$ 9. is evident for any category, and moreover 6. $\Leftrightarrow$ 9. by the fact that $\text{Gen}(A \cdot_C) \subseteq \sigma[A \cdot_C] \subseteq \sigma[C \cdot_C] = \mathcal{M}^C$. By Lemma 3.1 we have 3. $\Rightarrow$ 4. Now assuming 1. we conclude that A/B is C-Galois and that $_BA$ is flat (since 1. $\Leftrightarrow$ 5. $\Leftrightarrow$ 7., hence 1. $\Rightarrow$ 2. follows and we are done. $\square$

DEFINITION 3.4. ([21, Definition 2.4]) A left module P over a ring S is called a *weak generator*, if for any right S-module Y, $Y \otimes_S P = 0$ implies $Y = 0$. A right module P over a ring $\mathcal{R}$ is called a *quasiprogenerator* (resp. a *progenerator*), if $P_\mathcal{R}$ is finitely generated quasiprojective and generates each of its submodules (resp. $P_\mathcal{R}$ is finitely generated, projective and a generator). $P_\mathcal{R}$ is called *faithful* (resp. *balanced*), if the canonical morphism $\mathcal{R} \longrightarrow \text{End}(_{\text{End}(P_\mathcal{R})}P)^{\text{op}}$ is injective (resp. surjective).

THEOREM 3.5. *Assume that $_AC$ is flat. Then the following statements are equivalent:*

1. $\mathcal{M}^C$ *satisfies the strong structure theorem;*
2. $_BA$ *is faithfully flat and A/B is C-Galois.*

If $_AC$ is locally projective, then 1. and 2. are equivalent to:

3. $_BA$ *is faithfully flat and $\beta' := \Psi'_C$ is bijective;*
4. $_BA$ *is faithfully flat and for every $M \in \sigma[A \cdot_C]$, Ψ'_M is bijective;*
5. $A \cdot_C$ *is quasiprojective and generates each of its submodules, $_BA$ is a weak generator and $\sigma[C \cdot_C] = \sigma[A \cdot_C]$;*
6. $A \cdot_C$ *is a quasiprogenerator and $\sigma[C \cdot_C] = \sigma[A \cdot_C]$;*
7. $_BA$ *is a weak generator, Ψ'_M is an isomorphism for every $M \in \text{Gen}(A \cdot_C)$ and $\sigma[A \cdot_C] = \sigma[C \cdot_C]$;*
8. $\text{Hom}^C(A, -) : \mathcal{M}^C \longrightarrow \mathcal{M}_B$ *is an equivalence;*
9. A *is a progenerator in $\mathcal{M}^C$.*

Proof. 1. $\Leftrightarrow$2. is [4, Theorem 5.6]. Assume that $_AC$ is locally projective. Then 2. $\Leftrightarrow$ 3. follows by Lemma 2.1 and we obtain 1. $\Leftrightarrow$ 8. $\Leftrightarrow$ 9. by characterizations of progenerators in categories of type $\sigma[M]$ (see [25, 18.5, 46.2]). Moreover 4. $\Leftrightarrow$ 5. $\Leftrightarrow$ 6. $\Leftrightarrow$ 7. follows from [21, Theorem 2.5]. Obviously 3. $\Rightarrow$ 4. (note that 3. $\Leftrightarrow$ 2. $\Leftrightarrow$1.). Assume now 4. Then $_BA$ is faithfully flat and moreover Ψ'_C is bijective, since $C \in \sigma[A \cdot_C]$ by 6., i.e. 4. $\Rightarrow$ 3. and the proof is complete. $\square$

MORITA CONTEXTS FOR CORINGS AND EQUIVALENCES

Remark 3.6. Assume that $_AC$ is locally projective. Then $\mathrm{Im}(F) \subseteq {}^\square C$. In fact, we have for all $q \in Q$, $a \in A$, $g \in {}^*C$ and $c \in C$:

$$((qa) \cdot g)(c) = \sum g(c_1 q(c_2)a) = g(q(c)\mathbf{x}a) = q(c)g(\mathbf{x}a) = (qg(\mathbf{x}a)(c),$$

hence $qa \in {}^\square C$, with $\varrho(qa) = q \otimes_A \mathbf{x}a$.

PROPOSITION 3.7. *Assume that $_AC$ is locally projective and that there exists $\widehat{q} \in Q$ such that $\widehat{q}(\mathbf{x}) = 1_A$ (or, equivalently, $G : A \otimes_{\cdot c} Q \longrightarrow B$ is surjective). Then $\mathcal{M}^C$ satisfies the strong structure theorem if and only if $\mathrm{Im}(F) = {}^\square C$, and the following map is surjective, for every $M \in \mathcal{M}^C$,*

$$\varpi_M : M \otimes_{\cdot c} {}^\square C \longrightarrow M, \; m \otimes_{\cdot c} f \mapsto mf.$$

In this case $Q \otimes_B A \overset{F}{\cong} {}^\square C$ and $M \otimes_{\cdot c} {}^\square C \overset{\varpi_M}{\cong} M$ for every $M \in \mathcal{M}^C$.

Proof. For every $M \in \mathcal{M}^C$, we consider the commutative diagram

$$
\begin{array}{ccc}
M \otimes_{\cdot c} Q \otimes_B A & \xrightarrow{\;\xi_M \otimes id_A\;} & M^{\mathrm{co}C} \otimes_B A \\
{\scriptstyle id_M \otimes F}\Big\downarrow & & \Big\downarrow{\scriptstyle \Psi_M} \\
M \otimes_{\cdot c} {}^\square C & \xrightarrow[\;\varpi_M\;]{} & M
\end{array}
\tag{5}
$$

Assume that $\mathrm{Im}(F) = {}^\square C$ and that ϖ_M is surjective for every $M \in \mathcal{M}^C$. Then Ψ_M is obviously surjective. Let $K = \mathrm{Ke}(\Psi_M)$. Since Ψ_M is a morphism in $\mathcal{M}^C \simeq \sigma[\mathcal{C}_{\cdot c}]$, we have that $K \in \mathcal{M}^C$, hence $\Psi_K : K^{\mathrm{co}C} \otimes_B A \longrightarrow K$ is surjective. By Theorem 2.7, we have that $K \otimes_{\cdot c} Q \overset{\xi_K}{\cong} K^{\mathrm{co}C}$ and $A \otimes_{\cdot c} Q \overset{\xi_A}{\cong} B$, hence

$$K^{\mathrm{co}C} \simeq K \otimes_{\cdot c} Q = \mathrm{Ke}(\Psi_M) \otimes_{\cdot c} Q = \mathrm{Ke}(\Psi_M \otimes_{\cdot c} id_Q) = \mathrm{Ke}(id_{M^{\mathrm{co}C}}) = 0,$$

i.e. Ψ_M is bijective. By corollary 2.8, Φ_N is bijective for every $N \in \mathcal{M}_B$. Consequently $\mathcal{M}^C$ satisfies the strong structure theorem.

Conversely, assume that $\mathcal{M}^C$ satisfies the strong structure theorem. Note that F is the adjunction of $\Psi_{\square C}$, hence $Q \otimes_B A \overset{F}{\cong} {}^\square C$ and consequently ϖ_M is also bijective for every $M \in \mathcal{M}^C$ by the commutativity of (5). $\square$

Remarks 3.8. Assume that $_AC$ is locally projective.

1. $\varpi_A : A \otimes_{\cdot c} {}^\square C \longrightarrow A$ is surjective if and only if there exists $\widehat{g} \in {}^\square C$ with $\widehat{g}(\mathbf{x}) = 1_A$. This can be seen as follows. If ϖ_A is surjective, then there exist $\{(a_i, g_i)\}_{i=1}^k \subset A \times {}^\square C$ such that $\sum_{i=1}^k a_i \leftharpoonup g_i = 1_A$. Set $\widehat{g} := \sum_{i=1}^k a_i g_i \in {}^\square C$. Then $\widehat{g}(\mathbf{x}) = (\sum_{i=1}^k a_i g_i)(\mathbf{x}) = \sum_{i=1}^k g_i(\mathbf{x}a_i) = \sum_{i=1}^k a_i \leftharpoonup g_i = 1_A$. Conversely, assume that there exists $\widehat{g} \in {}^\square C$ with $\widehat{g}(\mathbf{x}) = 1_A$. Then for every $a \in A$ we have that $1_A \leftharpoonup (\widehat{g}a) = (\widehat{g}a)(\mathbf{x}) = \widehat{g}(\mathbf{x})a = a$, i.e. ϖ_A is surjective.

2. Assume that ϖ_A is surjective and take $M \in \mathcal{M}^C$. If Ψ_M is surjective, then ϖ_M is surjective, since

$$\varpi_M \circ (\Psi_M \otimes_{\cdot c} id_{\square C}) = \Psi_M \circ (id_{M^{\mathrm{co}C}} \otimes_B \varpi_A).$$

12   J. ABUHLAIL

THEOREM 3.9. *Assume that $_AC$ is finitely generated and projective. Then the following assertions are equivalent.*

1. $\mathcal{M}^C$ *satisfies the weak structure theorem;*
2. $_BA$ *is flat and A/B is C-Galois;*
3. $_BA$ *is flat and $\beta' := \Psi'_C$ is an isomorphism;*
4. $_BA$ *is flat and for every A-generated $M \in \mathcal{M}^C = \mathcal{M}_{\cdot C}$, the C-colinear morphism Ψ'_M is bijective;*
5. *for every $M \in \mathcal{M}^C$, the C-colinear morphism Ψ'_M is bijective;*
6. $_BA$ *is flat, $\mathcal{M}_{\cdot C} = \sigma[A_{\cdot C}]$ and $\mathrm{Hom}_{-\cdot C}(A, -) : \mathrm{Gen}(A_{\cdot C}) \longrightarrow \mathcal{M}_B$ is fully faithful;*
7. $\mathrm{Hom}_{-\cdot C}(A, -) : \mathcal{M}_{\cdot C} \longrightarrow \mathcal{M}_B$ *is faithful;*
8. $A_{\cdot C}$ *is a generator;*
9. $F : Q \otimes_B A \longrightarrow {}^*C$ *is surjective (bijective);*
10. (a) Q_B *is f.g. and projective;*
    (b) $\Omega : A \longrightarrow \mathrm{Hom}_{-B}(Q, B)$, $a \mapsto [q \mapsto a \leftharpoonup q]$ *is a bimodule isomorphism;*
    (c) $_{\cdot C}Q$ *is faithful;*
11. (a) $_BA$ *is finitely generated and projective;*
    (b) $\Lambda : {}^*C \longrightarrow \mathrm{End}(_BA)^{\mathrm{op}}$, $g \mapsto [a \mapsto a \leftharpoonup g]$ *is a ring isomorphism.*

Proof. The result follows from Theorems 2.10 and 3.3 and the fact that $\mathcal{M}^C = \mathcal{M}_{\cdot C} = \sigma[C_{\cdot C}]$ if $_AC$ is finitely generated and projective. $\qquad\square$

THEOREM 3.10. *(Morita, e.g. [16, 4.1.3, 4.3], [21, 2.6]). Let $\mathcal{R}$ be a ring, P a right $\mathcal{R}$-module, $\mathcal{S} := \mathrm{End}(P_\mathcal{R})$ and $P^* := \mathrm{Hom}_{-\mathcal{R}}(P, \mathcal{R})$.*
The following assertions are equivalent:

1. $P_\mathcal{R}$ *is a generator;*
2. $_\mathcal{S}P$ *is finitely generated and projective and $\mathcal{R}$ and $\mathrm{End}(_\mathcal{S}P)^{\mathrm{op}}$ are canonically isomorphic.*

The following assertions are equivalent:

1. $P_\mathcal{R}$ *is a faithful quasiprogenerator and $_\mathcal{S}P$ is finitely generated;*
2. $P_\mathcal{R}$ *is a progenerator;*
3. $_\mathcal{S}P$ *is a progenerator and $P_\mathcal{R}$ is faithfully balanced;*
4. $P_\mathcal{R}$ *and $_\mathcal{S}P$ are generators;*
5. $P_\mathcal{R}$ *and $_\mathcal{S}P$ are finitely generated and projective;*
6. $\mathrm{Hom}_{-\mathcal{R}}(P, -) : \mathcal{M}_\mathcal{R} \longrightarrow \mathcal{M}_\mathcal{S}$ *is an equivalence with inverse $\mathrm{Hom}_{-\mathcal{S}}(P^*, -)$;*
7. $- \otimes_\mathcal{R} P^* : \mathcal{M}_\mathcal{R} \longrightarrow \mathcal{M}_\mathcal{S}$ *is an equivalence with inverse $- \otimes_\mathcal{S} P$.*

As an immediate consequence of Theorems 3.5 and 3.10, we obtain

THEOREM 3.11. *Assume that $_AC$ is finitely generated and projective. Then the following statements are equivalent:*

1. $\mathcal{M}^C$ *satisfies the strong structure theorem;*
2. $_BA$ *is faithfully flat and A/B is C-Galois;*
3. $_BA$ *is faithfully flat and $\beta' := \Psi'_C$ is bijective;*
4. $_BA$ *is faithfully flat and for every $M \in \sigma[A_{\cdot C}]$, the map Ψ'_M is bijective;*
5. $A_{\cdot C}$ *is quasiprojective and generates each of its submodules, $_BA$ is a weak generator and $\mathcal{M}_{\cdot C} = \sigma[A_{\cdot C}]$;*

MORITA CONTEXTS FOR CORINGS AND EQUIVALENCES

6. $A_{\cdot C}$ is a quasiprogenerator and $\mathcal{M}_{\cdot C} = \sigma[A_{\cdot C}]$;
7. $_B A$ is a weak generator, Ψ'_M is an isomorphism for every $M \in \mathrm{Gen}(A_{\cdot C})$ and $\mathcal{M}_{\cdot C} = \sigma[A_{\cdot C}]$;
8. $A_{\cdot C}$ is a faithful quasiprogenerator and $_B A$ is finitely generated;
9. $_B A$ is a progenerator and $A_{\cdot C}$ is faithfully balanced;
10. $\mathrm{Hom}_{-\cdot C}(A, -) : \mathcal{M}_{\cdot C} \longrightarrow \mathcal{M}_B$ is an equivalence with inverse $\mathrm{Hom}_{-B}(Q, -)$;
11. $- \otimes_{\cdot C} Q : \mathcal{M}_{\cdot C} \longrightarrow \mathcal{M}_B$ is an equivalence with inverse $- \otimes_B A$;
12. $A_{\cdot C}$ and $_B A$ are generators;
13. $A_{\cdot C}$ and $_B A$ are finitely generated and projective;
14. $A_{\cdot C}$ is a progenerator.

4. CLEFT C-GALOIS EXTENSIONS

In what follows R is a *commutative* ring with $1_R \neq 0_R$ and $\mathcal{M}_R$ is the category of R-modules. For an R-coalgebra $(C, \Delta_C, \varepsilon_C)$ and an R-algebra (A, μ_A, η_A), we consider $(\mathrm{Hom}_R(C, A), \star) := \mathrm{Hom}_R(C, A)$ as an R-algebra with the *usual convolution product* $(f \star g)(c) := \sum f(c_1) g(c_2)$ and unit $\eta_A \circ \varepsilon_C$. The unadorned $- \otimes -$ means $- \otimes_R -$.

4.1. A right-right *entwining structure* (A, C, ψ) over R consists of an R-algebra (A, μ_A, η_A), an R-coalgebra $(C, \Delta_C, \varepsilon_C)$ and an R-linear map

$$\psi : C \otimes_R A \longrightarrow A \otimes_R C, \quad c \otimes a \mapsto \sum a_\psi \otimes c^\psi,$$

such that

$$\sum (a\tilde{a})_\psi \otimes c^\psi = \sum a_\psi \tilde{a}_\Psi \otimes c^{\psi \Psi}, \qquad \sum (1_A)_\psi \otimes c^\psi = 1_A \otimes c,$$
$$\sum a_\psi \otimes \Delta_C(c^\psi) = \sum a_{\psi \Psi} \otimes c_1^\Psi \otimes c_2^\psi, \qquad \sum a_\psi \varepsilon_C(c^\psi) = \varepsilon_C(c)a.$$

Let (A, C, ψ) be a right-right entwining structure. An *entwined module* corresponding to (A, C, ψ) is a right A-module, which is also a right C-comodule through ϱ_M, such that

$$\varrho_M(ma) = \sum m_{<0>} a_\psi \otimes m_{<1>}^\psi, \quad \text{for all } m \in M \text{ and } a \in A.$$

The category of right-right entwined modules and A-linear C-colinear morphisms is denoted by $\mathcal{M}_A^C(\psi)$. For $M, N \in \mathcal{M}_A^C(\psi)$ we denote by $\mathrm{Hom}_A^C(M, N)$ the set of A-linear C-colinear morphisms from M to N. With $\#_\psi^{\mathrm{op}}(C, A) := \mathrm{Hom}_R(C, A)$, we denote the A-ring with $(af)(c) = \sum a_\psi f(c^\psi)$, $(fa)(c) = f(c)a$, multiplication $(f \cdot g)(c) = \sum f(c_2)_\psi g(c_1^\psi)$ and unity $\eta_A \circ \varepsilon_C$ (see e.g. [1, Lemma 3.3]). Entwined modules were introduced by T. Brzeziński and S. Majid [3] as a generalization of the Doi-Koppinen modules presented in [12] and [17]. By a remark of M. Takeuchi (e.g. [4, Prop. 2.2]), we have an A-coring structure on $\mathcal{C} := A \otimes_R C$. The A-bimodule structure and comultiplication are given by the formulas

$$b(\tilde{a} \otimes c)a := \sum b\tilde{a}a_\psi \otimes c^\psi;$$

$$\Delta_{\mathcal{C}}(a \otimes c) = \sum (a \otimes c_1) \otimes_A (1_A \otimes c_2).$$

The counit is $\varepsilon_{\mathcal{C}} := id_A \otimes \varepsilon_C$. Moreover $\mathcal{M}_A^C(\psi) \simeq \mathcal{M}^{\mathcal{C}}$ and $\#_\psi^{\mathrm{op}}(C, A) \simeq {}^* \mathcal{C}$ as A-rings. If $_R C$ is flat (resp. finitely generated, projective), then $_A \mathcal{C}$ has the same property (see e.g. [1, Lemma 3.8]).

Inspired by [11, 3.1], we state the following definition:

4.2. Let (A, C, ψ) be a right-right entwining structure over R and consider the corresponding A-coring $\mathcal{C} := A \otimes_R C$. We say that (A, C, ψ) satisfies the *left α-condition*, if for every right A-module M the following map is injective:

$$\alpha_M^\psi : M \otimes_R C \longrightarrow \operatorname{Hom}_R(\#_\psi^{\mathrm{op}}(C, A), M), \ m \otimes c \mapsto [f \mapsto mf(c)].$$

This is equivalent to $_A C$ being locally projective. Let (A, C, ψ) be a right-right entwining structure satisfying the left α-condition, and let M be a right $\#_\psi^{\mathrm{op}}(C, A)$-module M and consider the canonical map $\rho_M : M \longrightarrow \operatorname{Hom}_R(\#_\psi^{\mathrm{op}}(C, A), M)$. Set $\operatorname{Rat}^C(M_{\#_\psi^{\mathrm{op}}(C,A)}) := (\rho_M^\psi)^{-1}(M \otimes_R C)$. We call M #-rational, if $\operatorname{Rat}^C(M_{\#_\psi^{\mathrm{op}}(C,A)}) = M$ and set $\varrho_M := (\alpha_M^\psi)^{-1} \circ \rho_M$. The category of #-rational right $\#_\psi^{\mathrm{op}}(C, A)$-modules will be denoted by $\operatorname{Rat}^C(M_{\#_\psi^{\mathrm{op}}(C,A)})$.

THEOREM 4.3. ([1, Theorem 3.10]) *Let (A, C, ψ) be a right-right entwining structure and consider the corresponding A-coring $\mathcal{C} := A \otimes_R C$.*

1. *If $_R C$ is flat, then $\mathcal{M}_A^C(\psi)$ is a Grothendieck category with enough injective objects.*

2. *If $_R C$ is locally projective, then*

$$\mathcal{M}_A^C(\psi) \simeq \operatorname{Rat}^C(M_{\#_\psi^{\mathrm{op}}(C,A)}) \simeq \sigma[(A \otimes_R C)_{\#_\psi^{\mathrm{op}}(C,A)}]. \tag{6}$$

3. *If $_R C$ is finitely generated and projective, then*

$$\mathcal{M}_A^C(\psi) \simeq \mathcal{M}_{\#_\psi^{\mathrm{op}}(C,A)}. \tag{7}$$

We fix a right-right entwining structure (A, C, ψ) and its corresponding coring $\mathcal{C} := A \otimes_R C$. Assume that $A \in \mathcal{M}_A^C(\psi) \simeq \mathcal{M}^C$ with

$$\varrho_A : A \longrightarrow A \otimes_R C, \ a \mapsto \sum a_{<0>} \otimes a_{<1>} = \sum 1_{<0>} a_\psi \otimes 1_{<1>}^\psi.$$

Then $\sum 1_{<0>} \otimes 1_{<1>} \in \mathcal{C}$ is a grouplike element and

$$Q \simeq \{q \in \operatorname{Hom}_R(C, A) \mid \sum q(c_2)_\psi \otimes c_1^\psi = \sum q(c) 1_{<0>} \otimes 1_{<1>} \text{ for all } c \in C\}.$$

For every $M \in \mathcal{M}_A^C(\psi)$, we set

$$M^{\mathrm{co}\mathcal{C}} := \{m \in M \mid \sum m_{<0>} \otimes m_{<1>} = \sum m 1_{<0>} \otimes 1_{<1>}\}.$$

Moreover we set $B := A^{\mathrm{co}\mathcal{C}}$.

Remark 4.4. Let $x \in C$ be a grouplike element. For every right C-comodule M we put $M^{\mathrm{co}C} := \{m \in M \mid \varrho_M(m) = m \otimes x\}$. If $\varrho_A(1_A) = 1_A \otimes x$, then we have $M^{\mathrm{co}\mathcal{C}} = M^{\mathrm{co}C}$ for every $M \in \mathcal{M}_A^C(\psi)$.

By [5, Cor. 3.4, 3.7] $- \otimes_R^c A : \mathcal{M}^C \longrightarrow \mathcal{M}_A^C(\psi)$ is a functor, which is left adjoint to the forgetful functor. Here, for every $N \in \mathcal{M}^C$, we consider the canonical right A-module $N \otimes_R^c A := N \otimes_R A$ with the C-coaction $n \otimes a \mapsto \sum n_{<0>} \otimes a_\psi \otimes n_{<1>}^\psi$.

PROPOSITION 4.5. *Let R be a QF ring. Assume that C is right semiperfect and that $_R C$ is locally projective (projective), and put $C^\square := \operatorname{Rat}(\cdot_C C^*)$.*
1. *The following statements are equivalent:*

(a) A is a generator in $\mathcal{M}_A^C(\psi)$;

(b) A generates $C^{\square} \otimes_R^c A$ in $\mathcal{M}_A^C(\psi)$;

(c) the map $\Psi'_{C^{\square} \otimes_R^c A} : \operatorname{Hom}_A^C(A, C^{\square} \otimes_R^c A) \otimes_B A \longrightarrow C^{\square} \otimes_R^c A$ is surjective (bijective).

2. The following statements are equivalent:

(a) A is a progenerator in $\mathcal{M}_A^C(\psi)$;

(b) $\Psi'_{C^{\square} \otimes_R^c A}$ is surjective (bijective) and $_BA$ is a weak generator.

Proof. By [20, 2.6], $C^{\square}$ is a generator in $\mathcal{M}^C$, hence $C^{\square} \otimes_R^c A$ is a generator in $\mathcal{M}_A^C(\psi)$ by the natural isomorphism $\operatorname{Hom}_A^C(C^{\square} \otimes_R^c A, M) \simeq \operatorname{Hom}^C(C^{\square}, M)$, for every $M \in \mathcal{M}_A^C(\psi)$. Then 1. follows from the above observation and Theorem 3.3. The implication
(a) $\Rightarrow$ (b) in 2. follows from Theorem 3.5.
(b) $\Rightarrow$ (a). By observation made above, $C^{\square} \otimes_R^c A$ is a generator in $\mathcal{M}_A^C(\psi)$, and the surjectivity of $\Psi'_{C^{\square} \otimes_R^c A}$ makes A a generator of $\mathcal{M}_A^C(\psi)$. So $_BA$ is flat by Theorem 3.3. The weak generator property makes $_BA$ faithfully flat, and we are done by Theorem 3.5. $\qquad\square$

DEFINITION 4.6. A *(total) integral* for C is a C-colinear morphism $\lambda : C \longrightarrow A$ (with $\sum 1_{<0>}\lambda(1_{<1>}) = 1_A$). We call the ring extension A/B *cleft*, if there exists a $\star$-invertible integral. We say that A has the *right normal basis property*, if there exists a left B-linear right C-colinear isomorphism $A \simeq B \otimes_R C$.

LEMMA 4.7. Let $\lambda \in \operatorname{Hom}_R(C, A)$ be $\star$-invertible with inverse $\overline{\lambda}$.

1. $\lambda \in \operatorname{Hom}^C(C, A)$ if and only if $\overline{\lambda} \in Q$.

2. If $\varrho(a) = \sum a_\psi \otimes x^\psi$ for some grouplike element $x \in C$, then there exists $\widehat{\lambda} \in Q$, such that $\sum 1_{<0>}\widehat{\lambda}(1_{<1>}) = \widehat{\lambda}(x) = 1_A$; in this case C has a total integral, namely the $\star$-inverse of $\widehat{\lambda}$.

Proof. Let $\lambda \in \operatorname{Hom}_R(C, A)$ be $\star$-invertible with inverse $\overline{\lambda}$.
1. If $\overline{\lambda} \in Q$, then we have for all $c \in C$:

$$
\begin{aligned}
\sum \lambda(c_1) \otimes c_2 &= \sum \lambda(c_1) 1_\psi \otimes c_2^\psi = \sum \lambda(c_1)\varepsilon(c_3) 1_\psi \otimes c_2^\psi \\
&= \sum \lambda(c_1)(\overline{\lambda}(c_3)\lambda(c_4))_\psi \otimes c_2^\psi = \sum \lambda(c_1)\overline{\lambda}(c_3)_\psi \lambda(c_4)_\Psi \otimes c_2^{\psi\Psi} \\
&= \sum \lambda(c_1)\overline{\lambda}(c_{22})_\psi \lambda(c_3)_\Psi \otimes c_{21}^{\psi\Psi} = \sum \lambda(c_1)\overline{\lambda}(c_2) 1_{<0>}\lambda(c_3)_\Psi \otimes 1_{<1>}^\Psi \\
&= \sum 1_{<0>}\lambda(c)_\Psi \otimes 1_{<1>}^\Psi = \sum \lambda(c)_{<0>} \otimes \lambda(c)_{<1>}
\end{aligned}
$$

i.e. $\lambda \in \operatorname{Hom}^C(C, A)$. Conversely, if λ is right C-colinear, then we have for all $c \in C$:

$$
\begin{aligned}
\sum \overline{\lambda}(c_2)_\psi \otimes c_1^\psi &= \sum \overline{\lambda}(c_1)\lambda(c_2)\overline{\lambda}(c_4)_\psi \otimes c_3^\psi \\
&= \sum \overline{\lambda}(c_1)\lambda(c_2)_{<0>}\overline{\lambda}(c_3)_\psi \otimes \lambda(c_2)_{<1>}^\psi \\
&= \sum \overline{\lambda}(c_1) 1_{<0>}\lambda(c_2)_\psi \overline{\lambda}(c_3)_\Psi \otimes 1_{<1>}^{\psi\Psi}
\end{aligned}
$$

$$= \sum \overline{\lambda}(c_1) 1_{<0>} (\lambda(c_2)\overline{\lambda}(c_3))_\psi \otimes 1^\psi_{<1>}$$

$$= \sum \overline{\lambda}(c) 1_{<0>} 1_\psi \otimes 1^\psi_{<1>}$$

$$= \sum \overline{\lambda}(c) 1_{<0>} \otimes 1_{<1>},$$

and it follows that $\overline{\lambda} \in Q$.

2. Assume that $\varrho(a) = \sum a_\psi \otimes x^\psi$ for some grouplike element $x \in C$. Let $\lambda \in \mathrm{Hom}^C(C, A)$ with $\overline{\lambda} \in Q$ (see (1)). Then $\widehat{\lambda} := \overline{\lambda}\lambda(x) \in Q$, since $\lambda(x) \in B$, and moreover $\sum 1_{<0>} \widehat{\lambda}(1_{<1>}) = \widehat{\lambda}(x) = \overline{\lambda}(x)\lambda(x) = (\overline{\lambda} \star \lambda)(x) = \varepsilon_C(x) 1_A = 1_A$. $\square$

PROPOSITION 4.8. *Assume that A/B is cleft.*

1. *$\mathcal{M}_A^C(\psi)$ satisfies the weak structure theorem; in particular A/B is C-Galois.*
2. *For every $M \in \mathcal{M}_A^C(\psi)$, the C-colinear morphism*

$$\gamma_M : M \longrightarrow M^{coC} \otimes_R C, \quad m \mapsto \sum m_{<0>} \overline{\lambda} \otimes m_{<1>}$$

   is an isomorphism.
3. *A has the right normal basis property.*
4. *If $_RC$ is faithfully flat, then $\mathcal{M}_A^C(\psi)$ satisfies the strong structure theorem.*

Proof. If A/B is cleft, then there exists a $\star$-invertible $\lambda \in \mathrm{Hom}^C(C, A)$ with inverse $\overline{\lambda} \in Q$, see Lemma 4.7 (1).

1. Let $M \in \mathcal{M}_A^C(\psi)$ and consider

$$\widetilde{\Psi}_M : M \longrightarrow M^{coC} \otimes_B A, \quad m \mapsto \sum m_{<0>} \overline{\lambda} \otimes \lambda(m_{<1>}).$$

Then we have for all $n \in M^{coC}$, $m \in M$ and $a \in A$:

$$(\widetilde{\Psi}_M \circ \Psi_M)(n \otimes a) = \widetilde{\Psi}_M(na)$$

$$= \sum (na_{<0>}) \overline{\lambda} \otimes_B \lambda(a_{<1>})$$

$$= \sum na_{<0><0>} \overline{\lambda}(a_{<0><1>}) \otimes_B \lambda(a_{<1>})$$

$$= \sum n \otimes_B a_{<0><0>} \overline{\lambda}(a_{<0><1>}) \lambda(a_{<1>})$$

$$= \sum n \otimes_B a_{<0>} \overline{\lambda}(a_{<1>1}) \lambda(a_{<1>2}) = n \otimes_B a$$

and

$$(\Psi_M \circ \widetilde{\Psi}_M)(m) = \sum (m_{<0>} \overline{\lambda}) \lambda(m_{<1>})$$

$$= \sum m_{<0><0>} \overline{\lambda}(m_{<0><1>}) \lambda(m_{<1>})$$

$$= \sum m_{<0>} \overline{\lambda}(m_{<1>1}) \lambda(m_{<1>2})$$

$$= \sum m_{<0>} \varepsilon_C(m_{<1>}) 1_A = m.$$

2. For every $M \in \mathcal{M}_A^C(\psi)$, the inverse of γ_M is

$$\widetilde{\gamma}_M : M^{coC} \otimes_R C \longrightarrow M, \quad n \otimes c \mapsto n\lambda(c).$$

MORITA CONTEXTS FOR CORINGS AND EQUIVALENCES

Indeed, for all $m \in M$, $n \in M^{coC}$ and $c \in C$, we have

$$(\tilde{\gamma}_M \circ \gamma_M)(m) = \sum (m_{<0>}\bar{\lambda})\lambda(m_{<1>}) = \sum m_{<0><0>}\bar{\lambda}(m_{<0><1>})\lambda(m_{<1>})$$
$$= \sum m_{<0>}\bar{\lambda}(m_{<1>1})\lambda(m_{<1>2}) = \sum m_{<0>}\varepsilon_C(m_{<1>}) = m,$$

and

$$(\gamma_M \circ \tilde{\gamma}_M)(n \otimes_B c) = \sum (n\lambda(c))_{<0>}\bar{\lambda} \otimes (n\lambda(c))_{<0>}$$
$$= \sum (n\lambda(c)_{<0>})\bar{\lambda} \otimes \lambda(c)_{<1>} = \sum (n\lambda(c_1))\bar{\lambda} \otimes c_2$$
$$= \sum n\lambda(c_1)_{<0>}\bar{\lambda}(\lambda(c_1)_{<1>}) \otimes c_2 = \sum n\lambda(c_{11})\bar{\lambda}(c_{12}) \otimes c_2 = n \otimes c.$$

3. By 2., the left B-linear right C-colinear map

$$\gamma_A : A \longrightarrow B \otimes_R C, a \mapsto \sum a_{<0>}\hookleftarrow\bar{\lambda} \otimes a_{<1>}$$

is an isomorphism with inverse $b \otimes c \mapsto b\lambda(c)$.

4. Assume that $_R C$ is faithfully flat. By 3., $A \simeq B \otimes_R C$ as left B-modules, hence $_B A$ is faithfully flat. By 1., A/B is C-Galois, and we are done by Theorem 3.5. $\square$

THEOREM 4.9. *The following statements are equivalent:*

1. A/B is cleft;
2. $\mathcal{M}_A^C(\psi)$ satisfies the weak structure theorem and A has the right normal basis property;
3. A/B is C-Galois and A has the right normal basis property;
4. $\Lambda : \#_\psi^{op}(C, A) \longrightarrow \mathrm{End}(_B A)^{op}, g \mapsto [a \mapsto a\hookleftarrow g]$ is a ring isomorphism and A has the right normal basis property.

If $_R C$ is faithfully flat, then 1.-4. are also equivalent to

5. $\mathcal{M}_A^C(\psi)$ satisfies the strong structure theorem and A has the right normal basis property.

Proof. 1. $\Rightarrow$ 2. This follows by Proposition 4.8.

2. $\Rightarrow$ 3. By assumption $\beta := \Psi_{A \otimes_R C}$ is an isomorphism.

3. $\Rightarrow$ 4. By assumption $A \otimes_B A \simeq A \otimes_R C$ as left A-modules, hence we have the canonical isomorphisms

$$\#_\psi^{op}(C, A) \simeq \mathrm{Hom}_{A-}(A \otimes_R C, A) \simeq \mathrm{Hom}_{A-}(A \otimes_B A, A)$$
$$\simeq \mathrm{Hom}_{B-}(A, \mathrm{End}(_A A)) \simeq \mathrm{End}(_B A).$$

4. $\Rightarrow$ 1. Assume that $\theta : B \otimes_R C \longrightarrow A$ is a left B-linear right C-colinear isomorphism and consider the right C-colinear morphism $\lambda : C \longrightarrow A, c \mapsto \theta(1_A \otimes c)$ and the left B-linear morphism $\delta := (id \otimes \varepsilon_C) \circ \theta^{-1} : A \longrightarrow B$. Define $\bar{\lambda} := \Lambda^{-1}(\delta) \in \#_\psi^{op}(C, A)$. Then we have for all $c \in C$:

$$\sum \lambda(c_1)\bar{\lambda}(c_2) = \sum \lambda(c)_{<0>}\bar{\lambda}(\lambda(c)_{<1>}) = \lambda(c)\hookleftarrow\bar{\lambda}$$
$$= \delta(\lambda(c)) = ((id \otimes \varepsilon_C) \circ \theta^{-1})(\lambda(c))$$
$$= ((id \otimes \varepsilon_C) \circ \theta^{-1})(\theta(1_A \otimes c)) = \varepsilon_C(c)1_A.$$

On the other hand, we have for all $a \in A$:

$$\Lambda(\overline{\lambda} \star \lambda)(a) = a \leftharpoonup (\overline{\lambda} \star \lambda) = \sum a_{<0>}(\overline{\lambda} \star \lambda)(a_{<1>})$$

$$= \sum a_{<0>} \overline{\lambda}(a_{<1>1}) \lambda(a_{<1>2}) = \sum a_{<0><0>} \overline{\lambda}(a_{<0><1>}) \lambda(a_{<1>})$$

$$= \sum (a_{<0>} \leftharpoonup \overline{\lambda}) \lambda(a_{<1>}) = \sum (a_{<0>} \leftharpoonup \Lambda^{-1}(\delta)) \lambda(a_{<1>})$$

$$= \sum \delta(a_{<0>}) \lambda(a_{<1>}) = \sum \delta(a_{<0>}) \theta(1_A \otimes a_{<1>})$$

$$= \sum \theta(\delta(a_{<0>}) \otimes a_{<1>}) = \theta(\theta^{-1}(a)) = a,$$

hence $\overline{\lambda} \star \lambda = \eta_A \circ \varepsilon_C$.

Now assume that $_RC$ is faithfully flat. 1. $\Rightarrow$ 5. follows by Proposition 4.8 4.; 5. $\Rightarrow$ 2. is trivial, and this finishes our proof. $\qquad\square$

Now we look at the special case where $\varrho(a) = \sum a_\psi \otimes x^\psi$, for some grouplike element $x \in C$. We obtain equivalent statements 1.-5. in Theorem 4.9 without any assumptions on C.

THEOREM 4.10. *Assume that* $\varrho(a) = \sum a_\psi \otimes x^\psi$ *for some grouplike element* $x \in C$. *The following statements are equivalent:*

1. *A/B is cleft;*
2. *$\mathcal{M}_A^C(\psi)$ satisfies the strong structure theorem and A has the right normal basis property;*
3. *$\mathcal{M}_A^C(\psi)$ satisfies the weak structure theorem and A has the right normal basis property;*
4. *A/B is C-Galois and A has the right normal basis property;*
5. *$\Lambda: \#_\psi^{\mathrm{op}}(C, A) \longrightarrow \mathrm{End}(_BA)^{op}$, $g \mapsto [a \mapsto a \leftharpoonup g]$ is a ring isomorphism and A has the right normal basis property.*

Proof. By Theorem 4.9 it remains to prove that Φ_N is an isomorphism for every $N \in \mathcal{M}_B$, if A/B is cleft. But in our special case there exists by Lemma 4.7 some $\widehat{\lambda} \in Q$ with $\sum 1_{<0>} \widehat{\lambda}(1_{<1>}) = 1_A$ and we are done by Corollary 2.8 (2). $\qquad\square$

Remark 4.11. Let (H, A, C) be a right-right (resp. a left-right) Doi-Koppinen structure. Then (A, C, ψ) is a right-right entwining structure with

$$\psi: C \otimes_R A \longrightarrow A \otimes_R C, \ c \otimes a \mapsto \sum a_{<0>} \otimes ca_{<1>}$$

(resp. a left-right entwining structure with

$$\psi: A \otimes_R C \longrightarrow A \otimes_R C, \ a \otimes c \mapsto \sum a_{<0>} \otimes a_{<1>}c).$$

If x is a grouplike element of C, then $A \in \mathcal{M}(H)_A^C$ with $\varrho(a) := \sum a_{<0>} \otimes xa_{<1>}$ (resp. $\varrho(a) = \sum a_{<0>} \otimes a_{<1>}x$) and we obtain [10, Theorem 1.5] (resp. [11, Theorem 2.5]) as special cases of Theorem 4.10.

REFERENCES

[1] J.Y. Abuhlail, Rational modules for corings, *Comm. Algebra* **31** (2003), 5793–5840.
[2] M. Beattie, S. Dăscălescu and Ş. Raianu, Galois extensions for co-Frobenius Hopf algebras, *J. Algebra* **198** (1997), 164–183.

[3] T. Brzeziński and S. Majid, Coalgebra bundles, *Comm. Math. Phys.* **191** (1998), 467–492.

[4] T. Brzeziński, The structure of corings. Induction functors, Maschke-type theorem, and Frobenius and Galois-type properties, Algebras Representation Theory **5** (2002), 389–410.

[5] T. Brzeziński, On modules associated to coalgebra Galois extensions, *J. Algebra* **215** (1999), 290–317.

[6] M. Cohen, D. Fischman and S. Montgomery, Hopf Galois extensions, smash products and Morita equivalence, *J. Algebra* **133** (1990), 351–372.

[7] S. Caenepeel, G. Militaru and Shenglin Zhu, "Frobenius and separable functors for generalized module categories and nonlinear equations", *Lect. Notes in Math.* **1787**, Springer Verlag, Berlin, 2002.

[8] S. Caenepeel, J. Vercruysse and Shuanhong Wang, Morita Theory for corings and cleft entwining structures, *J. Algebra*, to appear.

[9] S. Chase and M. Sweedler, "Hopf algebras and Galois theory", *Lect. Notes in Math.* **97**, Springer Verlag, Berlin, 1969.

[10] Y. Doi and A. Masuoka, Generalization of cleft comodule algebras, *Comm. Algebra* **20** (1992), 3703–3721.

[11] Y. Doi, Generalized smash products and Morita contexts for arbitrary Hopf algebras, in "Advances in Hopf algebras", J. Bergen and S. Montgomery, eds., *Lecture Notes Pure Appl. Math.* **158**, Dekker, New York, 1994.

[12] Y. Doi, Unifying Hopf modules, *J. Algebra* **153** (1992), 373-385.

[13] Y. Doi, On the structure of relative Hopf modules, *Comm. Algebra* **11** (1983), 243—255.

[14] Y. Doi and M. Takeuchi, Hopf-Galois extensions of algebras, the Miyashita-Ulbrich action, and Azumaya algebras, J. Algebra, **121** (1989), 488–516.

[15] Y. Doi and M. Takeuchi, Cleft comodule algebras for a bialgebra, *Comm. Algebra* **14** (1986), 801–817.

[16] C. Faith, Algebra I, Rings, Modules and Categories, Springer Verlag, Berlin, 1981.

[17] M. Koppinen, Variations on the smash product with applications to group-graded rings, *J. Pure Appl. Algebra* **104** (1995), 61–80.

[18] H. Kreimer and M. Takeuchi, Hopf Algebras and Galois extensions of an algebra, *Indiana Univ. Math. J.* **30** (1981), 675–692.

[19] C. Menini, A. Seidel, B. Torrecillas and R. Wisbauer, *A-H*-bimodules and equivalences, *Comm. Algebra* **29** (2001), 4619–4640.

[20] C. Menini, B. Torrecillas and R. Wisbauer, Strongly rational comodules and semiperfect Hopf algebras over QF Rings, *J. Pure Appl. Algebra* **155** (2001), 237–255.

[21] C. Menini and M. Zuccoli, Equivalence Theorems and Hopf-Galois extensions, *J. Algebra* **194** (1997), 245–274.

[22] H. J. Schneider, Principal homogeneous spaces for arbitrary Hopf algebras, *Israel J. Math.* **70** (1990), 167–195.

[23] M. E. Sweedler, The predual Theorem to the Jacobson-Bourbaki Theorem, *Trans. Amer. Math. Soc.* **213** (1975), 391–406.

[24] R. Wisbauer, On the category of comodules over corings, in "Mathematics and mathematics education (Bethlehem, 2000)", , S. Elaydi et al, eds., World Sci. Publishing, River Edge, NJ, 2002, 325-336.

[25] R. Wisbauer, Grundlagen der Modul- und Ringtheorie, Verlag Reinhard Fischer, Mnchen, 1988; Foundations of Module and Ring Theory, Gordon and Breach, Reading, 1991.

[26] B. Zimmermann-Huisgen, Pure submodules of direct products of free modules, *Math. Ann.* **224** (1976), 233–245.

Hopf Order Module Algebra Orders

FARAHAT S. ALY   Department of Mathematics, Faculty of Science
Al-Azhar University, Nasr City, Cairo 11884, Egypt
e-mail: *farahata@yahoo.com*

FREDDY VAN OYSTAEYEN   Department of Mathematics and Computer
Science, University of Antwerp. Middelheimlaan 1, B-2020 Antwerp, Belgium
e-mail: *fred.vanoystaeyen@ua.ac.be*

> ABSTRACT. We aim to study a theory for orders in Hopf module algebras with
> respect to their behaviour under actions by Hopf orders induced from the given
> action. The technique of Hopf filtrations and Hopf valuations play a central
> role.

1. INTRODUCTION

One of the few possibilities to extend some arithmetical aspects of number theory
to more abstract fields is via valuation theory and, as a first step, we may restrict
attention to discrete valuations. Then, in order to obtain some calculative tools in
algebras over a given field, it is useful to study orders and maximal orders in the
algebra over some discrete valuation ring of the field. This strategy is succesful in
connection with certain problems in the theory of the Brauer group of the field, in
fact the mother example is the calculation of the Brauer group of the rational field
in terms of its local components corresponding to the valuations of $\mathbb{Q}$. This paper
aims to introduce the same philosophy but with respect to the Brauer group of a
quantum group or more generally of a Hopf algebra.

Throughout, K is a field, $D = O_v$ is a discrete valuation of K and A is a finite
dimensional K-algebra which we assume to be central simple over K. We also fix
a Hopf algebra H over K. In fact we want to consider an H-Azumaya algebra A
over K as in [1], but as a first step in developing the theory we restrict to A being
Azumaya over K, i.e. central over K, equiped with a K-linear H-action making it
an H-module-algebra. How do the maximal order theory of A and the H-action
relate? Note that H cannot act on a D-order in A because H is a K-algebra; therefore we need the existence of a Hopf order over D in H, say $H(\xi)$, with notation
as in [1]. With respect to $H(\xi)$, we may look for $H(\xi)$-module algebra orders in A
over D and then try to relate these to the usual maximal D-orders containing them.

2000 *Mathematics Subject Classification.* 16W30, 16H05.
Key words and phrases. Order, Module algebra, valuation.

In Section 3, we establish the existence of nontrivial algebra orders in A via a constructive procedure based on $H(\xi)$-invariant D-lattices in A. In Section 4, we show how maximal $H(\xi)$-module orders relate to maximal D-orders.

We presented the case of $H(\xi)$-module-algebras, but it is clear that the $H(\xi)$-comodule-algebra version is also necessary in order to arrive at the $H(\xi)$-dimodule version necessary to deal with the properties related to the Brauer groups of the Hopf algebras $H(\xi) \subset H$ and the corresponding group morphism $\mathrm{Br}_D H(\xi) \to \mathrm{Br}_K H$. We hope to have made it clear that the new objects introduced in this paper, provide us with many interesting problems that deserve further investigation.

Section 2 is of a preliminary nature; we recall some definitions and some basic facts, in particular related to Hopf orders as in Section 2.

2. HOPF ORDERS AND HOPF MODULE ALGEBRAS

The field K and its valuation ring $O_v = D$ are fixed throughout. Let H be a finite dimensional Hopf algebra over K. The valuation v defines the valuation filtration fK on K, where $f_n K = \{\lambda \in K, v(\lambda) \le n\}$. A *filtration* on an arbitrary ring R is given by a family FR of additive subgroups $\{F_n R \mid n \in \mathbb{Z}\}$ such that: $F_n R F_m R \subset F_{n+m}R$, for all $n, m \in \mathbb{Z}, 1 \in F_0 R$ and $\cup_n F_n R = R$; such a filtration is separated if $\cap_n F_n R = 0$. A filtration FH on H is said to **extend** FK if $F_n H \cap K = f_n K$ for all $n \in \mathbb{Z}$. Since $f_n K f_m K = f_{n+m} K$, it follows that a filtration FH extending fK is a strong filtration, in the sense that $F_n H = F_m H . F_{n+m} H$ for all $n, m \in \mathbb{Z}$. From this it is also obvious that $F_n H = f_n K . F_0 H$, for all $n \in \mathbb{Z}$. Since $F_0 H$ is a torsionfree D-module, it is flat and therefore $\cap_n (f_n K F_0 H) = (\cap_n f_n K) F_0 H = 0$, hence FH is separated too. If $F_0 H$ is a Hopf subalgebra (over D) of H then we say that $F_0 H$ is a **Hopf order over** D in H. In this case, FH is a Hopf filtration of H in the sense of [1, Section 1]. Moreover, Theorem 2.3. of [1] states that the separated Hopf filtrations FH extending fK correspond bijectively to the set of Hopf valuation filtration functions $\xi : H \to \mathbb{Z} \cup \{-\infty\}$ (in [1], we deal with non-discrete valuations, so here we may set $\Gamma = \mathbb{Z}$) satisfying:

HV-1 $\xi(h) = -\infty$ if and only if $h = 0$;
HV-2 $\xi(1) = 0$;
HV-3 $\xi(\lambda h) = \xi(h) - v(\lambda)$ for all $h \in H$, $\lambda \in K$;
HV-4 $\xi(gh) \le \xi(g) + \xi(h)$, for all $g, h \in H$;
HV-5 $\xi(g + h) \le \max\{\xi(g), \xi(h)\}$ for all $g, h \in H$;
HV-6 $\xi(Sh) \le \xi(h)$ and $\xi(\varepsilon(h)) \le \xi(h)$ for all $h \in H$;
HV-7 $\xi(h) \ge \inf\{\max_{\Sigma}\{\xi(h_1) + \xi(h_2)\}\}$, where we use the Sweedler notation $\Delta(h) = \Sigma h_1 \otimes h_2$; $\max_{\Sigma}$ is taken over the terms in a fixed Sweedler decomposition of $\Delta(h)$, and the infimum is taken over all possible decompositions of $\Delta(h)$. In fact, HV-7 is an equality, see [1, Observation 2.1].

When FH corresponds to ξ in the above correspondence, then we write $G_0 H = H(\xi)$ for the corresponding Hopf order.

HOPF ORDER MODULE ALGEBRA ORDERS

In this paper, we consider a given fixed Hopf order $\mathcal{H} = H(\xi_0)$, with corresponding Hopf filtration FH on H extending the valuation filtration fK associated to v on K. Then $\mathcal{H}$ is a free D-module of finite rank $n = \dim_K H$ and we fix a D-basis B for $\mathcal{H}$. A K-algebra A is called a *left H-module-algebra* if there is a left H-action on A such that

1. A is a left H-module;
2. $h \cdot (ab) = \Sigma(h_1 \cdot a)(h_2 \cdot b)$, for all $h \in H$ and $a, b \in A$;
3. $h \cdot 1_A = \varepsilon(h)1_A$.

LEMMA 2.1. *Let H be a Hopf algebra, with antipode S, and A a left H-module algebra. For all $h \in H$ and $a \in A$, we have*

$$(h \cdot a)b = \Sigma h_1 \cdot (a(S(h_2) \cdot b)).$$

If the antipode S is bijective, with inverse $\overline{S}$, then

$$a(h \cdot b) = \Sigma h_2 \cdot ((\overline{S}(h_1) \cdot a)b).$$

Proof. This is an immediate verification, see for example [3, Lemma 6.1.3]. □

A *full D-lattice L* in A is a finitely generated D-module in A such that $K \otimes_D L = KL = A$. Since A is a left H-module-algebra we are interested in full D-lattices allowing an $\mathcal{H}$-action (induced by the H-action on A). A *full left $\mathcal{H}$-lattice* of A is a full D-lattice $\mathcal{L}$ that is a left $\mathcal{H}$-module, i.e. $\mathcal{H}.\mathcal{L} \subset \mathcal{L}$. A *$D$-order Λ* in A is a subring of A which is also a full D-lattice. A *left $\mathcal{H}$-order* in A is a D-order Λ that is a full left $\mathcal{H}$-lattice. In case there is given an action of H on the right, the right hand versions of the foregoing definitions may be given in a completely symmetric way.

PROPOSITION 2.2. *Consider a full D-lattice L in A and let*

$$\mathcal{L} = \mathcal{H} \rightharpoonup L = \mathcal{H} \cdot L = \left\{ \sum_{i=1}^d h_i \cdot . a_i \mid h_i \in \mathcal{H}, a_i \in L, i \in \{1, \dots, \alpha\} \right\}.$$

Then $\mathcal{L}$ is again a full D-lattice and a full left $\mathcal{H}$-lattice.

1. *$\lambda(\mathcal{L}) = \{a \in A, a\mathcal{L} \subset \mathcal{L}\}$ is a left $\mathcal{H}$-order in A.*
2. *If S is bijective, then $\rho(\mathcal{L}) = \{a \in A, La \subset \mathcal{L}\}$ is a left $\mathcal{H}$-order in A.*

Proof. Since both L and $\mathcal{H}$ are finitely generated D-modules, so is $\mathcal{L} = \mathcal{H} \rightharpoonup L$. Since $\mathcal{H}.(\mathcal{H} \cdot L) \subset \mathcal{H} \cdot L$ is clear from the fact that A is an H-module and $\mathcal{H}$ is a subring of H, it is also clear that $\mathcal{L}$ is a full left $\mathcal{H}$-lattice. Since $\mathcal{L}$ is a full D-lattice too, $\lambda(\mathcal{L})$ and $\rho(\mathcal{L})$ are D-orders (for classical results on orders and maximal orders we refer to [2]).

1) It remains to be verified that $\lambda(\mathcal{L})$ is a full left $\mathcal{H}$-lattice in A. Pick $h \in \mathcal{H}$, $a \in \lambda(\mathcal{L})$ and $b \in \mathcal{L}$; we obtain $(h \cdot a)b = \Sigma h_1 \cdot (a(S(h_2) \cdot b))$ (see Lemma 2.1). $b \in \mathcal{L}$, since $S(h_2).b \in \mathcal{L}$; it follows from $a \in \lambda(\mathcal{L})$ that $a(S(h_2) \cdot b) \in \mathcal{L}$, and using its $\mathcal{H}$-module structure again, we obtain that $(h \cdot a)b \in \mathcal{L}$, or $(h \cdot a) \in \lambda(\mathcal{L})$. $\lambda(\mathcal{L})$ is also a full left $\mathcal{H}$-lattice, thus an $\mathcal{H}$-order.

2) We have to verify that $\rho(\mathcal{L})$ is a full left $\mathcal{H}$-lattice. Pick $h \in \mathcal{H}$, $a \in \rho(\mathcal{L})$ and $b \in \mathcal{L}$; we obtain $b(h.a) = \Sigma h_2 \cdot ((\overline{S}(h_1) \cdot b)a)$. Now $\overline{S}(h_1) \cdot b \in \mathcal{L}$ and $(\overline{S}(h_1) \cdot b)a \in \mathcal{L}$ because $a \in \rho(\mathcal{L})$, then $b(h \cdot a) \in \mathcal{L}$, or $(h \cdot a) \in \rho(\mathcal{L})$ follows. Hence $\rho(\mathcal{L})$ is a D-order and a left $\mathcal{H}$-lattice, hence a left $\mathcal{H}$-order. □

Let A be an H-module algebra; recall that the multiplication $\mu_A : A \otimes A \to A$ is H-linear; the left H-action on $A \otimes A$ is given by the formula $h \cdot (a \otimes b) = \sum (h_1 \cdot a) \otimes (h_2 \cdot b)$. Also recall that the H-invariants of A are defined by

$$A^H = \{a \in A \mid h \cdot a = \varepsilon(h)a, \text{ for all } h \in H\}.$$

In particular, for any left $\mathcal{H}$-order Λ in A we have that $\mu : \Lambda \otimes_D \Lambda \to \Lambda$ is a morphism of $\mathcal{H}$-modules and the $\mathcal{H}$-invariants of Λ are defined by $\Lambda^{\mathcal{H}} = \{a \in \Lambda, h \cdot a = \varepsilon(h)a \text{ for all } h \in \mathcal{H}\}$. Obviously $\Lambda^{\mathcal{H}} = A^H \cap \Lambda$ because $K\mathcal{H} = H$. From now on, we assume that the antipode of H is bijective.

To any D-order Γ in A, there corresponds a filtration FA extending the valuation filtration fK, associated to v, in the following way: $F_n A = f_n K \cdot \Gamma$, $n \in \mathbb{Z}$. This is a strong filtration on A.

OBSERVATION 2.3. With notation as above, the following statements are equivalent:

1. The filtrations FH and FA define on A the structure of a filtered (left) H-module;
2. Γ is a full left $\mathcal{H}$-lattice i.e. an $\mathcal{H}$-order.

Indeed, both statements reduce to $F_0 H \cdot F_0 A \subset F_0 A$, i.e. $\mathcal{H}.\Gamma \subset \gamma$.

For any D-lattice L in A we define $L^{-1} = \{x \in A, LxL \subset L\}$.

PROPOSITION 2.4. *If $\mathcal{L}$ is a full $\mathcal{H}$-lattice in A, then $\mathcal{L}^{-1}$ is a full $\mathcal{H}$-lattice too.*

Proof. Clearly $\mathcal{L}^{-1}$ is a full D-lattice, so we only have to check that $\mathcal{H} \to \mathcal{L}^{-1} \subset \mathcal{L}^{-1}$. Take $\lambda, \mu \in \mathcal{L}$, $h \in \mathcal{H}$ and $x \in \mathcal{L}^{-1}$, and look at $\lambda(h \cdot x)\mu$. Applying Lemma 2.1, we compute

$$\lambda(h.x)\mu = \lambda \left[\Sigma h_1 \cdot (x(S(h_2) \cdot \mu)) \right]$$
$$= \Sigma \lambda \left[h_1 \cdot (x(S(h_2) \cdot \mu)) \right] = \Sigma h_{12} \cdot \left[((S^{-1}(h_{11}) \cdot \lambda)(x(S(h_2) \cdot \mu)) \right],$$

where $S^{-1}(h_{11}).\lambda \in \mathcal{L}$, $(S(h_2).\mu) \in \mathcal{L}$, hence $(S^{-1}(h_{11}).\lambda)(x(Sh_2) \cdot \mu)) \in \mathcal{L}\mathcal{L}^{-1}\mathcal{L} \subset \mathcal{L}$, and finally $\lambda(h.x)\mu \in \mathcal{H} \to \mathcal{L} = \mathcal{L}$ because $\mathcal{L}$ is an $\mathcal{H}$-lattice, thus $h.x \in \mathcal{L}^{-1}$ or $\mathcal{L}^{-1}$ is an $\mathcal{H}$-lattice too. $\square$

If $\mathcal{L}$ is a full left $\mathcal{H}$-lattice, then $\mathcal{L}$ is a $\lambda(\mathcal{L})$-$\rho(\mathcal{L})$-bimodule. If $\mathcal{J}$ is another full left $\mathcal{H}$-lattice such that $\rho(\mathcal{J}) = \lambda(\mathcal{L})$, then $\mathcal{J}\mathcal{L}$ is an $\mathcal{H}$-lattice and it is a $\lambda(\mathcal{J})$-$\rho(\mathcal{L})$-bimodule. It is clear that $\mathcal{L}^{-1}$ is a $\rho(\mathcal{L})$-$\lambda(\mathcal{L})$-bimodule such that $\mathcal{L}\mathcal{L}^{-1}$ is an ideal of $\lambda(\mathcal{L})$, $\mathcal{L}^{-1}\mathcal{L}$ is an ideal of $\rho(\mathcal{L})$, moreover these ideals are $\mathcal{H}$-submodules of $\lambda(\mathcal{L})$, resp. $\rho(\mathcal{L})$.

We are interested in the case where A is an H-Azumaya algebra over K. Let us start with an investigation of the case where A is just a central simple K-algebra that is an H-module algebra. Since the centre of A is K, $Z(\lambda(\mathcal{L})) = Z(\rho(\mathcal{L})) = D$ and $\lambda(\mathcal{L})\mathcal{L}$ as well as $\rho(\mathcal{L})$ is a (Noetherian) P.I. ring with Noetherian centre, hence they are finite D-modules and thus free D-modules of rank $\dim_K A$. In view of the remarks in the foregoing alinea, one may embark on a study of $\mathcal{H}$-normal ideals

HOPF ORDER MODULE ALGEBRA ORDERS

and the Hopf-Brandt groupoid, extending the classical theory to the setting of $\mathcal{H}$-orders but we do not go into this here; we restrict attention to the basic facts about maximal $\mathcal{H}$-orders.

3. MAXIMAL $\mathcal{H}$-ORDERS

D, K, H, A are as in Section 2. Consider a D-order Γ in A and look at $\mathcal{L} = \mathcal{H} \rightharpoonup \Gamma$. The results of Section 2 imply that $\lambda(\mathcal{L})$, resp. $\rho(\mathcal{L})$, is a left $\mathcal{H}$-order, and $\mathcal{L}$ is a $\lambda(\mathcal{L})$-$\rho(\mathcal{L})$-bimodule over D.

PROPOSITION 3.1. *With notation and assumptions as before, there exist maximal $\mathcal{H}$-orders in A and every $\mathcal{H}$-order is contained in a maximal $\mathcal{H}$-order.*

Proof. The existence of $\mathcal{H}$-orders in A follows from the construction in Proposition 2.2. Let $\mathcal{C}$ be the class of $\mathcal{H}$-orders in A, let $\mathcal{C}(\Lambda)$ be the subclass of those $\mathcal{H}$-orders containing a given $\mathcal{H}$-order Λ in A. A chain (i.e. a set ordered by inclusion) $\{\Lambda_\alpha \mid \alpha \in \mathcal{A}\} \subset \mathcal{C}$ defines $\Lambda' = \sum_{\alpha \in \mathcal{A}} \Lambda_\alpha = \cup_{\alpha \in \mathcal{A}} \Lambda_\alpha$. It is obvious that Λ' is a subring of A and in fact Λ' is a D-order. For each $\alpha \in \mathcal{A}$, the $\mathcal{H}$-action on Λ_α is induced by the H-action on A, so we may unambiguously (!) define an $\mathcal{H}$-structure on Λ' by defining $h.x$ for $\mathcal{L} \in H$ and $x \in \Lambda'$ by $h._\alpha x$, that is the $\mathcal{H}$-action as defined on Λ_α if $x \in \Lambda_\alpha, \alpha \in \mathcal{A}$. Therefore inductively ordered chains in $\mathcal{C}$ have a maximal element and we may conclude that maximal $\mathcal{H}$-orders exist. Applying the argument to $\mathcal{C}(\Lambda)$ yields the statement that every $\mathcal{H}$-order is contained in a maximal $\mathcal{H}$-order. $\qquad\square$

PROPOSITION 3.2. *If Γ is a maximal D-order in A containing an $\mathcal{H}$-order Λ, then:*
1. $\Lambda \subset \lambda(M)$, *where* $M = \mathcal{H} \rightharpoonup \Gamma$.
2. $\Lambda \subset \rho(M)$.

Proof. 1) Take $y \in \Lambda$, $h \in H$, $\gamma \in \Gamma$, and look at $y(h \cdot \gamma)$. We calculate using Lemma 2.1 that $y(h \cdot \gamma) = \Sigma h_2 \cdot ((\overline{S}(h_1), y)\gamma)$. Now, $\overline{S}(h_1) \cdot y \in \Lambda$. Since $\Lambda \subset \Gamma, \gamma \in \Gamma$ and Γ is a ring, we have that $(\overline{S}(h_1) \cdot y)\gamma \in \Gamma$, hence $y(h \cdot \gamma) \in H \rightharpoonup \Gamma$. Consequently, $\Lambda \subset \lambda(H \rightharpoonup \Gamma)$.
2) Take $x \in \Lambda, h \in H, \gamma \in \Gamma$, and look· at: $(h \cdot \gamma)x = \Sigma h_1 \cdot (\gamma(S(h_2) \cdot x))$ (again using Lemma 2.1). Then repeat the argument of the first part. $\qquad\square$

Remark 3.3. Looking at Example 3.4, it becomes clear that the connection between maximal orders and maximal $\mathcal{H}$-orders is far from obvious. The following facts are easily checked on Example 3.4:
1. A D-order in an H-module-algebra A is not necessarily an $\mathcal{H}$-order, if a Hopf order $\mathcal{H}$ over D is given in H.
2. If Γ is a D-order in A then $\mathcal{H} \rightharpoonup \Gamma$ need not be an $\mathcal{H}$-order in A (in fact not even a ring).
3. If Γ is a D-order in A then $\lambda(\mathcal{H} \rightharpoonup \Gamma)$ and $\rho(\mathcal{H} \rightharpoonup \Gamma)$ are $\mathcal{H}$-orders in A but none of them necessarily contains Γ.

EXAMPLE 3.4. Recall that Sweedler's fourdimensional Hopf algebra H_4 is generated as a K-algebra by the symbols g and h, subject to the relations $g^2 = 1, h^2 = 0$ and $gh + hg = 0$. H_4 is a Hopf algebra, with comultiplication, counit and antipode given by: $\Delta g = g \otimes g$, $\Delta(h) = 1 \otimes h + h \otimes g$, $S(g) = g$, $S(h) = gh$, $\varepsilon(g) = 1$,

$\varepsilon(h) = 0$. Let $\mathcal{H}$ be the Hopf order in H generated over D by 1, $f = g - 1$ and $\chi = \pi^{-m}h, g\chi$, where π is the generator of the maximal ideal of D (for detail and other examples on Hopf orders we refer to [1]). Let $\alpha, \beta, \gamma \in K$, and consider the quaternion algebra

$$A = \begin{pmatrix} \alpha\beta, \gamma \\ K \end{pmatrix} = K\{u, v; u^2 = \alpha, v^2 = \beta, uv + vu = \gamma\}.$$

Then the D-subalgebra Λ of A generated by u, v is a D-order in A. Consider the standard action of H on A making A into an H-module algebra:

$$g.1 = 1 \quad g.u = -u \quad g.v = -v \quad g.(uv) = uv$$
$$h.1 = 0 \quad h.u = 0 \quad h.v = 1 \quad h.(uv) = u$$

Then

$$M = \mathcal{H} \rightharpoonup \Lambda = (\pi^{-m}) + (\pi^{-m})u + Dv + Duv$$
$$\lambda(M) = D + Du + \pi^m v + \pi^m uv$$

It is clear that $M \not\subset \Lambda, \Lambda \not\subset \lambda(M)$ and $\mathcal{H} \rightharpoonup \lambda(M) = \lambda(M)$.

OBSERVATION 3.5. If a D-order Γ contains a left ideal L that is an $\mathcal{H}$-lattice, then Γ is contained in an $\mathcal{H}$-order. If Γ is a maximal D-order then it is also an $\mathcal{H}$-order and thus a maximal $\mathcal{H}$-order.

Proof. Since L is an $\mathcal{H}$-lattice, $\lambda(L)$ is an $\mathcal{H}$-order and since L is a left ideal for Γ, we have $\Gamma \subset \lambda(L)$. The other statement is obvious from this. $\square$

Notwithstanding the warning contained in Remark 3.3, we may obtain a first characterization of maximal $\mathcal{H}$-orders in terms of maximal orders.

THEOREM 3.6. *If Λ is a maximal $\mathcal{H}$-order in A, then there is a maximal order Γ in A such that $\Lambda = \lambda(\mathcal{H} \rightharpoonup \Gamma) = \rho(\mathcal{H} \rightharpoonup \Gamma)$.*

Proof. Pick a maximal order Γ containing Λ. It follows from Proposition 3.2 that, for $M = \mathcal{H} \rightharpoonup \Gamma$, we have: $\Lambda \subset \lambda(M)$ and $\Lambda \subset \rho(M)$. We know that $\lambda(M)$ and $\rho(M)$ are $\mathcal{H}$-orders, hence the maximality assumption on Λ leads to $\Lambda = \lambda(M) = \rho(M)$. Note that we obtain Λ from a normal ideal $\mathcal{H} \rightharpoonup \Gamma$ in A, i.e. $\mathcal{H} \rightharpoonup \Gamma$ is a Λ-bimodule over D. $\square$

4. CHANGE OF ORDER

We return to the notation introduced in Section 2 and look at the Hopf filtration FH extending fK associated to the valuation v; hence $\mathcal{H} = F_0H$ is the Hopf order we considered in H over $D = O_v$ and ξ_0 is the Hopf valuation function corresponding to FH. Let $\mathcal{L}_D(H)$ be the lattice of D-lattices in H. To FH we may associate a function $\rho : \mathcal{L}_D(H) \to \mathbb{Z} \cup \{\infty\}$, defined by $\rho(M) = -\min\{n, M \subset F_nH\} = \inf\{-\xi(h), h \in M\}$. Clearly we can determine ξ is we know ρ: $\xi(h) = -\rho(Dh)$, for every $h \in H$. So we can understand ξ in terms of the *Hopf pseudo-valuation* ρ.

DEFINITION 4.1. A Hopf pseudo-valuation extending the valuation $v : K \to \mathbb{Z} \cup \{\infty\}$ is a map $\rho : \mathcal{L}_D(H) \to \mathbb{Z} \cup \{\infty\}$ satisfying the following conditions
HP-1 $\rho(M) = \infty$ if and only if $M = 0$;

HP-2 for $n \in \mathbb{Z}$ we have $\rho(f_n K) = -n, \rho(D) = 0$;

HP-3 if $M \supset N$ in $\mathcal{L}_D(H)$ then $\rho(M) \leq \rho(N)$;

HP-4 for M, N in $\mathcal{L}_D(H)$: $\rho(MN) \geq \rho(M) + \rho(N)$ and equality holds when $M \subset K$;

HP-5 for M, N in $\mathcal{L}_D(H), \rho(M + N) = \min\{\rho(M), \rho(N)\}$;

HP-6 for $M \in \mathcal{L}_D(H), \rho(SM) = \rho(M)$;

HP-7 if in $H \otimes H$ there are M, M_1, M_2 in $\mathcal{L}_D(H)$ such that $\Delta(M) = \Sigma M_1 \otimes M_2$ (Sweedler notation), i.e. the space $\Delta(M)$ decomposes as a sum as indicated, then we have that $\rho(M) \leq \max\{\min_\Sigma\{\rho(M_1) + \rho(M_2)\}\}$, where $\min_\Sigma$ ranges over the factors in the decomposition of $\Delta(M)$ and then max is taken over the set of all such possible decomposition of $\Delta(M)$.

THEOREM 4.2. *There is a bijective correspondence between Hopf valuation filtration functions ξ satisfying HV-1,...,HV-7, (see Section 2) and Hopf pseudo-valuations ρ satisfying HP-1,...,HP-7. The correspondence is given by :*

$$\rho(M) = \inf\{-\xi(h), h \in M\}, \text{ for } M \in \mathcal{L}_D(H);$$
$$\xi(h) = -\rho(Dh), \text{ for } h \in H.$$

Proof. This is a straightforward verification. As an example, let us show how HP-4 follows from HV-4.

Look at $X = MN$ in $\mathcal{L}_D(H)$. Then,

$$-\rho(X) = \max\{\xi(x), x \in MN\}$$

For $x = \sum_i' m_i n_i$ with $m_i \in M, n_i \in N$, we have

$$\xi(x) \leq \max\{\xi(m_i) + \xi(n_i)\},$$

and, consequently,

$$\begin{aligned}
-\rho(X) &\leq \max_i\{\xi(m_i) + \xi(n_i)\} \\
&\leq \sup\{\xi(m_i).m_i \in M\} + \sup\{\xi(n_i), n_i \in N\} \\
&\leq -\rho(M) - \rho(N).
\end{aligned}$$

$\square$

If Λ is an $\mathcal{H}$-order in A, then $F_n A = (f_n K)\Lambda$ defines a separated filtration FA on A, extending the valuation filtration fK on K. This defines an algebra valuation filtration function $\chi : A \rightarrow \mathbb{Z} \cup \{-\infty\}$, defined by $\chi(a) = n$ if $a \in F_n A - F_{n-1} A$ and $\chi(b) = -\infty$ if and only if $b = 0$. Let $\mathcal{L}_D(A)$ denote the lattice of D-lattices contained in A, $D = O_v \subset K$. Then χ defines a pseudo-valuation on $\mathcal{L}_D(A)$, namely $\psi : \mathcal{L}_D(A) \rightarrow \mathbb{Z} \cup \{\infty\}$, defined by $\psi(L) = \inf\{-\chi(a), a \in L\}$. This may be compared to the early theory of pseudo-valuations of central simple algebras (in connection to generic crossed products) developed by the second author in [10].

If $\mathcal{H}$ acts on Λ, i.e. Λ is an $\mathcal{H}$-order, then FA makes A into a filtered H-module with respect to FH, in other words, $F_n H \cdot F_m A \subset F_{n+m} A$. In fact the latter is an equality for any $n, m \in \mathbb{Z}$ because the filtrations are strong filtrations. In Section 2, we introduced the following notation $\mathcal{H} = H(\xi_0) = F_0 H$, where FH is a Hopf filtration extending fK with corresponding Hopf valuation filtration function ξ_0. To a D-basis B of $\mathcal{H}$ we associate a Larson-type order $H_B(\xi)$, where ξ corresponds to the new filtration on H defined by $F_n^\xi H = f_n K H_B(\xi)$. We refer to [1] for the basic theory of such Hopf orders in arbitrary finite dimensional Hopf algebras over K. Here we just recall that $H_B(\xi)$ is the D-algebra generated by $\{\omega^{d_\xi(b)}b \mid b \in B\}$, with ω a generator of the maximal ideal of B. The classical examples are Larson's orders constructed in the group algebra KG, where $\mathcal{H} = DG$ and $H_B(\xi)$ is the Larson order constructed with respect to the D-basis $\{1, 1 - g \mid g \in G\}$.

We hope to characterize when an $\mathcal{H}$-order is also an $H_B(\xi)$-order.

THEOREM 4.3. *Consider $\mathcal{H}$ and $H_B(\xi)$ as before, and look at an $\mathcal{H}$-order Λ in A. Then Λ may be viewed as an $H_B(\xi)$-order (this structure also being induced from the H-action on A) if and only if for every $b \in B$: $\psi(b \cdot \Lambda) \geq \rho(Db)$, when ψ, ρ resp. are the pseudo-valuation functions corresponding to Λ, resp. $\mathcal{H}$.*

Proof. By definition, $H_B(\xi)$ is generated as a D-algebra by $\pi^{\xi(b)}b$ for $b \in B$, π a generator for the maximal ideal of D. Now $\psi(b \cdot \Lambda) \geq \rho(Db)$ if and only if $\psi(\pi^{\xi(b)}b.\Lambda) \geq 0$, if and only if $\pi^{\xi(b)}b, because of HP - 4.\Lambda \subset \Lambda$. From the fact that A is an H-module algebra, it then follows that $H_B(\xi) \cdot \Lambda \subset \Lambda$ and Λ is an $H_B|\xi|$-order. Clearly, the converse inclusion also holds, so $H_B(\xi) \cdot \Lambda = \Lambda$. Note in particular that this equality and HV-7 imply that $\Delta(\pi^{\xi(b)}b) = \pi^{\xi(b)}\Sigma b_1 \otimes b_2$ can be written as $\Sigma\pi^{\xi(b_1)+d_1}b_1 \otimes \pi^{\xi(b_2)}b_2$ with $d_1 \geq 0$. Moreover, the right hand side of HV-7 is an attained minimum, since $\mathbb{Z}$ is a discrete group, and $b = \Sigma b_1 \otimes b_2$ with $\xi(b) \geq \max\{\xi(b_1)+\xi(b_2)\}$. Consequently for $x, y \in \Lambda$ we may write $(\pi^{\xi(b)}b) \cdot (xy) = \Sigma(\pi^{d_1}(\pi^{\xi(b_1)}b_1) \cdot x)(\pi^{\xi(b_2)}b_2 \cdot y)$. $\qquad\square$

Let us give an example where different Hopf orders are used.

EXAMPLE 4.4. Let $H = K < x, y >$ be the Taft Hopf algebra; the Hopf algebra structure is given by the formulas

$$\Delta(x) = x \otimes x \qquad x^n = 1 \quad \varepsilon(x) = 1 \qquad S(x) = x^{n-1}$$
$$\Delta(y) = 1 \otimes y + y \otimes x \quad y^n = 0 \quad \varepsilon(x) = 0 \quad S(y) = -w^{-1}x^{n-1}y$$

and $xy = wyx$, where w is a primitive n-th root of unity. The following are Hopf orders in this Hopf algebra (cf. [1])

$$\mathcal{H}_1 = D + \sum_{i=1}^{n-1} D(x - 1)^i + \sum_{\substack{i=0 \\ j=2}}^{n-1}(\pi)^{-jn}(x - 1)^i y^j, \text{ where } n \neq p^s$$

$$\mathcal{H}_2 = D + \sum_{i=1}^{n-1}(\pi)^{-im}(x - 1)^i + \sum_{i=0, j=1}^{n-1}(\pi)^{-(im+jn)}(x - 1)^i y^j, n = p^s$$

and $\pi^{(p^s-p^{s-1})^m}|p, \mu^{-1} = \pi^m|(w - 1)$ (here the valuation ring D in K is assumed to dominate the p-adic valuation ring in $\mathbb{Q}$, p a prime). Now consider the generalized quaternion algebra :

$$A = K < u, v \mid u^n = \alpha, \ v^n = \beta, \ uv + vu = \gamma >,$$

HOPF ORDER MODULE ALGEBRA ORDERS

with $\alpha, \beta, \gamma \in D$. Clearly $\Gamma = D < u, v >$ is a D-order in A. Consider the H-action on A defined as follows:

$$x \rightharpoonup 1 = 1;$$
$$x^i \rightharpoonup u^j v^k = w^{i+j+k} u^i v^j, \text{ if } 1 \leq i \leq n;$$
$$y \rightharpoonup 1 = 0;$$
$$y \rightharpoonup u^i v^j = \delta u^i v^{j-1}, \text{ if } j \text{ odd and for some } \delta \in K;$$
$$y \rightharpoonup u^i v^j = 0, \text{ for } j \text{ even};$$
$$y^i \rightharpoonup u^j v^k = 0, \text{ if } 2 \leq i \text{ and for all } j, k.$$

Let (μ) be the D-ideal generated by μ. Then we obtain

$$\mathcal{H}_{1,2} \rightharpoonup \Gamma = \delta(\mu) + \sum_{j \text{ even}, i=1}^{n} \delta(\mu) u^i v^i + \sum_{j \text{ odd}, i=0}^{n} D u^i v^j.$$

Thus Γ is an $\mathcal{H}$-order if $\mu^{-1} | \delta$. If $\delta \notin D$ then Γ is not an $\mathcal{H}$-order; on the other hand

$$\lambda(\mathcal{H}_{1,2} \rightharpoonup \Gamma) = D + \sum_{\substack{j=\text{even} \\ i=1}}^{n} D u^i v^j + \sum_{\substack{j \text{ odd} \\ i=0}}^{n} \delta^{-1}(\mu^{-1}) u^i v^j$$

is an $\mathcal{H}_{1,2}$-order.

REFERENCES

[1] F. Aly, F. Van Oystaeyen, Hopf Filtrations and Larson-Type Orders in Hopf Algebras, *J. Algebra* **267** (2003), 756–772.

[2] R. G. Larson, Hopf Algebra Orders Determined by Group Valuations, *J. Algebra* **38** (1976), 414–452.

[3] R. G. Larson, Orders in Hopf Algebras, *J. Algebra* **22** (1972), 201–210.

[4] H. S. Li, F. Van Oystaeyen, "Zariskian Filtrations", *K-Monographs Math.* **3**, Kluwer Academic Publishers, Dordrecht, 1996.

[5] S. Montgomery, "Hopf algebras and their actions on rings", American Mathematical Society, Providence, 1993.

[6] I. Reiner, "Maximal Orders", Academic Press, 1975.

[7] O. Schilling, "The Theory of Valuations", *Math. Surveys* **4**, American Mathematical Society, Providence, 1950.

[8] M. E. Sweedler, "Hopf algebras", Benjamin, New York, 1969.

[9] E. Taft, The Order of the Antipode of a Finite Dimensional Hopf Algebra, *Proc. Nat. Acad. Sci. USA* **68** (1971), 2631–2633.

[10] F. Van Oystaeyen, On Pseudo-Places of Algebras, *Bull. Soc. Math. Belg.* **25** (1973), 139–159.

An alternative notion of Hopf Algebroid

GABRIELLA BÖHM   Research Institute for Particle and Nuclear Physics
Budapest, P.O.B. 49, H-1525 Budapest 114, Hungary
e-mail: *bgabr@rmki.kfki.hu*

> ABSTRACT. In [1] a new notion of Hopf algebroid has been introduced. It was
> shown to be inequivalent to the structure introduced under the same name
> in [18]. We review this new notion of Hopf algebroid. We prove that two
> Hopf algebroids are isomorphic as bialgebroids if and only if their antipodes
> are related by a 'twist' i.e. are deformed by the analogue of a character. A
> precise relation to weak Hopf algebras is given. After the review of the integral
> theory of Hopf algebroids we show how a right bialgebroid can be made a
> Hopf algebroid in the presence of a non-degenerate left integral. This can be
> interpreted as the 'half of the Larson-Sweedler theorem'. As an application
> we construct the Hopf algebroid symmetry of an abstract depth 2 Frobenius
> extension [2].

1. INTRODUCTION

Recently many authors introduced generalizations of bialgebras. These structures
are common in the feature that they do not need to be algebras rather bimodules
over some – possibly non-commutative – base ring L. In the paper [5] the notions of
Lu's bialgebroid [18] (axiomatized in a more compact form in [29]), Xu's bialgebroid
with anchor [33] and Takeuchi's $\times_L$-bialgebra [32] were shown to be equivalent. We
use the definition as follows:

DEFINITION 1.1. A *left bialgebroid* (or Takeuchi $\times_L$-bialgebra) $\mathcal{A}_L$ consists of the
data $(A, L, s_L, t_L, \gamma_L, \pi_L)$, where A and L are associative unital rings, called the
total ring and the base ring, and $s_L : L \to A$ and $t_L : L^{op} \to A$ are ring
homomorphisms such that the images of L in A commute and make A an L-L
bimodule:

$$l \cdot a \cdot l' := s_L(l)t_L(l')a. \tag{1}$$

The bimodule (1) is denoted by $_L A_L$. The triple $(_L A_L, \gamma_L, \pi_L)$ is a comonoid in
$_L \mathcal{M}_L$, the category of L-L bimodules. Introducing Sweedler's notation $\gamma_L(a) \equiv
a_{(1)} \otimes a_{(2)} \in A_L \otimes _L A$ the identities

$$a_{(1)}t_L(l) \otimes a_{(2)} = a_{(1)} \otimes a_{(2)}s_L(l) \tag{2}$$

$$\gamma_L(1_A) = 1_A \otimes 1_A \tag{3}$$

$$\gamma_L(ab) = \gamma_L(a)\gamma_L(b) \tag{4}$$

2000 *Mathematics Subject Classification.* 16W30, 13B02.

Key words and phrases. bialgebroid, Hopf algebroid, twist, integral.

Research supported by the Hungarian Scientific Research Fund, OTKA – T 034 512, FKFP –
0043/2001 and the Bolyai János Fellowship.

$$\pi_L(1_A) = 1_L \tag{5}$$

$$\pi_L(as_L \circ \pi_L(b)) = \pi_L(ab) = \pi_L(at_L \circ \pi_L(b)) \tag{6}$$

are required for all $l \in L$ and $a, b \in A$. The requirement (4) makes sense in view of (2).

With the help of the maps s_L and t_L we can introduce four commuting actions of L on A. They give rise to L-modules

$$_LA: \ l \cdot a = s_L(l)a \qquad A_L: \ a \cdot l = t_L(l)a$$
$$A^L: \ a \cdot l = as_L(l) \qquad {}^LA: \ l \cdot a = at_L(l).$$

One defines the bimodules $^LA^L$, ^LA_L and $_LA^L$ in the obvious way.

Throughout the paper it is a typical situation that the same ring A carries different L-module structures. In this situation the usual notation $A \underset{L}{\otimes} A$ is ambiguous. Our notation of bimodule tensor products is explained at the beginning of Section 2.

In [18] J. H. Lu introduced the notion of *Hopf algebroid* as a triple $(\mathcal{A}_L, S, \xi)$ consisting of a left bialgebroid $\mathcal{A}_L = (A, L, s_L, t_L, \gamma_L, \pi_L)$ such that A and L are algebras over a commutative ring k. It is equipped with an antipode $S : A \to A$. The S is required to be an anti-automorphism of the k-algebra A satisfying

$$S \circ t_L = s_L \tag{7}$$

$$m_A \circ (S \otimes_L \mathrm{id}_A) \circ \gamma_L = t_L \circ \pi_L \circ S \tag{8}$$

$$m_A \circ (\mathrm{id}_A \otimes_k S) \circ \xi \circ \gamma_L = s_L \circ \pi_L \tag{9}$$

where m_A is the multiplication in A and ξ is a section of the canonical projection $p_L : A \otimes_k A \to A_L \otimes {}_LA$ that is ξ is a map $A_L \otimes {}_LA \to A \otimes_k A$ satisfying $p_L \circ \xi = \mathrm{id}_{A \otimes_L A}$.

Following the result of [31, 17] – proving that an irreducible finite index depth 2 extension of von Neumann factors can be realized as a crossed product with a finite dimensional C^*-Hopf algebra – big effort has been made in order to make connection between more and more general kinds of extensions and of 'quantum symmetries' [25, 8, 9, 35, 24]. Allowing for *reducible* finite index D2 extensions of II_1 von Neumann factors in [19] and of von Neumann algebras with finite centers in [22, 23] the symmetry of the extension was shown to be described by a *finite dimensional C^*-weak Hopf algebra* introduced in [4, 21, 3]. A Galois correspondence has been established in [20] in the case of finite index finite depth extensions of II_1 factors. The infinite index D2 case has been treated in [10] for arbitrary von Neumann algebras endowed with a regular operator valued weight.

In the paper [14] depth 2 extensions of *rings* have been investigated. It was shown that there exists a canonical dual pair of (finite) bialgebroids associated to such a ring extension.

Studying the bialgebroids corresponding to a depth 2 *Frobenius* extension of rings one easily generalizes the formula describing the antipode in [31, 17, 24] to an anti-automorphism of the total ring satisfying the axioms (7-8). This supports the expectation that in the case of a depth 2 Frobenius extension of rings the canonical bialgebroids obtained in [14] can be made Hopf algebroids. However, no

AN ALTERNATIVE NOTION OF HOPF ALGEBROID

effort, made for checking the Lu-axioms [18] in this situation, brought success. (See however [12, 13] studying interesting subcases.) As a matter of fact the section ξ, appearing in the definition in [18], does not come naturally into this context.

This leads us to the introduction of an alternative notion of Hopf algebroid in [1] – the prototype of which is the symmetry of a depth 2 Frobenius extension of rings. In [1] three equivalent sets of axioms are formulated[1]. The first definition is analogous to the one in [18] in the aspect that a left bialgebroid $\mathcal{A}_L = (A, L, s_L, t_L, \gamma_L, \pi_L)$ is equipped with a bijective antipode map S which is a ring anti-automorphism and relates the left and right L-module structures as in (7). Also the antipode axiom (8) is identical. The main difference is that in the definition of [1] no reference to a section 'ξ' is needed. In its stead we deal with the maps

$$(S \otimes S) \circ \gamma_L^{op} \circ S^{-1} \quad \text{and} \quad (S^{-1} \otimes S^{-1}) \circ \gamma_L^{op} \circ S. \tag{10}$$

In the Hopf algebra case both are equal to the coproduct itself. In this more general case the images of A under γ_L and the maps (10) are however different. We require the two maps of (10) to be equal and be both a left and a right comodule map. This form of the axioms does not contain a second antipode axiom. This definition is cited at the end of Section 3 of this paper.

Analysing the consequences of this first definition one observes a hidden right bialgebroid [2] structure on A with the coproduct given by the equal maps (10). This lead us to the second 'symmetric' definition in [1] where the two antipode axioms have analogous forms using both the left and right bialgebroid structures of A in a symmetric way. We cite this definition in Definition 3.1 below.

The third definition in [1] is formulated without the explicit use of the antipode map. It borrows the philosophy of [27] where a Hopf algebroid like object (possibly without antipode), the so-called $\times_L$-*Hopf algebra* was introduced as a left bialgebroid s.t. the map

$$\alpha : {}^L A \underset{L^{op}}{\otimes} A_L \to A_L \otimes_L A, \qquad a \otimes b \mapsto a_{(1)} \otimes a_{(2)} b$$

is bijective. Requiring however the bijectivity of the maps α and its co-opposite β (leading to a bijective antipode in the bialgebra case) for a left bialgebroid does not imply the existence of an antipode. In order to have a definition which is equivalent to the other two both related bialgebroid structures are needed. This definition explicitly shows that the Hopf algebroids in the sense of [1] are $\times_L$-Hopf algebras.

It is proven in [1] that (weak) Hopf algebras are Hopf algebroids in the sense of [1]. Also some examples of Lu-Hopf algebroids [18, 5, 15, 7] are shown to be examples.

The two notions of Hopf algebroid – the one introduced in [18] and the one in [1] – were shown to be inequivalent by giving an example of the Hopf algebroid [1] that does not satisfy the axioms of [18]. This example is discussed in the Section 4 of this paper in more detail.

On the other hand no example of Lu-Hopf algebroid is known to us at the moment that does not satisfy the axioms of [1]. This leaves open the logical possibility that

---

[1]added in proof: The final version of [1] contains four equivalent definitions. A 'zeroth' one has been added later.

[2]For the definition of the right bialgebroid [14] see Definition 2.1 below.

34 G. BÖHM

the Lu-Hopf algebroid was a subcase of the one introduced in [1]. Until now we could neither prove nor exclude by examples this possibility.

In [1] the theory of non-degenerate integrals in a Hopf algebroid is developed. Though the axioms of the Hopf algebroid in [1] are by no means self-dual, it is proven in [1] that if there exists a non-degenerate left integral ℓ in a Hopf algebroid $\mathcal{A}$ then its dual (with respect to the base ring) also carries a Hopf algebroid structure which is unique upto an isomorphism of bialgebroids. The dual of the bialgebroid isomorphism class of $\mathcal{A}$ is defined as the bialgebroid isomorphism class of the Hopf algebroid constructed on the dual ring.

In this paper we present a review and also some new results on Hopf algebroids. In Section 2 we review some results about bialgebroids that were obtained in the papers [28, 32, 29, 14]. In Section 3 we repeat the definition of the Hopf algebroid given in [1] and cite some basic results without proofs. In Section 4 we generalize the notion of the *twist of a Hopf algebra* – introduced in [6] – to Hopf algebroids. In particular we prove that twisted (weak) Hopf algebras are Hopf algebroids in the sense of [1]. A most important example is the Connes-Moscovici algebra $\mathcal{H}_{FM}$ [7]. By twisting cocommutative Hopf algebras we construct examples that do not satisfy the Hopf algebroid axioms of [18]. We give a sufficient and necessary criterion on a Hopf algebroid under which it is a (twisted version) of weak Hopf algebra. In the Section 5 we review the integral theory of Hopf algebroids from [1] without proofs. The Section 6 deals with the question how can we make a right bialgebroid with a non-degenerate left integral into a Hopf algebroid. The result of Section 6 can be interpreted as the generalization of the 'easier half of the Larson-Sweedler theorem' [16].

2. BIALGEBROIDS

The total ring of a bialgebroid carries different module structures over the base ring. In this situation we make the following notational convention. In writing module tensor products we write out the (bi-) module factors explicitly. For the L-module tensor product of the bimodules $_L A^L$ and $_L A_L$, for example we write $_L A^{L'} \underset{L'}{\otimes} {_{L'}} A_L$, where L' stands for another copy of L, and it has been introduced to show explicitly which module structures are involved in the tensor product.

In order not to make the formulas more complicated than necessary we make a further simplification. In those situations in which it is clear from the tensor factors themselves over which ring the tensor product is taken, we do not denote it under the symbol $\otimes$. I.e. for the L-module tensor product of the right L-module A_L and the left L-module $_L A$, for example, we write $A_L \otimes_L A$.

The *bialgebroid* [18, 33, 29] or – what is equivalent to it – a Takeuchi $\times_L$-bialgebra [32] is a generalization of the bialgebra in the sense that it is not an algebra rather a bimodule over a non-commutative ring L. We use Definition 1.1 of the left bialgebroid. We use the name *left* bialgebroid as in [14] since the 'opposite structure' was called a *right bialgebroid* in [14]:

DEFINITION 2.1. A *right bialgebroid* $\mathcal{A}_R$ consists of the data $(A, R, s_R, t_R, \gamma_R, \pi_R)$. The A and R are associative unital rings, the total and base rings, respectively. The $s_R : R \to A$ and $t_R : R^{op} \to A$ are ring homomorphisms such that the images of R in A commute making A an $R - R$ bimodule:

$$r \cdot a \cdot r' := a s_R(r') t_R(r). \tag{11}$$

The bimodule (11) is denoted by ${}^R A^R$. The triple $({}^R A^R, \gamma_R, \pi_R)$ is a comonoid in ${}_R \mathcal{M}_R$. Introducing the Sweedler's notation $\gamma_R(a) \equiv a^{(1)} \otimes a^{(2)} \in A^R \otimes {}^R A$ the identities

$$s_R(r) a^{(1)} \otimes a^{(2)} = a^{(1)} \otimes t_R(r) a^{(2)},$$
$$\gamma_R(1_A) = 1_A \otimes 1_A,$$
$$\gamma_R(ab) = \gamma_R(a) \gamma_R(b),$$
$$\pi_R(1_A) = 1_R,$$
$$\pi_R (s_R \circ \pi_R(a) b) = \pi_R(ab) = \pi_R (t_R \circ \pi_R(a) b)$$

are required for all $r \in R$ and $a, b \in A$.

In addition to the bimodule ${}^R A^R$ we introduce also

$$_R A_R : \qquad r \cdot a \cdot r' := s_R(r) t_R(r') a.$$

If $\mathcal{A}_L = (A, L, s_L, t_L, \gamma_L, \pi_L)$ is a left bialgebroid then so is the co-opposite $(\mathcal{A}_L)_{cop} = (A, L^{op}, t_L, s_L, \gamma_L^{op}, \pi_L)$ – where $\gamma_L^{op} : A \to {}_L A \otimes A_L$ maps a to $a_{(2)} \otimes a_{(1)}$ –. The opposite $(\mathcal{A}_L)^{op} = (A^{op}, L, t_L, s_L, \gamma_L, \pi_L)$ is a right bialgebroid.

We use the terminology of homomorphisms of bialgebroids as introduced in [30]:

DEFINITION 2.2. A *left bialgebroid homomorphism* $\mathcal{A}_L \to \mathcal{A}'_{L'}$ is a pair of ring homomorphisms $(\Phi : A \to A', \phi : L \to L')$ such that

$$
\begin{aligned}
s'_L \circ \phi &= \Phi \circ s_L, \\
t'_L \circ \phi &= \Phi \circ t_L, \\
\pi'_L \circ \Phi &= \phi \circ \pi_L, \\
\gamma'_L \circ \Phi &= (\Phi \otimes \Phi) \circ \gamma_L.
\end{aligned}
$$

The last condition makes sense since by the first two conditions $\Phi \otimes \Phi$ is a well defined map $A_L \otimes {}_L A \to A'_{L'} \otimes {}_{L'} A'$. The pair (Φ, ϕ) is an *isomorphism of left bialgebroids* if it is a left bialgebroid homomorphism such that both Φ and ϕ are bijective.

A right bialgebroid homomorphism (isomorphism) $\mathcal{A}_R \to \mathcal{A}'_{R'}$ is a left bialgebroid homomorphism (isomorphism) $(\mathcal{A}_R)^{op} \to (\mathcal{A}'_{R'})^{op}$.

Let $\mathcal{A}_L$ be a left bialgebroid. The equation (1) describes two L-modules A_L and ${}_L A$. Their L-duals are the additive groups of L-module maps:

$$\mathcal{A}_* := \{ \phi_* : A_L \to L_L \} \quad \text{and} \quad {}_* \mathcal{A} := \{ {}_* \phi : {}_L A \to {}_L L \}$$

where ${}_L L$ stands for the left regular and L_L for the right regular L-module. Both $\mathcal{A}_*$ and ${}_* \mathcal{A}$ carry left A module structures via the transpose of the right regular action of A. For $\phi_* \in \mathcal{A}_*, {}_* \phi \in {}_* \mathcal{A}$ and $a, b \in A$ we have:

$$(a \rightharpoonup \phi_*)(b) = \phi_*(ba) \quad \text{and} \quad (a \rightharpoonup_* \phi)(b) = {}_* \phi(ba).$$

Similarly, in the case of a right bialgebroid $\mathcal{A}_R$ – denoting the left and right regular R-modules by ${}^R R$ and R^R, respectively, – the two R-dual additive groups

$$\mathcal{A}^* := \{\phi^* : A^R \to R^R\} \quad \text{and} \quad {}^*\!\mathcal{A} := \{{}^*\!\phi : {}^R A \to {}^R R\}$$

carry right A-module structures:

$$(\phi^* \leftharpoonup a)(b) = \phi^*(ab) \quad \text{and} \quad ({}^*\!\phi \leftharpoonup a)(b) = {}^*\!\phi(ab).$$

The comonoid structures can be transposed to give monoid (i.e. ring) structures to the duals. In the case of a left bialgebroid $\mathcal{A}_L$

$$(\phi_* \psi_*)(a) = \psi_* \left(s_L \circ \phi_*(a_{(1)}) a_{(2)} \right) \quad \text{and} \quad ({}_*\!\phi \, {}_*\!\psi)(a) = {}_*\!\psi \left(t_L \circ {}_*\!\phi(a_{(2)}) a_{(1)} \right) \quad (12)$$

for ${}_*\!\phi, {}_*\!\psi \in {}_*\!\mathcal{A}$, $\phi_*, \psi_* \in \mathcal{A}_*$ and $a \in A$. Similarly, in the case of a right bialgebroid $\mathcal{A}_R$

$$(\phi^* \psi^*)(a) = \phi^* \left(a^{(2)} t_R \circ \psi^*(a^{(1)}) \right) \quad \text{and} \quad ({}^*\!\phi \, {}^*\!\psi)(a) = {}^*\!\phi \left(a^{(1)} s_R \circ {}^*\!\psi(a^{(2)}) \right) \quad (13)$$

for $\phi^*, \psi^* \in \mathcal{A}^*$, ${}^*\!\phi, {}^*\!\psi \in {}^*\!\mathcal{A}$ and $a \in A$. In the case of a left bialgebroid $\mathcal{A}_L$ also the ring A has right $\mathcal{A}_*$- and right ${}_*\!\mathcal{A}$- module structures:

$$a \leftharpoonup \phi_* = s_L \circ \phi_*(a_{(1)}) a_{(2)} \quad \text{and} \quad a \leftharpoonup {}_*\!\phi = t_L \circ {}_*\!\phi(a_{(2)}) a_{(1)}$$

for $\phi_* \in \mathcal{A}_*$, ${}_*\!\phi \in {}_*\!\mathcal{A}$ and $a \in A$.

Similarly, in the case of a right bialgebroid $\mathcal{A}_R$ the ring A has left $\mathcal{A}^*$- and left ${}^*\!\mathcal{A}$ structures:

$$\phi^* \rightharpoonup a = a^{(2)} t_R \circ \phi^*(a^{(1)}) \quad \text{and} \quad {}^*\!\phi \rightharpoonup a = a^{(1)} s_R \circ {}^*\!\phi(a^{(2)})$$

for $\phi^* \in \mathcal{A}^*$, ${}^*\!\phi \in {}^*\!\mathcal{A}$ and $a \in A$.

It is also proven in [14] that if the L (R) module structure on A is finitely generated projective then the corresponding dual has also a bialgebroid structure.

3. HOPF ALGEBROID

Let $\mathcal{A}_L = (A, L, s_L, t_L, \gamma_L, \pi_L)$ be a left bialgebroid and $\mathcal{A}_R = (A, R, s_R, t_R, \gamma_R, \pi_R)$ a right bialgebroid such that the base rings are anti-isomorphic $R \simeq L^{op}$. Require that

$$s_L(L) = t_R(R) \quad \text{and} \quad t_L(L) = s_R(R) \quad (14)$$

as subrings of A. The requirement (14) implies that the coproduct γ_L is a quadro-module map ${}^R_L A^R_L \to {}^R_L A_{L'} \underset{L'}{\otimes} {}_{L'} A^R_L$ and γ_R is a quadro-module map ${}^R_L A^R_L \to {}^R_L A^{R'} \underset{R'}{\otimes} {}^{R'} A^R_L$. (The L' and R' denote another copy of L and R, respectively). This allows us to require that

$$(\gamma_L \otimes \mathrm{id}_A) \circ \gamma_R = (\mathrm{id}_A \otimes \gamma_R) \circ \gamma_L$$
$$(\gamma_R \otimes \mathrm{id}_A) \circ \gamma_L = (\mathrm{id}_A \otimes \gamma_L) \circ \gamma_R \quad (15)$$

as maps $A \to A_L \underset{R}{\otimes} {}_L A^R \underset{R}{\otimes} {}^R A$ and $A \to A^R \underset{R}{\otimes} {}^R A_L \underset{L}{\otimes} {}_L A$, respectively.

Let $S : A \to A$ be a bijection of additive groups such that S is a twisted isomorphism of bimodules ${}^L A_L \to {}_L A^L$ and ${}^R A_R \to {}_R A^R$ that is

$$S(t_L(l) a t_L(l')) = s_L(l') S(a) s_L(l) \quad \text{and} \quad S(t_R(r') a t_R(r)) = s_R(r) S(a) s_R(r') \quad (16)$$

for all $l, l' \in L$, $r, r' \in R$ and $a \in A$. The requirement (16) makes the expressions $S(a_{(1)})a_{(2)}$ and $a^{(1)}S(a^{(2)})$ meaningful. We require

$$S(a_{(1)})a_{(2)} = s_R \circ \pi_R(a) \text{ and } a^{(1)}S(a^{(2)}) = s_L \circ \pi_L(a) \tag{17}$$

for all a in A.

DEFINITION 3.1. The triple $\mathcal{A} = (\mathcal{A}_L, \mathcal{A}_R, S)$ satisfying (14-17) is a *Hopf algebroid*.

Throughout the paper we use the analogue of the Sweedler-Heyneman notation: $\gamma_L(a) = a_{(1)} \otimes a_{(2)}$ and $\gamma_R(a) = a^{(1)} \otimes a^{(2)}$.

For a Hopf algebroid $\mathcal{A}$ also the opposite $\mathcal{A}^{op} = (\mathcal{A}_R^{op}, \mathcal{A}_L^{op}, S^{-1})$ and the co-opposite $\mathcal{A}_{cop} = (\mathcal{A}_{L\,cop}, \mathcal{A}_{R\,cop}, S^{-1})$ are Hopf algebroids.

In Section 4 we will investigate the way in which (weak) Hopf algebra is a subcase.

The antipode of a Hopf algebra is a bialgebra anti-homomorphism. This property generalizes to Hopf algebroids as

PROPOSITION 3.2. *Both* $(S, \pi_R \circ s_L)$ *and* $(S^{-1}, \pi_R \circ t_L)$ *are left bialgebroid isomorphisms* $\mathcal{A}_L \to (\mathcal{A}_R)_{cop}^{op}$. *That is*

$$
\begin{array}{ll}
s_R \circ \pi_R \circ s_L = S \circ s_L & s_R \circ \pi_R \circ t_L = S^{-1} \circ s_L \\
t_R \circ \pi_R \circ s_L = S \circ t_L & t_R \circ \pi_R \circ t_L = S^{-1} \circ t_L \\
\pi_R \circ s_L \circ \pi_L = \pi_R \circ S & \pi_R \circ t_L \circ \pi_L = \pi_R \circ S^{-1} \\
S_{A \otimes_L A} \circ \gamma_L = \gamma_R \circ S & S_{A \otimes_R A}^{-1} \circ \gamma_L = \gamma_R \circ S^{-1}
\end{array}
$$

where $S_{A \otimes_L A}$ *is a map* $A_L \otimes_L A \to A^R \otimes^R A$, *it maps* $a \otimes b$ *to* $S(b) \otimes S(a)$. *Similarly,* $S_{A \otimes_R A}$ *is a map* $A^R \otimes^R A \to A_L \otimes_L A$, *it maps* $a \otimes b$ *to* $S(b) \otimes S(a)$.

The datum $(\mathcal{A}_L, \mathcal{A}_R, S)$ determining a Hopf algebroid is somewhat redundant. Indeed, suppose that we have given only a left bialgebroid $\mathcal{A}_L = (A, L, s_L, t_L, \gamma_L, \pi_L)$ and an anti-isomorphism S of the total ring A satisfying

$$S \circ t_L = s_L \tag{18}$$

$$m_A \circ (S \otimes \mathrm{id}_A) \circ \gamma_L = t_L \circ \pi_L \circ S \tag{19}$$

$$S_{A \otimes_L A} \circ \gamma_L \circ S^{-1} = S_{A \otimes_R A}^{-1} \circ \gamma_L \circ S \tag{20}$$

$$(\gamma_L \otimes \mathrm{id}_A) \circ \gamma_R = (\mathrm{id}_A \otimes \gamma_R) \circ \gamma_L \tag{21}$$

$$(\gamma_R \otimes \mathrm{id}_A) \circ \gamma_L = (\mathrm{id}_A \otimes \gamma_L) \circ \gamma_R \tag{22}$$

for m_A the multiplication in A and $\gamma_R: = S_{A \otimes_L A} \circ \gamma_L \circ S^{-1}$. Then the right bialgebroid $\mathcal{A}_R$ – together with which $(\mathcal{A}_L, \mathcal{A}_R, S)$ is a Hopf algebroid – can be reconstructed upto a trivial bialgebroid isomorphism. Namely, it follows from Proposition 3.2 that

$$\mathcal{A}_R = (A, R, S \circ s_L \circ \nu^{-1}, s_L \circ \nu^{-1}, S_{A \otimes_L A} \circ \gamma_L \circ S^{-1}, \nu \circ \pi_L \circ S^{-1})$$

for an arbitrary isomorphism $\nu : L^{op} \to R$.

4. TWIST OF THE HOPF ALGEBROID

It is clear that being given the left and right bialgebroids $\mathcal{A}_L$ and $\mathcal{A}_R$ satisfying the axioms (14) and (15) the antipode – if it exists – is unique. Indeed, if both S_1 and

38   G. BÖHM

S_2 make $(\mathcal{A}_L, \mathcal{A}_R, S_1)$ and $(\mathcal{A}_L, \mathcal{A}_R, S_2)$ Hopf algebroids then

$$
\begin{aligned}
S_2(a) &= s_R \circ \pi_R(a^{(1)})S_2(a^{(2)}) = S_1(a^{(1)}{}_{(1)})a^{(1)}{}_{(2)}S_2(a^{(2)}) \\
&= S_1(a_{(1)})a_{(2)}{}^{(1)}S_2(a_{(2)}{}^{(2)}) = S_1(a_{(1)})s_L \circ \pi_L(a_{(2)}) = S_1(a).
\end{aligned}
$$

There are some examples however in which only the left bialgebroid structure is naturally given and we have some ambiguity in the choice of the right bialgebroid structure and the corresponding antipode. (See for example the Hopf algebroid symmetry of a depth 2 Frobenius extension of rings in Section 3 of [1] or at the end of Section 6 below. In this example the ambiguity is nicely controlled by the Radon-Nykodim derivative relating the possible Frobenius maps.) In the following we address the question more generally: given a left bialgebroid $\mathcal{A}_L$ how are the possible antipodes satisfying the conditions (18-22) related?

DEFINITION 4.1. Let $(\mathcal{A}_L, S)$ be subject to the conditions (18-22). An invertible element g_* of $\mathcal{A}_*$ is called a *twist* of $(\mathcal{A}_L, S)$ if for all elements a, b of A

$$
\begin{aligned}
&i) \quad 1_A \leftharpoonup g_* = 1_A \\
&ii) \quad (a \leftharpoonup g_*)(b \leftharpoonup g_*) = ab \leftharpoonup g_* \\
&iii) \quad S(a_{(1)}) \leftharpoonup g_* \otimes a_{(2)} = S(a_{(1)}) \otimes a_{(2)} \leftharpoonup g_*^{-1}.
\end{aligned}
\qquad (23)
$$

The condition iii) is understood to be an identity in the product of the modules A^L and the left L-module on A:

$$
l \cdot a = s_L \circ g_*^{-1} \circ s_L(l)a
$$

where $g_* \circ s_L$ is an automorphism of L with inverse $g_*^{-1} \circ s_L$.

The *twist* of Definition 4.1 generalizes the notion of the character on a Hopf algebra. Clearly the twists of $(\mathcal{A}_L, S)$ form a group.

THEOREM 4.2. *Let $(\mathcal{A}_L, S)$ be subject to the conditions (18-22). Then $(\mathcal{A}_L, S')$ is subject to the conditions (18-22) if and only if there exists a twist g_* of $(\mathcal{A}_L, S)$ such that $S'(a) = S(a \leftharpoonup g_*)$ for all $a \in A$.*

Remark 4.3. For $(\mathcal{A}_L, S)$ a *Hopf algebra* the twisted antipode of the above form was introduced in [6]. In the view of Theorem 4.2 the twisted Hopf algebras in [6] are Hopf algebroids in the sense of Definition 3.1.

Proof. (of Theorem 4.2) *if part:* For a twist g_* the map $S_g(a): = S(a \leftharpoonup g_*)$ is bijective with inverse $S_g^{-1}(a) = S^{-1}(a) \leftharpoonup g_*^{-1}$. It is anti-multiplicative by ii) of (23). Using the property i) of (23)

$$
S_g \circ t_L(l) = S(t_L(l) \leftharpoonup g_*) = S((1_A \leftharpoonup g_*)t_L(l)) = S \circ t_L(l) = s_L(l).
$$

By $\gamma_L(a \leftharpoonup g_*) = (a_{(1)} \leftharpoonup g_*) \otimes a_{(2)}$ and $S \circ t_R \circ \pi_R = t_L \circ \pi_L \circ S$ we have

$$
S_g(a_{(1)})a_{(2)} = S(a_{(1)} \leftharpoonup g_*)a_{(2)} = s_R \circ \pi_R(a \leftharpoonup g_*) = t_L \circ \pi_L \circ S_g(a).
$$

In order to check the property (20) of $(\mathcal{A}_L, S_g)$ rewrite iii) of (23) into the equivalent form

$$
S(a_{(1)}) \otimes S_g(a_{(2)}) = S_g^{-1} \circ S^2(a_{(1)}) \otimes S(a_{(2)}).
\qquad (24)
$$

AN ALTERNATIVE NOTION OF HOPF ALGEBROID

Then using the fact that $(\mathcal{A}_L, S)$ satisfies (20) introduce $a^{(1)} \otimes a^{(2)} := S_{\mathcal{A} \otimes_L \mathcal{A}} \circ \gamma_L \circ S^{-1}(a) \equiv S_{\mathcal{A} \otimes_R \mathcal{A}}^{-1} \circ \gamma_L \circ S(a)$. By (24) we have

$$S_g(S_g^{-1}(a)_{(2)}) \otimes S_g(S_g^{-1}(a)_{(1)}) = S_g(S^{-1}(a)_{(2)}) \otimes S_g(S^{-1}(a)_{(1)} \leftharpoonup g_*^{-1})$$

$$= \; S_g(S^{-1}(a)_{(2)}) \otimes S(S^{-1}(a)_{(1)}) = S(S^{-1}(a)_{(2)}) \otimes S_g^{-1} \circ S^2(S^{-1}(a)_{(1)})$$

$$= \; a^{(1)} \otimes S_g^{-1} \circ S(a^{(2)}) = S_g^{-1} \circ S(a^{(1)} \leftharpoonup g_*) \otimes S_g^{-1} \circ S(a^{(2)})$$

$$= \; S_g^{-1}\left(S(a \leftharpoonup g_*)_{(2)}\right) \otimes S_g^{-1}\left(S(a \leftharpoonup g_*)_{(1)}\right)$$

$$= \; S_g^{-1}\left(S_g(a)_{(2)}\right) \otimes S_g^{-1}\left(S_g(a)_{(1)}\right).$$

The last condition (22) on $(\mathcal{A}_L, S_g)$ follows then easily by using the two forms of $S_{g \, \mathcal{A} \otimes_L \mathcal{A}} \circ \gamma_L \circ S_g^{-1}(a) \equiv \gamma_{gR}(a) = a^{(1)} \otimes S_g^{-1} \circ S(a^{(2)})$ and $\gamma_{gR}(a) = S_g \circ S^{-1}(a^{(1)}) \otimes a^{(2)}$, respectively:

$$(\gamma_L \otimes \mathrm{id}_A) \circ \gamma_{gR}(a) = a^{(1)}{}_{(1)} \otimes a^{(1)}{}_{(2)} \otimes S_g^{-1} \circ S(a^{(2)})$$

$$= \; a_{(1)} \otimes a_{(2)}{}^{(1)} \otimes S_g^{-1} \circ S(a_{(2)}{}^{(2)}) = (\mathrm{id}_A \otimes \gamma_{gR}) \circ \gamma_L(a)$$

$$(\mathrm{id}_A \otimes \gamma_L) \circ \gamma_{gR}(a) = S_g \circ S^{-1}(a^{(1)}) \otimes a^{(2)}{}_{(1)} \otimes a^{(2)}{}_{(2)}$$

$$= \; S_g \circ S^{-1}(a_{(1)}{}^{(1)}) \otimes a_{(1)}{}^{(2)} \otimes a_{(2)}(\gamma_{gR} \otimes \mathrm{id}_A) \circ \gamma_L(a).$$

only if part: Let $(\mathcal{A}_L, S)$ and $(\mathcal{A}_L, S')$ be subject to the conditions (18-22). By the considerations at the end of Section 3 we can construct the right bialgebroids $\mathcal{A}_R$ and $\mathcal{A}'_{R'}$ such that both $\mathcal{A} = (\mathcal{A}_L, \mathcal{A}_R, S)$ and $\mathcal{A}' = (\mathcal{A}_L, \mathcal{A}'_{R'}, S')$ are Hopf algebroids. Denoting the coproducts in $\mathcal{A}_R$ and $\mathcal{A}'_{R'}$ by $\gamma_R(a) = a^{(1)} \otimes a^{(2)}$ and $\gamma'_{R'}(a) = a^{(1)'} \otimes a^{(2)'}$, respectively, the maps

$$\alpha: \; {}^L A \otimes A_L \to A_L \otimes {}_L A, \qquad a \otimes b \mapsto a_{(1)} \otimes a_{(2)} b$$

$$\beta: \; A^L \otimes {}_L A \to {}_L A \otimes A_L, \qquad a \otimes b \mapsto a_{(2)} \otimes a_{(1)} b$$

are easily shown to be bijections with inverses

$$a^{(1)} \otimes S(a^{(2)}) b = \alpha^{-1}(a \otimes b) = a^{(1)'} \otimes S'(a^{(2)'}) b$$

$$a^{(2)} \otimes S^{-1}(a^{(1)}) b = \beta^{-1}(a \otimes b) = a^{(2)'} \otimes S'^{-1}(a^{(1)'}) b. \tag{25}$$

(This means that $\mathcal{A}_L$ and $(\mathcal{A}_L)_{cop}$ are $\times_L$-Hopf algebras in the sense of [27].)

We construct the twist $\pi_L \circ S^{-1} \circ S'$. It is an invertible element of $\mathcal{A}_*$ with inverse $\pi_L \circ S'^{-1} \circ S$:

$$[(\pi_L \circ S^{-1} \circ S')(\pi_L \circ S'^{-1} \circ S)](a)$$

$$= \; \pi_L \circ S'^{-1} \circ S\left(s_L \circ \pi_L \circ S^{-1} \circ S'(a_{(1)}) a_{(2)}\right)$$

$$= \; \pi_L \circ S'^{-1} \circ S\left(s_L \circ \pi_L \circ S^{-1}[S'(a)^{(2)'}] S'^{-1}[S'(a)^{(1)'}]\right)$$

$$= \; \pi_L \circ S'^{-1} \circ S\left(s_L \circ \pi_L \circ S^{-1}[S'(a)^{(2)}] S^{-1}[S'(a)^{(1)}]\right)$$

$$= \; \pi_L \circ S'^{-1}\left(S'(a)^{(1)} s_R \circ \pi_R(S'(a)^{(2)})\right) = \pi_L(a) \tag{26}$$

where in the third step (25) has been used. The relation $(\pi_L \circ S'^{-1} \circ S)(\pi_L \circ S^{-1} \circ S') = \pi_L$ follows by interchanging the roles of S and S'.

40   G. BÖHM

For all elements a, b in A we have

$$1_A \leftharpoonup \pi_L \circ S^{-1} \circ S' = s_L \circ \pi_L \circ S^{-1} \circ S'(1_A) = 1_A$$

$$(a \leftharpoonup \pi_L \circ S^{-1} \circ S')(b \leftharpoonup \pi_L \circ S^{-1} \circ S')$$

$$= s_L \circ \pi_L \circ S^{-1} \circ S'(a_{(1)})a_{(2)}s_L \circ \pi_L \circ S^{-1} \circ S'(b_{(1)})b_{(2)}$$

$$= s_L \circ \pi_L \left(S^{-1} \circ S'(a_{(1)})t_L \circ \pi_L \circ S^{-1} \circ S'(b_{(1)}) \right) a_{(2)}b_{(2)}$$

$$= s_L \circ \pi_L \left(S^{-1} \circ S'(a_{(1)})S^{-1} \circ S'(b_{(1)}) \right) a_{(2)}b_{(2)}$$

$$= s_L \circ \pi_L \circ S^{-1} \circ S'(a_{(1)}b_{(1)})a_{(2)}b_{(2)}$$

$$= ab \leftharpoonup \pi_L \circ S^{-1} \circ S'.$$

Using (25) we can show that

$$S(a \leftharpoonup \pi_L \circ S^{-1} \circ S') = S \left(s_L \circ \pi_L \circ S^{-1}[S'(a)^{(2)'}]S'^{-1}[S'(a)^{(1)'}] \right)$$

$$= S \left(s_L \circ \pi_L \circ S^{-1}[S'(a)^{(2)}]S^{-1}[S'(a)^{(1)}] \right)$$

$$= S'(a)^{(1)}s_R \circ \pi_R(S'(a)^{(2)}) = S'(a). \tag{27}$$

In order to check that $\pi_L \circ S^{-1} \circ S'$ satisfies iii) of (23) rewrite (25) into the equivalent forms:

$$a^{(1)} \otimes S'^{-1} \circ S(a^{(2)}) = a^{(1)'} \otimes a^{(2)'} = S' \circ S^{-1}(a^{(1)}) \otimes a^{(2)}$$

$$\Leftrightarrow \quad S^{-1}(S(a)_{(2)}) \otimes S'^{-1}(S(a)_{(1)}) = S'(S^{-1}(a)_{(2)}) \otimes S(S^{-1}(a)_{(1)})$$

$$\Leftrightarrow \quad S(a_{(2)}) \otimes S'^{-1} \circ S^2(a_{(1)}) = S'(a_{(2)}) \otimes S(a_{(1)}).$$

In view of (27), the last formula is equivalent to iii) of (23). This proves that $\pi_L \circ S^{-1} \circ S'$ is a twist relating S' to S. $\qquad\square$

Let $\mathcal{A} = (\mathcal{A}_L, \mathcal{A}_R, S)$ and $\mathcal{A}' = (\mathcal{A}'_{L'}, \mathcal{A}'_{R'}, S')$ be Hopf algebroids such that the underlying left bialgebroids $\mathcal{A}_L$ and $\mathcal{A}'_{L'}$ are isomorphic via the isomorphism ($\Phi : A \to A', \phi : L \to L'$). Then by Proposition 3.2 also the underlying right bialgebroids $\mathcal{A}_R$ and $\mathcal{A}'_{R'}$ are isomorphic and by Theorem 4.2 $S'(a') = \Phi \circ S \left(\Phi^{-1}(a') \leftharpoonup g_* \right)$ for a unique twist g_* of $(\mathcal{A}_L, S)$ and all $a' \in A'$. The Hopf algebroids $\mathcal{A}$ and $\mathcal{A}'$ are called *bialgebroid isomorphic* in the following.

Recall from [1] that a weak Hopf algebra $\mathbf{H} = (H, \Delta, \varepsilon, S)$ over a commutative ring k with bijective antipode S determines a Hopf algebroid $\mathcal{H} = (\mathcal{H}_L, \mathcal{H}_R, S)$ as follows:

$$\mathcal{H}_L = (H, L, \mathrm{id}_L, S^{-1}|_L, p_L \circ \Delta, \sqcap^L) \tag{28}$$

$$\mathcal{H}_R = (H, R, \mathrm{id}_R, S^{-1}|_R, p_R \circ \Delta, \sqcap^R) \tag{29}$$

where $\sqcap^L : H \to H$ is defined as $h \mapsto \varepsilon(1_{[1]}h)1_{[2]}$, $1_{[1]} \otimes 1_{[2]} = \Delta(1)$, $L : = \sqcap^L(H)$, and p_L is the canonical projection $H \otimes_k H \to H \otimes_L H$ the $L - L$-bimodule structure on H being given by

$$l \cdot h \cdot l' : = lS^{-1}(l')h.$$

Similarly, $\sqcap^R : H \to H$ is defined as $h \mapsto 1_{[1]}\varepsilon(h1_{[2]})$, $R : = \sqcap^R(H)$, and p_R is the canonical projection $H \otimes_k H \to H \otimes_R H$ the $R - R$-bimodule structure on H being given by

$$r \cdot h \cdot r' : = hr'S^{-1}(r).$$

In view of Theorem 4.2 we can obtain examples of Hopf algebroids by twisting weak Hopf algebras. The twists of the datum $(\mathcal{H}_L, S)$ are the characters on the weak Hopf algebra $\mathbf{H}$ or – if H is finite dimensional as a k-space – the group-like elements [3, 36] in the k-dual weak Hopf algebra $\hat{\mathbf{H}}$.

The twistings of cocommutative Hopf algebras are of particular interest as they provide examples of Hopf algebroids in the sense of Definition 3.1 that do *not* satisfy the Lu-Hopf algebroid axioms of [18]. (For [18]'s definition see the Introduction above.) If $\mathcal{H}_L$ is the left bialgebroid (28) corresponding to the *cocommutative k-Hopf algebra* $\mathbf{H}$ that is $\mathcal{H}_L = (H, L \equiv k, s_L \equiv \eta, t_L \equiv \eta, \gamma_L \equiv \Delta, \pi_L \equiv \varepsilon)$ – where $\eta : k \to H$ is the unit map $\lambda \mapsto \lambda 1_H$ – then the base ring is k itself, so the canonical projection is the identity map $p_L = \mathrm{id}_{H \otimes_k H}$. Then also $\xi = \mathrm{id}_{H \otimes_k H}$. Let χ be a character on $\mathbf{H}$ that is an algebra homomorphism $H \to k$, and $S_\chi : = (\chi \otimes S) \circ \Delta$ the twisted antipode. Then denoting $\Delta(h) = h_{(1)} \otimes h_{(2)}$ we have

$$s_L \circ \pi_L(h) = \varepsilon(h) 1_H$$

and

$$h_{(1)} S_\chi(h_{(2)}) = h_{(1)} \chi(h_{(2)}) S(h_{(3)})$$
$$= \chi(h_{(1)}) h_{(2)} S(h_{(3)}) = \chi(h_{(1)}) \varepsilon(h_{(2)}) 1_H = \chi(h) 1_H,$$

hence $(\mathcal{H}_L, S_\chi)$ – which satisfies the conditions (18-22) by Theorem 4.2, defines a Hopf algebroid in the sense of Definition 3.1 – is a Lu-Hopf algebroid (with the only possible section $\xi = \mathrm{id}_{H \otimes_k H}$) if and only if $\chi = \varepsilon$. It is easy, however, to find non-trivial characters on cocommutative Hopf algebras. In the simplest case of the group Hopf algebra kZ_2 we can set $\chi(t) = -t$ (where t is the second order generator of Z_2) if the characteristic of k is different from 2. This gives the Example presented in [1] proving that the two definitions – the one in [1] and the one in [18] – of Hopf algebroid are not equivalent.

We know from the above considerations that weak Hopf algebras with a bijective antipode provide examples of Hopf algebroids. In the following we identify the Hopf algebroids that arise in this way.

It is proven in [26, 29] that a left bialgebroid $\mathcal{A}_L = (A, L, s_L, t_L, \gamma_L, \pi_L)$ has a weak bialgebra structure *if and only if* A is an algebra over a commutative ring k and L is a separable k-algebra. Indeed let us fix a separability structure, that is a datum $(L, k, \delta : L \to L \otimes_k L, \psi : L \to k)$ where δ is a coassociative coproduct with counit ψ satisfying

$$(\mathrm{id}_L \otimes_k m_L) \circ (\delta \otimes_k \mathrm{id}_L) = \delta \circ m_L = (m_L \otimes_k \mathrm{id}_L) \circ (\mathrm{id}_L \otimes_k \delta)$$

and

$$m_L \circ \delta = \mathrm{id}_L.$$

m_L denotes the multiplication in L. The weak bialgebra structure on A, corresponding to the given separability structure, reads as

$$\Delta(a) : = t_L(e_i) a_{(1)} \otimes s_L(f_i) a_{(2)}, \qquad \varepsilon(a) = \psi \circ \pi_L(a) \qquad (30)$$

where $\sum_i e_i \otimes f_i = \delta(1_L)$ and the summation symbol is omitted.

42   G. BÖHM

On the other hand if $\mathbf{H} = (H, \Delta, \varepsilon)$ is a weak bialgebra over the commutative ring k then a separability structure $(L, k, \delta, \varepsilon)$ is given by

$$\delta(l) = l \sqcap^L (1_{[1]}) \otimes 1_{[2]} \equiv \sqcap^L(1_{[1]}) \otimes 1_{[2]}l. \tag{31}$$

This implies that the separability of the base algebra L is a necessary condition for the Hopf algebroid $\mathcal{A} = (\mathcal{A}_L, \mathcal{A}_R, S)$ to have a weak Hopf algebra structure. The different separability structures determine however different weak bialgebra structures (30). The given antipode S of the Hopf algebroid $\mathcal{A}$ cannot make all of them into a weak Hopf algebra. The following Theorem 4.4 gives a criterion on the separability structure (L, k, δ, ψ) under which the corresponding weak bialgebra (30) together with (a twist of) the antipode S becomes a weak Hopf algebra.

Let $\mathcal{A}_L = (A, L, s_L, t_L, \gamma_L, \pi_L)$ be a left bialgebroid such that A is an algebra over a commutative ring k and L is a separable k-algebra. Let us fix a separability structure (L, k, δ, ψ) and introduce the notation $\delta(1_L) = e_i \otimes f_i$ (summation on i is understood). Let Δ and ε be as in (30). Then we can equip the k-space $\hat{A}$ of k-linear maps $A \to k$ with an algebra structure with the multiplication

$$\phi\phi' := (\phi \otimes \phi') \circ \Delta \tag{32}$$

and unit ε. Now we are ready to formulate

THEOREM 4.4. *Let $\mathcal{A} = (\mathcal{A}_L, \mathcal{A}_R, S)$ be a Hopf algebroid such that the total ring A is an algebra over a commutative ring k and the base ring L of $\mathcal{A}_L$ is a separable k-algebra. Then fixing a separability structure (L, k, δ, ψ) the corresponding weak bialgebra (30) and the antipode S form a weak Hopf algebra if and only if $\psi \circ \pi_L \circ S = \psi \circ \pi_L$. Furthermore, there exists a twisted antipode S_g making the weak bialgebra (30) a weak Hopf algebra if and only if the element $\psi \circ \pi_L \circ S$ of the algebra $\hat{A}$ - defined in (32) - is invertible.*

Proof. if part: The separability structure (L, k, δ, ψ) defines an isomorphism of the algebras $\mathcal{A}_*$ - the L-dual algebra (12) of $\mathcal{A}_L$ - and $\hat{A}$ in (32):

$$\begin{aligned} \kappa: \quad & \hat{A} \to \mathcal{A}_* \quad & \phi \mapsto (a \mapsto \phi(t_L(e_i)a)f_i) \\ \kappa^{-1}: \quad & \mathcal{A}_* \to \hat{A} \quad & \phi_* \mapsto \psi \circ \phi_*. \end{aligned}$$

The element $\kappa(\psi \circ \pi_L \circ S)$ satisfies the properties *i)-iii)* of Definition 23:

$$1_A \leftharpoonup \kappa(\psi \circ \pi_L \circ S) = s_L(f_i)\psi \circ \pi_L \circ S \circ t_L(e_i) = s_L(1_L) = 1_A$$

$$\begin{aligned} &(a \leftharpoonup \kappa(\psi \circ \pi_L \circ S))(b \leftharpoonup \kappa(\psi \circ \pi_L \circ S)) \\ &\quad = s_L(f_i)a_{(2)}s_L(f_j)b_{(2)}\psi \circ \pi_L \circ S(t_L(e_i)a_{(1)})\psi \circ \pi_L \circ S(t_L(e_j)b_{(1)}) \\ &\quad = s_L(f_i)a_{(2)}b_{(2)}\psi \circ \pi_L \circ S \left(t_L[e_j\psi(f_j\pi_L \circ S\{t_L(e_i)a_{(1)}\})]b_{(1)} \right) \\ &\quad = s_L(f_i)a_{(2)}b_{(2)}\psi \circ \pi_L \circ S(t_L(e_i)a_{(1)}b_{(1)}) = ab \leftharpoonup \kappa(\psi \circ \pi_L \circ S) \end{aligned}$$

$$S(a_{(1)}) \leftharpoonup \kappa(\psi \circ \pi_L \circ S) \otimes a_{(2)} \leftharpoonup \kappa(\psi \circ \pi_L \circ S)$$
$$= s_L(f_i)S(a_{(1)})_{(2)}$$
$$\otimes s_L(f_j)a_{(2)(2)}\psi \circ \pi_L \circ S\left(t_L(e_i)S(a_{(1)})_{(1)}\right)\psi \circ \pi_L \circ S\left(t_L(e_j)a_{(2)(1)}\right)$$
$$= s_L(f_i)S(a_{(1)})_{(2)}s_L(f_j)$$
$$\otimes a_{(2)(2)}\psi \circ \pi_L \circ S\left(t_L(e_i)S(a_{(1)})_{(1)}\right)\psi \circ \pi_L \circ S\left(t_L(e_j)a_{(2)(1)}\right)$$
$$= s_L(f_i)S(a_{(1)}{}^{(1)})$$
$$\otimes a_{(2)(2)}\psi\left(f_j\pi_L \circ S\left[t_L(e_i)S(a_{(1)}{}^{(2)})\right]\right)\psi \circ \pi_L \circ S\left(t_L(e_j)a_{(2)(1)}\right)$$
$$= s_L(f_i)S(a_{(1)}{}^{(1)}) \otimes a_{(2)(2)}\psi \circ \pi_L \circ S\left(t_L(e_i)S(a_{(1)}{}^{(2)})a_{(2)(1)}\right)$$
$$= s_L(f_i)S(a^{(1)}) \otimes a^{(2)}{}_{(2)}\psi \circ \pi_L \circ S\left(t_L(e_i)S(a^{(2)}{}_{(1)(1)})a^{(2)}{}_{(1)(2)}\right)$$
$$= s_L(f_i)S(a^{(1)}) \otimes a^{(2)}{}_{(2)}\psi \circ \pi_L \circ S \circ t_L\left(\pi_L \circ S(a^{(2)}{}_{(1)})e_i\right)$$
$$= \psi\left(\pi_L \circ S(a_{(1)}{}^{(2)})e_i\right)s_L(f_i)S(a_{(1)}{}^{(1)}) \otimes a_{(2)}$$
$$= S \circ s_R \circ \pi_R(a_{(1)}{}^{(2)})S(a_{(1)}{}^{(1)}) \otimes a_{(2)} = S(a_{(1)}) \otimes a_{(2)}$$

hence, since $\psi \circ \pi_L \circ S \in \hat{A}$ is invertible by assumption, both $\kappa(\psi \circ \pi_L \circ S)$ and its inverse $g_* := \kappa(\psi \circ \pi_L \circ S)^{-1}$ are twists of (A_L, S) in the sense of Definition 23.

Using the standard properties [37] of the quasi-basis $e_i \otimes f_i$ of ψ one checks that the twisted antipode $S_g : a \mapsto S(a \leftharpoonup g_*)$ makes the weak bialgebra (30) a weak Hopf algebra. That is by definition
$$\psi \circ \pi_L \circ S_g = \kappa^{-1}(g_*)(\psi \circ \pi_L \circ S) = \psi \circ \pi_L.$$
Since $S_g \circ t_L = S \circ t_L = s_L$, we have
$$\psi\left(\pi_L(a)\pi_L \circ S_g^2 \circ t_L(l)\right) = \psi \circ \pi_L\left(S_g^2 \circ t_L(l)s_L \circ \pi_L(a)\right)$$
$$= \psi \circ \pi_L\left(t_L \circ \pi_L(a)s_L(l)\right) = \psi(l\pi_L(a)).$$
This implies that $e_i \otimes f_i = f_i \otimes \pi_L \circ S_g^{-2} \circ t_L(e_i)$ and hence $e_i \otimes s_L(f_i) = f_i \otimes S_g^{-1} \circ t_L(e_i)$. Then
$$m_A \circ (S_g \otimes_k \mathrm{id}_A) \circ \Delta(a) = S\left(s_L \circ g_*(a_{(1)})t_L(e_i)a_{(2)}\right)s_L(f_i)a_{(3)}$$
$$= s_R \circ \pi_R(a \leftharpoonup g_*) = t_L \circ \pi_L \circ S_g(a)$$
$$= \psi\left(\pi_L \circ S_g(a)e_i\right)t_L(f_i) = \psi \circ \pi_L\left(aS_g^{-1} \circ t_L(e_i)\right)t_L(f_i)$$
$$= t_L(e_i)\psi \circ \pi_L(as_L(f_i)),$$

$$m_A \circ (\mathrm{id}_A \otimes_k S_g) \circ \Delta(a) = t_L(e_i)a_{(1)}S(a_{(3)})S \circ s_L \circ g_*(s_L(f_i)a_{(2)})$$
$$= t_L(e_i)a_{(1)}t_L(e_k)\psi \circ \pi_L \circ$$
$$S\left(a_{(3)}{}^{(1)}t_L(f_k)\right)S\left(s_L \circ g_*(s_L(f_i)a_{(2)})a_{(3)}{}^{(2)}\right)$$
$$= t_L(e_i)a_{(1)}t_L(e_k)\psi \circ \pi_L \circ$$
$$S\left(s_L \circ g_*[s_L(f_i)a_{(2)}{}^{(1)}{}_{(1)}]a_{(2)}{}^{(1)}{}_{(2)}t_L(f_k)\right)S(a_{(2)}{}^{(2)})$$

$$= t_L(e_i)a_{(1)}t_L(e_k)\psi \circ \pi_L\left(s_L(f_i)a_{(2)}^{(1)}t_L(f_k)\right)S(a_{(2)}^{(2)})$$
$$= t_L \circ \pi_L(a^{(1)}{}_{(2)}t_L(f_k))a^{(1)}{}_{(1)}t_L(e_k)S(a^{(2)})$$
$$= s_L \circ \pi_L(a) = \psi \circ \pi_L(t_L(e_i)a)s_L(f_i),$$

$$S_g(t_L(e_i)a_{(1)})s_L(f_i)t_L(e_j)a_{(2)}S_g(s_L(f_j)a_{(3)}) = S_g(a_{(1)})s_L \circ \pi_L(a_{(2)}) = S_g(a),$$

for all $a \in A$. This finishes the proof of the if part.

only if part: If $\mathcal{H}_L$ is a left bialgebroid (28) corresponding to the weak Hopf algebra $\mathbf{H}$ then its base ring L is a separable k-algebra [3, 26]. The separability structure (31) is determined by the weak Hopf data in $\mathbf{H}$. Now if the datum $(\mathcal{H}_L, S')$ is obtained as a twist of the datum $(\mathcal{H}_L, S)$ then the twist element relating them is of the form

$$\sqcap^L \circ S^{-1} \circ S' = \kappa(\varepsilon \circ \sqcap^L \circ S^{-1} \circ S') = \kappa(\varepsilon \circ S^{-1} \circ S') = \kappa(\varepsilon \circ S') = \kappa(\varepsilon \circ \sqcap^L \circ S')$$

which is an invertible element of $\mathcal{H}_*$, by definition. Since κ is an algebra isomorphism this proves that $\varepsilon \circ \sqcap^L \circ S'$ is an invertible element of $\hat{H}$, hence the claim. $\square$

Remark 4.5. Proposition 4.4 implies that the Hopf algebroid $\mathcal{A} = (\mathcal{A}_L, \mathcal{A}_R, S)$ is obtained by a twist of a *Hopf algebra* over the commutative ring k if and only if the following conditions hold:

i) $L \simeq R^{op}$ is isomorphic to k;
ii) A is a k-algebra;
iii) $\pi_L \circ S : A \to k$ is invertible in $\hat{A}$, the dual k-algebra of A.

In particular it is a Hopf algebra if and only if the conditions $i)$ and $ii)$ hold and $\pi_L \circ S = \pi_L$.

5. THE THEORY OF INTEGRALS

The left/right *integrals* in a Hopf algebra $(H, \Delta, \varepsilon, S)$ are the invariants of left/right regular module. That is $\ell/\Upsilon \in H$ is a left/right integral if

$$h\ell = \varepsilon(h)\ell \quad / \quad \Upsilon h = \Upsilon \varepsilon(h)$$

for all $h \in H$. This notion has been generalized to a weak Hopf algebra $(H, \Delta, \varepsilon, S)$ as $\ell/\Upsilon \in H$ is a left/right integral if

$$h\ell = \sqcap^L(h)\ell \quad / \quad \Upsilon h = \Upsilon \sqcap^R(h)$$

for all $h \in H$ where $\sqcap^L(h) = \varepsilon(1_{[1]}h)1_{[2]}$, $\sqcap^R(h) = 1_{[1]}\varepsilon(h1_{[2]})$ and $1_{[1]} \otimes 1_{[2]} = \Delta(1_H)$.

It is then straightforward to generalize the notion of integrals to Hopf algebroids:

DEFINITION 5.1. The left integrals in a left bialgebroid $\mathcal{A}_L = (A, L, s_L, t_L, \gamma_L, \pi_L)$ are the elements $\ell \in A$ that satisfy

$$a\ell = s_L \circ \pi_L(a)\ell$$

for all $a \in A$. The right ideal of left integrals is denoted by $\mathcal{I}^L(\mathcal{A})$. The right integrals in the right bialgebroid $\mathcal{A}_R = (A, R, s_R, t_R, \gamma_R, \pi_R)$ are the elements $\Upsilon \in A$ for which

$$\Upsilon a = \Upsilon s_R \circ \pi_R(a)$$

for all $a \in A$. The left ideal of right integrals is denoted by $\mathcal{I}^R(A)$.

In a Hopf algebroid $\mathcal{A} = (\mathcal{A}_L, \mathcal{A}_R, S)$ the left/right integrals are the left/right integrals in $\mathcal{A}_L/\mathcal{A}_R$.

As a support of this definition [3, Lemma 3.2] generalizes as

LEMMA 5.2. *The following are equivalent:*
  i) $\ell \in \mathcal{I}^L(\mathcal{A})$;
  ii) $a\ell = t_L \circ \pi_L(a)\ell$, *for all* $a \in A$;
  iii) $S(\ell) \in \mathcal{I}^R(\mathcal{A})$;
  iv) $S^{-1}(\ell) \in \mathcal{I}^R(\mathcal{A})$;
  v) $S(a)\ell^{(1)} \otimes \ell^{(2)} = \ell^{(1)} \otimes a\ell^{(2)}$ *as elements of* $A^R \otimes^R A$, *for all* $a \in A$.

A left integral in a Hopf algebroid $\mathcal{A}$ is also a left integral in $\mathcal{A}_{cop}$ and it is a right integral in $\mathcal{A}^{op}$.

For the Hopf algebroid $\mathcal{A} = (\mathcal{A}_L, \mathcal{A}_R, S)$ we introduce the following notation: Let $\mathcal{A}^*$ and $^*\mathcal{A}$ denote the dual rings (13) of the right bialgebroid $\mathcal{A}_R$ and $_*\mathcal{A}$ and $\mathcal{A}_*$ denote the dual rings (12) of the left bialgebroid $\mathcal{A}_L$. We define the non-degeneracy of an integral as follows:

DEFINITION 5.3. The left integral $\ell \in \mathcal{I}^L(\mathcal{A})$ is non-degenerate if the maps

$$\ell_R : \ \mathcal{A}^* \ \to A, \qquad \phi^* \mapsto \phi^* \rightharpoonup \ell \tag{33}$$

and

$$_R\ell : \ ^*\mathcal{A} \ \to A \qquad {}^*\phi \mapsto {}^*\phi \rightharpoonup \ell \tag{34}$$

are bijective. A right integral $\Upsilon \in \mathcal{I}^R(\mathcal{A})$ is non-degenerate if the maps

$$_L\Upsilon : \ _*\mathcal{A} \to A, \qquad _*\phi \mapsto \Upsilon \leftharpoonup {}_*\phi \tag{35}$$

and

$$\Upsilon_L : \ \mathcal{A}_* \ \to A \qquad \phi_* \mapsto \Upsilon \leftharpoonup \phi_* \tag{36}$$

are bijective.

The identities

$$S(\phi^* \rightharpoonup a) = S(a) \leftharpoonup \pi_L \circ s_R \circ \phi^* \circ S^{-1}$$
$$S^{-1}(\phi^* \rightharpoonup a) = S^{-1}(a) \leftharpoonup \pi_L \circ t_R \circ \phi^* \circ S$$
$$S({}^*\phi \rightharpoonup a) = S(a) \leftharpoonup \pi_L \circ s_R \circ {}^*\phi \circ S^{-1}$$
$$S^{-1}({}^*\phi \rightharpoonup a) = S^{-1}(a) \leftharpoonup \pi_L \circ t_R \circ {}^*\phi \circ S$$

imply that ℓ is a non-degenerate left integral if and only if $S(\ell)$ is a non-degenerate right integral and if and only if $S^{-1}(\ell)$ is a non-degenerate right integral.

Let ℓ be a non-degenerate left integral in the Hopf algebroid $\mathcal{A}$. Introducing $\lambda^* := \ell_R^{-1}(1_A)$ and $^*\lambda := {}_R\ell^{-1}(1_A)$ we have

$$(\lambda^* \leftharpoonup S(a)) \rightharpoonup \ell = \ell^{(2)} t_R \circ \lambda^*(S(a)\ell^{(1)}) = a(\lambda^* \rightharpoonup \ell) = a,$$
$$({}^*\lambda \leftharpoonup S^{-1}(a)) \rightharpoonup \ell = \ell^{(1)} s_R \circ {}^*\lambda(S^{-1}(a)\ell^{(2)}) = a({}^*\lambda \rightharpoonup \ell) = a,$$

hence

$$\ell_R^{-1}(a) = \lambda^* \leftharpoonup S(a) \text{ and } {}_R\ell^{-1}(a) = {}^*\lambda \leftharpoonup S^{-1}(a). \tag{37}$$

Recall that in Definition 5.3 the non-degeneracy of a *left* integral is defined in terms of the duals $\mathcal{A}^*$ and $^*\mathcal{A}$ of the *right* bialgebroid $\mathcal{A}_R$. The explanation of this is that – in view of property v) in Lemma 5.2 – *this* notion of non-degeneracy is equivalent to the property that $(\lambda^*, \ell^{(1)} \otimes S(\ell^{(2)}))$ is a Frobenius system for the ring extension $s_R : R \to A$. This implies in particular that for a Hopf algebroid $\mathcal{A}$ possessing a non-degenerate left integral ℓ the modules A^R, ^RA, $_LA$ and A_L are all finitely generated projective. Hence by the results in [14], the corresponding duals $\mathcal{A}^*$ and $^*\mathcal{A}$ carry left- and $_*\mathcal{A}$ and $\mathcal{A}_*$ carry right bialgebroid structures. In addition to the maps ℓ_R and $_R\ell$ also the maps

$$\ell_L : \mathcal{A}_* \to A, \qquad \phi_* \mapsto \ell \leftharpoonup \phi_*$$

$$_L\ell : {}_*\mathcal{A} \to A, \qquad {}_*\phi \mapsto \ell \leftharpoonup {}_*\phi$$

turn out to be bijective. The map

$$\tilde{S} : a \mapsto \ell \leftharpoonup (a \rightharpoonup \pi_L \circ s_R \circ \lambda^*) \tag{38}$$

is an anti-automorphism of the ring A and we have the following isomorphisms of left bialgebroids:

$$
\begin{array}{ccc}
(\mathcal{A}_{*R})^{op}_{cop} & \xrightarrow{\ (_L\ell^{-1} \circ \tilde{S}^{-1} \circ \ell_L, \mathrm{id}_R)\ } & (_*\mathcal{A}_R)^{op}_{cop} \\[2mm]
{\scriptstyle (\ell_R^{-1} \circ \ell_L, \pi_R \circ s_L)} \downarrow & {\scriptstyle (_R\ell^{-1} \circ {}_L\ell, \pi_R \circ t_L)} & \downarrow \\[2mm]
\mathcal{A}^*_L & \xrightarrow{\ (_R\ell^{-1} \circ \tilde{S}^{-1} \circ \ell_R, \pi_R \circ S^{-1} \circ t_R)\ } & {}^*\mathcal{A}_L
\end{array}
$$

Let $\mathcal{A} = (\mathcal{A}_L, \mathcal{A}_R, S)$ be a Hopf algebroid with a non-degenerate left integral ℓ and let $\mathcal{A}' = (\mathcal{A}'_{L'}, \mathcal{A}'_{R'}, S')$ be a Hopf algebroid which is bialgebroid isomorphic to $\mathcal{A}$ via the isomorphism $(\Phi : A \to A', \phi : L \to L')$ of left bialgebroids. Then $\Phi(\ell)$ is a non-degenerate left integral in $\mathcal{A}'$.

The antipode $S_* := \ell_L^{-1} \circ \tilde{S} \circ \ell_L$, $\phi_* \mapsto (\ell \leftharpoonup \phi_*) \rightharpoonup \pi_L \circ s_R \circ \lambda^*$ makes the dual right bialgebroid $\mathcal{A}_{*R}$ into a Hopf algebroid called $\mathcal{A}^\ell_*$ possessing a two-sided non-degenerate integral $\pi_L \circ s_R \circ \lambda^*$. Clearly the bialgebroid isomorphism class of $\mathcal{A}^\ell_*$ does not depend on the choice of the non-degenerate integral ℓ. The *dual of the bialgebroid isomorphism class of $\mathcal{A}$* is then defined to be the bialgebroid isomorphism class of $\mathcal{A}^\ell_*$. This notion of duality is shown to be involutive and reproduces the duality of finite weak Hopf algebras [3] as follows:

Let $\mathbf{H} = (H, \Delta, \varepsilon, S)$ be a finite weak Hopf algebra over the commutative ring k and let $\hat{\mathbf{H}} = (\hat{H}, \hat{\Delta}, \hat{\varepsilon}, \hat{S})$ be its k-dual weak Hopf algebra [3]. The corresponding Hopf algebroids are denoted by $\mathcal{H} = (\mathcal{H}_L, \mathcal{H}_R, S)$ and $\hat{\mathcal{H}} = (\hat{\mathcal{H}}_{\hat{L}}, \hat{\mathcal{H}}_{\hat{R}}, \hat{S})$, respectively. Then the right bialgebroids $\mathcal{H}_{*R}$ and $\hat{\mathcal{H}}_{\hat{R}}$ are isomorphic via

$$\Phi : \mathcal{H}_* \to \hat{H}, \qquad \psi_* \mapsto \varepsilon \circ \psi_*, \tag{39}$$

$$\phi : L \to \hat{R}, \qquad l \mapsto \varepsilon_{[1]} \varepsilon_{[2]}(l), \tag{40}$$

where $\varepsilon_{[1]} \otimes \varepsilon_{[2]} = \hat{\Delta}(\varepsilon)$. This implies that the Hopf algebroids $\mathcal{H}^\ell_*$ and $\hat{\mathcal{H}}$ are bialgebroid isomorphic for any choice of the non-degenerate left integral ℓ. Making use of the separability structure (31) we can equip $\mathcal{H}_*$ with a weak bialgebra structure. Since the weak Hopf structure on a weak bialgebra is unique, there is a unique twist

AN ALTERNATIVE NOTION OF HOPF ALGEBROID

of S_* that makes this weak bialgebra into a weak Hopf algebra. The map Φ in (39) is a weak Hopf algebra isomorphism from it to $\hat{\mathbf{H}}$.

6. ANTIPODE FROM NON-DEGENERATE INTEGRAL

The Larson-Sweedler theorem [16] on Hopf algebras states that a finite dimensional bialgebra over a field is a Hopf algebra if and only if it has a non-degenerate left integral. It has been generalized to weak Hopf algebras in [36]. At the moment no generalization to Hopf algebroids is known. In this section we present a proof of the 'easy half' of the Larson-Sweedler theorem. Namely we show that if a finite *right* bialgebroid $\mathcal{A}_R$ – that is a right bialgebroid such that A^R and $^R A$ are finitely generated projective – has a non-degenerate *left* integral then it can be made a Hopf algebroid.

Notice that though we called it the 'easy half' of the Larson-Sweedler theorem, it is non-trivial in the sense that left integrals have not been defined in right bialgebroids until now. Recall that if $\mathcal{A}$ is a Hopf algebroid possessing a non-degenerate left integral ℓ then both maps ℓ_R and $_R\ell$ (33-34) are bijective and for all $a \in A$

$$\ell^{(1)} \otimes a\ell^{(2)} = S(a)\ell^{(1)} \otimes \ell^{(2)}.$$

The equations (37) imply that the antipode and its inverse can be written into the forms

$$\begin{aligned} S(a) &= (\tilde{\lambda} \leftharpoonup a) \rightharpoonup \ell \\ S^{-1}(a) &= (\lambda^* \leftharpoonup a) \rightharpoonup \ell \end{aligned}$$

with the help of $\lambda^* = \ell_R^{-1}(1_A)$ and $\tilde{\lambda} = {}_R\ell^{-1}(1_A)$. These formulae motivate

DEFINITION 6.1. An element ℓ of a finite right bialgebroid $\mathcal{A}_R$ is a *non-degenerate left integral* if

$$i) \quad \text{both maps } \ell_R \text{ and } {}_R\ell \text{ are bijective} \tag{41}$$

$$ii) \quad \ell^{(1)} \otimes a\ell^{(2)} = [(\tilde{\lambda} \leftharpoonup a) \rightharpoonup \ell]\ell^{(1)} \otimes \ell^{(2)} \tag{42}$$

$$a\ell^{(1)} \otimes \ell^{(2)} = \ell^{(1)} \otimes [(\lambda^* \leftharpoonup a) \rightharpoonup \ell]\ell^{(2)} \tag{43}$$

as elements of $A^R \otimes {}^R A$ for all $a \in A$, where $\lambda^* = \ell_R^{-1}(1_A)$ and $\tilde{\lambda} = {}_R\ell^{-1}(1_A)$.

The following Lemma is of technical use:

LEMMA 6.2. *Let $\mathcal{A}_R = (A, R, s_R, t_R, \gamma_R, \pi_R)$ be a finite right bialgebroid and $k \in A$ such that the map*

$$k_R : \mathcal{A}^* \rightarrow A, \qquad \phi^* \mapsto \phi^* \rightharpoonup k$$

is bijective. Set $\kappa^* := k_R^{-1}(1_A)$. Then

$$\kappa^* \rightharpoonup a = s_R \circ \kappa^*(a) \tag{44}$$

for all $a \in A$. Analogously, for $k \in A$ such that the map

$$_R k : {}^*\mathcal{A} \rightarrow A, \qquad {}^*\phi \mapsto {}^*\phi \rightharpoonup k$$

is bijective set $^*\kappa := {}_R k^{-1}(1_A)$. Then

$$^*\kappa \rightharpoonup a = t_R \circ {}^*\kappa(a) \tag{45}$$

for all $a \in A$.

Proof. Introduce the map $\hat{s}: R \to \mathcal{A}^*$ via $\hat{s}(r)(a) = r\pi_R(a)$. With its help

$$
\begin{aligned}
\phi^*\kappa^* &= k_R^{-1}(\phi^*\kappa^* \rightharpoonup k) = k_R^{-1}(\phi^* \rightharpoonup 1_A) = k_R^{-1}(\hat{s} \circ \phi^*(1_A) \rightharpoonup 1_A) \\
&= k_R^{-1}(\hat{s} \circ \phi^*(1_A)\kappa^* \rightharpoonup k) = \hat{s} \circ \phi^*(1_A)\kappa^*
\end{aligned}
$$

for all $\phi^* \in \mathcal{A}^*$. This implies that

$$
\phi^*(\kappa^* \rightharpoonup a) = (\phi^*\kappa^*)(a) = (\hat{s} \circ \phi^*(1_A)\kappa^*)(a) = \phi^*(1_A)\kappa^*(a) = \phi^*(s_R \circ \kappa^*(a)),
$$

for all $a \in A$ and $\phi^* \in \mathcal{A}^*$. Since A^R is finitely generated projective by assumption, this proves that $\kappa^* \rightharpoonup a = s_R \circ \kappa^*(a)$. The identity (45) follows by applying the same proof in $\mathcal{A}_{cop}$. $\qquad\square$

We are ready to formulate

THEOREM 6.3. *Let $\mathcal{A}_R$ be a finite right bialgebroid with non-degenerate left integral ℓ. Let $\lambda^*: = \ell_R^{-1}(1_A)$ and $\lambda: = {}_R\ell^{-1}(1_A)$. Then the map $S(a): = (\lambda \leftharpoonup a) \rightharpoonup \ell$ is an antipode making $\mathcal{A}_R$ into a Hopf algebroid. The element ℓ is a non-degenerate left integral in the resulting Hopf algebroid in the sense of Definition 5.3.*

Proof. We check that $(\mathcal{A}_R, S)$ satisfies the right analogues of the conditions (18-22) – what implies the claim. The conditions (42) and (43) imply that the map $S: = (\lambda \leftharpoonup a) \rightharpoonup \ell$ is bijective with inverse $S^{-1}(a): = (\lambda^* \leftharpoonup a) \rightharpoonup \ell$. It is anti-multiplicative by

$$
\begin{aligned}
S(b)S(a) &= S(b)S(a)\ell^{(1)}s_R \circ^*\lambda(\ell^{(2)}) = \ell^{(1)}s_R \circ^*\lambda(ab\ell^{(2)}) \\
&= S(ab)\ell^{(1)}s_R \circ^*\lambda(\ell^{(2)}) = S(ab).
\end{aligned}
$$

Also

$$
S \circ t_R(r) = \ell^{(1)}s_R \circ^*\lambda(t_R(r)\ell^{(2)}) = s_R(r).
$$

Using Lemma 6.2, we find

$$
\begin{aligned}
a^{(1)}S(a^{(2)}) &= a^{(1)}\ell^{(1)}s_R \circ^*\lambda(a^{(2)}\ell^{(2)}) = {}^*\lambda \rightharpoonup a\ell \\
&= t_R \circ^*\lambda(a\ell) = t_R \circ \pi_R((\lambda \leftharpoonup a) \rightharpoonup \ell) = t_R \circ \pi_R \circ S(a).
\end{aligned}
$$

Using the identities

$$
\begin{aligned}
\gamma_R \circ S(a) &= \ell^{(1)} \otimes (\lambda \leftharpoonup a) \rightharpoonup \ell^{(2)} && (46) \\
\gamma_R \circ S^{-1}(a) &= (\lambda^* \leftharpoonup a) \rightharpoonup \ell^{(1)} \otimes \ell^{(2)} && (47)
\end{aligned}
$$

one shows that

$$
\begin{aligned}
&S\left(S^{-1}(a)^{(2)}\right) \otimes S\left(S^{-1}(a)^{(1)}\right)^{(1)} \otimes S\left(S^{-1}(a)^{(1)}\right)^{(2)} \\
&= S(\ell^{(3)}) \otimes \ell^{(1)'} \otimes \ell^{(2)'}s_R \circ^*\lambda\left(\ell^{(2)}t_R \circ \lambda^*(a\ell^{(1)})\ell^{(3)'}\right) \\
&= S(\ell^{(3)}) \otimes \ell^{(1)'} \otimes (\lambda^* \rightharpoonup a\ell^{(1)})\ell^{(2)'}s_R \circ^*\lambda\left(\ell^{(2)}\ell^{(3)'}\right) \\
&= S(\ell^{(3)}) \otimes \ell^{(1)'} \otimes a^{(2)}\left(\lambda \rightharpoonup \ell^{(2)}t_R \circ \lambda^*(a^{(1)}\ell^{(1)})\ell^{(2)'}\right) \\
&= S(\ell^{(3)}) \otimes \ell^{(1)'}s_R \circ^*\lambda\left(\ell^{(2)}t_R \circ \lambda^*(a^{(1)}\ell^{(1)})\ell^{(2)'}\right) \otimes a^{(2)} \\
&= S\left(S^{-1}(a^{(1)})^{(2)}\right) \otimes S\left(S^{-1}(a^{(1)})^{(1)}\right) \otimes a^{(2)} && (48)
\end{aligned}
$$

AN ALTERNATIVE NOTION OF HOPF ALGEBROID

where $\ell^{(1)'} \otimes \ell^{(2)'} = \gamma_R(\ell) = \ell^{(1)} \otimes \ell^{(2)}$. Hence

$$S^2\left(S^{-1}(a)^{(2)}\right) S\left(S^{-1}(a)^{(1)}\right)^{(1)} \otimes S\left(S^{-1}(a)^{(1)}\right)^{(2)} = s_R \circ \pi_R(a^{(1)}) \otimes a^{(2)}. \quad (49)$$

Using (46), (47) and (49) one checks that

$$
\begin{aligned}
S&\left(S^{-1}(a)^{(2)}\right) \otimes S\left(S^{-1}(a)^{(1)}\right) \\
&= S\left(S^{-1}(a)^{(2)}\right) \otimes S^{-1}\left(S^2(S^{-1}(a)^{(1)})^{(1)} s_R \circ \pi_R[S^2(S^{-1}(a)^{(1)})^{(2)}]\right) \\
&= S^{-1} \circ t_R \circ \pi_R[S^2(S^{-1}(a)^{(1)})^{(2)}] S\left(S^{-1}(a)^{(2)}\right) \otimes S^{-1}\left(S^2(S^{-1}(a)^{(1)})^{(1)}\right) \\
&= S^{-1} \circ t_R \circ {}^*\lambda\left(S(S^{-1}(a)^{(1)})\ell^{(2)}\right) S\left(S^{-1}(a)^{(2)}\right) \otimes S^{-1}(\ell^{(1)}) \\
&= S^{-1}\left(S^2(S^{-1}(a)^{(2)})S(S^{-1}(a)^{(1)})^{(1)}\ell^{(2)} s_R \circ {}^*\lambda[S(S^{-1}(a)^{(1)})^{(2)}\ell^{(3)}]\right) \\
&\quad \otimes S^{-1}(\ell^{(1)}) \\
&= S^{-1}\left(s_R \circ \pi_R(a^{(1)})\ell^{(2)} s_R \circ {}^*\lambda(a^{(2)}\ell^{(3)})\right) \otimes S^{-1}(\ell^{(1)}) \\
&= S^{-1}\left(\ell^{(2)} s_R \circ {}^*\lambda(a\ell^{(3)})\right) \otimes S^{-1}(\ell^{(1)}) \\
&= S^{-1}\left(S(a)^{(2)}\right) \otimes S^{-1}\left(S(a)^{(1)}\right).
\end{aligned}
$$

Introducing $\gamma_L(a) := S\left(S^{-1}(a)^{(2)}\right) \otimes S\left(S^{-1}(a)^{(1)}\right) \equiv S^{-1}\left(S(a)^{(2)}\right) \otimes S^{-1}\left(S(a)^{(1)}\right)$ the identity

$$(\mathrm{id}_A \otimes \gamma_R) \circ \gamma_L = (\gamma_L \otimes \mathrm{id}_A) \circ \gamma_R$$

is proven by (48). The last axiom $(\gamma_R \otimes \mathrm{id}_A) \circ \gamma_L = (\mathrm{id}_A \otimes \gamma_L) \circ \gamma_R$ is checked similarly:

$$
\begin{aligned}
S^{-1}&\left(S(a)^{(2)}\right)^{(1)} \otimes S^{-1}\left(S(a)^{(2)}\right)^{(2)} \otimes S^{-1}\left(S(a)^{(1)}\right) \\
&= \ell^{(2)} t_R \circ \lambda^*\left(\ell^{(2)'} s_R \circ {}^*\lambda(a\ell^{(3)'})\ell^{(1)}\right) \otimes \ell^{(3)} \otimes S^{-1}\left(\ell^{(1)'}\right) \\
&= (\lambda^* \rightharpoonup a\ell^{(3)'})\ell^{(2)} t_R \circ \lambda^*\left(\ell^{(2)'}\ell^{(1)}\right) \otimes \ell^{(3)} \otimes S^{-1}\left(\ell^{(1)'}\right) \\
&= a^{(1)}\left(\lambda^* \rightharpoonup \ell^{(2)'} s_R \circ {}^*\lambda(a^{(2)}\ell^{(3)'})\ell^{(1)}\right) \otimes \ell^{(2)} \otimes S^{-1}\left(\ell^{(1)'}\right) \\
&= a^{(1)} \otimes \ell^{(2)} t_R \circ \lambda^*\left(\ell^{(2)'} s_R \circ {}^*\lambda(a^{(2)}\ell^{(3)'})\ell^{(1)}\right) \otimes S^{-1}\left(\ell^{(1)'}\right) \\
&= a^{(1)} \otimes S^{-1}\left(S(a^{(2)})^{(2)}\right) \otimes S^{-1}\left(S(a^{(2)})^{(1)}\right).
\end{aligned}
$$

$\square$

One defines a *non-degenerate right integral* in a left bialgebroid $\mathcal{A}_L = (A, L, s_L, t_L, \gamma_L, \pi_L)$ as a non-degenerate left integral in the right bialgebroid $(\mathcal{A}_L)^{op}$. That is as an element $\Upsilon \in A$ satisfying

    i)   *both maps $_L\Upsilon$ and Υ_L are bijective*

    ii) $\Upsilon_{(1)}a \otimes \Upsilon_{(2)} = \Upsilon_{(1)} \otimes \Upsilon_{(2)}[\Upsilon \leftharpoonup (a \rightharpoonup \rho_*)]$

           $\Upsilon_{(1)} \otimes \Upsilon_{(2)}a = \Upsilon_{(1)}[\Upsilon \leftharpoonup (a \rightharpoonup {}_*\rho)] \otimes \Upsilon_{(2)}$

50   G. BÖHM

for all $a \in A$ and $_*\rho := {}_L\Upsilon^{-1}(1_A)$ and $\rho_* := \Upsilon_L^{-1}(1_A)$. The application of Theorem 6.3 to the right bialgebroid $(\mathcal{A}_L)^{op}$ implies that a left bialgebroid $\mathcal{A}_L$ possessing a non-degenerate right integral Υ can be made a Hopf algebroid with the antipode $S(a) := \Upsilon \leftharpoonup (a \rightharpoonup \rho_*)$.

As an application of Theorem 6.3 we sketch a different derivation of the Hopf algebroid symmetry of an abstract depth 2 Frobenius extension – obtained in [2].

Let $\mathcal{C}$ be an additive 2-category closed under the direct sums and subobjects of 1-morphisms. By an abstract extension we mean a 1-morphism ι in $\mathcal{C}$ that posesses a left dual $\bar{\iota}$. This means the existence of 2-morphisms

$$ev_L \in \mathcal{C}^2(\bar{\iota} \times \iota, s_0(\iota)) \text{ and } coev_L \in \mathcal{C}^2(t_0(\iota), \iota \times \bar{\iota}) \tag{50}$$

satisfying the relations

$$(\iota \times ev_L) \circ (coev_L \times \iota) = \iota$$
$$(ev_L \times \bar{\iota}) \circ (\bar{\iota} \times coev_L) = \bar{\iota}$$

where $s_0(\iota)$ and $t_0(\iota)$ are the source and target 0-morphisms of ι, respectively, and $\times$ stands for the horizontal product and $\circ$ for the vertical product. The 1-morphism ι satisfies the *left depth 2* (or D2 for short) condition if $\iota \times \bar{\iota} \times \iota$ is a direct summand in a finite direct sum of copies of ι's [29]. This means the existence of finite sets of 2-morphisms $\beta_i \in \mathcal{C}^2(\iota \times \bar{\iota} \times \iota, \iota)$ and $\beta_i' \in \mathcal{C}^2(\iota, \iota \times \bar{\iota} \times \iota)$ satisfying $\sum_i \beta_i' \circ \beta_i = \iota \times \bar{\iota} \times \iota$. By the Theorem 3.5 in [29] in this case the ring of 2-morphisms: $A := \mathcal{C}^2(\iota \times \bar{\iota}, \iota \times \bar{\iota})$ carries a left bialgebroid structure $\mathcal{A}_L$ over the base $L := \mathcal{C}^2(\iota, \iota)$. The structural maps of $\mathcal{A}_L$ are explicitly given in [2]:

$$s_L(l) = l \times \bar{\iota}$$
$$t_L(l) = \iota \times [(ev_L \times \bar{\iota}) \circ (\bar{\iota} \times l \times \bar{\iota}) \circ (\bar{\iota} \times coev_L)]$$
$$\pi_L(a) = (\iota \times ev_L) \circ (a \times \iota) \circ (coev_L \times \iota)$$
$$\gamma_L(a) = (\iota \times ev_L \times \bar{\iota}) \circ (a \times \iota \times \bar{\iota}) \circ (\iota \times ev_L \times \bar{\iota} \times \iota \times \bar{\iota})$$
$$\circ (\iota \times \bar{\iota} \times \beta_i' \times \bar{\iota}) \circ (\iota \times \bar{\iota} \times coev_L) \otimes (\beta_i \times \bar{\iota}) \circ (\iota \times \bar{\iota} \times coev_L)$$

for $l \in L$ and $a \in A$. Summation over the index i is implicitly understood.

The 1-morphism ι satisfies the *right D2* condition if $\bar{\iota} \times \iota \times \bar{\iota}$ is a direct summand in a finite direct sum of copies of $\bar{\iota}$'s. This means the existence of finite sets of 2-morphisms $\hat{\beta}_i \in \mathcal{C}^2(\bar{\iota} \times \iota \times \bar{\iota}, \bar{\iota})$ and $\hat{\beta}_i' \in \mathcal{C}^2(\bar{\iota}, \bar{\iota} \times \iota \times \bar{\iota})$ satisfying $\sum_i \hat{\beta}_i' \circ \hat{\beta}_i = \bar{\iota} \times \iota \times \bar{\iota}$. In this case the ring $B := \mathcal{C}^2(\bar{\iota} \times \iota, \bar{\iota} \times \iota)$ carries a right bialgebroid structure $\mathcal{B}_R$ over the base $R := \mathcal{C}^2(\iota, \iota)$. The structural maps of $\mathcal{B}_R$ read as

$$s_R(r) = \bar{\iota} \times r$$
$$t_R(r) = [(ev_L \times \bar{\iota}) \circ (\bar{\iota} \times r \times \bar{\iota}) \circ (\bar{\iota} \times coev_L)] \times \iota$$
$$\pi_R(b) = (\iota \times ev_L) \circ (\iota \times b) \circ (coev_L \times \iota)$$
$$\gamma_R(b) = (\bar{\iota} \times \iota \times ev_L) \circ (\bar{\iota} \times \iota \times \hat{\beta}_i \times \iota) \circ (\bar{\iota} \times coev_L \times \iota \times \bar{\iota} \times \iota)$$
$$\circ (b \times \bar{\iota} \times \iota) \circ (\bar{\iota} \times coev_L \times \iota) \otimes (\bar{\iota} \times \iota \times ev_L) \circ (\hat{\beta}_i' \times \iota),$$

for $r \in R$ and $b \in B$. If ι satisfies both the left and the right D2 conditions, then the bialgebroids $\mathcal{A}_L$ and $\mathcal{B}_R$ are duals.

Let us assume that ι is a *Frobenius* 1-morphism that is its dual $\bar{\iota}$ is two-sided. This means the existence of further 2-morphisms

$$ev_R \in \mathcal{C}^2(\iota \times \bar{\iota}, t_0(\iota)) \qquad coev_R \in \mathcal{C}^2(s_0(\iota), \bar{\iota} \times \iota) \tag{51}$$

satisfying the relations

$$\begin{aligned}(ev_R \times \iota) \circ (\iota \times coev_R) &= \iota \\ (\bar{\iota} \times ev_R) \circ (coev_R \times \bar{\iota}) &= \bar{\iota}.\end{aligned}$$

Under this assumption the left- and right D2 conditions become equivalent (the 2-morphisms $\hat{\beta}_i$ and $\hat{\beta}_i'$ can be expressed in terms of β_i, β_i' and the 2-morphisms (50-51).) One checks that in the case of a D2 Frobenius 1-morphism ι the element

$$\Upsilon: = coev_L \circ ev_R \tag{52}$$

of A is a non-degenerate right integral. It leads to the antipode

$$\begin{aligned}S_A(a) = (\iota \times \bar{\iota} \times ev_R) &\circ (\iota \times \bar{\iota} \times \iota \times ev_L \times \bar{\iota}) \circ (\iota \times \bar{\iota} \times a \times \iota \times \bar{\iota}) \\ &\circ \ (\iota \times coev_R \times \bar{\iota} \times \iota \times \bar{\iota}) \circ (coev_L \times \iota \times \bar{\iota})\end{aligned}$$

obtained by different methods in [2]. Also the element

$$\ell: = coev_R \circ ev_L \tag{53}$$

of B is a non-degenerate left integral, leading to the antipode

$$\begin{aligned}S_B(b) = (\bar{\iota} \times \iota \times ev_L) &\circ (\bar{\iota} \times \iota \times \bar{\iota} \times ev_R \times \iota) \circ (\bar{\iota} \times \iota \times b \times \bar{\iota} \times \iota) \\ &\circ \ (\bar{\iota} \times coev_L \times \iota \times \bar{\iota} \times \iota) \circ (coev_R \times \bar{\iota} \times \iota).\end{aligned}$$

As a matter of fact $S_A(\Upsilon) = \Upsilon$ and $S_B(\ell) = \ell$, that is both Υ and ℓ turn out to be two-sided non-degenerate integrals.

We want to emphasize that the non-degenerate integrals (52) and (53) are not unique. Another choice of the non-degenerate integrals leads to other antipodes and other corresponding right and left bialgebroid structures on A and B, respectively.

The name 'abstract extension' is motivated by the most important example. For an extension $N \to M$ of rings the forgetful functor Φ of right modules $\mathcal{M}_M \to \mathcal{M}_N$ is a 1-morphism in the 2-category of categories. It posesses a left dual: the induction functor. The 1-morphism Φ is left/right D2 and Frobenius if and only if the extension $N \to M$ is left/right D2 and Frobenius, respectively. As it is shown in [14] in the case of a D2 ring extension $N \to M$ the above ring A is isomorphic to the endomorphism ring $\text{End}(_N M_N)$ and B is the center $(M \otimes_N M)^N$. Now for a Frobenius extension $N \to M$ let us *fix* a Frobenius system that is an $N - N$ bimodule map $\psi : \ _N M_N \to N$ and its quasi-basis $y_i \otimes x_i \in M_N \otimes \ _N M$. Then the non-degenerate integrals (52) and (53) are $\Upsilon = \psi$ and $\ell = y_i \otimes x_i$, respectively. The construction of the corresponding antipodes gives the Hopf algebroid structure on A that is discussed in the Section 3 of [1] on different grounds, and its dual – $\mathcal{A}_*^{\Upsilon}$ – on B.

REFERENCES

[1] G. Böhm, and K. Szlachányi, Hopf algebroids with bijective antipodes: axioms, integrals and duals, *J. Algebra*, to appear; preprint arXiv: math.QA/0302325.

[2] G. Böhm and K. Szlachányi, Hopf algebroid symmetry of abstract depth 2 Frobenius extensions, *Comm. Algebra*, to appear; preprint arXiv: math.QA/0305136.

[3] G. Böhm, F. Nill, and K. Szlachányi, Weak Hopf Algebras I: Integral Theory and C^*-structure, *J. Algebra* **221** (1999), 385–438

[4] G. Böhm and K. Szlachányi, A Coassociative C^*-Quantum Group with Non-Integral Dimensions, *Lett. Math. Phys.* **35** (1996), 437–456.

[5] T. Brzezinski and G. Militaru, Bialgebroids, $\times_R$-bialgebras and Duality, *J. Algebra* **251** (2002), 279–294

[6] A. Connes and H. Moscovici, Cyclic cohomology and Hopf algebra symmetry, Conference Moshé Flato Dijon 1999, *Lett. Math. Phys.* **52** (2000), 1–28.

[7] A. Connes and H. Moscovici, Differential cyclic cohomology and Hopf algebraic structures in transverse geometry, in "Essays on geometry and related topics", Monogr. Enseign. Math. **38**, Geneva 2001, 217–255.

[8] M-C. David, Paragroupe d'Adrian Ocneanu et algèbre de Kac, *Pacific J. Math.* **172** (1996), 331–363.

[9] M. Enock and R. Nest, Inclusions of Factors, Multiplicative Unitaries and Kac Algebras, *J. Funct. Anal.* **137** (1996), 466–543.

[10] M. Enock and J-M. Vallin, Inclusions of von Neumann algebras and quantum groupoids, *J. Funct. Anal.* **172** (2000), 249–300.

[11] M. Enock, Inclusions of von Neumann algebras and quantum groupoids II, *J. Funct. Anal.* **178** (2000), 156–225.

[12] L. Kadison, Hopf algebroids and H-separable extensions, *Proc. Amer. Math. Soc.* **131** (2003), 2993–3002.

[13] L. Kadison, Hopf algebroids and Galois extensions, *Bull. Belgian Math. Soc. - Simon Stevin*, to appear.

[14] L. Kadison and K. Szlachányi, Bialgebroid actions on depth two extensions and duality, *Adv. Math.* **179** (2003), 75–121.

[15] M. Khalkhali and B. Rangipour, On the cyclic cohomology of extended Hopf algebras, preprint arXiv: math.KT/0105105.

[16] R. Larson, M. Sweedler, An associative orthogonal bilinear form for Hopf algebras, *Amer. J. Math.* **91** (1969), 75–94.

[17] R. Longo, A duality for Hopf algebras and for subfactors I, *Commun. Math. Phys.* **159** (1994), 133–150.

[18] J. H. Lu, Hopf Algebroids and Quantum Groupoids, *Int. J. Math.* **7** (1996), 47–70.

[19] D. Nikshych and L. Vainerman, A characterization of depth 2 subfactors of II_1 factors, *J. Funct. Anal.* **171** (2000), 278–307.

[20] D. Nikshych and L. Vainerman, Galois correspondence for II_1 factors and quantum groupoids, *J. Funct. Anal.* **178** (2000), 113–142.

[21] F. Nill, Weak Bialgebras, preprint arXiv: math.QA/9805104.

[22] F. Nill, K. Szlachányi, and H-W. Wiesbrock, Weak Hopf Algebras and Reducible Jones Inclusions of Depth 2. I., preprint arXiv: math.QA/9806130.

[23] F. Nill, K. Szlachányi, H-W. Wiesbrock, Weak Hopf Algebras and Reducible Jones Inclusions of Depth 2. II., unpublished.

[24] F. Nill and H-W Wiesbrock, A comment on Jones Inclusions with finite Index, *Rev. Math. Phys.* **7** (1995), p.599.

[25] A. Ocneanu, A Galois Theory for Operator Algebras, Notes of a Lecture. Quantum Symmetry, Differential Geometry of Finite Graphs and Classification Subfactors, Lectures by A. Ocneanu given at the University of Tokyo, Notes taken by Y. Kawahigashi (1992).

[26] P. Schauenburg, Weak Hopf Algebras and Quantum Groupoids, preprint.

[27] P. Schauenburg, Duals and Doubles of Quantum Groupoids, *Contemp. Math.* **267** (2000), 273–299.

AN ALTERNATIVE NOTION OF HOPF ALGEBROID

[28] M. E. Sweedler, Groups of simple algebras, *Publ. Math., Inst. Hautes tud. Sci.* **44** (1974), 79–189.

[29] K. Szlachányi, Finite Quantum Groupoids and Inclusions of Finite Type, *Fields Inst. Comm.* **30** (2001), 393–407.

[30] K. Szlachányi, Galois actions by finite quantum groupoids, in "Proceedings of the 69 *ème recontre entre physiciens théoriciens et mathématiciens* Strasbourg 2002, L. Vainerman and V. Turaev (eds.), *IRMA Lectures in Mathematics and Theoretical Physics* **2**, de Gruyter, 2003.

[31] W. Szymanski, Finite Index Subfactors and Hopf Algebra Crossed Products, *Proc. Amer. Math. Soc.* **120** (1994), 519–528.

[32] M. Takeuchi, Groups of Algebras over $A \otimes \bar{A}$, *J. Math. Soc. Japan* **29** (1977), 459–492.

[33] P. Xu, Quantum Groupoids and Deformation Quantization, *C. R. Acad. Sci., Paris, Sr. I, Math.* **326** (1998), 289–294.

[34] P. Xu, Quantum Groupoids, *Comm. Math. Phys.* **216** (2001), 539–581.

[35] S. Yamagami, A note on On Ocneanu's approach to Jones' index theory, *Internat. J. Math.* **4** (1993), 859–871.

[36] P. Vecsernyés, Larson-Sweedler theorem and the role of grouplike elements in weak Hopf algebras, *J. Algebra* **270** (2003), 471–520.

[37] Y. Watatani, Index for C^*-subalgebras, *Mem. Amer. Math. Soc.* **424** (1990).

Topological Hopf Algebras, Quantum Groups and Deformation Quantization

PHILIPPE BONNEAU AND DANIEL STERNHEIMER
Laboratoire Gevrey de Mathématique physique, Université de Bourgogne
BP 47870, F-21078 Dijon Cedex, France.
e-mail *Philippe.Bonneau@u-bourgogne.fr, Daniel.Sternheimer@u-bourgogne.fr*

> ABSTRACT. After a presentation of the context and a brief reminder of defor-
> mation quantization, we indicate how the introduction of natural topological
> vector space topologies on Hopf algebras associated with Poisson Lie groups,
> Lie bialgebras and their doubles explains their dualities and provides a compre-
> hensive framework. Relations with deformation quantization and applications
> to the deformation quantization of symmetric spaces are described.

1. INTRODUCTION

1.1. Presentation of the context

The expression "quantum groups" is a name coined by Drinfeld (see [19]) in the first
half of the 80's which is superb, even if the notion is not necessarily quantum and
the objects are not really groups. But they are Hopf algebras and their theory can
be viewed as an avatar of deformation quantization [1] (see [17] for a recent review
which this presentation complements), applied to the quantization of Poisson-Lie
groups.
The philosophy underlying the role of *deformations in physics* has been consistently
put forward by Flato, almost since the definition of the deformation of rings and algebras by Gerstenhaber [27], and was eventually expressed by him in [25]. In short,
the passage from one level of physical theory to another, when a new fundamental
constant is imposed by experiments, can be understood (and might even have been
predicted) using deformation theory. The only question is, in which category do
we seek for deformations? Usually physics is rather conservative and if we start
e.g. with the category of associative or Lie algebras, we tend to deform in the same
category.
But there are important instances of generalizations of this principle. The most
elaborate is maybe noncommutative geometry, where the strategy is to formulate
the "undeformed" (commutative) geometry in terms of algebraic structures in such
a way that it becomes possible to "plug in" the deformation (noncommutativity)
in a quite natural, and mathematically rigorous, manner. We shall not elaborate
on that aspect here, refering e.g. to [12] for a presentation, to [13] for important

---

2000 *Mathematics Subject Classification.* Primary 54C40, 14E20; Secondary 46E25, 20C20.
Key words and phrases. Hopf algebras, topological vector spaces, quantum groups, deformation
quantization.

recent examples of noncommutative manifolds, and to [11, 14] for the basics and a relation with deformation quantization.

We shall concentrate on another prominent example: quantum groups. Instead of looking at the associative algebra of functions over a Poisson-Lie group or at the enveloping algebra, one makes full use of the Hopf algebra structure in both cases. In general both the product and the coproduct have to be (compatibly) deformed, but cohomological results ([20] and section 3.1) show that, when the Lie group is semi-simple, the deformation is always equivalent to a "preferred" one, that is, a deformation where only the product or the coproduct (resp.) is deformed. The group aspect is a special case of deformation quantization and we shall show that the enveloping algebra aspect can be seen as its dual, in the sense of topological vector spaces duality.

1.2. Deformation theory of algebras

A concise formulation of a Gerstenhaber deformation of an algebra (associative, Lie, bialgebra, etc.) is [27, 9]:

DEFINITION 1. A deformation of an algebra A over a field $\mathbb{K}$ is a $\mathbb{K}[[\nu]]$-algebra $\tilde{A}$ such that $\tilde{A}/\nu\tilde{A} \approx A$. Two deformations $\tilde{A}$ and $\tilde{A}'$ are said equivalent if they are isomorphic over $\mathbb{K}[[\nu]]$ and $\tilde{A}$ is said trivial if it is isomorphic to the original algebra A considered by base field extension as a $\mathbb{K}[[\nu]]$-algebra.

Whenever we consider a topology on A, $\tilde{A}$ is supposed to be topologically free. For associative (resp. Lie) algebras, Definition 1 tells us that there exists a new product $*$ (resp. bracket $[\cdot, \cdot]$) such that the new (deformed) algebra is again associative (resp. Lie). Denoting the original composition laws by ordinary product (resp. $\{\cdot, \cdot\}$) this means that, for $u, v \in A$ (we can extend this to $A[[\nu]]$ by $\mathbb{K}[[\nu]]$-linearity) we have:

$$u * v = uv + \sum_{r=1}^{\infty} \nu^r C_r(u, v) \tag{1}$$

$$[u, v] = \{u, v\} + \sum_{r=1}^{\infty} \nu^r B_r(u, v) \tag{2}$$

where the C_r are Hochschild 2-cochains and the B_r (skew-symmetric) Chevalley 2-cochains, such that for $u, v, w \in A$ we have $(u*v)*w = u*(v*w)$ and $\mathcal{S}[[u, v], w] = 0$, where $\mathcal{S}$ denotes summation over cyclic permutations.

For a (topological) *bialgebra* (an associative algebra A where we have in addition a coproduct $\Delta : A \longrightarrow A \otimes A$ and the obvious compatibility relations), denoting by $\otimes_\nu$ the tensor product of $\mathbb{K}[[\nu]]$-modules, we can identify $\tilde{A} \hat{\otimes}_\nu \tilde{A}$ with $(A \hat{\otimes} A)[[\nu]]$, where $\hat{\otimes}$ denotes the algebraic tensor product completed with respect to some topology (e.g. projective for Fréchet nuclear topology on A), we similarly have a deformed coproduct $\tilde{\Delta} = \Delta + \sum_{r=1}^{\infty} \nu^r D_r$, $D_r \in \mathcal{L}(A, A\hat{\otimes}A)$, satisfying $\tilde{\Delta}(u*v) = \tilde{\Delta}(u)*\tilde{\Delta}(v)$. In this context appropriate cohomologies can be introduced [30, 7]. There are natural additional requirements for Hopf algebras.

Equivalence means that there is an isomorphism $T_\nu = I + \sum_{r=1}^{\infty} \nu^r T_r$, $T_r \in \mathcal{L}(A, A)$ so that $T_\nu(u *' v) = (T_\nu u * T_\nu v)$ in the associative case, denoting by $*$ (resp. $*'$)

TOPOLOGICAL HOPF ALGEBRAS AND DEFORMATION QUANTIZATION   57

the deformed laws in $\tilde{A}$ (resp. $\tilde{A}'$); and similarly in the Lie, bialgebra and Hopf cases. In particular we see (for $r = 1$) that a deformation is trivial at order 1 if it starts with a 2-cocycle which is a 2-coboundary. More generally, exactly as above, we can show [1] ([30, 7] in the Hopf case) that if two deformations are equivalent up to some order t, the condition to extend the equivalence one step further is that a 2-cocycle (defined using the T_k, $k \leq t$) is the coboundary of the required T_{t+1} and therefore *the obstructions to equivalence lie in the 2-cohomology*. In particular, if that space is null, all deformations are trivial.

Unit. An important property is that a *deformation of an associative algebra with unit* (what is called a unital algebra) is again unital, and *equivalent to a deformation with the same unit*. This follows from a more general result of Gerstenhaber (for deformations leaving unchanged a subalgebra) and a proof can be found in [29].

Remark 1. In the case of (topological) *bialgebras* or *Hopf* algebras, *equivalence* of deformations has to be understood as an isomorphism of (topological) $\mathbb{K}[[\nu]]$-algebras, the isomorphism starting with the identity for the degree 0 in ν. A deformation is again said *trivial* if it is equivalent to that obtained by base field extension. For Hopf algebras the deformed algebras may be taken (by equivalence) to have the same unit and counit, but in general not the same antipode.

1.3. Deformation quantization and physics

Intuitively, classical mechanics is the limit of quantum mechanics when $\hbar = \frac{h}{2\pi}$ goes to zero. But how can this be realized when in classical mechanics the observables are functions over phase space (a Poisson manifold) and not operators? The deformation philosophy promoted by Flato shows the way: one has to look for deformations of algebras of classical observables, functions over Poisson manifolds, and realize there quantum mechanics in an *autonomous* manner.

What we call "deformation quantization" relates to (and generalizes) what in the conventional (operatorial) formulation are the Heisenberg picture and Weyl's quantization procedure. In the latter [51], starting with a classical observable $u(p,q)$, some function on phase space $\mathbb{R}^{2\ell}$ (with $p, q \in \mathbb{R}^\ell$), one associates an operator (the corresponding quantum observable) $\Omega(u)$ in the Hilbert space $L^2(\mathbb{R}^\ell)$ by the following general recipe:

$$u \mapsto \Omega_w(u) = \int_{\mathbb{R}^{2\ell}} \tilde{u}(\xi,\eta)\exp(i(P.\xi + Q.\eta)/\hbar)w(\xi,\eta)\, d^\ell\xi d^\ell\eta \tag{3}$$

where $\tilde{u}$ is the inverse Fourier transform of u, P_α and Q_α are operators satisfying the canonical commutation relations $[P_\alpha, Q_\beta] = i\hbar\delta_{\alpha\beta}$ ($\alpha, \beta = 1, ..., \ell$), w is a weight function and the integral is taken in the weak operator topology. What is called in physics normal (or antinormal) ordering corresponds to choosing for weight $w(\xi,\eta) = \exp(-\frac{1}{4}(\xi^2 \pm \eta^2))$. Standard ordering (the case of the usual pseudodifferential operators in mathematics) corresponds to $w(\xi,\eta) = \exp(-\frac{i}{2}\xi\eta)$ and the original Weyl (symmetric) ordering to $w = 1$. An inverse formula was found shortly afterwards by Eugene Wigner [52] and maps an operator into what mathematicians call its symbol by a kind of trace formula. For example Ω_1 defines an isomorphism of Hilbert spaces between $L^2(\mathbb{R}^{2\ell})$ and Hilbert-Schmidt operators on

$L^2(\mathbb{R}^\ell)$ with inverse given by

$$u = (2\pi\hbar)^{-\ell} \text{Tr}[\Omega_1(u) \exp((\xi.P + \eta.Q)/i\hbar)] \tag{4}$$

and if $\Omega_1(u)$ is of trace class one has $\text{Tr}(\Omega_1(u)) = (2\pi\hbar)^{-\ell} \int u\,\omega^\ell \equiv \text{Tr}_M(u)$, the "Moyal trace", where ω^ℓ is the (symplectic) volume dx on $\mathbb{R}^{2\ell}$. Looking for a direct expression for the symbol of a quantum commutator, Moyal found [41] what is now called the Moyal bracket:

$$M(u_1, u_2) = \nu^{-1}\sinh(\nu P)(u_1, u_2) = P(u_1, u_2) + \sum_{r=1}^{\infty}\frac{\nu^{2r}}{(2r+1)!}P^{2r+1}(u_1, u_2) \tag{5}$$

where $2\nu = i\hbar$, $P^r(u_1, u_2) = \Lambda^{i_1 j_1}\ldots\Lambda^{i_r j_r}(\partial_{i_1\ldots i_r}u_1)(\partial_{j_1\ldots j_r}u_2)$ is the rth power ($r \geq 1$) of the Poisson bracket bidifferential operator P, $i_k, j_k = 1,\ldots,2\ell$, $k = 1,\ldots,r$ and $(\Lambda^{i_k j_k}) = \binom{0\ -I}{I\ 0}$. To fix ideas we may assume here $u_1, u_2 \in \mathcal{C}^\infty(\mathbb{R}^{2\ell})$ and the sum is taken as a formal series. A corresponding formula for the symbol of a product $\Omega_1(u)\Omega_1(v)$ can be found in [31], and may now be written more clearly as a (Moyal) *star product*:

$$u_1 *_M u_2 = \exp(\nu P)(u_1, u_2) = u_1 u_2 + \sum_{r=1}^{\infty}\frac{\nu^r}{r!}P^r(u_1, u_2). \tag{6}$$

The formal series may be deduced (see e.g. [5]) from an integral formula of the type:

$$(u_1 * u_2)(x) = c_\hbar \int_{\mathbb{R}^{2\ell}\times\mathbb{R}^{2\ell}} u_1(x+y)u_2(x+z)e^{-\frac{i}{\hbar}\Lambda^{-1}(y,z)}dydz. \tag{7}$$

It was noticed, however after deformation quantization was introduced, that the composition of symbols of pseudodifferential operators (ordered, like differential operators, "first q, then p") is a star product.

One recognizes in (6) a special case of (1), and similarly for the bracket. So, via a Weyl quantization map, the algebra of quantized observables can be viewed as a deformation of that of classical observables.

But the deformation philosophy tells us more. Deformation quantization is not merely "a reformulation of quantizing a mechanical system" [18], e.g. in the framework of Weyl quantization: *The process of quantization itself is a deformation*. In order to show that explicitly it was necessary to treat in an *autonomous* manner significant physical examples, without recourse to the traditional operatorial formulation of quantum mechanics. That was achieved in [1] with the paradigm of the harmonic oscillator and more, including the angular momentum and the hydrogen atom. In particular what plays here the role of the unitary time evolution operator of a quantized system is the "star exponential" of its classical Hamiltonian H (expressed as a usual exponential series but with "star powers" of $tH/i\hbar$, t being the time, and computed as a distribution both in phase space variables and in time); in a very natural manner, the spectrum of the quantum operator corresponding to H is the support of the Fourier-Stieltjes transform (in t) of the star exponential (what Laurent Schwartz had called the spectrum of that distribution). Further examples were (and are still being) developed, in particular in the direction of field theory. That aspect of deformation theory has since 25 years or so been extended considerably. It now includes general symplectic and Poisson (finite dimensional) manifolds,

TOPOLOGICAL HOPF ALGEBRAS AND DEFORMATION QUANTIZATION

with further results for infinite dimensional manifolds, for "manifolds with singularities" and for algebraic varieties, and has many far reaching ramifications in both mathematics and physics (see e.g. a brief overview in [17]). As in quantization itself [51], symmetries (group theory) play a special role and an autonomous theory of star representations of Lie groups was developed, in the nilpotent and solvable cases of course (due to the importance of the orbit method there), but also in significant other examples. The presentation that follows can be seen as an extension of the latter, when one makes full use of the Hopf algebra structures and of the "duality" between the group structure and the set of its irreducible representations.

Finally one should mention that deformation theory and Hopf algebras are seminal in a variety of problems ranging from theoretical physics (see e.g. [15, 17]), including renormalization and Feynman integrals and diagrams, to algebraic geometry and number theory (see e.g. [35, 36]), including algebraic curves à la Zagier (cf. [16] and Connes' lectures at Collège de France, January to March 2003).

2. SOME TOPOLOGICAL HOPF ALGEBRAS

We shall now briefly review applications of the deformation theory of algebras in the context of Hopf algebras endowed with appropriate topologies and in the spirit of deformation quantization. That is, we shall consider Hopf algebras of functions on Poisson-Lie groups (or their topological duals) and their deformations, and show how this framework is a powerful tool to understand the standard examples of quantum groups, and more. In order to do so we first recall some notions on topological vector spaces and apply them to our context.

2.1. Well-behaved Hopf algebras

DEFINITION 2. A topological vector space (tvs) V is said *well-behaved* if V is either nuclear and Fréchet, or nuclear and dual of Fréchet [32, 50].

PROPOSITION 1. *If V is a well-behaved tvs and W a tvs, then*

$$(i)\ V^{**} \simeq V \qquad (ii)\ (V \hat{\otimes} V)^* \simeq V^* \hat{\otimes} V^* \qquad (iii)\ \mathsf{Hom}_{\mathbb{K}}(V, W) \simeq V^* \hat{\otimes} W$$

where V^* denotes the strong topological dual of V, $\hat{\otimes}$ the projective topological tensor product and the base field $\mathbb{K}$ is $\mathbb{R}$ or $\mathbb{C}$.

DEFINITION 3. $(A, \mu, \eta, \Delta, \epsilon, S)$ is a WB (well-behaved) Hopf algebra [9] if

- A is a well-behaved topological vector space.
- The multiplication $\mu : A \hat{\otimes} A \to A$, the coproduct $\Delta : A \to A \hat{\otimes} A$, the unit η, the counit ϵ, and the antipode S are continuous.
- $\mu, \eta, \Delta, \epsilon$ and S satisfy the usual axioms of a Hopf algebra.

COROLLARY 1. *If $(A, \mu, \eta, \Delta, \epsilon, S)$ is a WB Hopf algebra, then $(A^*, {}^t\Delta, {}^t\epsilon, {}^t\mu, {}^t\eta, {}^tS)$ is also a WB Hopf algebra.*

2.2. Examples of well-behaved Hopf algebras [9]

Let G be a semi-simple Lie group and $\mathfrak{g}$ its complexified Lie algebra. For simplicity we shall assume here G linear (i.e. with a faithful finite dimensional representation) but the same results hold, with some modification in the proofs, for any semi-simple Lie group.

Example 1

$\mathcal{C}^\infty(G)$, the algebra of the smooth functions on G, is a WB Hopf algebra (Fréchet and nuclear).

Example 2

$\mathcal{D}(G) = \mathcal{C}^\infty(G)^*$, the algebra of the compactly supported distributions on G, is a WB Hopf algebra (dual of Fréchet and nuclear). The product is the transposed map of the coproduct of $\mathcal{C}^\infty(G)$ that is, the convolution of distributions.

Example 3

$\mathcal{H}(G)$, the algebra of coefficient functions of finite dimensional representations of G (or polynomial functions on G) is a WB Hopf algebra, the Hopf structure being that induced from $\mathcal{C}^\infty(G)$.

A short description of that algebra is as follows: We take a set $\hat{G}$ of irreducible finite dimensional representations of G such that there is *one and only one* element for each equivalence class, and, if $\pi \in \hat{G}$, its contragredient $\check{\pi}$ is also in $\hat{G}$. We define $\quad C_\pi = \text{vect}\{\text{coefficient functions of } \pi\} \overset{Burnside}{\simeq} \text{End}(V_\pi)$ for $\pi \in \hat{G}$. Then

$$\mathcal{H}(G) \overset{alg.}{\simeq} \bigoplus_{\pi \in \hat{G}} C_\pi \overset{v.s.}{\simeq} \bigoplus_{\pi \in \hat{G}} \text{End}(V_\pi). \text{ So we take on } \mathcal{H}(G) \text{ the "direct sum" topology}$$

of $\bigoplus_{\pi \in \hat{G}} \text{End}(V_\pi)$. Then $\mathcal{H}(G)$ is dual of Fréchet and nuclear, that is, WB.

Example 4

Let $\mathcal{A}(G)$, the algebra of "generalized distributions", be defined by $\mathcal{A}(G) = \mathcal{H}(G)^*$ $\overset{alg.}{\simeq} \prod_{\pi \in \hat{G}} \text{End}(V_\pi)$. The (product) topology is Fréchet and nuclear, and therefore $\mathcal{A}(G)$ is WB.

2.3. Inclusions [3, 9]

We denote by $U\mathfrak{g}$ the universal enveloping algebra of $\mathfrak{g}$ and by $\mathbb{C}G$ the group algebra of G. All the following inclusions are inclusions of Hopf algebras. $\Subset$, $\Supset$, $\mathbb{U}$, $\mathbb{m}$ mean a *dense* inclusion.

$$
\begin{array}{ccccc|c}
U\mathfrak{g} & \Subset & \mathcal{A}(G) & \Supset & \mathbb{C}G & \mathcal{H}(G) \\
 & & \mathbb{U} & & & \mathbb{m} \qquad (*) \\
U\mathfrak{g} & \subset & \mathcal{D}(G) & \Supset & \mathbb{C}G & \mathcal{C}^\infty(G)
\end{array}
$$

$(*)$ is true if and only if G is linear, but comparable results can be obtained for G non linear.

3. TOPOLOGICAL QUANTUM GROUPS

We shall now deform the preceding topological Hopf algebras and indicate how this explains various models of quantum groups. For clarity of the exposition, throughout this Section and the remainder of the paper, we shall limit to a minimum the details concerning the Hopf algebra structures other than product and coproduct.

But whenever we write Hopf algebras and not only bialgebras, the relevant structures are included in the discussion and dealing with them is quite straightforward.

3.1. Quantization

THEOREM 1 ([20]). *Let $\mathfrak{g}$ be a semi-simple Lie algebra and $(\mathsf{U}\mathfrak{g}, \mu_0, \Delta_0)$ denote the usual Hopf structure on $\mathsf{U}\mathfrak{g}$.*

1. *If $(\mathsf{U}_t\mathfrak{g}, \mu_t)$ is a deformation (as an associative algebra) of $(\mathsf{U}\mathfrak{g}[[t]], \mu_0)$ then $\mathsf{U}_t\mathfrak{g} \overset{\varphi}{\simeq} \mathsf{U}\mathfrak{g}[[t]]$ (i.e. $\mathsf{U}\mathfrak{g}$ is rigid).*
2. *If $(\mathsf{U}\mathfrak{g}[[t]], \mu_0, \Delta_t)$ is a deformation (as a Hopf algebra) of $(\mathsf{U}\mathfrak{g}[[t]], \mu_0, \Delta_0)$ then*

$$\exists\, P_t \in (\mathsf{U}\mathfrak{g} \otimes \mathsf{U}\mathfrak{g})[[t]] \text{ such that } P_{t=0} = \mathsf{Id} \text{ and } \Delta_t(a) = P_t.\Delta_0(a).P_t^{-1},\ \forall a \in \mathsf{U}\mathfrak{g}.$$

An isomorphism φ (it is not unique!) appearing in item 1 above is called a *Drinfeld isomorphism*.

COROLLARY 2 ([9]). *Let G be a linear semi-simple Lie group and $\mathfrak{g}$ be its complexified Lie algebra.*

1. *If $\mathsf{U}_t\mathfrak{g}$ is a deformation of $\mathsf{U}\mathfrak{g}$ (a "quantum group") then $(\mathsf{U}_t\mathfrak{g}, \mu_t, \Delta_t) \simeq (\mathsf{U}\mathfrak{g}[[t]], \mu_0, P_t\Delta_0 P_t^{-1})$.*
2. *$\mathcal{A}_t(G) := (\mathcal{A}(G)[[t]], \mu_0, P_t \cdot \Delta_0 \cdot P_t^{-1})$ is a Hopf deformation of $\mathcal{A}(G)$ and $\mathsf{U}_t\mathfrak{g} \overset{\text{Hopf}}{\subset} \mathcal{A}_t(G)$.*
3. *$\mathcal{D}_t(G) := (\mathcal{D}(G)[[t]], \mu_0, P_t \cdot \Delta_0 \cdot P_t^{-1})$ is a Hopf deformation of $\mathcal{D}(G)$ and $\mathsf{U}_t\mathfrak{g} \overset{\text{Hopf}}{\subset} \mathcal{D}_t(G)$.*
4. *$\mathcal{C}_t^\infty(G) := \mathcal{D}_t(G)^*$ and $\mathcal{H}_t(G) := \mathcal{A}_t(G)^*$ are quantized algebras of functions. They are Hopf deformations of $\mathcal{C}^\infty(G)$ and $\mathcal{H}(G)$.*

Similar results hold in the non linear case [3] and for other WB Hopf algebras (e.g. constructed with infinite dimensional representations) [2].

Proof. (1) Direct consequence of Theorem 1.

(2) $P_t \in (\mathsf{U}\mathfrak{g} \otimes \mathsf{U}\mathfrak{g})[[t]] \subset (\mathcal{A}(G)\hat{\otimes}\mathcal{A}(G))[[t]]$. We obtain coassociativity by density: $\mathsf{U}\mathfrak{g} \in \mathcal{A}(G)$.

(3) By restriction of (2).

(4) By simple dualization from (2) and (3). $\qquad\qquad\qquad\qquad\qquad\qquad \square$

Remark 2. "Hidden group structure" in a quantum group. Here the deformations are *preferred*, that is, the product on $\mathcal{D}_t(G)$ and on $\mathcal{A}_t(G)$ (resp. the coproduct on $\mathcal{C}_t^\infty(G)$ and on $\mathcal{H}_t(G)$) is not deformed and the basic structure is still the product on the group G. So this approach gives an interpretation of the Tannaka-Krein philosophy in the case of quantum groups: it has often been noticed that, in the generic case, finite dimensional representations of a quantum group are (essentially) representations of its classical limit. So the algebras involved should be the same, which is justified by the above mentioned rigidity result of Drinfeld. This shows that the initial classical group is still there, acting as a kind of "hidden variables" in this quantum group theory, which is exactly what we see in this quantum group theory. This fact was implicit in Drinfeld's work. The Tannaka-Krein interpretation of the twisting of quasi-Hopf algebras can be found in Majid (see e.g. [38]). It was made explicit, within the framework exposed here, in [9].

3.2. Unification of models and generalizations

Drinfeld models

We call "Drinfeld model of quantum group" a deformation of $U\mathfrak{g}$ for $\mathfrak{g}$ simple, as given in [19]. We have seen in the preceding section that from any Drinfeld model $U_t\mathfrak{g}$ of a quantum group (which can be generalized to any deformation of the Hopf algebra $U\mathfrak{g}$), we obtain a deformation of $\mathcal{D}(G)$ and $\mathcal{A}(G)$ that contains $U_t\mathfrak{g}$ as a sub-Hopf algebra. So $\mathcal{D}_t(G)$ and $\mathcal{A}_t(G)$ are quantum group models that describe Drinfeld models. By duality, $C_t^\infty(G)$ and $\mathcal{H}_t(G)$ are "quantum group deformations" of $C^\infty(G)$ and $\mathcal{H}(G)$. The deformed product on $\mathcal{H}(G)$ is the restriction of that on $C^\infty(G)$. Furthermore, as we shall see, these deformations coincide with the usual "quantum algebras of functions". Let us look more in detail at $\mathcal{H}_t(G)$:

Faddeev-Reshetikhin-Takhtajan (FRT) models

In [24] quantized algebras of functions are defined in terms of generators and relations, the key relation being given by the star-triangle (Yang-Baxter) equation, $R(T \otimes \mathsf{Id})(\mathsf{Id} \otimes T) = (\mathsf{Id} \otimes T)(T \otimes \mathsf{Id})R$, for a given R-matrix $R \in \mathsf{End}(V \otimes V)$ and for $T \in \mathsf{End}(V)$, V being a finite dimensional vector space.

As our deformations are given by a twist P_t, it is not surprising, from a structural point of view [38] that, dually, we obtain in each case a Yang-Baxter relation and so a "FRT-type" quantized algebra of functions. Our Fréchet-topological context permits to write precisely such a construction for the infinite-dimensional Hopf algebras involved.

1. Linear case. If G is semi-simple and linear, there exists π a finite dimensional representation of G such that $\mathcal{H}(G) \simeq \mathbb{C}[\pi_{ij}; 1 \leqslant i,j \leqslant N]$ where the π_{ij} are the coefficient functions of π. Denote by $(\mathcal{H}_t(G), *)$ the deformation of $\mathcal{H}(G)$ obtained in this way and by T the matrix $[\pi_{ij}]$. Define $T_1 := T \otimes Id$ and $T_2 := Id \otimes T$. Then we have

PROPOSITION 2 ([9, 3]).

1. $\{\pi_{ij}; 1 \leqslant i,j \leqslant N\}$ *is a topological generator system of the* $\mathbb{C}[[t]]$-*algebra* $\mathcal{H}(G)_t$.
2. *There exists an invertible* $\mathcal{R} \in \mathcal{L}(V_\pi \otimes V_\pi)[[t]]$ *such that* $\mathcal{R} \cdot T_1 * T_2 = T_2 * T_1 \cdot \mathcal{R}$ *(so* $\mathcal{H}_t(G)$ *is a "quantum algebra of functions" of type FRT).*
3. *We recover every quantum group given in [24] by this construction.*

Proof. (Sketch)
1. Perform a precise study of the deformed tensor product of representations.
2. Since the deformations $\mathcal{A}_t(G)$ are given by a twist P_t, $\mathcal{A}_t(G)$ is quasi-cocommutative, i.e. there exists $R \in (\mathcal{A}(G)\hat{\otimes}\mathcal{A}(G))[[t]]$ such that $\sigma \circ \Delta_t(a) = R\Delta_t(a)R^{-1}$ with $\sigma(a \otimes b) = b \otimes a$. Standard computations give the result.
3. We want to follow the way used in [19] to link Drinfeld to FRT models. But the main point is that our deformations are obtained through a Drinfeld isomorphism. We therefore have to show:

TOPOLOGICAL HOPF ALGEBRAS AND DEFORMATION QUANTIZATION

- There exists a specific Drinfeld isomorphism deforming the standard representation of $\mathfrak{g}$ into the representation of $U_t\mathfrak{g}$ used in [19].
- Two Drinfeld isomorphisms give equivalent deformations.

$\square$

For instance, the FRT quantization of $SL(n)$ can be seen as a Hopf deformation of $\mathcal{H}(SU(n))$ (with non deformed coproduct). Moreover, this Hopf deformation extends to $C^\infty(G)$.

Remark 3.

1. This proposition justifies the terminology "deformation", often employed but never justified in these cases. See e.g. [28] where it is shown that relations of type $\mathcal{R}T_1T_2 = T_2T_1\mathcal{R}$ need not define a deformation, even if $\mathcal{R}$ is Yang-Baxter.
2. Starting from Drinfeld models, our construction produces FRT models also for e.g. $G = Spin(n)$ and for exceptional Lie groups. In addition, at least some multiparameter deformations [45] can be easily treated in this way [9].

2. Non-linear case.

PROPOSITION 3 ([3]). *If G is semi-simple with finite center, there exists a dense subalgebra of $(C_t^\infty(G), *)$ generated by the coefficient functions of a finite number of (possibly infinite dimensional) representations.*

Jimbo models

These are models [34] with generators $E_i^\pm$, K_i and K_i^{-1}. For $G = SU(2)$ [10] and $G = SL(2,\mathbb{C})$ [40] we realize $U_q\mathfrak{sl}(2)$ and $U_t\mathfrak{sl}(2,\mathbb{C})$ as dense sub-Hopf algebras of $\mathcal{A}(G)$, $\forall t \in \mathbb{C} \setminus 2\pi\mathbb{Q}$ (with $q = e^t$). For $\mathfrak{sl}(2)$ this gives the original model of Jimbo [34]. For the Lorentz algebra $\mathfrak{sl}(2,\mathbb{C})$ this unifies [40] all the models proposed so far in the literature for a quantum Lorentz group. We obtain here *convergent* deformations (not only formal).

For $\mathfrak{sl}(2,\mathbb{C})$ it was first proposed in [43] to consider the quantum double [19] of $U_q\mathfrak{su}(2)$ as q-deformed Lorentz group. It was known from [46] that in such cases the double, as an algebra, is the tensor product of two copies of $U_t\mathfrak{su}(2)$. See also [42, 47], and [39] for a dual version and another semi-direct product form.

Deformation quantization

From the main construction, using deformations of $U\mathfrak{g}$, we deduce the following general theorem:

THEOREM 2 ([3]). *Let G be a semi-simple connected Lie group with a Poisson-Lie structure. There exists a deformation $(C_t^\infty(G), *)$ of $C^\infty(G)$ such that $*$ is a (differential) star product.*

Remark 4.

1. When $\mathrm{Lie}(G)$ is the double of some Lie algebra, the same result holds.
2. $*$ is differential because $\Delta_t = P_t\Delta_0 P_t^{-1}$ (Δ_t is a twist), with $P_t \in (U\mathfrak{g} \times U\mathfrak{g})[[t]]$.
3. Since from any Drinfeld quantum group we obtain a star product, and since any FRT quantum group can be seen as a restriction of such a star product, we have showed that the data of a "semi-simple" quantum group is equivalent

to the data of a star product on $C^\infty(G)$ satisfying $\Delta(f*g) = \Delta(f)*\Delta(g)$. The functorial existence results of Etingof and Kazhdan [23] on the quantization of Lie bialgebras (see also [22]) show that the latter is true also for "non semi-simple" quantum groups.

4. Techniques similar to those indicated here can be applied to other q-algebras (more general quantum groups such as those in [26] and more recent examples, Yangians, etc.). In particular those used in the case of the Jimbo models should be applicable to q-algebras defined by generators and relations. That direction of research has not yet been developed.

4. TOPOLOGICAL QUANTUM DOUBLE

From now on we use the Sweedler notation for the coproducts [49]: in a coalgebra (H, Δ), $\Delta(x) = \sum_{(x)} x_{(1)} \otimes x_{(2)}$ and, by coassociativity, $(\mathrm{Id} \otimes \Delta)\Delta(x) = (\Delta \otimes \mathrm{Id})\Delta(x) = \sum_{(x)} x_{(1)} \otimes x_{(2)} \otimes x_{(3)}$.

In [19] Drinfeld defines the quantum double of $\mathsf{U}_t\mathfrak{g}$ (see also [48]). This can be adapted to the context of topological Hopf algebras [8].

4.1. Definitions

Let A be $\mathcal{D}(G), \mathcal{A}(G), \mathcal{D}_t(G)$ or $\mathcal{A}_t(G)$. If $A = (A, \mu, \Delta, S)$ then $A^* = (A^*, {}^t\Delta, {}^t\mu, {}^tS)$. Define $A^0 = A^{*\ co-op} = (A^*, {}^t\Delta, {}^t\mu^{op}, {}^tS^{op})$, where $\mu^{op}(x \otimes y) := \mu(y \otimes x)$ and S^{op} is the antipode compatible with μ^{op} and Δ.

If we consider the vector space $A^* \otimes A$, Drinfeld [19] defines the quantum double as follows :

i) $D(A) \simeq A^0 \otimes A$ as coalgebras,

ii) $(f \otimes Id_A).(Id_{A^0} \otimes b) = f \otimes b$,

iii) $(Id_{A^0} \otimes e_s).(e^t \otimes Id_A) = \Delta_s^{kjn} \mu_{plk}^t S_n^{\prime p} (e^l \otimes Id_A) (Id_{A^0} \otimes e_j)$, where $\{e_s\}$ is a basis of A and $\{e^t\}$ the dual basis.

The Drinfeld double was expressed [37] in a Sweedler form for dually paired Hopf algebras as an example of a theory of 'double smash products'. Adapting that formulation to our topological context we can now define the double as:

DEFINITION 4. The double of A, $D(A)$, is the topological Hopf algebra $(A^* \overline{\otimes} A, \mu_D, {}^t\mu^{op} \otimes \Delta, {}^tS^{op} \otimes S)$ with

$$\mu_D((f \otimes a) \otimes (g \otimes b)) = \sum_{(a)} f < g, S^{op}(a_{(3)})\,?\,a_{(1)} > \otimes a_{(2)} b$$

$$= \sum_{(a)(g)} < g_{(1)}, a_{(1)} > < {}^tS^{op}(g_{(3)}), a_{(3)} > fg_{(2)} \otimes a_{(2)}b$$

where $< , >$ denotes the pairing A^*/A, "?" stands for a variable in A and $\overline{\otimes}$ is the completed inductive tensor product.

As topological vector spaces we have $D(A) = A^* \overline{\otimes} A$. Thus $D(A)^* = A \hat{\otimes} A^*$ and $D(A)^{**} = D(A)$. So $D(A)$ is "almost self dual" (it is self dual up to a completion) and is reflexive.

4.2. Extension theory

- If A is cocommutative then the product μ_D of $D(A)$ is the *smash product* $\vec{\mu}$ on $A^0 \overline{\otimes} A$

$$\vec{\mu}\left((f \otimes a) \otimes (g \otimes b)\right) = \sum_{(a)} f(a_{(1)} \rightharpoonup g) \otimes a_{(2)}b$$

where $\rightharpoonup$ denotes the coadjoint action of A on A^0, defined by $< a \rightharpoonup f, b > = \sum_{(a)} < f, S(a_{(1)})ba_{(2)} >$. This product is the "zero class" of an extension theory, defined by Sweedler [49], classified by a space of 2-cohomology $H^2_{sw}(A, A^0)$. The products are of the form, for τ a 2-cocycle,

$$\vec{\mu}_\tau\left((f \otimes a) \otimes (g \otimes b)\right) = \sum_{(a)(b)} f(a_{(1)} \rightharpoonup g)\tau(a_{(2)} \otimes b_{(2)}) \otimes a_{(3)}b_{(2)}.$$

- The coproduct of $D(A)$ is a smash coproduct for the trivial co-action. We can dualize the theory and, putting the two things together, we obtain an extension theory for bialgebras which is classified by a cohomology space $H^2_{bisw}(A^0, A)$.

Question : Are there other possible definitions of the double as an extension of A^0 by A?
Answer : NO, for $A = \mathcal{D}(G)$ [8], because $H^2_{bisw}\left(\mathcal{D}(G), \mathcal{C}^\infty(G)\right) = \{0\}$.

5. CROSSED PRODUCTS AND DEFORMATION QUANTIZATION

In this section we shall see that the Hopf algebra techniques presented in the preceding sections can be useful not only to understand quantum groups, but also to develop very nice formulas in deformation quantization itself.

In order to shed light on the general definition which follows, we return to the simplest case of deformation quantization: the Moyal product on $\mathbb{R}^2$. We look at $\mathbb{R}^2$ as $T^*\mathbb{R} \equiv \mathbb{R} \times \mathbb{R}^*$ and therefore can write $\mathcal{C}^\infty(\mathbb{R}^2) \simeq \mathcal{C}^\infty(\mathbb{R}) \hat{\otimes} \mathcal{C}^\infty(\mathbb{R}^*)$. We consider first two functions of a special kind in this algebra: $u(x) = u(x_1, x_2) = f(x_1)P(x_2)$ and $v(x) = v(x_1, x_2) = g(x_1)Q(x_2)$ where $f, g \in \mathcal{C}^\infty_0(\mathbb{R})$ and P, Q are polynomials in $\mathrm{Pol}(\mathbb{R}^*) \simeq S\mathbb{R}$. We can then write is the usual coproduct on the symmetric algebra $S\mathbb{R}$ as $\Delta(P)(x_2, y_2) = P(x_2 + y_2)(\overset{\text{notation}}{=} \sum_{(P)} P_{(1)}(x_2)P_{(2)}(y_2))$.

We now look at Formula (7) for the Moyal star product on $\mathbb{R}^2$ and perform on it some formal calculations (we do not discuss the convergence of the integrals involved). Up to a constant (depending on $\hbar$) we get:

$$(u * v)(x) = \int_{\mathbb{R}^2 \times \mathbb{R}^2} u(x+y)v(x+z)e^{-\frac{i}{\hbar}\Lambda^{-1}(y,z)}dydz$$

$$= \int_{\mathbb{R}^2 \times \mathbb{R}^2} f(x_1+y_1)P(x_2+y_2)g(x_1+z_1)Q(x_2+z_2) \times$$

$$e^{-\frac{i}{\hbar}(y_1 z_2 - y_2 z_1)}dy_1 dy_2 dz_1 dz_2$$

$$= \int_{\mathbb{R}^2} f(x_1+y_1)Q(x_2+z_2)e^{-\frac{i}{\hbar}y_1 z_2}dy_1 dz_2 \times$$

$$\int_{\mathbb{R}^2} g(x_1+z_1)P(x_2+y_2)e^{\frac{i}{\hbar}y_2 z_1}dy_2 dz_1$$

$$= \sum_{(P)(Q)} (\partial^+_{Q_{(1)}} f)(x_1)Q_{(2)}(x_2).(\partial^-_{P_{(1)}} g)(x_1)P_{(2)}(x_2) \quad \text{(up to a constant)}$$

with $\partial^\pm_{Q_{(1)}} = Q_{(1)}(\mp i\hbar\partial_{x_1})$ (the same for P), considering that $F_\hbar^\mp\left(\alpha F_\hbar^\pm(h)(\alpha)\right)(x)$
$= \mp i\hbar\partial_x h(x)$ for $h \in \mathcal{C}_0^\infty(\mathbb{R})$ with $F_\hbar^\pm(h)(\alpha)$ defined as $\int_\mathbb{R} h(x)e^{\mp\frac{i}{\hbar}x\alpha}dx$. This
suggests the following small generalization of the smash product:

DEFINITION 5. Let B be a cocommutative bialgebra and C a B-bimodule algebra
[i.e. C is both a left B-module algebra and a right B-module algebra such that
$(a \rightharpoonup f) \leftharpoonup b = a \rightharpoonup (f \leftharpoonup b)$]. We define the L-R smash product on $C \otimes B$ by

$$(f \otimes a) \star (g \otimes b) = \sum_{(a)} (f \leftharpoonup b_{(1)})(a_{(1)} \rightharpoonup g) \otimes a_{(2)} b_{(2)}.$$

PROPOSITION 4. *The L-R smash product is associative.*

5.1. Relation with usual deformation quantization

Let G be a Lie group, T^*G its cotangent bundle, $\mathfrak{g} = \text{Lie}(G)$. We have

$$\mathcal{C}^\infty(T^*G) \simeq \mathcal{C}^\infty(G \times \mathfrak{g}^*) \simeq \mathcal{C}^\infty(G)\hat{\otimes}\mathcal{C}^\infty(\mathfrak{g}^*) \supset \mathcal{C}^\infty(G) \otimes \text{Pol}(\mathfrak{g}^*) \simeq \mathcal{C}^\infty(G) \otimes \mathsf{Sg}.$$

We define a deformation of $\mathcal{C}^\infty(G) \otimes \mathsf{Sg}$ by a L-R smash product:

- We deform Sg by the "parametrized version" of Ug:

$$\mathsf{Ug}[[t]] = \frac{\mathsf{Tg}}{< xy - yx - t[x,y] >}.$$

This is a Hopf algebra with Δ, ϵ and S as for Ug.

- Let $\{X_i \; ; \; i = 1, \ldots, n\}$ be a basis of $\mathfrak{g}$ and $\overrightarrow{X}_i$ (resp. $\overleftarrow{X}_i$) be the left (resp.

  right) invariant vector fields on G associated with X_i. For $\lambda \in [0,1]$ we
  consider the following actions of $B = \mathsf{Ug}[[t]]$ on $C = \mathcal{C}^\infty(G)$:

  1. $(X_i \rightharpoonup f)(x) = t(\lambda - 1)(\overrightarrow{X}_i \cdot f)(x)$

  2. $(f \leftharpoonup X_i)(x) = t\lambda(\overleftarrow{X}_i \cdot f)(x)$.

LEMMA 1. *These actions define on $\mathcal{C}^\infty(G)$ a B-bimodule algebra structure.*

DEFINITION 6. We denote by $\star_\lambda$ the L-R smash product on $\mathcal{C}^\infty(G) \otimes \text{Pol}(\mathfrak{g}^*)$ given
by this B-bimodule algebra structure on $\mathcal{C}^\infty(G)$.

TOPOLOGICAL HOPF ALGEBRAS AND DEFORMATION QUANTIZATION

PROPOSITION 5. *For $G = \mathbb{R}^n$, $\star_{1/2}$ is the Moyal (Weyl ordered) star product, $\star_0$ is the standard ordered star product and in general $\star_\lambda$ is called λ-ordered star product on $\mathbb{R}^{2n}$* [44].

Remark 5. For a general Lie group G, $\star_\lambda$ gives in the generic case new deformation quantization formulas on T^*G. It would be interesting to study the properties of these $\star_\lambda$ for a noncommutative G and their relations with the star products that are known. In particular $\star_{1/2}$ is formally different from the star product on $C^\infty(T^*G)$ given by S. Gutt in [33] but preliminary calculations seem to indicate that, in a neighborhood of the unit of G, they are equivalent by a symplectomorphism.

5.2. Application to the quantization of symmetric spaces

DEFINITION 7 ([4]). A *symplectic symmetric space* is a triple (M, ω, s), where (M, ω) is a smooth connected symplectic manifold and $s : M \times M \to M$ is a smooth map such that:

(i) for all x in M, the partial map $s_x : M \to M : y \mapsto s_x(y) := s(x, y)$ is an involutive symplectic diffeomorphism of (M, ω) called the *symmetry* at x.

(ii) For all x in M, x is an isolated fixed point of s_x.

(iii) For all x and y in M, one has $s_x s_y s_x = s_{s_x(y)}$.

Two symplectic symmetric spaces (M, ω, s) and (M', ω', s') are *isomorphic* if there exists a symplectic diffeomorphism $\varphi : (M, \omega) \to (M', \omega')$ such that $\varphi s_x = s'_{\varphi(x)} \varphi$.

DEFINITION 8. Let $(\mathfrak{g}, \sigma)$ be an *involutive algebra*, that is, $\mathfrak{g}$ is a finite dimensional real Lie algebra and σ is an involutive automorphism of $\mathfrak{g}$. Let Ω be a skewsymmetric bilinear form on $\mathfrak{g}$. Then the triple $(\mathfrak{g}, \sigma, \Omega)$ is called a *symplectic triple* if the following properties are satisfied:

1. Let $\mathfrak{g} = \mathfrak{k} \oplus \mathfrak{p}$ where $\mathfrak{k}$ (resp. $\mathfrak{p}$) is the $+1$ (resp. -1) eigenspace of σ. Then $[\mathfrak{p}, \mathfrak{p}] = \mathfrak{k}$ and the representation of $\mathfrak{k}$ on $\mathfrak{p}$, given by the adjoint action, is faithful.

2. Ω is a Chevalley 2-cocycle for the trivial representation of $\mathfrak{g}$ on $\mathbb{R}$ such that $\forall X \in \mathfrak{k}$, $i(X)\Omega = 0$. Moreover, the restriction of Ω to $\mathfrak{p} \times \mathfrak{p}$ is nondegenerate.

The dimension of $\mathfrak{p}$ defines the *dimension* of the triple. Two such triples $(\mathfrak{g}_i, \sigma_i, \Omega_i)$ $(i = 1, 2)$ are *isomorphic* if there exists a Lie algebra isomorphism $\psi : \mathfrak{g}_1 \to \mathfrak{g}_2$ such that $\psi \circ \sigma_1 = \sigma_2 \circ \psi$ and $\psi^*\Omega_2 = \Omega_1$.

PROPOSITION 6 ([4]). *There is a bijective correspondence between the isomorphism classes of simply connected symplectic symmetric spaces (M, ω, s) and the isomorphism classes of symmetric triples $(\mathfrak{g}, \sigma, \Omega)$.*

DEFINITION 9. A symplectic symmetric space (M, ω, s) is called an *elementary solvable* symplectic symmetric space if its associated triple $(\mathfrak{g}, \sigma, \Omega)$ is of the following type:

1. The Lie algebra $\mathfrak{g}$ is a split extension of Abelian Lie algebras $\mathfrak{a}$ and $\mathfrak{b}$:

$$\mathfrak{b} \longrightarrow \mathfrak{g} \overrightarrow{\longleftarrow} \mathfrak{a}.$$

2. The automorphism σ preserves the splitting $\mathfrak{g} = \mathfrak{b} \oplus \mathfrak{a}$.

3. There exists $\xi \in \mathfrak{k}^*$ such that $\Omega(X, Y) = \delta\xi = <\xi, [X, Y]_{\mathfrak{g}}>$ (Chevalley 2-coboundary).

For such an elementary solvable symplectic symmetric space there exists a global Darboux chart such that $(M, \omega) \simeq (\mathfrak{p} = \mathfrak{l} \oplus \mathfrak{a}, \Omega)$ [5]. So we have

$$\mathcal{C}^\infty(M) \quad \simeq \quad \mathcal{C}^\infty(\mathfrak{p}) \simeq \mathcal{C}^\infty(\mathfrak{l}) \hat{\otimes} \mathcal{C}^\infty(\mathfrak{a}) \simeq \mathcal{C}^\infty(\mathfrak{l}) \hat{\otimes} \mathcal{C}^\infty(\mathfrak{l}^*)$$
$$\underset{\mathfrak{a} \simeq \mathfrak{l}^*}{\supset} \quad \mathcal{C}^\infty(\mathfrak{l}) \otimes \mathsf{Pol}(\mathfrak{l}^*) \underset{\mathfrak{l} \text{ abelian}}{\simeq} \quad \mathcal{C}^\infty(\mathfrak{l}) \otimes U\mathfrak{l}$$

One can now define $\star_{1/2}$ (Moyal) on $\mathcal{C}^\infty(M) \simeq \mathcal{C}^\infty(\mathfrak{l} \oplus \mathfrak{a})$ or, using our preceding construction, on $\mathcal{C}^\infty(\mathfrak{l}) \otimes U\mathfrak{l}$.

In order to have an *invariant* star product on M under the action of G (such that $\mathfrak{g} = \mathrm{Lie}(G)$) P. Bieliavsky [5] defines an integral transformation $S : \mathcal{C}^\infty(\mathfrak{l}) \to \mathcal{C}^\infty(\mathfrak{l})$ and then an invariant star product $\star_S$ by, for $T := S \otimes \mathsf{Id}$,

$$(f \otimes a) \star_S (g \otimes b) := T^{-1}(T(f \otimes a) \star_{1/2} T(g \otimes b)).$$

Let us define $\quad f \bullet_S g := S^{-1}(Sf.Sg), \quad a \overset{S}{\rightharpoonup} f := S^{-1}(a \rightharpoonup Sf) \quad$ and $\quad f \overset{S}{\leftharpoonup} a := S^{-1}(Sf \leftharpoonup a).$

PROPOSITION 7 ([6]). *$\star_S$ is the L-R smash product of $(\mathcal{C}^\infty(\mathfrak{l}), \bullet_S)$ by $U\mathfrak{l}$ with the $U\mathfrak{l}$-bimodule structure given by $\overset{S}{\rightharpoonup}$ and $\overset{S}{\leftharpoonup}$.*

Remark 6. Since we were dealing with quantum groups in the first sections, we want to stress that the homogeneous (symmetric) spaces involved here are strictly different from those appearing in the quantum group approach of quantized homogeneous spaces [21]. Indeed, in the latter, the spaces come from Poisson-Lie groups, so that the Poisson bracket has to be singular; therefore this bracket (and a fortiori a star product deforming this bracket) cannot be invariant (otherwise it would be zero everywhere). Here the Poisson brackets are invariant and regular.

Acknowledgements

This survey owes a lot to the insight shown by Moshé Flato in pushing forward the deformation quantization program, including in its aspects related to quantum groups where the inputs of Georges Pinczon and Murray Gerstenhaber were, as can be seen here, very important. Thanks are also due to the referee for a number of valuable comments.

REFERENCES

[1] François Bayen, Moshé Flato, Christian Frønsdal, André Lichnerowicz and Daniel Sternheimer. Deformation theory and quantization I, II. *Ann. Phys.* (NY) (1978) **111**, 61–110, 111–151.

[2] Frédéric Bidegain. A candidate for a noncompact quantum group. *Lett. Math. Phys.*, **36** (1996), 157–167.

[3] Frédéric Bidegain and Georges Pinczon. A star-product approach to noncompact quantum groups. *Lett. Math. Phys.* **33** (1995), 231–240 (**hep-th/9409054**). Quantization of Poisson-Lie groups and applications. *Comm. Math. Phys.*, **179** (1996), 295–332.

[4] Pierre Bieliavsky. *Espaces symétriques symplectiques*. Ph. D. thesis, Université Libre de Bruxelles 1995.

[5] Pierre Bieliavsky. Strict quantization of solvable symmetric spaces, *J. Sympl. Geom.* **1** (2002), no. 2, 269–320.

TOPOLOGICAL HOPF ALGEBRAS AND DEFORMATION QUANTIZATION   69

[6] Pierre Bieliavsky, Philippe Bonneau and Yoshiaki Maeda. Universal Deformation Formulae, Symplectic Lie groups and Symmetric Spaces, math.QA/0308189. Universal Deformation Formulae for Three-Dimensional Solvable Lie groups, math.QA/0308188.

[7] Philippe Bonneau. Cohomology and associated deformations for not necessarily co-associative bialgebras. *Lett. Math. Phys.*, **26** (1992), 277–283.

[8] Philippe Bonneau. Topological quantum double. *Rev. Math. Phys.*, **6** (1994), 305–318.

[9] Philippe Bonneau, Moshé Flato, Murray Gerstenhaber, and Georges Pinczon. The hidden group structure of quantum groups: strong duality, rigidity and preferred deformations. *Comm. Math. Phys.*, **161** (1994), 125–156.

[10] Philippe Bonneau, Moshé Flato, and Georges Pinczon. A natural and rigid model of quantum groups. *Lett. Math. Phys.*, **25** (1992), 75–84.

[11] Alain Connes. *Noncommutative Geometry*, Academic Press, San Diego 1994.

[12] Alain Connes. Noncommutative geometry—year 2000. GAFA 2000 (Tel Aviv, 1999). *Geom. Funct. Anal.* (2000) Special Volume, Part II, 481–559 (math.QA/0011193).

[13] Alain Connes and Michel Dubois-Violette. Noncommutative finite-dimensional manifolds. I. Spherical manifolds and related examples. *Comm. Math. Phys.* **230** (2002), 539–579 (math.QA/0107070). Moduli space and structure of noncommutative 3-spheres, *Lett. Math. Phys.* (2003), in press (math.QA/0308275).

[14] Alain Connes, Moshé Flato and Daniel Sternheimer. Closed star-products and cyclic cohomology, *Lett. Math. Phys.* **24** (1992), 1–12.

[15] Alain Connes and Dirk Kreimer. Lessons from Quantum Field Theory — Hopf Algebras and Spacetime Geometries. *Lett. Math. Phys.* **48** (1999), 85–96.

[16] Alain Connes and Henri Moscovici. Modular Hecke Algebras and their Hopf Symmetry, math.QA/0301089. Rankin-Cohen Brackets and the Hopf Algebra of Transverse Geometry, math.QA/0304316.

[17] Giuseppe Dito and Daniel Sternheimer. Deformation Quantization: Genesis, Developments and Metamorphoses, in *Deformation quantization* (G.Halbout ed.), IRMA Lectures in Math. Theoret. Phys. **1**, pp. 9–54 . Walter de Gruyter, Berlin 2002 (math.QA/0201168).

[18] Michael R. Douglas and Nikita A. Nekrasov. Noncommutative Field Theory. *Rev.Mod.Phys.* **73** (2001), 977–1029 (hep-th/0106048).

[19] Vladimir G. Drinfeld. Quantum groups, in *Proceedings of the International Congress of Mathematicians*, Vol. 1–2 (Berkeley, Calif.; 1986), pp. 798–820, Amer. Math. Soc. Providence, RI 1987.

[20] Vladimir G. Drinfeld. Almost cocommutative Hopf algebras. *Algebra i Analiz*, **1** (1989), 30–46.

[21] Vladimir G. Drinfeld. On Poisson homogeneous spaces of Poisson-Lie groups. *Teoret. Mat. Fiz.*, **95** (1993), 226–227; translation in *Theoret. and Math. Phys.*, **95** (1993), 524–525.

[22] Benjamin Enriquez. A cohomological construction of quantization functors of Lie bialgebras. math.QA/0212325 (2002). Benjamin Enriquez and Pavel Etingof. On the invertibility of quantization functors. math.QA/0306212 (2003).

[23] Pavel Etingof and David Kazhdan. Quantization of Lie bialgebras I. *Selecta Math. (N.S.)*, **2** (1996), 1–41; II, III *ibid.* **4** (1998), 213–231, 233–269; IV, V *ibid.* **6** (2000), 79–104, 105–130; Quantization of Poisson algebraic groups and Poisson homogeneous spaces, in *Symétries quantiques* (Les Houches, 1995), 935–946, North-Holland, Amsterdam 1998.

[24] Ludwig D. Faddeev, Nicolai Yu. Reshetikhin, and Leon A. Takhtajan. Quantization of Lie groups and Lie algebras, in *Algebraic analysis, Vol. I*, pp. 129–139. Academic Press, Boston, MA 1988.

[25] Moshé Flato. Deformation view of physical theories, *Czechoslovak J. Phys.* **B32** (1982), 472–475.

[26] Christian Frønsdal. Generalization and exact deformations of quantum groups. *Publ. Res. Inst. Math. Sci.* **33** (1997), 91–149.

[27] Murray Gerstenhaber. On the deformation of rings and algebras, *Ann. Math.* **79** (1964), 59–103; and (IV), *ibid.* **99** (1974), 257–276.

[28] Murray Gerstenhaber, Anthony Giaquinto and Samuel D. Schack. Quantum symmetry, in *Lect. Notes in Math.* **1510**, 9–46, Springer 1991.

70 PH. BONNEAU AND D. STERNHEIMER

[29] Murray Gerstenhaber and Samuel D. Schack. Algebraic cohomology and deformation theory, in *Deformation Theory of Algebras and Structures and Applications* (M. Hazewinkel and M. Gerstenhaber Eds.), NATO ASI Ser. C **247**, 11–264, Kluwer Acad. Publ., Dordrecht 1988.

[30] Murray Gerstenhaber and Samuel D. Schack. Bialgebra cohomology, deformations, and quantum groups. *Proc. Natl. Acad. Sci. USA*, **87** (1990) 478–481.

[31] Hip J. Groenewold. On the principles of elementary quantum mechanics. *Physica* **12** (1946), 405–460.

[32] Alexander Grothendieck. Produits tensoriels topologiques et espaces nucléaires. *Mem. Amer. Math. Soc.*, No 16 (140p.), 1955.

[33] Simone Gutt. An explicit ∗-product on the cotangent bundle of a Lie group. *Lett. Math. Phys.*, **7** (1983), 249–258.

[34] Michio Jimbo. A q-difference algebra of $\mathcal{U}(\mathfrak{g})$ and the Yang-Baxter equation. *Lett. Math. Phys.* **10** (1985), 63–69.

[35] Maxim Kontsevich. Deformation quantization of algebraic varieties. in *EuroConférence Moshé Flato 2000*, Part III (Dijon), *Lett. Math. Phys.* **56** (2001), 271–294.

[36] Maxim Kontsevich and Don Zagier. Periods, in *Mathematics unlimited—2001 and beyond*, 771–808, Springer, Berlin 2001.

[37] Shahn Majid. Physics for algebraists: noncommutative and noncocommutative Hopf algebras by a bicrossproduct construction. *J. Algebra* **130** (1968), 17–64.

[38] Shahn Majid. Tannaka-Krein theorem for quasi-Hopf algebras and other results, in *Deformation theory and quantum groups with applications to mathematical physics* (Amherst, MA, 1990), *Contemp. Math.*, **134** (1992), 219–232.

[39] Shahn Majid. Braided matrix structure of the Sklyanin algebra and of the quantum Lorentz group. *Comm. Math. Phys.* **156** (1993), 607–638. See also: *A quantum groups primer*, London Mathematical Society Lecture Note Series **292**, x+169 pp., Cambridge University Press, Cambridge, 2002.

[40] Christiane Martin and Mohamed Zouagui. A noncommutative Hopf structure on $C^\infty[SL(2,\mathbf{C})]$ as a quantum Lorentz group. *J. Math. Phys.*, **37** (1996), 3611–3629.

[41] Jose E. Moyal. Quantum mechanics as a statistical theory, *Proc. Cambridge Phil. Soc.* **45** (1949), 99–124.

[42] Oleg Ogievetsky, William B. Schmidke, Julius Wess, and Bruno Zumino. Six generator q-deformed Lorentz algebra. *Lett. Math. Phys.* **23** (1991), 233–240.

[43] Piotr Podleś and Stanisław L. Woronowicz. Quantum deformation of Lorentz group. *Comm. Math. Phys.* **130** (1990), 381–431.

[44] Markus J. Pflaum. Deformation quantization on cotangent bundles, in *Coherent states, differential and quantum geometry* (Białowieża 1997), *Rep. Math. Phys.*, **43** (1999), 291–297.

[45] Nicolai Yu. Reshetikhin. Multiparameter quantum groups and twisted quasitriangular Hopf algebras. *Lett. Math. Phys.* **20** (1990), 331–335.

[46] Nicolai Yu. Reshetikhin and Michael A. Semenov-Tian-Shansky. Quantum R-matrices and factorization problems. *J. Geom. Phys.* **5**(4) (1989), 533–550.

[47] William B. Schmidke, Julius Wess and Bruno Zumino. A q-deformed Lorentz algebra. *Z. Phys. C* **52** (1991), 471–476.

[48] Michael A. Semenov-Tian-Shansky. Poisson Lie groups, quantum duality principle, and the quantum double, in *Mathematical aspects of conformal and topological field theories and quantum groups (South Hadley, MA, 1992)*, *Contemp. Math.* **175**, 219–248. Amer. Math. Soc., Providence, RI 1994.

[49] Moss E. Sweedler. Cohomology of algebras over Hopf algebras. *Trans. Am. Math. Soc.* **133** (1968), 205–239.

[50] François Trèves. *Topological vector spaces, distributions and kernels*, xvi+624 pp., Academic Press, New York–London 1967.

[51] Hermann Weyl. *The theory of groups and quantum mechanics*, Dover, New-York 1931. *Gruppentheorie und Quantenmechanik*, Reprint of the second edition [Hirzel, Leipzig 1931], xi+366 pp. Wissenschaftliche Buchgesellschaft, Darmstadt 1977.

[52] Eugene P. Wigner. Quantum corrections for thermodynamic equilibrium, *Phys. Rev.* **40** (1932), 749–759.

On Coseparable and Biseparable Corings

TOMASZ BRZEZIŃSKI Department of Mathematics, University of Wales Swansea, Singleton Park, Swansea SA2 8PP, U.K.
e-mail: *T.Brzezinski@swansea.ac.uk*

LARS KADISON Department of Mathematics and Statistics, University of New Hampshire, Kingsbury Hall, Durham, NH 03824, USA
e-mail: *kadison@math.unh.edu*

ROBERT WISBAUER Department of Mathematics, Heinrich-Heine University, D-40225 Düsseldorf, Germany
e-mail: *wisbauer@math.uni-duesseldorf.de*

> ABSTRACT. A relationship between coseparable corings and separable non-unital rings is established. In particular it is shown that a coseparable A-coring C has an associative A-balanced product. A Morita context is constructed for a coseparable coring with a grouplike element. Biseparable corings are defined, and a conjecture relating them to Frobenius corings is proposed.

1. INTRODUCTION

Corings were introduced by Sweedler in [22] as a generalisation of coalgebras and a means for dualising the Jacobson-Bourbaki theorem. Recently, corings have resurfaced in the theory of Hopf-type modules, in particular it has been shown in [5] that the category of entwined modules is an example of a category of comodules of a coring. Since entwined modules appear to be the most general of Hopf-type modules studied since the mid-seventies, the theory of corings provides one with a uniform and general approach to studying all such modules. This simple observation renewed interest in general theory of corings.

Corings appear naturally in the theory of ring extensions. Indeed, they provide an equivalent description of certain types of extensions (cf. [6]). In this paper we study properties of corings associated to extensions. In particular, we study *coseparable corings* introduced by Guzman [13] (and recently studied in [12] from a different point of view) and we reveal an intriguing duality between such corings and a non-unital generalisation of separable ring extensions. We also show that to any grouplike element in a coseparable coring one can associate a Morita context. This leads to a pair of adjoint functors. One of these functors turns out to be fully faithful. Furthermore, we introduce the notion of a *biseparable coring* and study

---

2000 *Mathematics Subject Classification.* 16W30, 13B02.
Key words and phrases. coring, A-ring, Morita context, ring extension, Frobenius coring, coseparable, cosplit, depth two.

its relationship to Frobenius corings introduced in [6]. This allows us to consider a conjecture from [9], concerning biseparable and Frobenius extensions in a new framework.

Our paper is organised as follows. In the next section, apart from recalling some basic facts about corings and comodules, we introduce a non-unital generalisation of separable extensions, which we term *separable A-rings*. We show that any coseparable coring is an example of such a separable A-ring, and conversely, that every separable A-ring leads to a non-unital coring. We then proceed in Section 3 to construct a Morita context associated to a grouplike element in a coseparable coring. We consider some examples coming from ring extensions and bialgebroids. Finally in Section 4 we introduce the notion of *biseparable corings*. These are closely related to biseparable extensions, and may serve as a means for settling the question put forward in [9] of whether biseparable extensions are Frobenius.

Throughout the paper, A denotes an associative ring with unit 1_A, and we use the standard notation for right (resp. left) A-modules $\mathbf{M}_A$ (resp. $_A\mathbf{M}$), bimodules, such as $\mathrm{Hom}_A(-,-)$ for right A-module maps, $_A\mathrm{Hom}(-,-)$ for left A-module maps etc. For any (A,A)-bimodule M the centraliser of A in M is denoted by M^A, i.e., $M^A := \{m \in M \mid \forall a \in A,\ am = ma\}$.

2. COSEPARABLE A-CORINGS AND SEPARABLE A-RINGS

2.1. Coseparable corings

We begin by recalling the definition of a coring from [22]. An (A,A)-bimodule $\mathcal{C}$ is said to be a *non-counital A-coring* if there exists an (A,A)-bimodule map $\Delta_\mathcal{C} : \mathcal{C} \to \mathcal{C} \otimes_A \mathcal{C}$ rendering the following diagram commutative

$$
\begin{array}{ccc}
\mathcal{C} & \xrightarrow{\ \Delta_\mathcal{C}\ } & \mathcal{C} \otimes_A \mathcal{C} \\
{\scriptstyle \Delta_\mathcal{C}}\downarrow & & \downarrow{\scriptstyle I_\mathcal{C} \otimes \Delta_\mathcal{C}} \\
\mathcal{C} \otimes_A \mathcal{C} & \xrightarrow{\ \Delta_\mathcal{C} \otimes I_\mathcal{C}\ } & \mathcal{C} \otimes_A \mathcal{C} \otimes_A \mathcal{C} .
\end{array}
$$

The map $\Delta_\mathcal{C}$ is termed a *coproduct*. Given a non-counital A-coring $\mathcal{C}$ with a coproduct $\Delta_\mathcal{C}$, an (A,A)-bimodule map $\epsilon_\mathcal{C} : \mathcal{C} \to A$ such that

$$(\epsilon_\mathcal{C} \otimes I_\mathcal{C}) \circ \Delta_\mathcal{C} = I_{A \otimes_A \mathcal{C}} = I_\mathcal{C}, \quad (I_\mathcal{C} \otimes \epsilon_\mathcal{C}) \circ \Delta_\mathcal{C} = I_{\mathcal{C} \otimes_A A} = I_\mathcal{C}.$$

is called a *counit* of $\mathcal{C}$. A non-counital A-coring with a counit is called an *A-coring*. If $\mathcal{C}$ is an (non-counital) A-coring, a right A-module M is called a *non-counital right $\mathcal{C}$-comodule* if there exists a right A-module map $\varrho^M : M \to M \otimes_A \mathcal{C}$ rendering the following diagram commutative

$$
\begin{array}{ccc}
M & \xrightarrow{\ \varrho^M\ } & M \otimes_A \mathcal{C} \\
{\scriptstyle \varrho^M}\downarrow & & \downarrow{\scriptstyle \varrho^M \otimes I_\mathcal{C}} \\
M \otimes_A \mathcal{C} & \xrightarrow{\ I_M \otimes \Delta_\mathcal{C}\ } & M \otimes_A \mathcal{C} \otimes_A \mathcal{C} .
\end{array}
$$

The map ϱ^M is called a *$\mathcal{C}$-coaction*. If, in addition, a $\mathcal{C}$-coaction satisfies the condition

$$(I_M \otimes \epsilon_\mathcal{C}) \circ \varrho^M = I_{M \otimes_A A} = I_M,$$

then M is called a *right C-comodule*. Similarly one defines left C-comodules, and (C, C)-bicomodules. Given right C-comodules M, N, a right A-linear map $f : M \to N$ is called a *morphism of right C-comodules* provided the following diagram

$$
\begin{array}{ccc}
M & \xrightarrow{\ f\ } & N \\
\varrho^M \downarrow & & \downarrow \varrho^N \\
M \otimes_A C & \xrightarrow{\ f \otimes I_C\ } & N \otimes_A C
\end{array}
$$

is commutative. The category of right C-comodules is denoted by $\mathbf{M}^C$. We use Sweedler notation to denote the action of a coproduct or a coaction on elements,

$$
\Delta_C(c) = \sum c_{(1)} \otimes c_{(2)}, \qquad \varrho^M(m) = \sum m_{(0)} \otimes m_{(1)}.
$$

An immediate example of a left and right C-comodule is provided by C itself. In both cases coaction is given by the coproduct Δ_C. Also, for any right (resp. left) A-module M, the tensor product $M \otimes_A C$ (resp. $C \otimes_A M$) is a right (resp. left) C-comodule with the coaction $I_M \otimes \Delta_C$ (resp. $\Delta_C \otimes I_M$). This defines a functor which is the right adjoint of a forgetful functor from the category of C-comodules to the category of A-modules. This functor can be defined for non-counital corings and non-counital comodules, and adjointness holds for corings with a counit.
In particular $C \otimes_A C$ is a (C, C)-bicomodule, and Δ_C is a (C, C)-bicomodule map, and following [13] we have

DEFINITION 2.1. A (non-counital) coring C is said to be *coseparable* if there exists a (C, C)-bicomodule splitting of the coproduct Δ_C.

Although Definition 2.1 makes sense for non-counital corings, it is much more meaningful in the case of corings with a counit. In this case (C, C)-bicomodule splittings of Δ_C, $\pi : C \otimes_A C \to C$ are in bijective correspondence with (A, A)-bimodule maps $\gamma : C \otimes_A C \to A$ such that for all $c, c' \in C$,

$$
\sum \gamma(c \otimes c'_{(1)}) c'_{(2)} = \sum c_{(1)} \gamma(c_{(2)} \otimes c'), \qquad \sum \gamma(c_{(1)} \otimes c_{(2)}) = \epsilon_C(c).
$$

Such a map γ is termed a *cointegral* in C, and the first of the above equations is said to express a *colinearity* of a cointegral. The correspondence is given by $\gamma = \epsilon_C \circ \pi$ and $\pi(c \otimes c') = \sum c_{(1)} \gamma(c_{(2)} \otimes c')$. Furthermore, C is a coseparable A-coring if and only if the forgetful functor $\mathbf{M}^C \to \mathbf{M}_A$ is separable (cf. [5, Theorem 3.5]).
Corings appear naturally in the context of ring extensions. A ring extension $B \to A$ determines the *canonical* Sweedler A-coring $C := A \otimes_B A$ with coproduct $\Delta_C : C \to C \otimes_A C$ given by $\Delta_C(a \otimes a') = a \otimes 1_A \otimes a'$ and counit $\epsilon_C : C \to A$ given by $\epsilon_C(a \otimes a') = aa'$ for all $a, a' \in A$. Recall from [19] that an extension $B \to A$ is said to be *split* if there exists a (B, B)-bimodule map $E : A \to B$ such that $E(1_A) = 1_B$. The map E is known as a *conditional expectation*. The canonical Sweedler coring associated to a split ring extension is coseparable. A cointegral γ coincides with the splitting map E via the natural isomorphisms

$$
{}_B\mathrm{Hom}_B(A, B) \subset {}_B\mathrm{Hom}_B(A, A) \cong {}_A\mathrm{Hom}_A(A \otimes_B A \otimes_A A \otimes_B A, A)
$$

(cf. [5, Corollary 3.7]).

2.2. Separable A-rings

Corings can be seen as a dualisation of A-*rings* and coseparable corings turn out to be closely related to a generalisation of separable extensions of rings. In this subsection we describe this generalisation.

DEFINITION 2.2. An (A, A)-bimodule B is called an A-*ring* provided there exists an (A, A)-bimodule map $\mu : B \otimes_A B \to B$ rendering commutative the diagram

$$
\begin{array}{ccc}
B \otimes_A B \otimes_A B & \xrightarrow{I_B \otimes \mu} & B \otimes_A B \\
{\scriptstyle \mu \otimes I_B} \downarrow & & \downarrow {\scriptstyle \mu} \\
B \otimes_A B & \xrightarrow{\mu} & B.
\end{array}
$$

This means that μ is associative. Note that an A-ring is necessarily a (non-unital) ring in the usual sense. Equivalently, an A-ring can be defined as a ring and an (A, A)-bimodule B with product that is an A-balanced (A, A)-bimodule map.

Note further that the notion of an A-ring in Definition 2.2 is a non-unital generalisation of ring extensions. Indeed, it is only natural to call an (A, A)-bilinear map $\iota : A \to B$ a *unit* (for (B, μ)) if it induces a commutative diagram

$$
\begin{array}{ccc}
B & \xrightarrow{I_B \otimes \iota} & B \otimes_A B \\
{\scriptstyle \iota \otimes I_B} \downarrow & {\scriptstyle =} & \downarrow {\scriptstyle \mu} \\
B \otimes_A B & \xrightarrow{\mu} & B.
\end{array}
$$

If this holds then $\iota(1_A) = 1_B$ is a unit of B in the usual sense. One can then easily show that ι is a ring map, hence a unital A-ring is simply a ring extension.

DEFINITION 2.3. Given an A-ring B, a right A-module M is said to be a *right B-module* provided there exists a right A-module map $\varrho_M : M \otimes_A B \to M$ making the following diagram

$$
\begin{array}{ccc}
M \otimes_A B \otimes_A B & \xrightarrow{\varrho_M \otimes I_B} & M \otimes_A B \\
{\scriptstyle I_M \otimes \mu} \downarrow & & \downarrow {\scriptstyle \varrho_M} \\
M \otimes_A B & \xrightarrow{\varrho_M} & M
\end{array}
$$

commute. The map ϱ_M is called a *right B-action*. On elements the action is denoted by a dot in a standard way, i.e., $m \cdot b = \varrho_M(m \otimes b)$. Remember that for all $a \in A$, $(ma) \cdot b = m \cdot (ab)$.

A morphism $f : M \to N$ between two B-modules is an A-linear map which makes the following diagram

$$
\begin{array}{ccc}
M \otimes_A B & \xrightarrow{f \otimes I_B} & N \otimes_A B \\
{\scriptstyle \varrho_M} \downarrow & & \downarrow {\scriptstyle \varrho_N} \\
M & \xrightarrow{f} & N
\end{array}
$$

ON COSEPARABLE AND BISEPARABLE CORINGS

commute. A right B-module M is said to be *firm* provided the induced map $M \otimes_B B \to M$, $m \otimes b \mapsto m \cdot b$ is a right B-module isomorphism. The category of firm right B-modules is denoted by $\mathbf{M}_B$.

Obviously left B-modules are defined in a symmetric way. Similarly one defines (B, B)-bimodules, (A, B)-bimodules etc.

Dually to the definition of coseparable corings we can define separable A-rings.

DEFINITION 2.4. An A-ring B is said to be *separable* if the product map $\mu : B \otimes_A B \to B$ has a (B, B)-bimodule section $\delta : B \to B \otimes_A B$.

If B is a separable A-ring then clearly μ is surjective and the induced map $B \otimes_B B \to B$ is an isomorphism. Therefore B is a firm left and right B-module, i.e., B is a firm ring.

Note that if B has a unit $\iota : A \to B$ then B is a separable A-ring if and only if B is a separable extension of A. Thus Definition 2.4 extends the notion of a separable extension to non-unital rings. Note, however, that in general this is not an extension, since there is no (ring) map $A \to B$.

Remark 2.5. In consistency with A-corings, we use the terminology of [3] in Definition 2.2. In [19, 11.7] A-rings are termed *multiplicative A-bimodules*. Following [23] one might call a separable A-ring (as defined in Definition 2.4) an *A-ring with a splitting map*.

2.3. Coseparable A-corings are separable A-rings

The main result of this section is contained in the following

THEOREM 2.6. *If C is a coseparable A-coring then C is a separable A-ring.*

Proof. Let $\pi : C \otimes_A C \to C$ be a bicomodule retraction of the coproduct Δ_C, and let $\gamma = \epsilon_C \circ \pi$ be the corresponding cointegral. We claim that C is an associative A-ring with product $\mu = \pi$. Indeed, since the alternative expressions for product are $cc' = \sum \gamma(c \otimes c'_{(1)}) c'_{(2)} = \sum c_{(1)} \gamma(c_{(2)} \otimes c')$, for all $c, c', c'' \in C$ we obtain, using the left A-linearity of γ and Δ_C,

$$(cc')c'' = \sum (\gamma(c \otimes c'_{(1)}) c'_{(2)}) c'' = \sum \gamma(c \otimes c'_{(1)}) \gamma(c'_{(2)} \otimes c''_{(1)}) c''_{(2)}.$$

On the other hand, the colinearity and right A-linearity of γ, and the left A-linearity of Δ_C imply

$$\begin{aligned} c(c'c'') &= \sum c(\gamma(c' \otimes c''_{(1)}) c''_{(2)}) = \sum \gamma(c \otimes \gamma(c' \otimes c''_{(1)}) c''_{(2)}) c''_{(3)} \\ &= \sum \gamma(c \otimes c'_{(1)} \gamma(c'_{(2)} \otimes c''_{(1)})) c''_{(2)} = \sum \gamma(c \otimes c'_{(1)}) \gamma(c'_{(2)} \otimes c''_{(1)}) c''_{(2)}. \end{aligned}$$

This explicitly proves that the product in C is associative. Clearly this product is (A, A)-bilinear. Note that Δ_C is a (C, C)-bimodule map since

$$c \Delta_C(c') = \sum cc'_{(1)} \otimes c'_{(2)} = \sum \pi(c \otimes c'_{(1)}) \otimes c'_{(2)} = \Delta_C \circ \pi(c \otimes c') = \Delta_C(cc'),$$

from the right colinearity of π. Similarly for the left C-linearity. Finally π is split by Δ_C since π is a retraction of Δ_C. This proves that C is a separable A-ring. $\square$

PROPOSITION 2.7. *Let C be a coseparable A-coring with a cointegral γ. View C as an A-ring with product π as in Theorem 2.6. Then any right C-comodule M is a firm right C-module with the action $\varrho_M = (I_M \otimes \gamma) \circ (\varrho^M \otimes I_C)$.*

Proof. Take any $m \in M$ and $c, c' \in C$. Then explicitly the action reads $m \cdot c = \sum m_{(0)} \gamma(m_{(1)} \otimes c)$, and we can compute

$$
\begin{aligned}
(m \cdot c) \cdot c' &= \sum (m_{(0)} \gamma(m_{(1)} \otimes c)) \cdot c' = \sum m_{(0)} \gamma(m_{(1)} \gamma(m_{(2)} \otimes c) \otimes c') \\
&= \sum m_{(0)} \gamma(\gamma(m_{(1)} \otimes c_{(1)}) c_{(2)} \otimes c') = \sum m_{(0)} \gamma(m_{(1)} \otimes c_{(1)}) \gamma(c_{(2)} \otimes c') \\
&= \sum m_{(0)} \gamma(m_{(1)} \otimes c_{(1)} \gamma(c_{(2)} \otimes c')) = \sum m_{(0)} \gamma(m_{(1)} \otimes cc') = m \cdot (cc'),
\end{aligned}
$$

as required. We used the following properties of a cointegral: colinearity to derive the third equality and A-bilinearity to derive the fourth and fifth equalities. Obviously the action is right A-linear. Thus M is a C-module. We need to show that it is firm.

Note that $M \otimes_C C$ is defined as a cokernel of the following right C-linear map

$$
\lambda : M \otimes_A C \otimes_A C \to M \otimes_A C, \quad m \otimes c \otimes c' \mapsto mc \otimes c' - m \otimes cc'.
$$

Since γ is a cointegral, ϱ_M is a right C-linear retraction of ϱ^M, hence, in particular it is a surjection and there is the following sequence of right C-module maps

$$
M \otimes_A C \otimes_A C \xrightarrow{\lambda} M \otimes_A C \xrightarrow{\varrho_M} M \longrightarrow 0.
$$

We need to show that this sequence is exact. Clearly the associativity of the action of C on M implies that $\varrho_M \circ \lambda = 0$, so that $\mathrm{Im}\,\lambda \subseteq \ker \varrho_M$. Furthermore, for all $m \in M$ and $c \in C$,

$$
\begin{aligned}
(\varrho^M \circ \varrho_M - \lambda \circ (I_M \otimes \Delta_C))(m \otimes c) &= \sum m_{(0)} \otimes m_{(1)} \gamma(m_{(2)} \otimes c) \\
&\quad - \sum m \cdot c_{(1)} \otimes c_{(2)} + \sum m \otimes \pi(c_{(1)} \otimes c_{(2)}) \\
&= \sum m_{(0)} \gamma(m_{(1)} \otimes c_{(1)}) \otimes c_{(2)} \\
&\quad - \sum m_{(0)} \gamma(m_{(1)} \otimes c_{(1)}) \otimes c_{(2)} + m \otimes c \\
&= m \otimes c,
\end{aligned}
$$

where we used the colinearity of a cointegral. This implies that $\ker \varrho_M \subseteq \mathrm{Im}\,\lambda$, i.e., the above sequence is exact as required. $\qquad\square$

As an example of a coseparable coring one can take the canonical Sweedler coring associated to a split ring extension $B \to A$. In this case the product in $A \otimes_B A$ comes out as

$$
(a \otimes a')(a'' \otimes a''') = aE(a'a'') \otimes a''', \qquad \forall a, a', a'', a''' \in A
$$

where E is a splitting map. This is known as the *E-multiplication*. Since a comodule of the canonical coring is a *descent datum* for a ring extension $B \to A$, Proposition 2.7 implies that every descent datum is a firm module of the A-ring $A \otimes_B A$ with the E-multiplication (cf. [5, Example 2.1], [8, Section 25] or [10, Section 4.8] for a discussion of the correspondence between descent data and comodules of a canonical coring).

Theorem 2.6 has the following (part-) converse.

PROPOSITION 2.8. *Let B be a separable A-ring. Then B is a coseparable noncounital coring.*

Proof. Let $\Delta : B \xrightarrow{\prime} B \otimes_A B$ be a (B,B)-bimodule map splitting the product μ in B. The B-linearity of Δ implies that the following diagram

$$
\begin{array}{ccccc}
B \otimes_A B \otimes_A B & \xleftarrow{\ I_B \otimes \Delta\ } & B \otimes_A B & \xrightarrow{\ \Delta \otimes I_B\ } & B \otimes_A B \otimes_A B \\
{\scriptstyle \mu \otimes I_B} \big\downarrow & & {\scriptstyle \mu} \big\downarrow & & \big\downarrow {\scriptstyle I_B \otimes \mu} \\
B \otimes_A B & \xleftarrow{\ \Delta\ } & B & \xrightarrow{\ \Delta\ } & B \otimes_A B
\end{array}
$$

is commutative. For all $b \in B$ we write $(\Delta \otimes I_B) \circ \Delta(b) = \sum b_{(1)(1)} \otimes b_{(1)(2)} \otimes b_{(2)}$ and $(I_B \otimes \Delta) \circ \Delta(b) = \sum b_{(1)} \otimes b_{(2)(1)} \otimes b_{(2)(2)}$, and use the above diagram to obtain

$$
\begin{aligned}
\Delta(b) = (\Delta \circ \mu \circ \Delta)(b) &= (I_B \otimes \mu) \circ (\Delta \otimes I_B) \circ \Delta(b) = \sum b_{(1)(1)} \otimes \mu(b_{(1)(2)} \otimes b_{(2)}) \\
&= (\mu \otimes I_B) \circ (I_B \otimes \Delta) \circ \Delta(b) = \sum \mu(b_{(1)} \otimes b_{(2)(1)}) \otimes b_{(2)(2)}.
\end{aligned}
$$

Using these identities we can compute

$$
\begin{aligned}
(I_B \otimes \Delta) \circ \Delta(b) &= \sum b_{(1)(1)} \otimes (\Delta \circ \mu)(b_{(1)(2)} \otimes b_{(2)}) \\
&= \sum b_{(1)(1)} \otimes ((\mu \otimes I_B) \circ (I_B \otimes \Delta))(b_{(1)(2)} \otimes b_{(2)}) \\
&= \sum b_{(1)(1)} \otimes \mu(b_{(1)(2)} \otimes b_{(2)(1)}) \otimes b_{(2)(2)},
\end{aligned}
$$

and

$$
\begin{aligned}
(\Delta \otimes I_B) \circ \Delta(b) &= \sum (\Delta \circ \mu)(b_{(1)} \otimes b_{(2)(1)}) \otimes b_{(2)(2)} \\
&= \sum ((I_B \otimes \mu) \circ (\Delta \otimes I_B))(b_{(1)} \otimes b_{(2)(1)}) \otimes b_{(2)(2)} \\
&= \sum b_{(1)(1)} \otimes \mu(b_{(1)(2)} \otimes b_{(2)(1)}) \otimes b_{(2)(2)},
\end{aligned}
$$

i.e., $(\Delta \otimes I_B) \circ \Delta = (I_B \otimes \Delta) \circ \Delta$. This proves that B is a non-counital A-coring with coproduct Δ.

Next note that the above diagram can also be understood as a statement that μ is a (B,B)-bicomodule map. Since μ is a retraction for Δ, B is a coseparable non-counital coring as required. This completes the proof. $\qquad\square$

3. MORITA CONTEXTS FOR SEPARABLE CORINGS

Although the Morita theory is usually developed for rings with unit, it can be extended to firm rings without units (cf. [4, Exercise 4.1.4]). Recall that a right module M of a non-unital ring R is said to be *firm* if the map $M \otimes_R R \to M$ induced from the R-product in M is an R-module isomorphism. Similarly one defines left firm modules. A non-unital ring is a firm ring if it is firm as a left and right R-module.

DEFINITION 3.1. Given a pair of firm non-unital rings R, S, a Morita context consists of a firm (R, S)-bimodule V and a firm (S, R)-bimodule W and a pair of bimodule maps

$$
\sigma : W \otimes_R V \to S, \qquad \tau : V \otimes_S W \to R,
$$

such that the following diagrams

commute. A Morita context is denoted by $(R, S, V, W, \tau, \sigma)$. A Morita context is said to be *strict* provided σ and τ are isomorphisms.

The Morita theory for non-unital rings can be developed along the same lines as the usual Morita theory [18]. The aim of this section is to show that there is a Morita context associated to any coseparable coring with a grouplike element. To construct such a context we employ techniques developed in recent papers [1], [11]. First recall that an element g of an A-coring $\mathcal{C}$ is said to be a *grouplike element*, provided $\Delta_\mathcal{C}(g) = g \otimes g$ and $\epsilon_\mathcal{C}(g) = 1$. Obviously, not every coring has grouplike elements. The results of this section are contained in the following

THEOREM 3.2. *Let $\mathcal{C}$ be a coseparable A-coring with a cointegral γ and a grouplike element g. View $\mathcal{C}$ as a separable A-ring as in Theorem 2.6 and let $\mathbf{M}_\mathcal{C}$ denote the category of firm right modules of the A-ring $\mathcal{C}$ (cf. Definition 2.3). For any $M \in \mathbf{M}_\mathcal{C}$ define*

$$M_{g,\gamma}^\mathcal{C} = \{m \in M \mid \forall c \in \mathcal{C}, \ m \cdot c = m\gamma(g \otimes c)\}.$$

Then:

(1) $B = A_{g,\gamma}^\mathcal{C} = \{b \in A \mid \forall c \in \mathcal{C}, \ \gamma(gb \otimes c) = b\gamma(g \otimes c)\}$, is a subring of A.

(2) The assignment $(-)_{g,\gamma}^\mathcal{C} : \mathbf{M}_\mathcal{C} \to \mathbf{M}_B$, $M \mapsto M_{g,\gamma}^\mathcal{C}$ is a covariant functor which has a left adjoint $- \otimes_B A : \mathbf{M}_B \to \mathbf{M}_\mathcal{C}$.

(3) $Q = \mathcal{C}_{g,\gamma}^\mathcal{C}$ is a firm left ideal in $\mathcal{C}$ and hence a $(\mathcal{C}, B)$-bimodule.

(4) For every $M \in \mathbf{M}_\mathcal{C}$, the additive map

$$\omega_M : M \otimes_\mathcal{C} Q \to M_{g,\gamma}^\mathcal{C}, \qquad m \otimes q \mapsto m \cdot q,$$

is bijective.

(5) Define two maps

$$\sigma : Q \otimes_B A \to \mathcal{C}, \ q \otimes a \mapsto qa \quad \text{and} \quad \tau : A \otimes_\mathcal{C} Q \to B, \ a \otimes q \mapsto \gamma(ga \otimes q).$$

Then $(B, \mathcal{C}, A, Q, \tau, \sigma)$ is a Morita context, in which τ is surjective (hence an isomorphism).

Proof. (1) First note that A is a right $\mathcal{C}$-comodule with the coaction $a \mapsto 1_A \otimes ga$. Thus by Proposition 2.7, A is a firm $\mathcal{C}$-module with the action $a \cdot c = \gamma(ga \otimes c)$, for all $a \in A$ and $c \in \mathcal{C}$. Therefore the definition of B makes sense and takes the form stated. Obviously $1_A \in B$. Furthermore, for all $b, b' \in B$ and $c \in \mathcal{C}$,

$$\gamma(gbb' \otimes c) = \gamma(gb \otimes b'c) = b\gamma(g \otimes b'c) = bb'\gamma(g \otimes c),$$

so that $bb' \in B$ as required. Alternatively, we note that $B \cong \mathrm{End}_\mathcal{C}(A)$ and is therefore a ring.

ON COSEPARABLE AND BISEPARABLE CORINGS

(2) We note that $M_{g,\gamma}^{\mathcal{C}} \cong \text{Hom}_{\mathcal{C}}(A, M)$ via $f \mapsto f(1_A)$ has the left adjoint $-\otimes_B A$. In more detail, take any $M \in \mathbf{M}_{\mathcal{C}}$, $m \in M_{g,\gamma}^{\mathcal{C}}$, $b \in B$ and $c \in \mathcal{C}$, and use the definitions of B and $M_{g,\gamma}^{\mathcal{C}}$ to compute

$$(mb) \cdot c = m \cdot (bc) = m\gamma(g \otimes bc) = mb\gamma(g \otimes c).$$

This shows that $M_{g,\gamma}^{\mathcal{C}}$ is a right B-module, hence $(-)_{g,\gamma}^{\mathcal{C}}$ is a functor as stated. In the opposite direction, for any right B-module N, $N \otimes_B A$ is a firm right $\mathcal{C}$-module with the action $(n \otimes a) \cdot c = n \otimes \gamma(ga \otimes c)$. Note that this action is well-defined by the construction of B. More precisely, A is a left B-module and a firm right $\mathcal{C}$-module (since it is a right $\mathcal{C}$-comodule). It is a $(B, \mathcal{C})$-bimodule, since for every $b \in B$, $a \in A$ and $c \in \mathcal{C}$,

$$(ba) \cdot c = \gamma(gba \otimes c) = \gamma(gb \otimes ac) = b\gamma(g \otimes ac) = b(a \cdot c),$$

by (1). The above action is simply induced from the action of $\mathcal{C}$ on A and thus well-defined. Therefore there is a functor as required. Now, one can easily check that the unit and counit of the adjunction are given by

$$\eta_N : N \to (N \otimes_B A)_{g,\gamma}^{\mathcal{C}}, \qquad n \mapsto n \otimes 1_A,$$

$$\epsilon_M : M_{g,\gamma}^{\mathcal{C}} \otimes_B A \to M, \qquad m \otimes a \mapsto ma,$$

for all $N \in \mathbf{M}_B$ and $M \in \mathbf{M}_{\mathcal{C}}$.

(3) Note that $Q \cong \text{Hom}_{\mathcal{C}}(A, \mathcal{C})$ and is therefore a natural $(\mathcal{C}, B)$-bimodule. In more detail, $(cq)c' = c(qc') = cq\gamma(g \otimes c')$ for any $c, c' \in \mathcal{C}$ and $q \in Q$, so cq is an element of Q, hence Q is a left $\mathcal{C}$-ideal. By (2) Q is also a right B-module, and since the product in $\mathcal{C}$ is an (A, A)-bimodule map, Q is a $(\mathcal{C}, B)$-bimodule. We only need to show that Q is firm as a left $\mathcal{C}$-module. This can be shown by the same technique as in the proof of Proposition 2.7. $\mathcal{C} \otimes_{\mathcal{C}} Q$ is defined as a cokernel of the following left $\mathcal{C}$-linear map

$$\lambda : \mathcal{C} \otimes_A \mathcal{C} \otimes_A Q \to \mathcal{C} \otimes_A Q, \qquad \lambda = I_{\mathcal{C}} \otimes \pi - \pi \otimes I_Q,$$

where π is the product map in $\mathcal{C}$ (i.e., the splitting of $\Delta_{\mathcal{C}}$) corresponding to the cointegral γ. Observe that the product map $\pi : \mathcal{C} \otimes_A Q \to Q$ is a surjection. Indeed, first note that since for all $c \in \mathcal{C}$, $\pi(g \otimes c) = g\gamma(g \otimes c)$ by the relationship between π and γ, the grouplike element g is in Q. For any $q \in Q$ take $q \otimes g \in \mathcal{C} \otimes_A Q$. Then $\pi(q \otimes g) = q\gamma(g \otimes g) = q$, by the properties of the cointegral γ. Thus π is a surjection as claimed.

Consider the following sequence of left $\mathcal{C}$-module maps

$$\mathcal{C} \otimes_A \mathcal{C} \otimes_A Q \xrightarrow{\lambda} \mathcal{C} \otimes_A Q \xrightarrow{\pi} Q \longrightarrow 0.$$

We need to show that this sequence is exact. Clearly the associativity of π implies that $\pi \circ \lambda = 0$, so that $\text{Im}\lambda \subseteq \ker \pi$. Furthermore,

$$\Delta_{\mathcal{C}} \circ \pi - \lambda \circ (\Delta_{\mathcal{C}} \otimes I_Q) = (I_{\mathcal{C}} \otimes \pi) \circ (\Delta_{\mathcal{C}} \otimes I_Q) - (I_{\mathcal{C}} \otimes \pi) \circ (\Delta_{\mathcal{C}} \otimes I_Q) + (\pi \circ \Delta_{\mathcal{C}}) \otimes I_Q = I_{\mathcal{C}} \otimes I_Q,$$

where we used the colinearity of π and the fact that π is a splitting of $\Delta_{\mathcal{C}}$. This implies that $\ker \pi \subseteq \text{Im}\lambda$, i.e., the above sequence is exact as required.

(4) Note that ω_M is the natural map $M \otimes_{\mathcal{C}} \text{Hom}_{\mathcal{C}}(A, \mathcal{C}) \to \text{Hom}_{\mathcal{C}}(A, M) \cong M_{g,\gamma}^{\mathcal{C}}$, $m \otimes f \mapsto [a \mapsto f(1)m]$, and hence it is well-defined. Explicitly, for all $m \in M$, $q \in Q$

and $c \in C$ we compute $(m \cdot q) \cdot c = m \cdot (qc) = m \cdot q\gamma(g \otimes c)$, as required. We need to show that ω_M is bijective. Consider a map

$$\theta_M : M_{g,\gamma}^C \to M \otimes_C Q, \qquad m \mapsto m \otimes g.$$

This is well-defined, since as shown in the proof of (3), $g \in Q$. Take any $m \in M_{g,\gamma}^C$. Then

$$\omega_M(\theta_M(m)) = m \cdot g = m\gamma(g \otimes g) = m,$$

by the definition of $M_{g,\gamma}^C$ and properties of a cointegral. Conversely, for any simple tensor $m \otimes q \in M \otimes_C Q$,

$$\theta_M(\omega_M(m \otimes q)) = m \cdot q \otimes g = m \otimes qg = m \otimes q\gamma(g \otimes g) = m \otimes q,$$

again by the definition of Q and properties of çointegrals.

(5) Note that σ is the evaluation mapping $\operatorname{Hom}_C(A, C) \otimes_B A \to C$, while τ is the canonical map $A \otimes_C \operatorname{Hom}_C(A, C) \to \operatorname{End}_C(A)$. In more detail, we show that the maps σ and τ are well-defined as bimodule maps. Obviously, σ is left C-linear. Take any $q \in Q$, $a \in A$ and $c \in C$ and compute

$$\sigma(q \otimes a \cdot c) = q\gamma(ga \otimes c) = q\gamma(g \otimes ac) = \pi(q \otimes ac) = \pi(qa \otimes c) = (qa)c.$$

This shows that σ is (C, C)-bilinear as required. Note that $\tau = \omega_A$, and since $B = A_{g,\gamma}^C$ it is well defined and surjective. Clearly τ is right B-linear. An easy computation which involves the definition of B confirms that τ is (B, B)-bilinear. Next we need to check the commutativity of diagrams in Definition 3.1. The commutativity of the second diagram follows immediately from A-linearity of the cointegral. Now take any $a \in A$, and $q, q' \in Q$ and compute

$$\sigma(q \otimes a)q' = (qa)q' = q(aq') = q\gamma(g \otimes aq') = q\tau(a \otimes q'),$$

where the definition of Q was used to derive the third equality. Thus there is a Morita context as required. A standard argument in Morita theory confirms that τ is an isomorphism (cf. [2, II (3.4) Theorem]). $\square$

COROLLARY 3.3. *With the assumptions and notation as in Theorem 3.2 we have:*

(1) Q is a subring of C with a right unit g.

(2) The functor $- \otimes_B A : \mathbf{M}_B \to \mathbf{M}_C$ is fully faithful, i.e., the unit of adjunction η is an isomorphism.

(3) A_C and $_CQ$ are direct summands of C^n for some (finite) $n \in \mathbb{N}$.

(4) A and Q are generators as left resp. right B-modules.

Proof. (1) follows immediately from the proof of Theorem 3.2(3), while (2)-(4) follow from Morita theory with surjective τ, and can be proven by the same methods as in the unital case (cf. [2, Ch. II.3]). In particular, in the case of (3) the "dual bases" of A and Q can be constructed as follows. Let $\{a_i \in A, q^i \in Q\}_{i=1,\ldots,n}$ be such that $1_B = \sum_i \tau(a_i \otimes q^i)$. Define $\sigma^i = \sigma(q^i \otimes -) \in \operatorname{Hom}_C(A, C)$. Then for every $a \in A$,

$$\sum_i a_i \cdot \sigma^i(a) = \sum_i a_i \cdot \sigma(q^i \otimes a) = \sum_i \tau(a_i \otimes q^i)a = a,$$

so that $\{a_i, \sigma^i\}_{i=1,\ldots,n}$ is a dual basis for A_C. Similarly a dual basis for $_CQ$ can be constructed as $\{q^i, \sigma_i\}$ with $\sigma_i = \sigma(-\otimes a_i)$. $\square$

ON COSEPARABLE AND BISEPARABLE CORINGS

One can easily find a sufficient condition for the Morita context of Theorem 3.2 to be strict.

PROPOSITION 3.4. *Let $\mathcal{C}$ be a coseparable A-coring with a grouplike element g. Then the Morita context $(B, \mathcal{C}, A, Q, \tau, \sigma)$ in Theorem 3.2 is strict, provided g is a left unit in $\mathcal{C}$.*

Proof. In this case $\gamma(g \otimes c) = \epsilon_{\mathcal{C}}(c)$, hence B and Q are characterised by relations $\epsilon_{\mathcal{C}}(bc) = b\epsilon_{\mathcal{C}}(c)$ and $qc = q\epsilon_{\mathcal{C}}(c)$, respectively, for all $c \in \mathcal{C}$. We need to show that σ is an isomorphism. For any $c \in \mathcal{C}$, $\sigma(g \otimes \epsilon_{\mathcal{C}}(c)) = g\epsilon_{\mathcal{C}}(c) = gc = c$, since $g \in Q$. Thus σ is surjective. Suppose now that $\sum_i q_i \otimes a_i \in \ker \sigma$, i.e., $\sum_i q_i a_i = 0$. This implies that $\sum_i \epsilon_{\mathcal{C}}(q_i) a_i = 0$, so that

$$0 = g \otimes_B \sum_i \epsilon_{\mathcal{C}}(q_i) a_i = \sum_i g\epsilon_{\mathcal{C}}(q_i) \otimes a_i = \sum_i gq_i \otimes a_i = \sum_i q_i \otimes a_i.$$

Here we used that for all $q \in Q$, $\epsilon_{\mathcal{C}}(q) = \gamma(g \otimes q) \in B$, the fact that $g \in Q$ and that g is a left unit in $\mathcal{C}$. This completes the proof. $\square$

Finally, we consider two examples of Theorem 3.2.

EXAMPLE 3.5. Consider a split extension $\bar{B} \overset{\iota}{\longrightarrow} A$ with splitting map $E : A \to \bar{B}$. Then the canonical Sweedler coring $\mathcal{C} = A \otimes_{\bar{B}} A$ is an A-ring with the E-multiplication, $1_A \otimes 1_A \in \mathcal{C}$ is a grouplike element, and the Morita context constructed in Theorem 3.2 comes out as follows. The ring B is just

$$B = \iota(\bar{B}),$$

while the $(\mathcal{C}, B)$-bimodule $Q \subset A \otimes_B A$ is

$$Q \cong A$$

via $a \mapsto a \otimes 1_A$ and the left module action of the A-ring $\mathcal{C}$ on A given by $c \cdot a = \sum c^1 E(c^2 a)$ (suppressing a possible summation in $c = \sum c^1 \otimes c^2 \in A \otimes_B A$). The module $A_{\mathcal{C}}$ is similarly given by $a \cdot c = \sum E(ac^1) c^2$ for each $a \in A, c \in \mathcal{C}$. The Morita maps read $\sigma(a \otimes a') = a \otimes a'$ and $\tau(a \otimes a') = E(aa')$. This context is obviously strict.

The proof of this involves applying the theorem, noting that

$$B = \{b \in A \mid \forall a \in A, \ 1_A E(ba) = bE(a)\} = \iota(\bar{B})$$

since $\supseteq$ is clear and $\subseteq$ follows from letting $a = 1_A$. Next one notes that

$$Q = \{q = \sum_i q_i \otimes \bar{q}_i \in A \otimes_B A \mid \forall a \in A, \ \sum_i q_i E(\bar{q}_i a) \otimes 1_A = qE(a)\} = A \otimes 1_A$$

since $\supseteq$ is clear and $\subseteq$ follows from taking $a = 1_A$.

EXAMPLE 3.6. Hopf algebroids over a noncommutative base k-algebra A, where k is a commutative ring, provide examples of A-corings with grouplike elements; in particular, the canonical bialgebroids $\text{End}_k A$ and $A \otimes_k A^{\text{op}}$ do (cf. [16] [7] for the definition and examples of Hopf algebroids). They can be extended to ring extensions via an algebraic formulation of depth two for subfactors [15]: a ring extension $B \to A$ is of *depth two* (D2) if $A \otimes_B A$ is isomorphic to a direct summand of $A \oplus \cdots \oplus A$ as a (B, A)-bimodule, and similarly as an (A, B)-bimodule. The two conditions are equivalent respectively to the existence of finitely many elements $c_j, b_i \in (A \otimes_B A)^B$ and $\gamma_j, \beta_i \in {}_B\mathrm{End}_B(A)$ such that for all $a, a' \in A$,

$$a \otimes a' = \sum_i b_i \beta_i(a) a' = \sum_j a \gamma_j(a') c_j. \tag{1}$$

Denoting the centraliser A^B of a D2 extension $B \to A$ by R, the following R-coring structure for $\mathcal{C} := {}_B\mathrm{End}_B(A)$ is considered in [15]. The (R, R)-bimodule structure is $r \alpha r' = r \alpha(-) r'$ ($\alpha \in \mathcal{C}$). The coproduct is given most simply by noting [15, 3.10]: $\mathcal{C} \otimes_R \mathcal{C} \cong {}_B\mathrm{Hom}_B(A \otimes_B A, A)$ via

$$\alpha \otimes \beta \longmapsto [a \otimes a' \mapsto \alpha(a)\beta(a')].$$

Then

$$\Delta_{\mathcal{C}}(\alpha)(a \otimes a') = \alpha(aa'),$$

with counit

$$\epsilon_{\mathcal{C}}(\alpha) = \alpha(1_A).$$

We also have the alternative formulae for the coproduct [15, Eqs. (66), (68)]:

$$\Delta_{\mathcal{C}}(\alpha) = \sum_j \gamma_j \otimes_R c_j^1 \alpha(c_j^2 -) = \sum_i \alpha(-b_i^1) b_i^2 \otimes_R \beta_i,$$

where $c_j = \sum c_j^1 \otimes c_j^2$ and $b_i = \sum b_i^1 \otimes b_i^2$ is a notation suppressing a possible summation index. We note the grouplike element I_A.

Suppose the D2 extension $B \to A$ is separable with separability element $e = \sum e^1 \otimes e^2 \in A \otimes_B A$ (summation index suppressed). Then the R-coring $\mathcal{C}$ is coseparable with cointegral $\gamma : \mathcal{C} \otimes_R \mathcal{C} \to R$ given by $\gamma(\alpha \otimes \beta) = \sum \alpha(e^1)\beta(e^2)$. The corresponding R-ring structure on $\mathcal{C}$ is given by ($x \in A$)

$$(\alpha * \beta)(x) = \sum \alpha(xe^1)\beta(e^2) = \sum \alpha(e^1)\beta(e^2 x)$$

with the R-bimodule structure above.

The Morita context in the theorem applied to $\mathcal{C}$ turns out as follows:

PROPOSITION 3.7. *The centre Z of A and the non-unital ring $(\mathcal{C}, *)$ are related by the Morita context $(Z, \mathcal{C}, {}_Z R_{\mathcal{C}}, {}_{\mathcal{C}} R_Z, \tau, \sigma)$ where ${}_Z R_{\mathcal{C}}$ is given by $zr \cdot \alpha = \sum z e^1 r \alpha(e^2)$, ${}_{\mathcal{C}} R_Z$ by $\alpha \cdot rz = \sum \alpha(e^1) r e^2 z$,*

$$\sigma : R \otimes_Z R \to \mathcal{C}, \quad r \otimes r' \longmapsto \lambda(r)\rho(r') = \rho(r')\lambda(r),$$

where $\lambda(r), \rho(r) \in \mathcal{C}$ denote left and right multiplication by $r \in R$, respectively, and

$$\tau : R \otimes_{\mathcal{C}} R \xrightarrow{\cong} \mathcal{C}, \quad r \otimes r' \longmapsto \sum e^1 r r' e^2.$$

The Morita context is strict if $B \to A$ is H-separable.

Proof. We check that γ is a cointegral. Take any $\alpha \in \mathcal{C}$ and compute

$$\sum \gamma(\alpha_{(1)} \otimes \alpha_{(2)}) = \sum \alpha(e^1 e^2) = \epsilon_{\mathcal{C}}(\alpha).$$

Thus γ is normalised. Furthermore, for all $\alpha, \beta \in \mathcal{C}$,

$$\sum \gamma(\alpha \otimes \beta_{(1)})\beta_{(2)} = \sum_i \alpha(e^1)\beta(e^2 b_i^1) b_i^2 \beta_i = \sum \alpha(e^1)\beta(e^2 -).$$

On the other hand

$$\sum \alpha_{(1)}\gamma(\alpha_{(2)}\otimes\beta) = \sum_j \gamma_j(-)c_j^1\alpha(c_j^2 e^1)\beta(e^2) = \sum \alpha(-e^1)\beta(e^2),$$

so that γ is colinear and hence a cointegral.

The subring of R in Theorem 3.2 is

$$R_{I_A,\gamma}^{\mathcal{C}} = \{r \in R \mid \forall a \in A, \sum e^1 r\alpha(e^2) = \sum re^1\alpha(e^2)\} = Z,$$

since $\supseteq$ is clear, and $\subseteq$ follows from taking $\alpha = I_A$ and observing $\sum e^1 re^2 \in Z$. The Morita context bimodule

$$Q = \{q \in \mathcal{C} \mid \forall \alpha \in \mathcal{C}, \sum q(-e^1)\alpha(e^2) = \sum q(-)e^1\alpha(e^2)\} = \lambda(R),$$

since $\supseteq$ is clear, and $\subseteq$ follows from taking $\alpha = I_A$, whence $q = \sum q(-e^1)e^2 = \sum \lambda(q(e^1)e^2) \in \lambda(R)$.

That $(Z, \mathcal{C}, Q, R, \sigma, \tau)$ is a Morita context is now straightforward; τ being epi by an old lemma of Hirata and Sugano [14].

Recall that $B \to A$ is H-separable (after Hirata) if there are (Casimir) elements $e_i \in (A\otimes_B A)^A$ and $r_i \in R$ (the centraliser) such that $1_A\otimes 1_A = \sum_i e_i r_i (= \sum_i r_i e_i)$ (a very strong version of Eqs. (1) above). It is well-known that A is a separable extension of B. Moreover,

$$R\otimes_Z R^{\mathrm{op}} \xrightarrow{\cong} {}_B\mathrm{End}_B(A),$$

via $r\otimes r' \mapsto \lambda(r)\rho(r')$ with inverse $\alpha \mapsto \sum_i \alpha(e_i^1)e_i^2\otimes r_i$. Whence σ is an isomorphism if we begin with an H-separable extension. $\qquad\square$

If A is a separable algebra over a (commutative) ground ring B, then the proposition shows that the center Z and $\mathrm{End}_B(A)$ (with the exotic multiplication above) are related by a Morita context, which is strict if $Z = B1_A$, i.e. A is Azumaya.

As a third example, we may instead work with the dual bialgebroid in [15] and prove that a split D2 extension $B \to A$ has a coseparable R-coring structure on $(A\otimes_B A)^B$ which is essentially a restriction of $\mathcal{C}$ in Example 3.5.

4. ARE BISEPARABLE CORINGS FROBENIUS?

In this section we will show that a one-sided, slightly stronger version of the problem in [9] is equivalent to the problem if cosplit, coseparable corings with a condition of finite projectivity are Frobenius. Given the techniques developed for corings and the many examples coming from entwined structures [5], we expect this equivalence to be useful in solving this problem.

As recalled in Section 2, an A-coring $\mathcal{C}$ is coseparable if the forgetful functor $F : \mathbf{M}^{\mathcal{C}} \to \mathbf{M}_A$ is separable (cf. [17] for the definition of a separable functor). Dually, we say that $\mathcal{C}$ is *cosplit* if the functor $-\otimes_A \mathcal{C}$ is a separable functor from the category of right A-modules $\mathbf{M}_A$ into the category of right $\mathcal{C}$-comodules $\mathbf{M}^{\mathcal{C}}$. (Recall that F is the left adjoint of $-\otimes_A \mathcal{C}$ [13].)

An A-coring $\mathcal{C}$ determines two ring extensions $\iota^* : A \to \mathcal{C}^*$ and ${}^*\iota : A \to {}^*\mathcal{C}$ where $\mathcal{C}^* := \mathrm{Hom}_A(\mathcal{C}, A)$ and ${}^*\mathcal{C} := {}_A\mathrm{Hom}(\mathcal{C}, A)$, i.e., the right and left duals of $\mathcal{C}$. The ring structure on $\mathcal{C}^*$ is given by $(\xi\xi')(c) = \sum \xi(\xi'(c_{(1)})c_{(2)})$ $(\xi, \xi' \in \mathcal{C}^*, c \in \mathcal{C})$ with unity $\epsilon_{\mathcal{C}}$ and the natural Abelian group structure, while the ring structure on $^*\mathcal{C}$ is given by $(\xi\xi')(c) = \sum \xi'(c_{(1)}\xi(c_{(2)}))$. The mappings ι^* and $^*\iota$ are given by $\iota^*(a) = \epsilon_\mathcal{C}(a-)(= a\epsilon_\mathcal{C})$ and $^*\iota(a) = \epsilon_\mathcal{C}(-a)$. We note by short calculations that the induced (A, A)-bimodule structures on $\mathcal{C}^*$ and $^*\mathcal{C}$ coincide with the usual structures, which we recall are given by $(a\xi a')(c) := a\xi(a'c)$ for $\xi \in \mathcal{C}^*$ and $a, a' \in A$ and $(a\xi a')(c) = \xi(ca)a'$ for $\xi \in {}^*\mathcal{C}$. Also note that $^*\iota$ and ι^* are monomorphisms if $\epsilon_\mathcal{C} : \mathcal{C} \to A$ is surjective.

Recall from [9] that a ring extension $B \to A$ is biseparable if it is split, separable and the natural modules A_B and $_BA$ are finitely generated projective. We will say that $B \to A$ is *left or right biseparable* if $B \to A$ is split, separable but only one of $_BA$ or A_B, respectively, need be finitely generated projective. This motivates the following

DEFINITION 4.1. An A-coring $\mathcal{C}$ is said to be *biseparable* if $\mathcal{C}_A$ and $_A\mathcal{C}$ are finitely generated projective and $\mathcal{C}$ is cosplit as well as coseparable.

PROPOSITION 4.2. *If $B \to A$ is a biseparable extension, then the canonical Sweedler A-coring $\mathcal{C} := A \otimes_B A$ is a biseparable coring.*

Proof. Since $B \to A$ is separable, the induction functor $- \otimes_A \mathcal{C}$ from $\mathbf{M}_A$ into $\mathbf{M}^\mathcal{C}$ is separable by [5, Corollary 3.4]. Since $B \to A$ is split, and A_B is a projective generator (therefore faithfully flat), the forgetful functor $F : \mathbf{M}^\mathcal{C} \to \mathbf{M}_A$ is a separable functor by [5, Corollary 3.7]. It follows by definition that the canonical coring $\mathcal{C}$ is cosplit and coseparable.

Finally we note that $_BA$ finitely generated projective implies $_AA\otimes_BA$ finitely generated projective. Similarly, $\mathcal{C}_A$ is finitely generated projective, and we conclude that $\mathcal{C}$ is biseparable. $\qquad\square$

Recall that an A-coring $\mathcal{C}$ is said to be *Frobenius* if the forgetful functor $F : \mathbf{M}^\mathcal{C} \to \mathbf{M}_A$ is a Frobenius functor (has the same left and right adjoint), i.e. $- \otimes_A \mathcal{C}$ is also a left adjoint of F [5, 6]. Motivated by the question in [9] let us make

CONJECTURE 4.3. *A biseparable A-coring $\mathcal{C}$ is Frobenius.*

PROPOSITION 4.4. *If Conjecture 4.3 is true, then biseparable extensions are Frobenius.*

Proof. Given a biseparable extension $B \to A$, its canonical coring $\mathcal{C} = A\otimes_B A$ is biseparable by the previous proposition. If $\mathcal{C}$ is then a Frobenius coring by hypothesis, it follows from [6, Theorem 2.7] that $B \to A$ is a Frobenius extension, since $_BA$ is faithfully flat. $\qquad\square$

We now proceed to establish a converse to this proposition.

PROPOSITION 4.5. *If $\mathcal{C}$ is a cosplit A-coring, then $\iota^* : A \to \mathcal{C}^*$ and $^*\iota : A \to {}^*\mathcal{C}$ are both split extensions.*

Proof. By [5, Theorem 3.3], $\mathcal{C}$ is cosplit if and only if there is $e \in \mathcal{C}^A$ such that $\epsilon_\mathcal{C}(e) = 1_A$. (In other words, $\epsilon_\mathcal{C} : \mathcal{C} \to A$ is a split (A, A)-epimorphism.) We now define a "conditional expectation" or bimodule projection $E^* : \mathcal{C}^* \to A$, respectively $^*E : {}^*\mathcal{C} \to A$ simply by

$$E^*(\xi) = \xi(e), \quad {}^*E(\xi') = \xi'(e) \quad (\xi \in \mathcal{C}^*, \xi' \in {}^*\mathcal{C}).$$

ON COSEPARABLE AND BISEPARABLE CORINGS 85

Note that $E^*(\epsilon_{\mathcal{C}}) = 1_A = {}^*E(1 \cdot_{\mathcal{C}})$ and

$$E^*(a\xi a') = a\xi(a'e) = a\xi(ea') = a\xi(e)a' = aE^*(\xi)a'$$

for $a, a' \in A$, whence E^* and similarly ${}^*E$ give (A, A)-bimodule splittings of ι^* and ${}^*\iota$. □

For example, the canonical Sweedler coring $\mathcal{C}$ of a ring extension $B \to A$ is cosplit if and only if $B \to A$ is a separable extension. Now $\mathcal{C}^* \cong \mathrm{End}_B(A)$ as rings via $\xi \mapsto \xi(-\otimes 1_A)$ with inverse

$$f \longmapsto [a \otimes a' \mapsto f(a)a'].$$

Since ι^* corresponds to the left regular representation $\lambda : A \to \mathrm{End}_B(A)$, we recover results by Müller and Sugano that λ is a split extension if $B \to A$ is separable.

PROPOSITION 4.6. *Let $\mathcal{C}$ be an A-coring. If ${}_A\mathcal{C}$ is finitely generated projective, then $\mathcal{C}$ is coseparable if and only if ${}^*\iota : A \to {}^*\mathcal{C}$ is a separable extension. If $\mathcal{C}_A$ is finitely generated projective, then $\mathcal{C}$ is a coseparable A-coring if and only if $\iota^* : A \to \mathcal{C}^*$ is a separable extension.*

Proof. We will prove the first statement, the second follows similarly. If ${}_A\mathcal{C}$ is finitely generated projective, then the category of right comodules $\mathbf{M}^{\mathcal{C}}$ is isomorphic to the category $\mathbf{M}_{*\mathcal{C}}$ of right modules over ${}^*\mathcal{C}$ [5, Lemma 4.3]. Recall that given a coaction $\varrho^M : M_A \to M \otimes_A \mathcal{C}$, we define an action of $\xi \in {}^*\mathcal{C}$ on $m \in M \in \mathbf{M}^{\mathcal{C}}$ by $m \cdot \xi = \sum m_{(0)} \xi(m_{(1)})$. It is trivial to check that $(M, \cdot) \in \mathbf{M}_{*\mathcal{C}}$. Inversely, given dual bases $\{\xi_i \in {}^*\mathcal{C}\}$ and $\{c_i \in \mathcal{C}\}$ such that $c = \sum_i \xi_i(c)c_i$ for each $c \in \mathcal{C}$, and right action of ${}^*\mathcal{C}$ on $M \in \mathbf{M}_{*\mathcal{C}}$, we define a coaction

$$\varrho^M(m) = \sum_i m \cdot \xi_i \otimes_A c_i.$$

It is easily checked that $(M, \varrho^M) \in \mathbf{M}^{\mathcal{C}}$, and that the two operations are natural and inverses to one another, so that $\mathbf{M}_{*\mathcal{C}} \cong \mathbf{M}^{\mathcal{C}}$.
Now $\mathcal{C}$ is coseparable if and only if the forgetful functor $F : \mathbf{M}^{\mathcal{C}} \to \mathbf{M}_A$ is a separable functor. Since

$$m \cdot {}^*\iota(a) = \sum m_{(0)} \epsilon_{\mathcal{C}}(m_{(1)})a = ma$$

for each $a \in A, m \in M \in \mathbf{M}^{\mathcal{C}}$, F corresponds under the isomorphism of categories above to the forgetful functor $G : \mathbf{M}_{*\mathcal{C}} \to \mathbf{M}_A$ induced by ${}^*\iota : A \to {}^*\mathcal{C}$. But G is a separable functor if and only if ${}^*\iota$ is a separable extension [17]. □

PROPOSITION 4.7. *Suppose $\mathcal{C}$ is an A-coring which is reflexive as a left and right A-module. Then $\mathcal{C}$ is a Frobenius coring if and only if ${}^*\iota : A \to {}^*\mathcal{C}$ is a Frobenius extension if and only if $\iota^* : A \to \mathcal{C}^*$ is a Frobenius extension.*

Proof. The proof is quite similar to the proof of Proposition 4.6 (cf. [5, Theorem 4.1]). If ${}^*\iota$ is Frobenius, it follows that ${}^*\mathcal{C}_A$ is finitely generated projective, so ${}_A({}^*\mathcal{C})^*$ is finitely generated projective, whence by reflexivity ${}_A\mathcal{C}$ is finitely generated projective. Then the categories $\mathbf{M}_{*\mathcal{C}}$ and $\mathbf{M}^{\mathcal{C}}$ are isomorphic. But the forgetful functor $G : \mathbf{M}_{*\mathcal{C}} \to \mathbf{M}_A$ has equal left and right adjoint if ${}^*\iota$ is Frobenius, in which case F is Frobenius and $\mathcal{C}$ is a Frobenius coring. The other case is entirely similar.

Conversely, if C is a Frobenius A-coring, then both C_A and $_AC$ are finitely generated projective [6, Corollary 2.3]. The rest follows from the functorial definitions of Frobenius extension and coring applied to either isomorphism of left or right module and comodule categories. $\square$

THEOREM 4.8. *Biseparable corings are Frobenius if and only if left or right biseparable extensions are Frobenius.*

Proof. Without one-sidedness, we saw $\Rightarrow$ in Proposition 4.4. ($\Leftarrow$) Suppose C is a biseparable A-coring. Then $^*\iota : A \to {}^*C$ and $\iota^* : A \to C^*$ are split, separable extensions by Propositions 4.5 and 4.6. Since $_AC$ and C_A are finitely generated projective, it follows that $^*C_A$ and $_AC^*$ are finitely generated projective. If either left or right biseparable extensions are Frobenius, then either ι^* or $^*\iota$ is a Frobenius extension. In either case, Proposition 4.7 shows C to be a Frobenius coring. $\square$

We note the following special "depth one" case for which there is a solution to our conjecture. If C is a centrally projective A-bimodule, i.e., as (A, A)-bimodules $C \oplus W \cong \oplus^n A$ for some (A, A)-bimodule W, and C is moreover biseparable, then C is Frobenius by a classical result of Sugano:

PROPOSITION 4.9. *If C is a centrally projective, cosplit, coseparable A-coring, then C is Frobenius.*

Proof. We easily obtain $C^* \oplus W^* \cong \oplus^n A$ as (A, A)-bimodules, whence $\iota^* : A \to C^*$ is a centrally projective separable extension by Proposition 4.6, and monomorphism since ϵ_C is surjective. By [21, Theorem 2] ι^* is a Frobenius extension. Then by Proposition 4.7, C is a Frobenius coring. $\square$

Acknowledgements

Tomasz Brzeziński would like to thank the Engineering and Physical Sciences Research Council for an Advanced Fellowship. Lars Kadison thanks Dmitri Nikshych and the University of New Hampshire Department of Mathematics and Statistics for their generosity.

REFERENCES

[1] J.Y. Abuhlail. Morita contexts for corings and equivalences, in "Hopf algebras in noncommutative geometry and physics", S. Caenepeel and F. Van Oystaeyen, eds., *Lecture Notes Pure Appl. Math.*, Dekker, New York, to appear.
[2] H. Bass "Algebraic K-Theory", Benjamin, New York, 1968.
[3] G.M. Bergman and A.O. Hausknecht, "Cogroups and Co-rings in Categories of Associative Rings", Amer. Math. Soc., Providence R.I., 1996.
[4] A.J. Berrick and M.A. Keating, "Categories and Modules with K-Theory in View", Cambridge Univ. Press, Cambridge, 2000.
[5] T. Brzeziński, The structure of corings. Induction functors, Maschke-type theorem, and Frobenius and Galois-type properties, *Algebras Representation Theory* 5 (2002), 389–410.
[6] T. Brzeziński, Towers of corings, *Comm. Algebra* 31 (2003), 2015–2026.
[7] T. Brzeziński and G. Militaru, Bialgebroids, $\times_A$-bialgebras and duality, *J. Algebra* 251 (2002), 279–294.
[8] T. Brzeziński and R. Wisbauer, "Corings and Comodules", *London Math. Soc. Lect. Note Ser.* 309, Cambridge University Press, Cambridge, 2003.

ON COSEPARABLE AND BISEPARABLE CORINGS

[9] S. Caenepeel and L. Kadison. Are biseparable extensions Frobenius? *K-Theory* **24** (2001), 361–383.

[10] S. Caenepeel, G. Militaru and Shenglin Zhu, "Frobenius and separable functors for generalized module categories and nonlinear equations", *Lect. Notes in Math.* **1787**, Springer Verlag, Berlin, 2002.

[11] S. Caenepeel, J. Vercruysse and S. Wang, Morita theory for corings and cleft entwining structures, *J. Algebra*, in press. Preprint arXiv: math.RA/0206198, 2002.

[12] J. Gómez-Torrecillas and A. Louly, Coseparable corings, *Comm. Algebra* **31** (2003), 4455–4471.

[13] F. Guzman, Cointegrations, relative cohomology for comodules and coseparable corings, *J. Algebra* **126** (1989), 211–224.

[14] K. Hirata and K. Sugano, On semisimple extensions and separable extensions over noncommutative rings, *J. Math. Soc. Japan* **18** (1966), 360–373.

[15] L. Kadison and K. Szlachányi, Bialgebroid actions on depth two extensions and duality, *Adv. Math.* **179** (2003), 75–121.

[16] J.H. Lu, Hopf algebroids and quantum groupoids, *Int. J. Math.* **7** (1996), 47–70.

[17] C. Nastasescu, M. Van den Bergh and F. Van Oystaeyen, Separable functors applied to graded rings, *J. Algebra* **123** (1989), 397–413.

[18] D. Quillen, Module theory over non-unital rings, Preprint 1997.

[19] R.S. Pierce, "Associative Algebras", Springer Verlag, Berlin, 1982.

[20] M.D. Rafael, Separable functors revisited, *Comm. Algebra* **18** (1990), 1445–1459.

[21] K. Sugano. Separable extensions and Frobenius extensions, *Osaka J. Math.* **7** (1970), 291–299.

[22] M. Sweedler, The predual theorem to the Jacobson-Bourbaki theorem, *Trans. Amer. Math. Soc.*, **213** (1975), 391–406.

[23] J.L. Taylor, A bigger Brauer group, *Pacific J. Math.* **103** (1962), 163—203.

More Properties of Yetter-Drinfeld Modules over Quasi-Hopf Algebras

DANIEL BULACU   Faculty of Mathematics, University of Bucharest
Str. Academiei 14, RO-70109 Bucharest, Romania
e-mail: *dbulacu@al.math.unibuc.ro*

STEFAAN CAENEPEEL   Faculty of Applied Sciences, Vrije Universiteit
Brussel, VUB, Pleinlaan 2, B-1050 Brussels, Belgium
e-mail: *scaenepe@vub.ac.be*

FLORIN PANAITE   Institute of Mathematics of the Romanian Academy
PO-Box 1-764, RO-70700 Bucharest, Romania
e-mail: *florin.panaite@imar.ro*

> ABSTRACT. We generalize various properties of Yetter-Drinfeld modules over
> Hopf algebras to quasi-Hopf algebras. The dual of a finite dimensional Yetter-
> Drinfeld module is again a Yetter-Drinfeld module. The algebra H_0 in the
> category of Yetter-Drinfeld modules that can be obtained by modifying the
> multiplication in a proper way is quantum commutative. We give a Structure
> Theorem for Hopf modules in the category of Yetter-Drinfeld modules, and
> deduce the existence and uniqueness of integrals from it.

1. INTRODUCTION

The motivation for studying Yetter-Drinfeld modules over quasi-Hopf algebras is
the same as for Hopf algebras. It is well known that for any finite dimensional Hopf
algebra H the category of Yetter-Drinfeld modules ${}_H\mathcal{YD}^H$ is isomorphic to the category of modules over the quantum double $D(H)$. From a categorical point of view,
the quantum double $D(H)$ arises by considering the center $\mathcal{Z}({}_H\mathcal{M})$ of the monoidal
category ${}_H\mathcal{M}$ of left H-modules. More precisely, one has $\mathcal{Z}({}_H\mathcal{M}) \simeq {}_{D(H)}\mathcal{M}$ if H
is finite dimensional. Actually, the category of Yetter-Drinfeld modules appears
as an intermediate step in the proof of this isomorphism: one first proves that
$\mathcal{Z}({}_H\mathcal{M}) \simeq {}_H\mathcal{YD}^H$, and then ${}_H\mathcal{YD}^H \simeq {}_{D(H)}\mathcal{M}$, where the finite dimensionality is
not needed in the proof of the first isomorphism, see [16] for full detail.

2000 *Mathematics Subject Classification.* 16W30.

Key words and phrases. quasi-Hopf algebra, Yetter-Drinfeld module, quantum double, braided
monoidal category.

This research was supported by the bilateral project "Hopf Algebras in Algebra, Topology,
Geometry and Physics" of the Flemish and Romanian governments. The third author was also
partially supported by the programmes SCOPES and EURROMMAT. The first and third author
wish to the Vrije Universiteit Brussel for its warm hospitality during their visit there.

Quasi-bialgebras and quasi-Hopf algebras were introduced by Drinfeld [13]; a categorical interpretation is the following: a quasi-bialgebra H is an algebra with the additional structure that is needed to make the category of left H-modules, with the tensor product over k as tensor product and k as unit object into a monoidal category. The difference with a usual bialgebra is that we do not require that the associativity isomorphism coincides with the associativity in the category of vector spaces. A quasi-Hopf algebra is a quasi-bialgebra with additional structure making the category of finite dimensional H-modules into a monoidal category with duality. The center construction $\mathcal{Z}(\mathcal{C})$ can be applied to any monoidal category $\mathcal{C}$. Majid [19] computed the center of the category of left modules over a quasi-Hopf algebra H, and introduced the category of Yetter-Drinfeld modules over H. Hausser and Nill [14], [15] constructed the quantum double $D(H)$ of a finite dimensional quasi-Hopf algebra H, and proved that $_H\mathcal{YD}^H \simeq {}_{D(H)}\mathcal{M}$. Recently, Schauenburg [22] gave the equivalence between the category of Yetter-Drinfeld modules $_H^H\mathcal{YD}$ and the category $_H^H\mathcal{M}_H^H$ of Hopf bimodules. In [5], the relation between Yetter-Drinfeld modules and Radford's biproduct is studied. In [4], the rigidity of the category of Yetter-Drinfeld modules is investigated, as well as the relations between left, left-right, right-left and right Yetter-Drinfeld modules.

In this paper, which can be seen as a sequel to [4], we continue our investigations of properties of Yetter-Drinfeld modules. In Section 3, we show that the linear dual of a finite dimensional right-left Yetter-Drinfeld module is a left-right Yetter-Drinfeld module.

It was shown in [7], [5] that the multiplication on H can be modified in such a way that we obtain an algebra in the category of left Yetter-Drinfeld modules. The main result of Section 4 is that H_0 is quantum commutative.

In Section 5, we will generalize Doi's results [12] about Hopf modules in the category of Yetter-Drinfeld modules to our situation: we give a Structure Theorem for Hopf modules in the category of Yetter-Drinfeld modules over a quasi-Hopf algebras, and we use this result to obtain the existence and uniqueness of integrals for a finite dimensional braided Hopf algebra in $_H^H\mathcal{YD}$. We apply this to the braided Hopf algebra considered in Section 4, in the case where H is finite dimensional and quasitriangular.

2. PRELIMINARY RESULTS

2.1. Quasi-Hopf algebras

We work over a commutative field k. All algebras, linear spaces etc. will be over k; unadorned $\otimes$ means $\otimes_k$. Following Drinfeld [13], a quasi-bialgebra is a fourtuple $(H, \Delta, \varepsilon, \Phi)$, where H is an associative algebra with unit, Φ is an invertible element in $H \otimes H \otimes H$, and $\Delta : H \to H \otimes H$ and $\varepsilon : H \to k$ are algebra homomorphisms satisfying the identities

$$(id \otimes \Delta)(\Delta(h)) = \Phi(\Delta \otimes id)(\Delta(h))\Phi^{-1}, \tag{1}$$

$$(id \otimes \varepsilon)(\Delta(h)) = h \otimes 1, \quad (\varepsilon \otimes id)(\Delta(h)) = 1 \otimes h, \tag{2}$$

YETTER-DRINFELD MODULES OVER QUASI-HOPF ALGEBRAS

for all $h \in H$, and Φ has to be a normalized 3-cocycle, in the sense that

$$(1 \otimes \Phi)(id \otimes \Delta \otimes id)(\Phi)(\Phi \otimes 1) = (id \otimes id \otimes \Delta)(\Phi)(\Delta \otimes id \otimes id)(\Phi), \quad (3)$$

$$(id \otimes \varepsilon \otimes id)(\Phi) = 1 \otimes 1 \otimes 1. \quad (4)$$

The map Δ is called the coproduct or the comultiplication, ε the counit and Φ the reassociator. As for Hopf algebras [23] we use the notation $\Delta(h) = \sum h_1 \otimes h_2$. Since Δ is only quasi-coassociative we adopt the further notation

$$(\Delta \otimes id)(\Delta(h)) = \sum h_{(1,1)} \otimes h_{(1,2)} \otimes h_2, \quad (id \otimes \Delta)(\Delta(h)) = \sum h_1 \otimes h_{(2,1)} \otimes h_{(2,2)},$$

for all $h \in H$. We will denote the tensor components of Φ by capital letters, and the ones of Φ^{-1} by small letters, namely

$$\Phi = \sum X^1 \otimes X^2 \otimes X^3 = \sum T^1 \otimes T^2 \otimes T^3 = \sum V^1 \otimes V^2 \otimes V^3 = \cdots$$

$$\Phi^{-1} = \sum x^1 \otimes x^2 \otimes x^3 = \sum t^1 \otimes t^2 \otimes t^3 = \sum v^1 \otimes v^2 \otimes v^3 = \cdots$$

A quasi-bialgebra H is called a quasi-Hopf algebra if there exists an anti-automorphism S of the algebra H and $\alpha, \beta \in H$ such that:

$$\sum S(h_1)\alpha h_2 = \varepsilon(h)\alpha \quad \text{and} \quad \sum h_1 \beta S(h_2) = \varepsilon(h)\beta, \quad (5)$$

$$\sum X^1 \beta S(X^2)\alpha X^3 = 1 \quad \text{and} \quad \sum S(x^1)\alpha x^2 \beta S(x^3) = 1, \quad (6)$$

for all $h \in H$. It is shown in [9] that the condition that the antipode is bijective follows automatically from the other axioms in the case where H is finite dimensional. Observe that the antipode of a quasi-Hopf algebra is determined uniquely up to a transformation $\alpha \mapsto U\alpha$, $\beta \mapsto \beta U^{-1}$, $S(h) \mapsto US(h)U^{-1}$, where $U \in H$ is invertible. The axioms for a quasi-Hopf algebra imply that $\varepsilon(\alpha)\varepsilon(\beta) = 1$, so, by rescaling α and β, we may assume without loss of generality that $\varepsilon(\alpha) = \varepsilon(\beta) = 1$ and $\varepsilon \circ S = \varepsilon$. The identities (2-4) also imply that

$$(\varepsilon \otimes id \otimes id)(\Phi) = (id \otimes id \otimes \varepsilon)(\Phi) = 1 \otimes 1 \otimes 1. \quad (7)$$

Together with a quasi-Hopf algebra $H = (H, \Delta, \varepsilon, \Phi, S, \alpha, \beta)$ we also have H^{op}, H^{cop} and $H^{\mathrm{op,cop}}$ as quasi-Hopf algebras, where "op" means opposite multiplication and "cop" means opposite comultiplication. The reassociators of these three quasi-Hopf algebras are $\Phi_{\mathrm{op}} = \Phi^{-1}$, $\Phi_{\mathrm{cop}} = (\Phi^{-1})^{321}$, $\Phi_{\mathrm{op,cop}} = \Phi^{321}$, the antipodes are $S_{\mathrm{op}} = S_{\mathrm{cop}} = (S_{\mathrm{op,cop}})^{-1} = S^{-1}$, and the elements α, β are $\alpha_{\mathrm{op}} = S^{-1}(\beta)$, $\beta_{\mathrm{op}} = S^{-1}(\alpha)$, $\alpha_{\mathrm{cop}} = S^{-1}(\alpha)$, $\beta_{\mathrm{cop}} = S^{-1}(\beta)$, $\alpha_{\mathrm{op,cop}} = \beta$ and $\beta_{\mathrm{op,cop}} = \alpha$.

Recall next that the definition of a quasi-Hopf algebra is "twist coinvariant", in the following sense. An invertible element $F \in H \otimes H$ is called a *gauge transformation* or *twist* if $(\varepsilon \otimes id)(F) = (id \otimes \varepsilon)(F) = 1$. If H is a quasi-Hopf algebra and $F = \sum F^1 \otimes F^2 \in H \otimes H$ is a gauge transformation with inverse $F^{-1} = \sum G^1 \otimes G^2$, then we can define a new quasi-Hopf algebra H_F by keeping the multiplication, unit, counit and antipode of H and replacing the comultiplication, antipode and the elements α and β by

$$\Delta_F(h) = F\Delta(h)F^{-1}, \quad (8)$$

$$\Phi_F = (1 \otimes F)(id \otimes \Delta)(F)\Phi(\Delta \otimes id)(F^{-1})(F^{-1} \otimes 1), \quad (9)$$

$$\alpha_F = \sum S(G^1)\alpha G^2, \quad \beta_F = \sum F^1 \beta S(F^2). \quad (10)$$

It is well-known that the antipode of a Hopf algebra is an anti-coalgebra morphism. The corresponding statement for a quasi-Hopf algebra is the following: there exists a gauge transformation $f \in H \otimes H$ such that

$$f\Delta(S(h))f^{-1} = \sum(S \otimes S)(\Delta^{\text{cop}}(h)), \tag{11}$$

for all $h \in H$, where $\Delta^{\text{cop}}(h) = \sum h_2 \otimes h_1$. The element f can be computed explicitly. First set

$$\sum A^1 \otimes A^2 \otimes A^3 \otimes A^4 = (\Phi \otimes 1)(\Delta \otimes id \otimes id)(\Phi^{-1}), \tag{12}$$

$$\sum B^1 \otimes B^2 \otimes B^3 \otimes B^4 = (\Delta \otimes id \otimes id)(\Phi)(\Phi^{-1} \otimes 1) \tag{13}$$

and then define $\gamma, \delta \in H \otimes H$ by

$$\gamma = \sum S(A^2)\alpha A^3 \otimes S(A^1)\alpha A^4 \text{ and } \delta = \sum B^1\beta S(B^4) \otimes B^2\beta S(B^3). \tag{14}$$

Then f and f^{-1} are given by the formulas

$$f = \sum(S \otimes S)(\Delta^{\text{op}}(x^1))\gamma\Delta(x^2\beta S(x^3)), \tag{15}$$

$$f^{-1} = \sum\Delta(S(x^1)\alpha x^2)\delta(S \otimes S)(\Delta^{\text{op}}(x^3)). \tag{16}$$

Moreover, f satisfies the following relations:

$$f\Delta(\alpha) = \gamma, \quad \Delta(\beta)f^{-1} = \delta. \tag{17}$$

Furthermore the corresponding twisted reassociator (see (9)) is given by

$$\Phi_f = \sum(S \otimes S \otimes S)(X^3 \otimes X^2 \otimes X^1). \tag{18}$$

In a Hopf algebra H, we obviously have the identity

$$\sum h_1 \otimes h_2 S(h_3) = h \otimes 1, \text{ for all } h \in H.$$

We will need the generalization of this formula to the quasi-Hopf algebra setting. Following [14, 15], we define

$$p_R = \sum p_R^1 \otimes p_R^2 = \sum x^1 \otimes x^2\beta S(x^3), \tag{19}$$

$$q_R = \sum q_R^1 \otimes q_R^2 = \sum X^1 \otimes S^{-1}(\alpha X^3)X^2, \tag{20}$$

$$p_L = \sum p_L^1 \otimes p_L^2 = \sum X^2 S^{-1}(X^1\beta) \otimes X^3, \tag{21}$$

$$q_L = \sum q_L^1 \otimes q_L^2 = \sum S(x^1)\alpha x^2 \otimes x^3. \tag{22}$$

We then have, for all $h \in H$,

$$\sum\Delta(h_1)p_R[1 \otimes S(h_2)] = p_R(h \otimes 1), \tag{23}$$

$$\sum[1 \otimes S^{-1}(h_2)]q_R\Delta(h_1) = (h \otimes 1)q_R, \tag{24}$$

$$\sum\Delta(h_2)p_L[S^{-1}(h_1) \otimes 1] = p_L(1 \otimes h), \tag{25}$$

$$\sum[S(h_1) \otimes 1]q_L\Delta(h_2) = (1 \otimes h)q_L, \tag{26}$$

and

$$(q_R \otimes 1)(\Delta \otimes id)(q_R)\Phi^{-1} = \sum [1 \otimes S^{-1}(X^3) \otimes S^{-1}(X^2)]$$
$$[1 \otimes S^{-1}(f^2) \otimes S^{-1}(f^1)](id \otimes \Delta)(q_R\Delta(X^1)), \tag{27}$$

where $f = \sum f^1 \otimes f^2$ is the twist defined in (15).

A quasi-Hopf algebra H is quasitriangular if there exists an element $R \in H \otimes H$ such that

$$(\Delta \otimes id)(R) = \sum \Phi_{312}R_{13}\Phi_{132}^{-1}R_{23}\Phi, \tag{28}$$

$$(id \otimes \Delta)(R) = \sum \Phi_{231}^{-1}R_{13}\Phi_{213}R_{12}\Phi^{-1}, \tag{29}$$

$$\Delta^{\text{cop}}(h)R = R\Delta(h), \text{ for all } h \in H, \tag{30}$$

$$(\varepsilon \otimes id)(R) = (id \otimes \varepsilon)(R) = 1. \tag{31}$$

Here we used the following notation: if σ is a permutation of $\{1,2,3\}$, then we write $\Phi_{\sigma(1)\sigma(2)\sigma(3)} = \sum X^{\sigma^{-1}(1)} \otimes X^{\sigma^{-1}(2)} \otimes X^{\sigma^{-1}(3)}$; R_{ij} means R acting non-trivially on the i-th and j-th tensor factors of $H \otimes H \otimes H$.

It is shown in [10] that R is invertible. Furthermore, the element

$$u = \sum S(R^2p^2)\alpha R^1p^1, \tag{32}$$

with $p_R = \sum p^1 \otimes p^2$ defined as in (19), is invertible in H, and

$$u^{-1} = \sum X^1R^2p^2S(S(X^2R^1p^1)\alpha X^3), \tag{33}$$

$$\varepsilon(u) = 1 \text{ and } S^2(h) = uhu^{-1}, \tag{34}$$

for all $h \in H$. Consequently the antipode S is bijective, so, as in the Hopf algebra case, the assumptions about invertibility of R and bijectivity of S can be dropped. Moreover, the R-matrix $R = \sum R^1 \otimes R^2$ satisfies the identity (see [1], [15], [10]):

$$f_{21}Rf^{-1} = (S \otimes S)(R) \tag{35}$$

where $f = \sum f^1 \otimes f^2$ is the twist defined in (15), and $f_{21} = \sum f^2 \otimes f^1$.

2.2. Monoidal categories

A monoidal or tensor category is a sixtuple $(\mathcal{C}, \otimes, \underline{1}, a, l, r)$, where $\mathcal{C}$ is a category, $\otimes$ is a functor $\mathcal{C} \times \mathcal{C} \to \mathcal{C}$ (called the tensor product), $\underline{1}$ is an object of $\mathcal{C}$, and

$$a_{U,V,W} : (U \otimes V) \otimes W \to U \otimes (V \otimes W)$$
$$l_V : V \cong V \otimes \underline{1} \, ; \, r_V : V \cong \underline{1} \otimes V$$

are natural isomorphisms satisfying certain coherence conditions, see for example [16, 18, 20]. An object V of a monoidal category $\mathcal{C}$ has a left dual if there exists an object V^* and morphisms $\text{ev}_V : V^* \otimes V \to \underline{1}$, $\text{coev}_V : \underline{1} \to V \otimes V^*$ in $\mathcal{C}$ such that

$$l_V^{-1} \circ (id_V \otimes ev_V) \circ a_{V,V^*,V} \circ (coev_V \otimes id_V) \circ r_V = id_V, \tag{36}$$

$$r_{V^*}^{-1} \circ (ev_V \otimes id_{V^*}) \circ a_{V^*,V,V^*}^{-1} \circ (id_{V^*} \otimes coev_V) \circ l_{V^*} = id_{V^*}. \tag{37}$$

$\mathcal{C}$ is called a rigid monoidal category if every object of $\mathcal{C}$ has a dual.

A braided monoidal category is a monoidal category equipped with a commutativity natural isomorphism $c_{U,V} : U \otimes V \to V \otimes U$, compatible with the unit and the associativity.

In a braided monoidal category, we can define algebras, coalgebras, bialgebras and Hopf algebras. For example, a bialgebra $(B, \underline{m}, \underline{\eta}, \underline{\Delta}, \underline{\varepsilon})$ consists of $B \in \mathcal{C}$, a multiplication $\underline{m} : B \otimes B \to B$ which is associative up to the natural isomorphism a, and a unit $\underline{\eta} : \underline{1} \to B$ such that $\underline{m} \circ (\underline{\eta} \otimes id) = \underline{m} \circ (id \otimes \underline{\eta}) = id$. The properties of the comultiplication $\underline{\Delta}$ and the counit $\underline{\varepsilon}$ are similar. In addition, $\underline{\Delta} : B \to B \otimes B$ has to be an algebra morphism, where $B \otimes B$ is an algebra with multiplication $\underline{m}_{B \otimes B}$, defined as the composition

$$
\begin{array}{ll}
(B \otimes B) \otimes (B \otimes B) & \xrightarrow{a} & B \otimes (B \otimes (B \otimes B)) \\
& \xrightarrow{id \otimes a^{-1}} & B \otimes ((B \otimes B) \otimes B) \\
& \xrightarrow{id \otimes c \otimes id} & B \otimes ((B \otimes B) \otimes B) \\
& \xrightarrow{id \otimes a} & B \otimes (B \otimes (B \otimes B)) \\
& \xrightarrow{a^{-1}} & (B \otimes B) \otimes (B \otimes B) \\
& \xrightarrow{m \otimes m} & B \otimes B
\end{array}
\tag{38}
$$

A Hopf algebra B is a bialgebra with a morphism $\underline{S} : B \to B$ in $\mathcal{C}$ (the antipode) satisfying the usual axioms $\underline{m} \circ (\underline{S} \otimes id) \circ \underline{\Delta} = \underline{\eta} \circ \underline{\varepsilon} = \underline{m} \circ (id \otimes \underline{S}) \circ \underline{\Delta}$. It is known, see e.g. [21], that the antipode $\underline{S}$ of a Hopf algebra B in a braided monoidal category $\mathcal{C}$ is an antialgebra and anticoalgebra morphism, in the sense that

$$
\underline{S} \circ \underline{m} = \underline{m} \circ (\underline{S} \otimes \underline{S}) \circ c_{B,B} \quad \text{and} \quad \underline{\Delta} \circ \underline{S} = c_{B,B} \circ (\underline{S} \otimes \underline{S}) \circ \underline{\Delta}.
\tag{39}
$$

Recall also that an algebra A in a braided monoidal category $\mathcal{C}$ is called quantum commutative if $\underline{m} \circ c_{A,A} = \underline{m}$.

Assume that $(H, \Delta, \varepsilon, \Phi)$ is a quasi-bialgebra, and let U, V, W be left H-modules. We define a left H-action on $U \otimes V$ by

$$
h \cdot (u \otimes v) = \sum h_1 \cdot u \otimes h_2 \cdot v.
$$

We have isomorphisms $a_{U,V,W} : (U \otimes V) \otimes W \to U \otimes (V \otimes W)$ in ${}_H\mathcal{M}$ given by

$$
a_{U,V,W}((u \otimes v) \otimes w) = \Phi \cdot (u \otimes (v \otimes w)).
\tag{40}
$$

The counit $\varepsilon : H \to k$ makes $k \in {}_H\mathcal{M}$, and the natural isomorphisms $\lambda : k \otimes H \to H$ and $\rho : H \otimes k \to H$ are in ${}_H\mathcal{M}$. With this structures, $({}_H\mathcal{M}, \otimes, k, a, \lambda, \rho)$ is a monoidal category.

If H is a quasi-Hopf algebra then the category of finite dimensional left H-modules is rigid; the left dual of V is V^* with the H-module structure given by $(h \cdot \varphi)(v) = \varphi(S(h) \cdot v)$, for all $v \in V$, $\varphi \in V^*$, $h \in H$ and with

$$
\mathrm{ev}_V(\varphi \otimes v) = \varphi(\alpha \cdot v), \quad \mathrm{coev}_V(1) = \sum_{i=1}^{n} \beta \cdot v_i \otimes v^i,
\tag{41}
$$

where $\{v_i\}$ is a basis in V with dual basis $\{v^i\}$.

Now let H be a quasitriangular quasi-Hopf algebra, with R-matrix $R = \sum R^1 \otimes R^2$. For two left H-modules U and V, we define

$$
c_{U,V} : U \otimes V \to V \otimes U
$$

by
$$c_{U,V}(u \otimes v) = \sum R^2 \cdot v \otimes R^1 \cdot u \tag{42}$$
and then $({}_H\mathcal{M}, \otimes, k, a, \lambda, \rho, c)$ is a braided monoidal category (cf. [16] or [20]).

3. YETTER-DRINFELD MODULES AND THE QUASI-YANG-BAXTER EQUATION

From [19], we recall the notion of Yetter-Drinfeld module over a quasi-bialgebra.

DEFINITION 3.1. Let H be a quasi-bialgebra with reassociator Φ. A left H-module M together with a left H-coaction
$$\lambda_M : M \to H \otimes M, \ \lambda_M(m) = \sum m_{(-1)} \otimes m_{(0)}$$
is called a left Yetter-Drinfeld module if the following equalities hold, for all $h \in H$ and $m \in M$:
$$\sum X^1 m_{(-1)} \otimes (X^2 \cdot m_{(0)})_{(-1)} X^3 \otimes (X^2 \cdot m_{(0)})_{(0)}$$
$$= \sum X^1 (Y^1 \cdot m)_{(-1)_1} Y^2 \otimes X^2 (Y^1 \cdot m)_{(-1)_2} Y^3 \otimes X^3 \cdot (Y^1 \cdot m)_{(0)} \tag{43}$$
$$\sum \varepsilon(m_{(-1)}) m_{(0)} = m \tag{44}$$
$$\sum h_1 m_{(-1)} \otimes h_2 \cdot m_{(0)} = \sum (h_1 \cdot m)_{(-1)} h_2 \otimes (h_1 \cdot m)_{(0)}. \tag{45}$$

The category of left Yetter-Drinfeld H-modules and k-linear maps that intertwine the H-action and H-coaction is denoted by ${}^H_H\mathcal{YD}$. In [19] it is shown that ${}^H_H\mathcal{YD}$ is a prebraided monoidal category. The forgetful functor ${}^H_H\mathcal{YD} \to {}_H\mathcal{M}$ is monoidal, and the coaction on the tensor product $M \otimes N$ of two Yetter-Drinfeld modules M and N is given by
$$\lambda_{M \otimes N}(m \otimes n) = \sum X^1 (x^1 Y^1 \cdot m)_{(-1)} x^2 (Y^2 \cdot n)_{(-1)} Y^3 \tag{46}$$
$$\otimes \ X^2 \cdot (x^1 Y^1 \cdot m)_{(0)} \otimes X^3 x^3 \cdot (Y^2 \cdot n)_{(0)}. \tag{47}$$
The braiding is given by
$$c_{M,N}(m \otimes n) = \sum m_{(-1)} \cdot n \otimes m_{(0)}. \tag{48}$$
This braiding is invertible if H is a quasi-Hopf algebra [5], and its inverse is then given by
$$c_{M,N}^{-1}(n \otimes m) = \sum y_1^3 X^2 \cdot (x^1 \cdot m)_{(0)}$$
$$\otimes S^{-1}(S(y^1)\alpha y^2 X^1 (x^1 \cdot m)_{(-1)} x^2 \beta S(y_2^3 X^3 x^3)) \cdot n. \tag{49}$$
Let (H, R) be a quasitriangular quasi-bialgebra. It is well-known (see for example [16]) that R satisfies the so-called quasi-Yang-Baxter equation in $H \otimes H \otimes H$:
$$R_{12}\Phi_{312}R_{13}\Phi_{132}^{-1}R_{23}\Phi = \Phi_{321}R_{23}\Phi_{231}^{-1}R_{13}\Phi_{213}R_{12}.$$
On the other hand, if H is a bialgebra and M is a left-right Yetter-Drinfeld module over H, with structures
$$H \otimes M \to M, \quad h \otimes m \mapsto h \cdot m;$$
$$M \to M \otimes H, \quad m \mapsto \sum m_{(0)} \otimes m_{(1)},$$

then the map $R_M : M \otimes M \to M \otimes M$, $R_M(m \otimes n) = \sum n_{(1)} \cdot m \otimes n_{(0)}$ is a solution in $\text{End}(M \otimes M \otimes M)$ of the quantum Yang-Baxter equation

$$R_{12}R_{13}R_{23} = R_{23}R_{13}R_{12},$$

see for instance [17].
We will show a similar result for quasi-bialgebras; first we define left-right Yetter-Drinfeld modules over quasi-bialgebras as follows

$$_H\mathcal{Y}D^H = {}^{H^{\text{cop}}}_{H^{\text{cop}}}\mathcal{Y}D.$$

This is stated more explicitly in the next definition.

DEFINITION 3.2. Let H be a quasi-bialgebra. A k-linear space M with a left H-action $h \otimes m \mapsto h \cdot m$, and a right H-coaction $M \to M \otimes H$, $m \mapsto \sum m_{(0)} \otimes m_{(1)}$ is called a left-right Yetter-Drinfeld module if the following relations hold, for all $m \in M$ and $h \in H$:

$$\sum (x^2 \cdot m_{(0)})_{(0)} \otimes (x^2 \cdot m_{(0)})_{(1)} x^1 \otimes x^3 m_{(1)}$$
$$= \sum x^1 \cdot (y^3 \cdot m)_{(0)} \otimes x^2 (y^3 \cdot m)_{(1)_1} y^1 \otimes x^3 (y^3 \cdot m)_{(1)_2} y^2 \quad (50)$$

$$\sum \varepsilon(m_{(1)}) m_{(0)} = m \quad (51)$$

$$\sum h_1 \cdot m_{(0)} \otimes h_2 m_{(1)} = \sum (h_2 \cdot m)_{(0)} \otimes (h_2 \cdot m)_{(1)} h_1. \quad (52)$$

PROPOSITION 3.3. Let H be a quasi-bialgebra and $M \in {}_H\mathcal{Y}D^H$. The map $R = R_M : M \otimes M \to M \otimes M$, $R(m \otimes n) = \sum n_{(1)} \cdot m \otimes n_{(0)}$, is a solution of the quasi-Yang-Baxter equation

$$R_{12}\Phi_{312}R_{13}\Phi_{132}^{-1}R_{23}\Phi = \Phi_{321}R_{23}\Phi_{231}^{-1}R_{13}\Phi_{213}R_{12} \quad (53)$$

on $\text{End}(M \otimes M \otimes M)$.

We considered R_{12}, Φ_{312}, etc. as elements in $\text{End}(M \otimes M \otimes M)$ by left multiplication, for example $R_{12}(l \otimes m \otimes n) = \sum R^1 \cdot l \otimes R^2 \cdot m \otimes n$, $\Phi_{312}(l \otimes m \otimes n) = \sum X^2 \cdot l \otimes X^3 \cdot m \otimes X^1 \cdot n$ etc.

Proof. ${}_H\mathcal{Y}D^H$ is a prebraided category, hence the result is a consequence of the fact (see [16]) that the braiding satisfies the categorical version of the Yang-Baxter equation. A direct proof is also possible. For all $l, m, n \in M$, we compute that

$$R_{12}\Phi_{312}R_{13}\Phi_{132}^{-1}R_{23}\Phi(l \otimes m \otimes n)$$
$$= \sum (Y^3 x^3 (X^3 \cdot n)_{(1)} X^2 \cdot m)_{(1)} Y^2 (x^2 \cdot (X^3 \cdot n)_{(0)})_{(1)} x^1 X^1 \cdot l$$
$$\otimes (Y^3 x^3 (X^3 \cdot n)_{(1)} X^2 \cdot m)_{(0)} \otimes Y^1 \cdot (x^2 \cdot (X^3 \cdot n)_{(0)})_{(0)}$$
$$(50) = \sum (Y^3 x^3 (y^3 X^3 \cdot n)_{(1)_2} y^2 X^2 \cdot m)_{(1)} Y^2 x^2 (y^3 X^3 \cdot n)_{(1)_1} y^1 X^1 \cdot l$$
$$\otimes (Y^3 x^3 (y^3 X^3 \cdot n)_{(1)_2} y^2 X^2 \cdot m)_{(0)} \otimes Y^1 x^1 \cdot (y^3 X^3 \cdot n)_{(0)}$$
$$= \sum (n_{(1)_2} \cdot m)_{(1)} n_{(1)_1} \cdot l \otimes (n_{(1)_2} \cdot m)_{(0)} \otimes n_{(0)}$$
$$(52) = \sum n_{(1)_2} m_{(1)} \cdot l \otimes n_{(1)_1} \cdot m_{(0)} \otimes n_{(0)}$$

YETTER-DRINFELD MODULES OVER QUASI-HOPF ALGEBRAS   97

and

$$\Phi_{321} R_{23} \Phi_{231}^{-1} R_{13} \Phi_{213} R_{12}(l \otimes m \otimes n)$$

$$= \sum Y^3 x^3 (X^3 \cdot n)_{(1)} X^2 m_{(1)} \cdot l \otimes Y^2 (x^2 \cdot (X^3 \cdot n)_{(0)})_{(1)} x^1 X^1 \cdot m_{(0)}$$

$$\otimes Y^1 \cdot (x^2 \cdot (X^3 \cdot n)_{(0)})_{(0)}$$

$$(50) \quad = \sum Y^3 x^3 (y^3 X^3 \cdot n)_{(1)_2} y^2 X^2 m_{(1)} \cdot l \otimes Y^2 x^2 (y^3 X^3 \cdot n)_{(1)_1} y^1 X^1 \cdot m_{(0)}$$

$$\otimes Y^1 x^1 \cdot (y^3 X^3 \cdot n)_{(0)}$$

$$= \sum n_{(1)_2} m_{(1)} \cdot l \otimes n_{(1)_1} \cdot m_{(0)} \otimes n_{(0)}$$

and (53) follows. $\square$

We will now present a generalization of [17, Prop. 4.4.2], stating that the dual M^* of a finite dimensional right-left Yetter-Drinfeld module is a left-right Yetter-Drinfeld module and that $R_{M^*} = R_M^*$.
First we define right-left Yetter-Drinfeld modules for quasi-bialgebras as follows:

$$^H \mathcal{YD}_H = {}_{H^{\mathrm{op,cop}}} \mathcal{YD}^{H^{\mathrm{op,cop}}}.$$

More explicitely:

DEFINITION 3.4. Let H be a quasi-bialgebra. A k-linear space M with a right H-action $m \otimes h \mapsto m \cdot h$, and a left H-coaction $M \to H \otimes M$, $m \mapsto \sum m_{(-1)} \otimes m_{(0)}$ is called a right-left Yetter-Drinfeld module if the following relations hold, for all $m \in M$ and $h \in H$:

$$\sum m_{(-1)} x^1 \otimes x^3 (m_{(0)} \cdot x^2)_{(-1)} \otimes (m_{(0)} \cdot x^2)_{(0)}$$

$$= \sum y^2 (m \cdot y^1)_{(-1)_1} x^1 \otimes y^3 (m \cdot y^1)_{(-1)_2} x^2 \otimes (m \cdot y^1)_{(0)} \cdot x^3 \quad (54)$$

$$\sum \varepsilon(m_{(-1)}) m_{(0)} = m \quad (55)$$

$$\sum m_{(-1)} h_1 \otimes m_{(0)} \cdot h_2 = \sum h_2 (m \cdot h_1)_{(-1)} \otimes (m \cdot h_1)_{(0)}. \quad (56)$$

For $M \in {}^H \mathcal{YD}_H$, we consider the map

$$R_M : M \otimes M \to M \otimes M, \quad R_M(m \otimes n) = \sum m \cdot n_{(-1)} \otimes n_{(0)}.$$

If we consider M as an object in $_{H^{\mathrm{op,cop}}} \mathcal{YD}^{H^{\mathrm{op,cop}}}$, then we obtain the same map R_M, so R_M is also a solution of the corresponding quasi-Yang-Baxter equation, which is obtained after replacing Φ by $\Phi_{op,cop} = \Phi^{321}$).
Now let M be a finite dimensional right-left Yetter-Drinfeld module. Then M^* is a left H-module, with action given by $(h \cdot m^*)(m) = m^*(m \cdot h)$, for all $h \in H, m \in M, m^* \in M^*$. We also define a k-linear map $M^* \to M^* \otimes H$, $m^* \mapsto \sum m^*_{(0)} \otimes m^*_{(1)}$, by the condition

$$\sum m^*_{(0)}(m) m^*_{(1)} = \sum m^*(m_{(0)}) m_{(-1)} \quad (57)$$

for all $m \in M$. We can prove now the following result.

PROPOSITION 3.5. Let H be a quasi-bialgebra, M a finite dimensional right-left Yetter-Drinfeld module. Then
(i) $M^* \in_H \mathcal{YD}^H$;

(ii) $R_{M^*} = R_M^*$.

Proof. (*i*) We prove that (50), (51), (52) are satisfied. For $m^* \in M^*$ and $m \in M$, we compute:

$$\sum (x^2 \cdot m_{(0)}^*)_{(0)}(m)(x^2 \cdot m_{(0)}^*)_{(1)} x^1 \otimes x^3 m_{(1)}^*$$

$$(57) \quad = \quad \sum (x^2 \cdot m_{(0)}^*)(m_{(0)}) m_{(-1)} x^1 \otimes x^3 m_{(1)}^*$$

$$= \quad \sum m_{(0)}^*(m_{(0)} \cdot x^2) m_{(-1)} x^1 \otimes x^3 m_{(1)}^*$$

$$(54) \quad = \quad \sum m^*((m \cdot y^1)_{(0)} \cdot x^3) y^2 (m \cdot y^1)_{(-1)_1} x^1 \otimes y^3 (m \cdot y^1)_{(-1)_2} x^2$$

$$= \quad \sum (x^3 \cdot m^*)((m \cdot y^1)_{(0)}) y^2 (m \cdot y^1)_{(-1)_1} x^1 \otimes y^3 (m \cdot y^1)_{(-1)_2} x^2$$

$$(57) \quad = \quad \sum (x^3 \cdot m^*)_{(0)} (m \cdot y^1) y^2 (x^3 \cdot m^*)_{(1)_1} x^1 \otimes y^3 (x^3 \cdot m^*)_{(1)_2} x^2$$

$$= \quad \sum (y^1 \cdot (x^3 \cdot m^*)_{(0)})(m) y^2 (x^3 \cdot m^*)_{(1)_1} x^1 \otimes y^3 (x^3 \cdot m^*)_{(1)_2} x^2$$

so obtain (50). Now we compute:

$$\sum \varepsilon(m_{(1)}^*) m_{(0)}^*(m) = \sum \varepsilon(m_{(0)}^*(m) m_{(1)}^*)$$

$$(57) \quad = \quad \sum \varepsilon(m^*(m_{(0)}) m_{(-1)}) = \sum m^*(\varepsilon(m_{(-1)}) m_{(0)}) = m^*(m),$$

using (55) at the last step. Thus (51) holds. For $h \in H$, we compute:

$$\sum (h_1 \cdot m_{(0)}^*)(m) h_2 m_{(1)}^* = \sum m_{(0)}^*(m \cdot h_1) h_2 m_{(1)}^*$$

$$(57) \quad = \quad \sum m^*((m \cdot h_1)_{(0)}) h_2 (m \cdot h_1)_{(-1)}$$

$$(56) \quad = \quad \sum m^*(m_{(0)} \cdot h_2) m_{(-1)} h_1$$

$$= \quad \sum (h_2 \cdot m^*)(m_{(0)}) m_{(-1)} h_1$$

$$= \quad \sum (h_2 \cdot m^*)_{(0)}(m)(h_2 \cdot m^*)_{(1)} h_1$$

and (52) follows.

(*ii*) We identify $(M \otimes M)^* = M^* \otimes M^*$, and we prove that R_{M^*} and R_M^* coincide as maps $M^* \otimes M^* \to M^* \otimes M^*$. For $m, n \in M$ and $m^*, n^* \in M^*$, we compute:

$$R_{M^*}(m^* \otimes n^*)(m \otimes n) = \sum (n_{(1)}^* \cdot m^*)(m) n_{(0)}^*(n)$$

$$= \quad \sum m^*(m \cdot n_{(1)}^*) n_{(0)}^*(n)$$

$$(57) \quad = \quad \sum m^*(m \cdot n_{(-1)}) n^*(n_{(0)})$$

$$= \quad (m^* \otimes n^*)(R_M(m \otimes n))$$

$$= \quad R_M^*(m^* \otimes n^*)(m \otimes n),$$

as needed. $\qquad \square$

4. THE QUANTUM COMMUTATIVITY OF H_0

Let H be a Hopf algebra. It is well-known that H is an algebra in the monoidal category $^H_H\mathcal{YD}$, with left action and coaction given by

$$h \triangleright h' = \sum h_1 h' S(h_2), \quad \lambda(h) = \sum h_1 \otimes h_2.$$

Moreover, H is quantum commutative as an algebra in $^H_H\mathcal{YD}$, see for example [11]. We will now prove a similar result for quasi-Hopf algebras. Let H be a quasi-Hopf algebra. In [7], a new multiplication on H was introduced; this multiplication is given by the formula

$$h \circ h' = \sum X^1 h S(x^1 X^2) \alpha x^2 X^3_1 h' S(x^3 X^3_2) \tag{58}$$

for all $h, h' \in H$. β is a unit for this multiplication $\circ$. Let H_0 be the k-linear space H, with multiplication $\circ$, and left H-action given by

$$h \triangleright h' = \sum h_1 h' S(h_2). \tag{59}$$

Then H_0 is a left H-module algebra. In H_0, we also define a left H-coaction, as follows

$$\begin{aligned}
\lambda_{H_0}(h) &= \sum h_{(-1)} \otimes h_{(0)} \\
&= \sum X^1 Y^1_1 h_1 g^1 S(q^2 Y^2_2) Y^3 \otimes X^2 Y^2_1 h_2 g^2 S(X^3 q^1 Y^2_1), \tag{60}
\end{aligned}$$

where $f^{-1} = \sum g^1 \otimes g^2$ and $q_R = \sum q^1 \otimes q^2$ are the elements defined by (16) and (19). Then H_0 is an algebra in $^H_H\mathcal{YD}$, see [5] for details. In Proposition 4.2, we will show that H_0 is quantum commutative. But first we need the following formulas, which are of independent interest. Recall that $q_R = \sum q^1 \otimes q^2$, q_L, $f = \sum f^1 \otimes f^2$ and $f^{-1} = \sum g^1 \otimes g^2$ are defined by (20), (22), (15) and (16).

LEMMA 4.1. *Let H be a quasi-Hopf algebra. Then we have*

$$\sum q^1 y^1 \otimes S(q^2 y^2) y^3 = 1 \otimes \alpha, \tag{61}$$

$$\Phi(\Delta \otimes id)(f^{-1}) = \sum g^1 S(X^3) f^1 \otimes g^2_1 G^1 S(X^2) f^2 \otimes g^2_2 G^2 S(X^1), \tag{62}$$

$$\sum S(g^1) \alpha g^2 = S(\beta), \quad \sum f^1 \beta S(f^2) = S(\alpha), \tag{63}$$

$$\sum S(q^2_2 X^3) f^1 \otimes S(q^1 X^1 \beta S(q^2_1 X^2) f^2) = (id \otimes S)(q_L). \tag{64}$$

Proof. (61) and (62) are a direct consequence of (19) and (18). (63) has been proved in [6, Lemma 2.6] and [10, Lemma 2.5]. We are left to prove (64). Using (27), we obtain:

$$(id \otimes \Delta)(q) = \sum (1 \otimes S^{-1}(x^3 g^2) \otimes S^{-1}(x^2 g^1))(q \otimes 1)(\Delta \otimes id)(q) \Phi^{-1}(id \otimes \Delta)(\Delta(x^1))$$

and, using the formula (see [8])

$$(\Delta \otimes id)(q) \Phi^{-1} = \sum Y^1 \otimes q^1 Y^2_1 \otimes S^{-1}(Y^3) q^2 Y^2_2,$$

we obtain

$$(id \otimes \Delta)(q) = \sum Q^1 Y^1 x^1_1 \otimes S^{-1}(x^3 g^2) Q^2 q^1 Y^2_1 x^1_{(2,1)} \otimes S^{-1}(Y^3 x^2 g^1) q^2 Y^2_2 x^1_{(2,2)} \tag{65}$$

where $q_R = \sum q^1 \otimes q^2 = \sum Q^1 \otimes Q^2$. Now we compute

$$\sum S(q_2^2 X^3)f^1 \otimes S(q^1 X^1 \beta S(q_1^2 X^2)f^2)$$

$$(65) \quad = \quad \sum S(q^2 Y_2^2 x_{(2,2)}^1 X^3)Y^3 x^2 \otimes S(Q^1 Y^1 x_1^1 X^1 \beta S(Q^2 q^1 Y_1^2 x_{(2,1)}^1 X^2)x^3)$$

$$(1) \quad = \quad \sum S(q^2 Y_2^2 X^3 x_2^1)Y^3 x^2 \otimes S(Q^1 Y^1 X^1 x_{(1,1)}^1 \beta S(Q^2 q^1 Y_1^2 X^2 x_{(1,2)}^1)x^3)$$

$$(5) \quad = \quad \sum S(q^2 Y_2^2 X^3 x^1)Y^3 x^2 \otimes S(Q^1 Y^1 X^1 \beta S(Q^2 q^1 Y_1^2 X^2)x^3)$$

$$(3) \quad = \quad \sum S(q^2 y^2 X_1^3 Y^2 x^1)y^3 X_2^3 Y^3 x^2 \otimes S(Q^1 X^1 Y_1^1 \beta S(Q^2 q^1 y^1 X^2 Y_2^1)x^3)$$

$$(5,61) \quad = \quad \sum S(X_1^3 x^1)\alpha X_2^3 x^2 \otimes S(Q^1 X^1 \beta S(Q^2 X^2)x^3)$$

$$(5,7) \quad = \quad \sum S(x^1)\alpha x^2 \otimes S(Q^1 \beta S(Q^2)x^3)$$

$$(19,6) \quad = \quad \sum S(x^1)\alpha x^2 \otimes S(x^3),$$

as needed. $\qquad\qquad\qquad\qquad\qquad\qquad\qquad\qquad\qquad\qquad\qquad\qquad\qquad\qquad\qquad\square$

We can prove now the main result of this Section.

PROPOSITION 4.2. *Let H be a quasi-Hopf algebra. Then H_0 is quantum commutative as an algebra in ${}_H^H \mathcal{YD}$, that is, for all $h, h' \in H$:*

$$h \circ h' = \sum (h_{(-1)} \triangleright h') \circ h_{(0)}.$$

Proof. For all $h, h' \in H$ we compute:

$$\sum (h_{(-1)} \triangleright h') \circ h_{(0)}$$

$$(60) \quad = \quad \sum (X^1 Y_1^1 h_1 g^1 S(q^2 Y_2^2)Y^3 \triangleright h') \circ X^2 Y_2^1 h_2 g^2 S(X^3 q^1 Y_1^2)$$

$$(59,58) \quad = \quad \sum Z^1 X_1^1 Y_{(1,1)}^1 h_{(1,1)} g_1^1 S(q^2 Y_2^2)_1 Y_1^3 h'$$

$$\quad S(x^1 Z^2 X_2^1 Y_{(1,2)}^1 h_{(1,2)} g_2^1 S(q^2 Y_2^2)_2 Y_2^3)$$

$$\quad \alpha x^2 Z_1^3 X^2 Y_2^1 h_2 g^2 S(x^3 Z_2^3 X^3 q^1 Y_1^2)$$

$$(3,5) \quad = \quad \sum Z^1 Y_{(1,1)}^1 h_{(1,1)} g_1^1 S(q^2 Y_2^2)_1 Y_1^3 h'$$

$$\quad S(Z^2 Y_{(1,2)}^1 h_{(1,2)} g_2^1 S(q^2 Y_2^2)_2 Y_2^3)$$

$$\quad \alpha Z^3 Y_2^1 h_2 g^2 S(q^1 Y_1^2)$$

$$(11) \quad = \quad \sum Z^1 [Y^1 h S(Y^2)]_{(1,1)} g_1^1 S(q^2)_1 Y_1^3 h'$$

$$\quad S(Z^2 [Y^1 h S(Y^2)]_{(1,2)} g_2^1 S(q^2)_2 Y_2^3)$$

$$\quad \alpha Z^3 [Y^1 h S(Y^2)]_2 g^2 S(q^1)$$

$$(1,5) \quad = \quad \sum Y^1 h S(Y^2) Z^1 g_1^1 S(q^2)_1 Y_1^3 h' S(Z^2 g_2^1 S(q^2)_2 Y_2^3)\alpha Z^3 g^2 S(q^1)$$

$$(62) \quad = \quad \sum Y^1 h S(Y^2)g^1 S(X^3)f^1 S(q^2)_1 Y_1^3 h'$$

$$\quad S(g_1^2 G^1 S(X^2)f^2 S(q^2)_2 Y_2^3)\alpha g_2^2 G^2 S(q^1 X^1)$$

$$(5,63) \quad = \quad \sum Y^1 h S(X^3 Y^2) f^1 S(q^2)_1 Y_1^3 h' S(q^1 X^1 \beta S(X^2) f^2 S(q^2)_2 Y_2^3)$$

$$(11) \quad = \quad \sum Y^1 h S(q_2^2 X^3 Y^2) f^1 Y_1^3 h' S(q^1 X^1 \beta S(q_1^2 X^2) f^2 Y_2^3)$$

$$(64) \quad = \quad \sum Y^1 h S(x^1 Y^2) \alpha x^2 Y_1^3 h' S(x^3 Y_2^3)$$

$$(58) \quad = \quad h \circ h'.$$

$\square$

5. HOPF MODULES IN $_H^H \mathcal{YD}$. INTEGRALS

Let H be a quasi-Hopf algebra. The aim of this Section is to define the space of integrals of a finite dimensional braided Hopf algebra in $_H^H \mathcal{YD}$, and to prove, following [24], [12], that it is an object of $_H^H \mathcal{YD}$, and that it has dimension 1. We will apply our results to the braided Hopf algebra associated to H, in the case where H is a quasitriangular quasi-Hopf algebra.

Let A be an algebra in a monoidal category $\mathcal{C}$. Recall that a right A-module M is an object $M \in \mathcal{C}$ together with a morphism $\underline{\omega}_M : M \otimes A \to M$ in $\mathcal{C}$ such that $\underline{\omega}_M \circ (id_M \otimes \underline{\eta}) = l_M^{-1}$ and the following diagram is commutative:

$$
\begin{array}{ccccc}
(M \otimes A) \otimes A & \xrightarrow{\underline{\omega}_M \otimes id_A} & M \otimes A & \xrightarrow{\underline{\omega}_M} & M \\
{\scriptstyle a_{M,A,A}} \downarrow & & & & \uparrow {\scriptstyle \underline{\omega}_M} \\
M \otimes (A \otimes A) & \xrightarrow{\hspace{1cm} id_M \otimes \underline{m} \hspace{1cm}} & & & M \otimes A.
\end{array}
$$

Clearly A itself is a right A-module, by right multiplication. Right comodules over a coalgebra C in $\mathcal{C}$ can be defined in a similar way: we need $N \in \mathcal{C}$ together with a morphism $\underline{\rho}_N : N \to N \otimes C$ in $\mathcal{C}$ such that $(id_N \otimes \underline{\varepsilon}) \circ \underline{\rho}_N = l_N$ and the following diagram is commutative:

$$
\begin{array}{ccccc}
N & \xrightarrow{\underline{\rho}_N} & N \otimes C & \xrightarrow{\underline{\rho}_N \otimes id_C} & (N \otimes C) \otimes C \\
{\scriptstyle \underline{\rho}_N} \downarrow & & & & \downarrow {\scriptstyle a_{N,C,C}} \\
N \otimes C & \xrightarrow{\hspace{1cm} id_N \otimes \underline{\Delta} \hspace{1cm}} & & & N \otimes (C \otimes C).
\end{array}
$$

C itself is a right C-comodule via the comultiplication $\underline{\Delta}$.
From [3], [21], [24], we recall the following.

DEFINITION 5.1. Let B be a bialgebra in a braided category $\mathcal{C}$. A right B-Hopf module is a triple $(M, \underline{\omega}_M, \underline{\rho}_M)$, where $(M, \underline{\omega}_M)$ is a right B-module and $(M, \underline{\rho}_M)$ is a right B-comodule such that $\underline{\rho}_M : M \to M \otimes B$ is right B-linear. The B-module structure $\underline{\omega}_{M \otimes B} : (M \otimes B) \otimes B \to M \otimes B$ on $M \otimes B$ is given by the following composition:

$$(M \otimes B) \otimes B \xrightarrow{\ id_{M \otimes B} \otimes \underline{\Delta}\ } (M \otimes B) \otimes (B \otimes B)$$
$$\xrightarrow{\ a_{M,B,B \otimes B}\ } M \otimes (B \otimes (B \otimes B))$$
$$\xrightarrow{\ id_M \otimes a_{B,B,B}^{-1}\ } M \otimes ((B \otimes B) \otimes B)$$
$$\xrightarrow{\ id_M \otimes (c_{B,B} \otimes id_B)\ } M \otimes ((B \otimes B) \otimes B) \qquad (66)$$
$$\xrightarrow{\ id_M \otimes a_{B,B,B}\ } M \otimes (B \otimes (B \otimes B))$$
$$\xrightarrow{\ a_{M,B,B \otimes B}^{-1}\ } (M \otimes B) \otimes (B \otimes B)$$
$$\xrightarrow{\ \underline{\omega}_M \otimes \underline{m}\ } M \otimes B$$

$\mathcal{M}_B^B$ will denote the category of right B-Hopf modules and morphisms in $\mathcal{C}$ preserving the B-action and the corresponding B-coaction.

We can consider algebras, coalgebras, bialgebras and Hopf algebras in the braided category ${}_H^H\mathcal{Y}D$ over a quasi-Hopf algebra H. More precisely, an algebra B in ${}_H^H\mathcal{Y}D$ is an object $B \in {}_H^H\mathcal{Y}D$ such that

- B is a left H-module algebra, i.e. B has a multiplication $\underline{m}$ and a usual unit 1_B satisfying the following conditions:

$$(ab)c = \sum (X^1 \cdot a)[(X^2 \cdot b)(X^3 \cdot c)], \qquad (67)$$

$$h \cdot (ab) = \sum (h_1 \cdot a)(h_2 \cdot b), \quad h \cdot 1_B = \varepsilon(h)1_B, \qquad (68)$$

for all $a, b, c \in B$ and $h \in H$.

- B is a quasi-comodule algebra, that is, the multiplication $\underline{m}$ and the unit $\underline{\eta}$ of B intertwine the H-coaction λ_B. By (47) this means:

$$\lambda_B(bb') = \sum X^1 (x^1 Y^1 \cdot b)_{(-1)} x^2 (Y^2 \cdot b')_{(-1)} Y^3$$
$$\otimes [X^2 \cdot (x^1 Y^1 \cdot b)_{(0)}][X^3 x^3 \cdot (Y^2 \cdot b')_{(0)}], \qquad (69)$$

for all $b, b' \in B$, and

$$\lambda_B(1_B) = 1_H \otimes 1_B. \qquad (70)$$

$M \in {}_H^H\mathcal{Y}D$ is a right B-module if there exists a morphism $\underline{\omega}_M : M \otimes B \to M$ in ${}_H^H\mathcal{Y}D$ (we will denote $\underline{\omega}_M(m \otimes b) := m \leftarrow b$) such that

$$m \leftarrow 1_B = m, \quad (m \leftarrow b) \leftarrow b' = \sum (X^1 \cdot m) \leftarrow [(X^2 \cdot b)(X^3 \cdot b')] \qquad (71)$$

for all $m \in M$, $b, b' \in B$. The fact that $\underline{\omega}_M$ is a morphism in ${}_H^H\mathcal{Y}D$ means (see (47))

$$h \cdot (m \leftarrow b) = \sum (h_1 \cdot m) \leftarrow (h_2 \cdot b), \qquad (72)$$

$$\lambda_M(m \leftarrow b) = \sum X^1 (x^1 Y^1 \cdot m)_{(-1)} x^2 (Y^2 \cdot b)_{(-1)} Y^3$$
$$\otimes [X^2 \cdot (x^1 Y^1 \cdot m)_{(0)}] \leftarrow [X^3 x^3 \cdot (Y^2 \cdot b)_{(0)}] \qquad (73)$$

for all $m \in M$, $b \in B$.

Similarly, $B \in {}_H^H\mathcal{Y}D$ is a coalgebra if

YETTER-DRINFELD MODULES OVER QUASI-HOPF ALGEBRAS

- B is a left H-module coalgebra, i.e. B has a comultiplication $\underline{\Delta}_B : B \to B \otimes B$ (we will denote $\underline{\Delta}(b) = \sum b_{\underline{1}} \otimes b_{\underline{2}}$) and a usual counit $\underline{\varepsilon}_B$ such that:

$$\sum X^1 \cdot b_{(\underline{1},\underline{1})} \otimes X^2 \cdot b_{(\underline{1},\underline{2})} \otimes X^3 \cdot b_{\underline{2}} = \sum b_{\underline{1}} \otimes b_{(\underline{2},\underline{1})} \otimes b_{(\underline{2},\underline{2})}, \tag{74}$$

$$\underline{\Delta}_B(h \cdot b) = \sum h_1 \cdot b_{\underline{1}} \otimes h_2 \cdot b_{\underline{2}}, \quad \underline{\varepsilon}_B(h \cdot b) = \varepsilon(h)\underline{\varepsilon}_B(b), \tag{75}$$

for all $h \in H$, $b \in B$, where we use the same notation for the quasi-coassociativity of $\underline{\Delta}_B$ as in Section 2.

- B is a quasi-comodule coalgebra, i.e. the comultiplication $\underline{\Delta}_B$ and the counit $\underline{\varepsilon}_B$ intertwine the H-coaction λ_B. Explicitly, for all $b \in B$ we must have that:

$$\sum b_{(-1)} \otimes b_{(0)_{\underline{1}}} \otimes b_{(0)_{\underline{2}}} = \sum X^1(x^1Y^1 \cdot b_{\underline{1}})_{(-1)}x^2(Y^2 \cdot b_{\underline{2}})_{(-1)}Y^3$$
$$\otimes X^2 \cdot (x^1Y^1 \cdot b_{\underline{1}})_{(0)} \otimes X^3x^3 \cdot (Y^2 \cdot b_{\underline{2}})_{(0)}, \tag{76}$$

and

$$\sum \varepsilon_B(b_{(0)})b_{(-1)} = \varepsilon_B(b)1. \tag{77}$$

A right B-comodule in $^H_H\mathcal{YD}$ is an object $M \in {}^H_H\mathcal{YD}$ together with a morphism $\varrho_M : M \to M \otimes B$ in $^H_H\mathcal{YD}$ (we will denote $\varrho_M(m) = \sum m_{(0)} \otimes m_{(1)}$ for all $m \in M$) such that the following relations hold, for all $m \in M$:

$$\sum X^1 \cdot m_{(0,0)} \otimes X^2 \cdot m_{(0,1)} \otimes X^3 \cdot m_{(1)} = \sum m_{(0)} \otimes m_{(1)_{\underline{1}}} \otimes m_{(1)_{\underline{2}}}, \tag{78}$$

$$\sum \underline{\varepsilon}(m_{(1)})m_{(0)} = m, \tag{79}$$

where we will denote

$$(\varrho_M \otimes id_B)(\varrho_M(m)) = \sum m_{(0,0)} \otimes m_{(0,1)} \otimes m_{(1)} \text{ etc.}$$

The fact that ϱ_M is a morphism in $^H_H\mathcal{YD}$ means that (see (47))

$$\varrho_M(h \cdot m) = \sum h_1 \cdot m_{(0)} \otimes h_2 \cdot m_{(1)}, \tag{80}$$

and

$$\sum m_{(-1)} \otimes m_{(0)_{(0)}} \otimes m_{(0)_{(1)}} = \sum X^1(x^1Y^1 \cdot m_{(0)})_{(-1)}x^2(Y^2 \cdot m_{(1)})_{(-1)}Y^3$$
$$\otimes X^2 \cdot (x^1Y^1 \cdot m_{(0)})_{(0)} \otimes X^3x^3 \cdot (Y^2 \cdot m_{(1)})_{(0)}, \tag{81}$$

for all $h \in H$ and $m \in M$.

Now, a bialgebra $B \in {}^H_H\mathcal{YD}$ is an algebra and a coalgebra in $^H_H\mathcal{YD}$ such that $\underline{\Delta}_B$ is an algebra morphism, i.e. $\underline{\Delta}_B(1_B) = 1_B \otimes 1_B$ and, by (38) and (48), for all $b, b' \in B$ we have that:

$$\Delta_B(bb') = \sum [y^1X^1 \cdot b_{\underline{1}}][y^2Y^1(x^1X^2 \cdot b_{\underline{2}})_{(-1)}x^2X_1^3 \cdot b_{\underline{1}}']$$
$$\otimes [y_1^3Y^2 \cdot (x^1X^2 \cdot b_{\underline{2}})_{(0)}][y_2^3Y^3x^3X_2^3 \cdot b_{\underline{2}}']. \tag{82}$$

If $B \in {}^H_H\mathcal{YD}$ is a bialgebra then $M \in {}^H_H\mathcal{YD}$ is a right B-Hopf module if M is a right B-module (as above, we will denote $\underline{\omega}_M(m \otimes b) = m \leftarrow b$) and a right B-comodule

104   D. BULACU, S.CAENEPEEL AND F. PANAITE

such that the right B-coaction on M, $\underline{\rho}_M : M \to M \otimes B$, is right B-linear, which means that the following relation holds, for all $m \in M$ and $b \in B$ (see (66)):

$$\underline{\rho}_M(m \leftharpoonup b) = \sum (y^1 X^1 \cdot m_{(\underline{0})}) \leftharpoonup [y^2 Y^1 (x^1 X^2 \cdot m_{(\underline{1})})_{(-1)} x^2 X_1^3 \cdot b_{\underline{1}}]$$
$$\otimes [y_1^3 Y^2 \cdot (x^1 X^2 \cdot m_{(\underline{1})})_{(0)}][y_2^3 Y^3 x^3 X_2^3 \cdot b_{\underline{2}}]. \tag{83}$$

Finally, a bialgebra B in $^H_H \mathcal{YD}$ is a braided Hopf algebra if there exists a morphism $\underline{S} : B \to B$ in $^H_H \mathcal{YD}$ such that $\sum \underline{S}(b_{\underline{1}})b_{\underline{2}} = \sum b_{\underline{1}} \underline{S}(b_{\underline{2}}) = \varepsilon(b)1_B$, for all $b \in B$. Since $\underline{S}$ is a morphism in $^H_H \mathcal{YD}$, we have that

$$\underline{S}(h \cdot b) = h \cdot \underline{S}(b) \quad \text{and} \quad \sum \underline{S}(b)_{(-1)} \otimes \underline{S}(b)_{(0)} = \sum b_{(-1)} \otimes \underline{S}(b_{(0)}), \tag{84}$$

for all $h \in H$, $b \in B$. Also, by (39) and (48) we obtain that

$$\underline{S}(bb') = \sum [b_{(-1)} \cdot \underline{S}(b')] \underline{S}(b_{(0)}) \quad \text{and} \quad \underline{\Delta}(\underline{S}(b)) = \sum b_{\underline{1}_{(-1)}} \cdot \underline{S}(b_{\underline{2}}) \otimes \underline{S}(b_{\underline{1}_{(0)}}), \tag{85}$$

for all $b, b' \in B$.

The first step to prove the existence and uniqueness of integrals in a finite dimensional braided Hopf algebra is the structure theorem for Hopf modules. To this end we need first the following result.

LEMMA 5.2. *Let H be a quasi-bialgebra, B a bialgebra in $^H_H \mathcal{YD}$ and $N \in {}^H_H \mathcal{YD}$. Then $N \otimes B \in \mathcal{M}^B_B$ with following action $\underline{\omega}_{N \otimes B} : (N \otimes B) \otimes B \to N \otimes B$ and coaction $\underline{\rho}_{N \otimes B} : N \otimes B \to (N \otimes B) \otimes B$ given by*

$$(n \otimes b) \prec b' = \sum X^1 \cdot n \otimes [(X^2 \cdot b)(X^3 \cdot b')], \tag{86}$$

$$\underline{\rho}_{N \otimes B}(n \otimes b) := \sum x^1 \cdot n \otimes x^2 \cdot b_{\underline{1}} \otimes x^3 \cdot b_{\underline{2}}, \tag{87}$$

for all $n \in N$ and $b, b' \in B$.

Proof. $^H_H \mathcal{YD}$ is a braided category, so $N \otimes B \in {}^H_H \mathcal{YD}$. It is not hard to see that (1) and (67) imply that $\underline{\omega}_{N \otimes B}$ is left H-linear. It intertwines also the corresponding H-coaction. Indeed, by (47), the left H-coaction on $(N \otimes B) \otimes B$ is given by

$$\lambda_{(N \otimes B) \otimes B}((n \otimes b) \otimes b')$$
$$= \sum Z^1 X^1 (x^1 Y^1 y_1^1 T_1^1 \cdot n)_{(-1)} x^2 (Y^2 y_2^1 T_2^1 \cdot b)_{(-1)} Y^3 y^2 (T^2 \cdot b')_{(-1)} T^3$$
$$\otimes Z_1^2 X^2 \cdot (x^1 Y^1 y_1^1 T_1^1 \cdot n)_{(0)} \otimes Z_2^2 X^3 x^3 \cdot (Y^2 y_2^1 T_2^1 \cdot b)_{(0)} \otimes Z^3 y^3 \cdot (T^2 \cdot b')_{(0)},$$

YETTER-DRINFELD MODULES OVER QUASI-HOPF ALGEBRAS

for all $n \in N$, $b, b' \in B$. Therefore:

$$(id_H \otimes \underline{\omega}_{N \otimes B}) \circ \lambda_{(N \otimes B) \otimes B}((n \otimes b) \otimes b')$$

$$(86) = \sum Z^1 X^1 (x^1 Y^1 y_1^1 T_1^1 \cdot n)_{(-1)} x^2 (Y^2 y_2^1 T_2^1 \cdot b)_{(-1)}$$
$$Y^3 y^2 (T^2 \cdot b')_{(-1)} T^3 \otimes W^1 Z_1^2 X^2 \cdot (x^1 Y^1 y_1^1 T_1^1 \cdot n)_{(0)}$$
$$\otimes [(W^2 Z_2^2 X^3 x^3 \cdot (Y^2 y_2^1 T_2^1 \cdot b)_{(0)})][(W^3 Z^3 y^3 \cdot (T^2 \cdot b')_{(0)})]$$

$$(3, 45, 67) = \sum Z^1 (X_1^1 x^1 Y^1 y_1^1 T_1^1 \cdot n)_{(-1)} X_2^1 x^2 (Y^2 y_2^1 T_2^1 \cdot b)_{(-1)}$$
$$Y^3 y^2 (T^2 \cdot b')_{(-1)} T^3 \otimes Z^2 \cdot (X_1^1 x^1 Y^1 y_1^1 T_1^1 \cdot n)_{(0)}$$
$$\otimes Z^3 \cdot [(X^2 x^3 \cdot (Y^2 y_2^1 T_2^1 \cdot b)_{(0)})(X^3 y^3 \cdot (T^2 \cdot b')_{(0)})]$$

$$(3) \text{ twice}, (45) = \sum Z^1 (x^1 Y^1 T_1^1 \cdot n)_{(-1)} x^2 X^1 (y^1 Y^2 T_2^1 \cdot b)_{(-1)} y^2$$
$$(Y_1^3 T^2 \cdot b')_{(-1)} Y_2^3 T^3 \otimes Z^2 \cdot (x^1 Y^1 T_1^1 \cdot n)_{(0)}$$
$$\otimes Z^3 x^3 \cdot [(X^2 \cdot (y^1 Y^2 T_2^1 \cdot b)_{(0)})(X^3 y^3 \cdot (Y_1^3 T^2 \cdot b')_{(0)})]$$

$$(3, 69) = \sum Z^1 (x^1 Y^1 T^1 \cdot n)_{(-1)} x^2 [(Y_1^2 T^2 \cdot b)(Y_2^2 T^3 \cdot b')]_{(-1)} Y^3$$
$$\otimes Z^2 \cdot (x^1 Y^1 T^1 \cdot n)_{(0)} \otimes Z^3 x^3 \cdot [(Y_1^2 T^2 \cdot b)(Y_2^2 T^3 \cdot b')]_{(0)}$$

$$(47, 86) = \sum \lambda_{N \otimes B}(T^1 \cdot n \otimes (T^2 \cdot b)(T^3 \cdot b'))$$
$$= \lambda_{N \otimes B} \circ \underline{\omega}_{N \otimes B}((n \otimes b) \otimes b')$$

for all $n \in N$ and $b, b' \in B$. In a similar way, it can be proved that the map $\rho_{N \otimes B}$ is a morphism in $^H_H \mathcal{YD}$, we leave it to the reader to verify the details.

Using (67) and (3), it easily follows that $N \otimes B$ is a right B-module. Also, it is not hard to see that (74), (75) and (3) imply that $N \otimes B$ is a right B-comodule. It remains only to show that $\rho_{N \otimes B}$ is right B-linear. By (66), we have that the right B-module structure of $(N \otimes B) \otimes B$ is given by

$$[(n \otimes b) \otimes b'] \bullet b''$$
$$= \sum [Z^1 y_1^1 X_1^1 \cdot n \otimes (Z^2 y_2^1 X_2^1 \cdot b)(Z^3 y^2 Y^1 (x^1 X^2 \cdot b')_{(-1)} x^2 X_1^3 \cdot b_1'')]$$
$$\otimes [y_1^3 Y^2 \cdot (x^1 X^2 \cdot b')_{(0)}][y_2^3 Y^3 x^3 X_2^3 \cdot b_2''],$$

for all $n \in N$ and $b, b', b'' \in B$. This allows us to compute, for any $n \in N$ and $b, b' \in B$, that:

$$\rho_{N \otimes B}(n \otimes b) \bullet b' = \sum [(z^1 \cdot n \otimes z^2 \cdot b_{\underline{1}}) \otimes z^3 \cdot b_{\underline{2}}] \bullet b'$$
$$= \sum [Z^1 y_1^1 X_1^1 z^1 \cdot n$$
$$\otimes (Z^2 y_2^1 X_2^1 z^2 \cdot b_{\underline{1}})(Z^3 y^2 Y^1 (x^1 X^2 z^3 \cdot b_{\underline{2}})_{(-1)} x^2 X_1^3 \cdot b_{\underline{1}}')]$$
$$\otimes [y_1^3 Y^2 \cdot (x^1 X^2 z^3 \cdot b_{\underline{2}})_{(0)}][y_2^3 Y^3 x^3 X_2^3 \cdot b_{\underline{2}}']$$

$$
\begin{aligned}
(3) \quad &= \sum [Z^1 y_1^1 z^1 X^1 \cdot n \otimes (Z^2 y_2^1 z^2 T^1 X_1^2 \cdot b_{\underline{1}}) \\
&\quad (Z^3 y^2 Y^1 (x^1 z_1^3 T^2 X_2^2 \cdot b_{\underline{2}})_{(-1)} x^2 z_{(2,1)}^3 T_1^3 X_1^3 \cdot b_{\underline{1}}')] \\
&\quad \otimes [y_1^3 Y^2 \cdot (x^1 z_1^3 T^2 X_2^2 \cdot b_{\underline{2}})_{(0)}][y_2^3 Y^3 x^3 z_{(2,2)}^3 T_2^3 X_2^3 \cdot b_{\underline{2}}'] \\
(1,45) \quad &= \sum [Z^1 y_1^1 z^1 X^1 \cdot n \otimes (Z^2 y_2^1 z^2 T^1 X_1^2 \cdot b_{\underline{1}}) \\
&\quad (Z^3 y^2 Y^1 z_{(1,1)}^3 (x^1 T^2 X_2^2 \cdot b_{\underline{2}})_{(-1)} x^2 T_1^3 X_1^3 \cdot b_{\underline{1}}')] \\
&\quad \otimes [y_1^3 Y^2 z_{(1,2)}^3 \cdot (x^1 T^2 X_2^2 \cdot b_{\underline{2}})_{(0)}][y_2^3 Y^3 z_2^3 x^3 T_2^3 X_2^3 \cdot b_{\underline{2}}'] \\
(1,3,67) \quad &= \sum \{y^1 X^1 \cdot n \otimes y^2 \cdot [(z^1 T^1 X_1^2 \cdot b_{\underline{1}}) \\
&\quad (z^2 Y^1 (x^1 T^2 X_2^2 \cdot b_{\underline{2}})_{(-1)} x^2 T_1^3 X_1^3 \cdot b_{\underline{1}}')]\} \\
&\quad \otimes y^3 \cdot \{[z_1^3 Y^2 \cdot (x^1 T^2 X_2^2 \cdot b_{\underline{2}})_{(0)}][z_2^3 Y^3 x^3 T_2^3 X_2^3 \cdot b_{\underline{2}}']\} \\
(75,82) \quad &= \sum \{y^1 X^1 \cdot n \otimes y^2 \cdot [(X^2 \cdot b)(X^3 \cdot b')]_{\underline{1}}\} \otimes y^3 \cdot [(X^2 \cdot b)(X^3 \cdot b')]_{\underline{2}} \\
(87,86) \quad &= \sum \rho_{N \otimes B}(X^1 \cdot n \otimes (X^2 \cdot b)(X^3 \cdot b')) = \rho_{N \otimes B}((n \otimes b) \prec b'),
\end{aligned}
$$

as needed. $\qquad\square$

Our next result is the Fundamental Theoreom for Hopf modules in the braided monoidal category $_H^H \mathcal{YD}$, generalizing [12, Theorem 1].

THEOREM 5.3. *Let H be a quasi-Hopf algebra, B a Hopf algebra in $_H^H\mathcal{YD}$ and $M \in \mathcal{M}_B^B$.*

(i) *$M^{coB} = \{m \in M \mid \rho_M(m) = m \otimes 1_B\} \in {}_H^H\mathcal{YD}$.*

(ii) *For all $m \in M$, we have that $P(m) = \sum m_{(0)} \leftarrow \underline{S}(m_{(1)}) \in M^{coB}$.*

(iii) *$\rho_M(n \leftarrow b) = \sum (x^1 \cdot n) \leftarrow (x^2 \cdot b_{\underline{1}}) \otimes x^3 \cdot b_{\underline{2}}$ and $P(n \leftarrow b) = \varepsilon(b)n$, for all $n \in M^{coB}$ and $b \in B$.*

(iv) *The map*
$$ F: \; M^{coB} \otimes B \to M, \; F(n \otimes b) = n \leftarrow b, $$
is an isomorphism of Hopf modules in $_H^H\mathcal{YD}$, with inverse G given by
$$ G(m) = \sum P(m_{(0)}) \otimes m_{(1)}. $$

Proof. (i) If $n \in M^{coB}$, then $\rho_M(h \cdot n) = \sum h_1 \cdot n \otimes h_2 \cdot 1_B = h \cdot n \otimes 1_B$, by (72) and (67). This shows that M^{coB} is an H-submodule of M. On the other hand, for any $n \in N$ we have

$$
\begin{aligned}
\sum n_{(-1)} &\otimes n_{(0)_{(0)}} \otimes n_{(0)_{(1)}} \\
(81) \quad &= \sum X^1 (x^1 Y^1 \cdot n)_{(-1)} x^2 (Y^2 \cdot 1_B)_{(-1)} Y^3 \\
&\quad \otimes X^2 \cdot (x^1 Y^1 \cdot n)_{(0)} \otimes X^3 x^3 \cdot (Y^2 \cdot 1_B)_{(0)} \\
(67) \text{ twice, } (70) \quad &= \sum n_{(-1)} \otimes n_{(0)} \otimes 1_B.
\end{aligned}
$$

Thus, $\rho_M(n) = \sum n_{(-1)} \otimes n_{(0)} \in H \otimes M^{coB}$ which means that M^{coB} is a left H-quasi-subcomodule of M. It follows from the above arguments that $M^{coB} \in {}_H^H\mathcal{YD}$.

YETTER-DRINFELD MODULES OVER QUASI-HOPF ALGEBRAS

(ii) For any $m \in M$, we have that

$$
\begin{aligned}
\rho_M(P(m)) &= \sum \rho_M(m_{(0)} \leftharpoonup \underline{S}(m_{(1)})) \\
(83) \quad &= \sum (y^1 X^1 \cdot m_{(0,0)}) \leftharpoonup [y^2 Y^1 (x^1 X^2 \cdot m_{(0,1)})_{(-1)} x^2 X_1^3 \cdot \underline{S}(m_{(1)})_{\underline{1}}] \\
&\qquad [y_1^3 Y^2 \cdot (x^1 X^2 \cdot m_{(0,1)})_{(0)}][y_2^3 Y^3 x^3 X_2^3 \cdot \underline{S}(m_{(1)})_{\underline{2}}] \\
(78, 84, 85) \quad &= \sum (y^1 \cdot m_{(0)}) \leftharpoonup \\
&\qquad [y^2 Y^1 (x^1 \cdot m_{(1)_{\underline{1}}})_{(-1)} x^2 \underline{S}(m_{(1)_{(2,1)}})_{(-1)} \cdot \underline{S}(m_{(1)_{(2,2)}})] \\
&\qquad \otimes y^3 \cdot \{[Y^2 \cdot (x^1 \cdot m_{(1)_{\underline{1}}})_{(0)}][Y^3 x^3 \cdot \underline{S}(m_{(1)_{(2,1)}})_{(0)}]\} \\
(69) \quad &= \sum (y^1 \cdot m_{(0)}) \leftharpoonup \\
&\qquad y^2 [(x^1 \cdot m_{(1)_{\underline{1}}})(x^2 \cdot \underline{S}(m_{(1)_{(2,1)}}))]_{(-1)} x^3 \cdot \underline{S}(m_{(1)_{(2,2)}}) \\
&\qquad \otimes y^3 \cdot [(x^1 \cdot m_{(1)_{\underline{1}}})(x^2 \cdot \underline{S}(m_{(1)_{(2,1)}}))]_{(0)} \\
(84, 74, 67) \quad &= \sum (y^1 \cdot m_{(0)}) \leftharpoonup (y^2 \cdot \underline{S}(m_{(1)})) \otimes y^3 \cdot 1_B = P(m) \otimes 1_B.
\end{aligned}
$$

(iii) For all $n \in N$ and $b \in B$, we compute, using (83),

$$
\begin{aligned}
\rho_M(n \leftharpoonup b) &= \sum (y^1 X^1 \cdot n) \leftharpoonup [y^2 Y^1 (x^1 X^2 \cdot 1_B)_{(-1)} x^2 X_1^3 \cdot b_{\underline{1}}] \\
&\qquad \otimes [y_1^3 Y^2 \cdot (x^1 X^2 \cdot 1_B)_{(0)}][y_2^3 Y^3 x^3 X_2^3 \cdot b_{\underline{2}}] \\
(67, 70) \quad &= \sum (y^1 \cdot n) \leftharpoonup (y^2 \cdot b_{\underline{1}}) \otimes y^3 \cdot b_{\underline{2}}.
\end{aligned}
$$

For all $n \in M^{coB}$, we find

$$
\begin{aligned}
P(n \leftharpoonup b) &= \sum [(y^1 \cdot n) \leftharpoonup (y^2 \cdot b_{\underline{1}})] \leftharpoonup \underline{S}(y^3 \cdot b_{\underline{2}}) \\
(71, 84) \quad &= \sum n \leftharpoonup b_{\underline{1}} \underline{S}(b_{\underline{2}}) = \underline{\varepsilon}(b) n \leftharpoonup 1_B = \underline{\varepsilon}(b) n.
\end{aligned}
$$

(iv) By (i) and Lemma 5.2, we obtain that $M^{coB} \otimes B \in \mathcal{M}_B^B$. It follows from (72) that F is left H-linear. It also intertwines the corresponding left H-coaction by (47) and (73). Now we will prove that F and G are inverses. For all $m \in M$, we have

$$
\begin{aligned}
FG(m) &= \sum P(m_{(0)}) \leftharpoonup m_{(1)} \\
(71) \quad &= \sum (X^1 \cdot m_{(0,0)}) \leftharpoonup [(X^2 \cdot \underline{S}(m_{(0,1)}))(X^3 \cdot m_{(1)})] \\
(84, 78, 79) \quad &= \sum m_{(0)} \leftharpoonup \underline{S}(m_{(1)_{\underline{1}}}) m_{(1)_{\underline{2}}} = m \leftharpoonup 1_B = m.
\end{aligned}
$$

Similarly, for any $n \in M^{coB}$ and $b \in B$, we compute

$$
\begin{aligned}
GF(n \otimes b) &= \sum P((n \leftharpoonup b)_{(0)}) \otimes (n \leftharpoonup b)_{(1)} \\
(iii) \quad &= \sum P((x^1 \cdot n) \leftharpoonup (x^2 \cdot b_{\underline{1}})) \otimes x^3 \cdot b_{\underline{2}} \\
(iii), (75) \quad &= \sum P(n) \otimes b = n \otimes b.
\end{aligned}
$$

We are left to show that F is a morphism in $\mathcal{M}_B^B$. It is not hard to see that (86) and (71) imply that F is right B-linear. Also, (iii) implies that

$$\rho_M \circ F(n \otimes b) = (F \otimes id_B) \circ \rho_{M^{c \circ B} \otimes B}(n \otimes b) = \sum (x^1 \cdot n) \leftarrow (x^2 \cdot b_{\underline{1}}) \otimes x^3 \cdot b_{\underline{2}},$$

for all $n \in N$ and $b \in B$, and this finishes the proof. $\qquad\square$

Let H be a quasi-Hopf algebra, and let ${}_H^H \mathcal{YD}^{\text{fd}}$ be the category of finite dimensional left Yetter-Drinfeld modules over H. If $M \in {}_H^H \mathcal{YD}^{\text{fd}}$, then $M^* \in {}_H^H \mathcal{YD}^{\text{fd}}$ (cf. [4]). The action and coaction are given by

$$(h \cdot m^*)(m) = m^*(S(h) \cdot m) \tag{88}$$

$$\lambda_{M^*}(m^*) = \sum m^*_{(-1)} \otimes m^*_{(0)} = \sum_{i=1}^{n} \langle m^*, f^2 \cdot (g^1 \cdot {}_i m)_{(0)} \rangle$$
$$S^{-1}(f^1 (g^1 \cdot {}_i m)_{(-1)} g^2) \otimes {}^i m \tag{89}$$

for all $h \in H$, $m^* \in M^*$, $m \in M$. Here $f = \sum f^1 \otimes f^2$ is the twist defined in (15), $({}_i m)_{i=\overline{1,n}}$ is a basis of M and $({}^i m)_{i=\overline{1,n}}$ its dual basis. Moreover, ${}_H^H \mathcal{YD}^{f.d.}$ is a rigid monoidal category. For each object $M \in {}_H^H \mathcal{YD}^{\text{fd}}$, the evaluation and coevaluation maps (ev_M and $coev_M$, respectively) are given by (41).

In addition, if $B \in {}_H^H \mathcal{YD}^{\text{fd}}$ is a Hopf algebra, then B^* is a Hopf algebra in ${}_H^H \mathcal{YD}^{f.d.}$. The structure is the following.

- The multiplication and unit are given by

$$(\varphi * \psi)(b) = \langle \varphi, f^2 \tilde{q}_2^2 Y^3 S^{-1} (\tilde{q}^1 Y^1 (p^1 \cdot b_{\underline{2}})_{(-1)} p^2) \cdot b_{\underline{1}} \rangle$$
$$\langle \psi, f^1 \tilde{q}_1^2 Y^2 \cdot (p^1 \cdot b_{\underline{2}})_{(0)} \rangle, \tag{90}$$

$$1_{B^*} = \underline{\varepsilon} \tag{91}$$

for all $\varphi, \psi \in B^*$, $b \in B$, where $q_L = \sum \tilde{q}^1 \otimes \tilde{q}^2$ and $p_R = \sum p^1 \otimes p^2$ are the elements defined in (21) and (19).

- the comultiplication and counit are given by the formulas

$$\underline{\Delta}_{B^*}(\varphi) = \sum_{i,j=1}^{n} \langle \varphi, [(g^1 \cdot {}_j b)_{(-1)} g^2 \cdot {}_i b] (g^1 \cdot {}_j b)_{(0)} \rangle {}^i b \otimes {}^j b \tag{92}$$

$$\underline{\varepsilon}_{B^*}(\varphi) = \varphi(1_B), \tag{93}$$

for any $\varphi \in B^*$, where $f^{-1} = \sum g^1 \otimes g^2$ was defined in (16), $({}_i b)_{i=\overline{1,n}}$ is a basis of B and $({}^i b)_{i=\overline{1,n}}$ the corresponding dual basis of B^*.

- the antipode is given by

$$\underline{S}_{B^*} = \underline{S}^*, \quad \text{i. e.} \quad \underline{S}_{B^*}(\varphi) = \varphi \circ \underline{S}, \tag{94}$$

for all $\varphi \in B^*$.

PROPOSITION 5.4. *Let* $B \in {}_H^H \mathcal{YD}^{\text{fd}}$ *a Hopf algebra. Then* B^* *is a right B-Hopf module, with structure:*

$$\langle \varphi \leftarrow b, b' \rangle = \sum \langle \varphi, [(U^1 \cdot b)_{(-1)} U^2 \cdot b'] \underline{S}((U^1 \cdot b)_{(0)}) \rangle, \tag{95}$$

YETTER-DRINFELD MODULES OVER QUASI-HOPF ALGEBRAS

$$\underline{\rho}_{B^*}(\varphi) = \sum_{i=1}^n (S(\tilde{p}^1) \cdot_i b)_{(-1)} \cdot [{}^i b * (\tilde{p}^2 \cdot \varphi)] \otimes (S(\tilde{p}^1) \cdot_i b)_{(0)}, \tag{96}$$

for all $\varphi \in B^*$, $b, b' \in B$, where

$$U = \sum U^1 \otimes U^2 := \sum g^1 S(q^2) \otimes g^2 S(q^1), \tag{97}$$

$p_L = \sum \tilde{p}^1 \otimes \tilde{p}^2$, $q_R = \sum q^1 \otimes q^2$ and $f^{-1} = \sum g^1 \otimes g^2$ are the elements defined by (21), (19) and (16), and $\{{}_i b\}_{i=\overline{1,n}}$ is a basis of B with corresponding dual basis $\{{}^i b\}_{i=\overline{1,n}}$. Moreover,

$$B^{*coB} = \{\Lambda \in B^* \mid \sum (\tilde{p}^1 \cdot \varphi) * (\tilde{p}^2 \cdot \Lambda) = \varphi(1_B)\Lambda \text{ for all } \varphi \in B^*\}.$$

Proof. If B is a Hopf algebra in a braided rigid monoidal category $\mathcal{C}$, then B^* is a right Hopf B-module, as follows.

- the right B-module structure $\leftharpoonup: B^* \otimes B \to B$ on B^* is the composition

$$
\begin{array}{ll}
B^* \otimes B & \xrightarrow{\ l_{B^* \otimes B}\ } & (B^* \otimes B) \otimes \underline{1} \\
\xrightarrow{\ (id_{B^*} \otimes \underline{S}) \otimes coev_B\ } & (B^* \otimes B) \otimes (B \otimes B^*) \\
\xrightarrow{\ a_{B^* \otimes B, B, B^*}^{-1}\ } & ((B^* \otimes B) \otimes B) \otimes B^* \\
\xrightarrow{\ a_{B^*, B, B} \otimes id_{B^*}\ } & (B^* \otimes (B \otimes B)) \otimes B^* \\
\xrightarrow{\ (id_{B^*} \otimes \underline{m}) \otimes id_{B^*}\ } & (B^* \otimes B) \otimes B^* \\
\xrightarrow{\ ev_B \otimes id_{B^*}\ } & \underline{1} \otimes B^* \\
\xrightarrow{\ r_{B^*}^{-1}\ } & B^*
\end{array} \tag{98}
$$

- the right B-comodule structure $\rho_{B^*} : B^* \to B^* \otimes B$ on B^* is the composition

$$
\begin{array}{ll}
B^* & \xrightarrow{\ r_{B^*}\ } & \underline{1} \otimes B^* & \xrightarrow{\ coev_B \otimes id_{B^*}\ } & (B \otimes B^*) \otimes B^* \\
\xrightarrow{\ a_{B, B^*, B^*}\ } & B \otimes (B^* \otimes B^*) & \xrightarrow{\ id_B \otimes m_{B^*}\ } & B \otimes B^* \\
\xrightarrow{\ c_{B, B^*}\ } & B^* \otimes B.
\end{array} \tag{99}
$$

Let $\gamma = \sum \gamma^1 \otimes \gamma^2$ and $f^{-1} = \sum g^1 \otimes g^2$ be the elements defined in (14) and (16). By (98), we have, for all $\varphi \in B^*$ and $b, b' \in B$:

$$
\begin{aligned}
\langle \varphi &\leftharpoonup b, b' \rangle \\
&= \sum \langle \varphi, S(X^1 p_1^1)\alpha \cdot [((X^2 p_2^1 \cdot \underline{S}(b))_{(-1)} X^3 p^2 \cdot b')(X^2 p_2^1 \cdot \underline{S}(b))_{(0)}] \rangle \\
(67, 45) \quad &= \sum \langle \varphi, [(S(X^1 p_1^1)_1 \alpha_1 X^2 p_2^1 \cdot \underline{S}(b))_{(-1)} S(X^1 p_1^1)_2 \alpha_2 X^3 p^2 \cdot b'] \\
&\quad (S(X^1 p_1^1)_1 \alpha_1 X^2 p_2^1 \cdot \underline{S}(b))_{(0)} \rangle
\end{aligned}
$$

$$(17,11) \quad = \quad \sum \langle \varphi, [(g^1 S(X_2^1 p_{(1,2)}^1) \gamma^1 X^2 p_2^1 \cdot \underline{S}(b))_{(-1)} g^2 S(X_1^1 p_{(1,1)}^1) \gamma^2 X^3 p^2 \cdot b')]$$
$$(g^1 S(X_2^1 p_{(1,2)}^1) \gamma^1 X^2 p_2^1 \cdot \underline{S}(b))_{(0)} \rangle$$

$$(14,3,5) \quad = \quad \sum \langle \varphi, g^1 S(Y^2 p_{(1,2)}^1) \alpha Y^3 p_2^1 \cdot \underline{S}(b))_{(-1)} g^2 S(Y^1 p_{(1,1)}^1) \alpha p^2 \cdot b']$$
$$(g^1 S(Y^2 p_{(1,2)}^1) \alpha Y^3 p_2^1 \cdot \underline{S}(b))_{(0)} \rangle$$

$$(1,5,6) \quad = \quad \sum \langle \varphi, [(g^1 S(Y^2) \alpha Y^3 \cdot \underline{S}(b))_{(-1)} g^2 S(Y^1) \cdot b']$$
$$(g^1 S(Y^2) \alpha Y^3 \cdot \underline{S}(b))_{(0)} \rangle$$

$$(19,97,84) \quad = \quad \sum \langle \varphi, [\underline{S}(U^1 \cdot b)_{(-1)} U^2 \cdot b'] \underline{S}(U^1 \cdot b)_{(0)} \rangle$$

$$(84) \quad = \quad \sum \langle \varphi, [(U^1 \cdot b)_{(-1)} U^2 \cdot b'] \underline{S}((U^1 \cdot b)_{(0)}) \rangle$$

which is just (95). (96) follows easily by (99), the details are left to the reader. Finally, by (99) we have

$$\Lambda \in B^{*co(B)} \iff \varrho_{B^*}(\Lambda) = \Lambda \otimes 1_B$$
$$\iff c_{B,B^*}^{-1} \circ \varrho_{B^*}(\Lambda) = c_{B,B^*}^{-1}(\Lambda \otimes 1_B)$$
$$\iff \sum_{i=1}^n S(\tilde{p}^1) \cdot {}_i b \otimes {}^i b * (\tilde{p}^2 \cdot \Lambda) = 1_B \otimes \Lambda$$
$$\iff \sum (\tilde{p}^1 \cdot \varphi) * (\tilde{p}^2 \cdot \Lambda) = \varphi(1_B)\Lambda, \quad \text{for all } \varphi \in B^*.$$

$\square$

We define the space of left integrals by $I_l(B^*) = B^{*co(B)}$. From the Fundamental Theorem for Hopf modules, we then obtain.

COROLLARY 5.5. *Let H be a quasi-Hopf algebra and B a finite dimensional Hopf algebra in ${}_H^H \mathcal{YD}$. Then $I_l(B^*) \otimes B \simeq B^*$ as right B-Hopf modules. In particular, $dim_k(I_l(B^*)) = 1$.*

Now, let H be a quasi-Hopf algebra and H_0 the H-module algebra described in Section 4. If (H, R) is quasitriangular, then H_0 is a Hopf algebra in ${}_H^H \mathcal{YD}$, see [5]. The additional structure is the following.

$$\lambda_{H_0}(h) = \sum R^2 \otimes R^1 \triangleright h, \tag{100}$$

$$\underline{\Delta}(h) = \sum h_{\underline{1}} \otimes h_{\underline{2}} \tag{101}$$

$$= \sum x^1 X^1 h_1 g^1 S(x^2 R^2 y^3 X_2^3) \otimes x^3 R^1 \triangleright y^1 X^2 h_2 g^2 S(y^2 X_1^3), \tag{102}$$

$$\underline{\varepsilon}(h) = \varepsilon(h), \tag{103}$$

$$\underline{S}(h) = \sum X^1 R^2 p^2 S(q^1(X^2 R^1 p^1 \triangleright h)S(q^2)X^3), \tag{104}$$

for all $h \in H$, where $R = \sum R^1 \otimes R^2$ and $f^{-1} = \sum g^1 \otimes g^2$, $p_R = \sum p^1 \otimes p^2$ and $q_R = \sum q^1 \otimes q^2$ are the elements defined by (16), (19) and (20). By the above arguments, if H is a finite dimensional Hopf algebra, then H_0^* is also a Hopf algebra

YETTER-DRINFELD MODULES OVER QUASI-HOPF ALGEBRAS

in $_H^H\mathcal{YD}$, with structure

$$(\varphi * \Psi)(h) = \sum \langle \varphi, f^2 \overline{R}^2 \triangleright h_{\underline{1}} \rangle \langle \Psi, f^1 \overline{R}^1 \triangleright h_{\underline{2}} \rangle \tag{105}$$

$$= \sum \langle \varphi, f^2 \triangleright Y^2 \overline{R}^2 X^1 x_1^1 h_1 g^1 S(Y^3 x^3) \rangle$$

$$\langle \Psi, f^1 Y^1 \overline{R}^1 \triangleright X^2 x_2^1 h_2 g^2 S(X^3 x^2) \rangle, \tag{106}$$

$$1_{H_0^*} = \underline{\varepsilon}, \tag{107}$$

$$\underline{\Delta}_{H_0^*}(\varphi) = \sum_{i,j=1}^n \langle \varphi, (R^2 g^2 \triangleright {}_i e)(R^1 g^1 \triangleright {}_j e) \rangle {}^i e \otimes {}^j e, \tag{108}$$

$$\underline{\varepsilon}_{H_0^*}(\varphi) = \varphi(\beta), \tag{109}$$

$$\underline{S}_{H_0^*}(\varphi) = v \circ \underline{S}, \tag{110}$$

for all $h \in H$ and $\varphi \in H^*$, where $R^{-1} = \sum \overline{R}^1 \otimes \overline{R}^2$, $\{{}_ie\}_{i=\overline{1,n}}$ is a basis of H and $\{{}^ie\}_{i=\overline{1,n}}$ the corresponding dual basis of H^*. In this particular case we have

$$I_l(H_0^*) = \{\Lambda \in H^* \mid \sum \Lambda(S(\tilde{p}^2) f^1 \overline{R}^1 \triangleright h_{\underline{2}}) S(\tilde{p}^1) f^2 \overline{R}^2 \triangleright h_{\underline{1}} = \Lambda(b)\beta, \text{ for all } h \in H\}.$$

REFERENCES

[1] D. Altschuler and A. Coste, Quasi-quantum groups, knots, three-manifolds, and topological field theory, *Comm. Math. Phys.* **150** (1992), 83–107.

[2] N. Andruskiewitsh and M. Graña, Braided Hopf algebras over abelian finite groups, *Bol. Acad. Ciencias (Còrdoba)* **63** (1999), 45–78.

[3] Y. Bespalov, T. Kerler and V. Lyubashenko, Integrals for braided Hopf algebras, *J. Pure Appl. Algebra* **148** (2000), 113–164.

[4] D. Bulacu, S. Caenepeel and F. Panaite, Yetter-Drinfeld categories over quasi-Hopf algebras, in preparation.

[5] D. Bulacu and E. Nauwelaerts, Radford's biproduct for quasi-Hopf algebras and bosonization, *J. Pure Appl. Algebra* **174** (2002), 1–42.

[6] D. Bulacu and E. Nauwelaerts, Relative Hopf modules for (dual) quasi-Hopf algebras, *J. Algebra* **229** (2000), 632–659.

[7] D. Bulacu, F. Panaite and F. Van Oystaeyen, Quasi-Hopf algebra actions and smash products, *Comm. Algebra* **28** (2000), 631–651.

[8] D. Bulacu and S. Caenepeel, The quantum double for quasitriangular quasi-Hopf algebras, *Comm. Algebra* **31** (2003), 1403–1425.

[9] D. Bulacu and S. Caenepeel, Integrals for (dual) quasi-Hopf algebras. Applications, *J. Algebra* **266** (2003), 552–583.

[10] D. Bulacu and E. Nauwelaerts, Quasitriangular and ribbon quasi-Hopf algebras, *Comm. Algebra* **31** (2003), 1–16.

[11] S. Caenepeel, F. Van Oystaeyen and Y. H. Zhang, Quantum Yang-Baxter module algebras, *K-Theory* **8** (1994), 231–255.

[12] Y. Doi, Hopf Modules in Yetter-Drinfeld categories, *Comm. Algebra* **26** (1998), 3057–3070.

[13] V. G. Drinfeld, Quasi-Hopf algebras, *Leningrad Math. J.* **1** (1990), 1419–1457.

[14] F. Hausser and F. Nill, Diagonal crossed products by duals of quasi-quantum groups, *Rev. Math. Phys.* **11** (1999), 553–629.

[15] F. Hausser and F. Nill, Doubles of quasi-quantum groups, *Comm. Math. Phys.* **199** (1999), 547–589.

[16] C. Kassel, "Quantum Groups", *Graduate Texts Math.* **155**, Springer Verlag, Berlin, 1995.

[17] L. A. Lambe and D. E. Radford, Algebraic aspects of the quantum Yang-Baxter equation, *J. Algebra* **154** (1992), 228–288.

112   D. BULACU, S.CAENEPEEL AND F. PANAITE

[18] S. Mac Lane, Categories for the working mathematician, second edition, *Graduate Texts Math.* **5**, Springer Verlag, Berlin, 1997.
[19] S. Majid, Quantum double for quasi-Hopf algebras, *Lett. Math. Phys.* **45** (1998), 1–9.
[20] S. Majid, "Foundations of quantum group theory", Cambridge Univ. Press, Cambridge, 1995.
[21] S. Majid, Algebras and Hopf algebras in braided categories, in "Advances in Hopf Algebras", *Lect. Notes Pure Appl. Math.* **158**, Dekker, New York, 1994, 55–105.
[22] P. Schauenburg, Hopf modules and the double of a quasi-Hopf algebra, preprint 2002.
[23] M. E. Sweedler, "Hopf algebras", Benjamin, New York, 1969.
[24] M. Takeuchi, Finite Hopf algebras in braided tensor categories, *J. Pure Appl. Algebra* **138** (1999), 59–82.

Rationality Properties for Morita Contexts associated to Corings

STEFAAN CAENEPEEL AND JOOST VERCRUYSSE
Faculty of Applied Sciences, Vrije Universiteit Brussel, VUB
Pleinlaan 2, B-1050 Brussels, Belgium
e-mail: *scaenepe@vub.ac.be, joost.vercruysse@vub.ac.be*

SHUANHONG WANG   Department of Mathematics
Henan Normal University, Henan, Xinxiang 453002, China
e-mail: *shuanhwang2002@yahoo.com*

> ABSTRACT. Given an A-coring C with a fixed grouplike element, we can construct a Morita context connecting the dual of the coring with the ring of coinvariants of A. In this paper, we discuss the image of one of the two connecting maps, and show that it is contained in the rational part of the dual of the coring, at least if the coring is locally projective. We apply our result to entwined modules, and this leads to the introduction of factorizable entwined modules.

1. INTRODUCTION

During the past decades, several variations of the notion of Hopf module have been proposed, with applications in various directions, for example relative Hopf modules (in connection with Hopf Galois theory), Yetter-Drinfeld modules (in connection with quantum groups), Long dimodules (in connection with the Brauer group). Doi [11] and Koppinen [16] gave a unification of all these types of modules, nowadays usually called Doi-Hopf modules, and their construction was generalized later by Brzeziński and Majid [6], who introduced entwined modules. In a mathematical review [19], Takeuchi observed that entwined modules and all their special cases can be considered as comodules over a coring, a notion that goes back to Sweedler [18]. This idea has been worked out by Brzeziński in [5], see also [15], [21], [22], and it turned out that many interesting properties of Hopf modules and their generalizations can be at the same time generalized and reformulated more elegantly using the language of corings. For example, Galois corings generalize Hopf Galois extensions, and they can be introduced in such a way that the connection with descent theory is clarified (see [22] or [7, Sec. 4.8]).

Hopf Galois extensions were first considered by Chase and Sweedler [10], and their

---

2000 *Mathematics Subject Classification.* 16W30.

Key words and phrases. Galois coring, Morita context, rational module, factorizable entwined module, co-Frobenius coring.

Research supported by the bilateral project "Hopf Algebras in Algebra, Topology, Geometry and Physics" of the Flemish and Chinese governments.

work reveals that there is a close connection with Morita theory: to a comodule algebra, they associate a Morita context that is strict in case the comodule algebra in question is a Hopf Galois extension. This Morita context has been generalized by Doi [12]. In [8] and, independently in [1], it was discussed how Doi's Morita context can be generalized to corings. Of course one then wants to investigate when this context is strict. For one of the two connecting maps, there is no problem to find necessary and sufficient conditions for its surjectivity (see [8]), but for the other one, having values in the dual coring, a satisfactory answer can be given only in the situation where the coring is finitely generated and projective over the groundring A. In fact we will prove in this paper that surjectivity of this second map μ implies that the coring $\mathcal{C}$ is finitely generated projective (Corollary 5.2). The aim of this paper is to show that - under the condition that $\mathcal{C}$ is locally projective, the image of the map μ is contained in the rational part of the dual of $\mathcal{C}$ (Proposition 5.1). If $\mathcal{C}$ satisfies the Weak or Strong Structure Theorem, then μ is surjective onto the rational part.

This paper is organized as follows. In Section 2, we recall some generalities about corings, entwined modules and Doi-Hopf modules. In Section 3, we introduce a relative version of local projectivity, and relate it to the so-called α-condition. In Section 4, we discuss rationality properties of modules over a suitable subring $\mathcal{R}$ of the dual of a coring; in case the coring is $\mathcal{R}$-locally projective, we can introduce the $\mathcal{R}$-rational part of such a module. Sections 3 and 4 provide the necessary machinery to state and prove our main results in Section 5. We apply our results to entwined modules in Section 6, and this leads us to the introduction of what we called factorizable entwining structures. In Section 7, we prove that there is an injective map between the two connecting modules in the Morita context, under the condition that the coring $\mathcal{C}$ is co-Frobenius, which means that there exists a $^{*}\mathcal{C}$-linear map from $\mathcal{C}$ to its left dual.

2. PRELIMINARY RESULTS

Corings

Let A be a ring. Recall that an A-coring $\mathcal{C}$ is an A-bimodule together with two A-bilinear maps $\Delta_{\mathcal{C}} : \mathcal{C} \to \mathcal{C} \otimes_A \mathcal{C}$, $\Delta_{\mathcal{C}}(c) = c_{(1)} \otimes_A c_{(2)}$ and $\varepsilon_{\mathcal{C}} : \mathcal{C} \to A$, such that

$$\Delta_{\mathcal{C}}(c_{(1)}) \otimes_A c_{(2)} = c_{(1)} \otimes_A \Delta(c_{(2)}) \text{ and } c_{(1)}\varepsilon_{\mathcal{C}}(c_{(2)}) = \varepsilon_{\mathcal{C}}(c_{(1)})c_{(2)} = c$$

for all $c \in \mathcal{C}$. A right $\mathcal{C}$-comodule M is a right A-module together with a map $\rho^r : M \to M \otimes_A \mathcal{C}$, $\rho^r(m) = m_{[0]} \otimes m_{[1]}$ such that

$$\rho^r(m_{[0]}) \otimes_A m_{[1]} = m_{[0]} \otimes_A \Delta_{\mathcal{C}}(m_{[1]})$$

and

$$m_{[0]}\varepsilon_{\mathcal{C}}(m_{[1]}) = m$$

for all $m \in M$. ρ^r is called a right $\mathcal{C}$-coaction on M. Left $\mathcal{C}$-comodules are introduced in a similar way. A right A-linear map $f : M \to N$ between two right $\mathcal{C}$-comodules is called right $\mathcal{C}$-colinear if f preserves the coaction, that is

$$f(m_{[0]}) \otimes_A m_{[1]} = f(m)_{[0]} \otimes_A f(m)_{[1]}$$

for all $m \in M$. The category of right C-comodules and C-colinear maps is denoted by $\mathcal{M}^C$.

The left dual $^*C = {}_A\mathrm{Hom}(C, A)$ of a coring C is a ring with multiplication

$$(f \# g)(c) = g(c_{(1)} f(c_{(2)}))$$

and unit ε_C. We have a ring homomorphism $i : A \to {}^*C$, $i(a)(c) = \varepsilon_C(c)a$. It is easy to verify that

$$(i(a) \# f)(c) = f(ca) \qquad \text{and} \qquad (f \# i(a))(c) = f(c)a$$

for all $a \in A$, $f \in {}^*C$ and $c \in C$.

An element $x \in C$ is called grouplike if

$$\Delta_C(x) = x \otimes_A x \text{ and } \varepsilon_C(x) = 1. \tag{1}$$

The set of grouplike elements of C is denoted by $G(C)$. Grouplike elements correspond bijectively to right (or left) C-coactions on A:

$$\begin{aligned} G(C) &\cong \{\rho^r : A \to A \otimes_A C \cong C \mid \rho^r \text{ makes } A \text{ into a right } C\text{-comodule}\} \\ &\cong \{\rho^\ell : A \to C \otimes_A A \cong C \mid \rho^\ell \text{ makes } A \text{ into a left } C\text{-comodule}\} \end{aligned}$$

The right and left coactions ρ^r and ρ^ℓ corresponding to $x \in G(C)$ are given by the formulas

$$\rho^r(a) = xa \text{ and } \rho^\ell(a) = ax$$

(where we identify $C \otimes_A A$ and C).

If $G(C) \neq \emptyset$, then $i : A \to {}^*C$ is injective, since for every grouplike element x, the map $\chi : {}^*C \to A$, $\chi(f) = f(x)$ is a left inverse of i. In this case, ε_C is surjective, with right inverse ρ^r (or ρ^ℓ).

For a right C-comodule M, we define the set of coinvariants as follows:

$$M^{coC} = \{m \in M \mid \rho(m) = m \otimes_A x\} = \mathrm{Hom}^C(A, M).$$

$B = A^{coC} = \{b \in A \mid bx = xb\}$ is a subring of A and it is easy to see that M^{coC} is a B-submodule of M. C is a right C-comodule (the coaction is the comultiplication), and $C^{coC} \cong A$ as a right B-module: ρ^ℓ is a right B-linear map from A to C^{coC}, with inverse ε_C.

Recall that we have a functor $F : \mathcal{M}^C \to \mathcal{M}_{{}^*C}$, $F(M) = M$ with $m \cdot f = m_{[0]} f(m_{[1]})$. For every $M \in \mathcal{M}_{{}^*C}$, we define

$$M^{{}^*C} = \{m \in M \mid m \cdot f = mf(x), \text{ for all } f \in {}^*C\}.$$

In particular, $A^{{}^*C} = B'$ is a subring of A. Obviously $M^{coC} \subset M^{{}^*C}$.

A Morita context associated to a coring

Let (C, x) be a coring with a fixed grouplike element. Following [8], we define

$$Q = \{q \in {}^*C \mid c_{(1)} q(c_{(2)}) = q(c)x, \text{ for all } c \in C\} = {}^C\mathrm{Hom}(C, A)$$

and

$$Q \subset Q' = ({}^*C)^{{}^*C}.$$

Q is a $({}^*C, B)$-bimodule, and A is a $(B, {}^*C)$-bimodule, with right $^*C$-action $a \leftharpoonup f = f(xa)$. Consider the bimodule maps

$$\mu : Q \otimes_B A \to {}^*C, \ \mu(q \otimes_B a) = q \# a$$

$$\tau: A \otimes_{\cdot C} Q \to B, \ \tau(a \otimes_{\cdot C} q) = a \leftharpoonup q = q(xa)$$

Then $(B, {}^*C, A, Q, \tau, \mu)$ is a Morita context. In a similar way, Q' is a $({}^*C, B')$-bimodule, A is a $(B', {}^*C)$-bimodule, and we can consider the maps

$$\mu': \ Q' \otimes_{B'} A \to {}^*C \text{ and } \tau': \ A \otimes_{\cdot C} Q' \to B'$$

defined in the same way as μ and τ. Then $(B', {}^*C, A, Q', \tau', \mu')$ is also a Morita context, and it was shown in [8] that the two contexts coincide if C is finitely generated projective as a left A-module. From [8] we also recall the following result.

THEOREM 2.1. *With notation as above, the following statements are equivalent:*

1. *τ is surjective (and, a fortiori, bijective);*
2. *there exists $\Lambda \in Q$ such that $\Lambda(x) = 1$;*
3. *for every right C-comodule M, the map*

$$\omega_M: \ M \otimes_{\cdot C} Q \to M^{coC}, \quad \omega_M(m \otimes_{\cdot C} q) = m \cdot q$$

   is bijective.

Consider a right A-submodule $\mathcal{R}$ of ${}^*C$ that is closed under multiplication, and let $\widetilde{Q} = Q \cap \mathcal{R}$. Then we have a Morita context $(B, \mathcal{R}, A, \widetilde{Q}, \widetilde{\mu}, \widetilde{\tau})$ with $\widetilde{\mu}$ and $\widetilde{\tau}$ the obvious restrictions of the original maps μ and τ. It is possible that $\mathcal{R}$ has no unit. For Morita contexts between rings without unit, see [3].

Galois corings

Let $(\mathcal{C}, x)$ be an A-coring with a fixed grouplike element. Let

$$B = A^{coC} = \{b \in A \mid bx = xb\}.$$

We have a pair of adjoint functors (F, G) between the categories $\mathcal{M}_B$ and $\mathcal{M}^C$, namely, for $N \in \mathcal{M}_B$ and $M \in \mathcal{M}^C$,

$$F(N) = N \otimes_B A \text{ and } G(M) = M^{coC}.$$

The unit and counit of the adjunction are

$$\eta_N: \ N \to (N \otimes_B A)^{coC}, \ \eta_N(n) = n \otimes_B 1$$
$$\varepsilon_M: \ M^{coC} \otimes_B A \to M, \ \varepsilon_M(m \otimes_B a) = ma$$

We say that $(\mathcal{C}, x)$ satisfies the *Weak Structure Theorem* if ε_M is an isomorphism for all $M \in \mathcal{M}^C$, that is, $G = \bullet^{coC}$ is a fully faithful functor. $(\mathcal{C}, x)$ satisfies the *Strong Structure Theorem* if, in addition, all η_N are isomorphisms, or F is fully faithful, and therefore (F, G) is an equivalence of categories.

The canonical coring associated to $i: \ B \to A$ is $\mathcal{D} = A \otimes_B A$, with structure maps

$$\Delta_{\mathcal{D}}: \ A \otimes_B A \to (A \otimes_B A) \otimes_A (A \otimes_B A) \cong A \otimes_B A \otimes_B A \text{ and } \varepsilon_{\mathcal{D}}: \ A \otimes_B A \to A$$

given by

$$\Delta_{\mathcal{D}}(a \otimes_B b) = (a \otimes_B 1) \otimes_A (1 \otimes_B b) = a \otimes_B 1 \otimes_B b$$
$$\varepsilon_{\mathcal{D}}(a \otimes_B b) = ab$$

$1 \otimes_B 1$ is a grouplike element. If A is faithfully flat as a B-module, then $(\mathcal{D}, 1 \otimes_B 1)$ satisfies the Strong Structure Theorem. We have a canonical coring morphism

$$\text{can}: \ \mathcal{D} \to \mathcal{C}; \ \text{can}(a \otimes_B b) = axb$$

We say that $(\mathcal{C}, x)$ is a *Galois coring* if can is an isomorphism of corings. In this situation, we obviously have an isomorphism between the categories $\mathcal{M}^{\mathcal{C}}$ and $\mathcal{M}^{\mathcal{D}}$, and if $(\mathcal{D}, 1 \otimes_B 1)$ satisfies the Strong, resp. Weak Structure Theorem (for example if A/B is faithfully flat, resp. A/B is flat), then $(\mathcal{C}, x)$ also satisfies the Strong resp. Weak Structure Theorem. If $(\mathcal{C}, x)$ satisfies the Weak Structure Theorem, then $(\mathcal{C}, x)$ is Galois (see [8, Proposition 1.1]).

Entwined modules

Let k be a commutative ring, A a k-algebra, C a k-coalgebra, and $\psi : C \otimes A \to A \otimes C$ a k-linear map satisfying the following four conditions:

$$(ab)_\psi \otimes c^\psi = a_\psi b_\Psi \otimes c^{\psi\Psi} \tag{2}$$

$$(1_A)_\psi \otimes c^\psi = 1_A \otimes c \tag{3}$$

$$a_\psi \otimes \Delta_C(c^\psi) = a_{\psi\Psi} \otimes c^\Psi_{(1)} \otimes c^\psi_{(2)} \tag{4}$$

$$\varepsilon_C(c^\psi)a_\psi = \varepsilon_C(c)a \tag{5}$$

Here we used the sigma notation

$$\psi(c \otimes a) = a_\psi \otimes c^\psi = a_\Psi \otimes c^\Psi$$

We then call (A, C, ψ) a (right-right) entwining structure. To an entwining structure (A, C, ψ), we can associate an A-coring $\mathcal{C} = A \otimes C$. The structure maps are given by the formulas

$$a'(b \otimes c)a = a'ba_\psi \otimes c^\psi$$
$$\Delta_{\mathcal{C}}(a \otimes c) = (a \otimes c_{(1)}) \otimes_A (1 \otimes c_{(2)})$$
$$\varepsilon_{\mathcal{C}}(a \otimes c) = a\varepsilon_C(c)$$

An entwined module M is a k-module together with a right A-action and a right C-coaction, in such a way that

$$\rho^r(ma) = m_{[0]}a_\psi \otimes m^\psi_{[1]}$$

for all $m \in M$ and $a \in A$. The category $\mathcal{M}(\psi)^C_A$ of entwined modules and A-linear C-colinear maps is isomorphic to the category of right C-comodules.

Doi-Hopf modules

Let H be a Hopf algebra, A a right H-comodule algebra, and C a right H-module coalgebra. We call (H, A, C) a right-right Doi-Hopf structure. We associate an entwining structure (A, C, ψ) to (H, A, C) as follows, with ψ defined by $\psi(c \otimes a) = a_{[0]} \otimes ca_{[1]}$. The corresponding entwined modules are called Doi-Hopf modules. They have to satisfy the compatibility relation

$$\rho(ma) = m_{[0]}a_{[0]} \otimes m_{[1]}a_{[1]}$$

for all $m \in M$ and $a \in A$.

Factorization structures and the smash product

Let A and S be k-algebras, and $R : S \otimes A \to A \otimes S$ a k-linear map. We will write

$$R(s \otimes a) = a_R \otimes s_R = a_r \otimes s_r$$

(summation understood). $A\#_R S$ will be the k-module $A \otimes S$, with multiplication

$$(a\#s)(b\#t) = ab_R\#s_R t \tag{6}$$

It is straightforward to verify that this multiplication is associative with unit $1_A\#1_S$ if and only if

$$\begin{aligned}
R(s \otimes 1_A) &= 1_A \otimes s \tag{7}\\
R(1_S \otimes a) &= a \otimes 1_S \tag{8}\\
R(st \otimes a) &= a_{Rr} \otimes s_r t_R \tag{9}\\
R(s \otimes ab) &= a_R b_r \otimes s_{Rr} \tag{10}
\end{aligned}$$

for all $a, b \in A$ and $s, t \in S$. We then call (A, S, R) a factorization structure, and $A\#_R S$ the smash product of A and S.

3. $\mathcal{R}$-LOCALLY PROJECTIVE MODULES

Let A be a ring and P a left A-module. Then $^*P = {}_A\mathrm{Hom}(P, A)$ is a right A-module, with action given by $(f \cdot a)(p) = f(p)a$, for $f \in {}^*P$, $p \in P$ and $a \in A$. Let $\mathcal{R}$ be an abelian group, and $\varphi : \mathcal{R} \to {}^*P$ a morphism of abelian groups. For $M \in \mathcal{M}_A$, we consider the map

$$\xi_{M,\mathcal{R}} : M \otimes_A P \to \mathrm{Hom}_{\mathbf{Z}}(\mathcal{R}, M), \ \xi_{M,\mathcal{R}}(m \otimes_A p)(r) = m \cdot \varphi(r)(p)$$

Since we will only be interested in $\mathcal{R}$ as acting on P, it will be no restriction to assume that φ is injective, and consider $\mathcal{R}$ as a subgroup of $^*P$.
If $\mathcal{R}$ is a right A-submodule of $^*P$, then $\mathrm{Im}\,(\xi_{M,\mathcal{R}}) \subset \mathrm{Hom}_A(\mathcal{R}, M)$ and $\xi_{M,\mathcal{R}}$ factors as the composition of a map

$$\xi'_{M,\mathcal{R}} : M \otimes_A P \to \mathrm{Hom}_A(\mathcal{R}, M)$$

followed by the natural inclusion $\mathrm{Hom}_A(\mathcal{R}, M) \subset \mathrm{Hom}_{\mathbf{Z}}(\mathcal{R}, M)$.

DEFINITION 3.1. Let $\mathcal{R}$ be an additive subgroup of $^*P$. We say that P satisfies the α-condition for $\mathcal{R}$ iff $\xi_{M,\mathcal{R}}$ is injective for all $M \in \mathcal{M}_A$. If P satisfies the α-condition for $^*P$ we just say that P satisfies the α-condition.

More specifically, the α-condition for $\mathcal{R}$ is equivalent to the following statement: if $\sum_i m_i f(p_i) = 0$ for all $f \in \mathcal{R}$, then $\sum_i m_i \otimes_A p_i = 0$.
If $\mathcal{S}$ is an abelian subgroup of $\mathcal{R}$, then obviously

$$\ker \xi_{M,\mathcal{R}} \subset \ker \xi_{M,\mathcal{S}}, \quad \mathrm{Im}\,\xi_{M,\mathcal{S}} \subset \mathrm{Im}\,\xi_{M,\mathcal{R}}$$

and the α-condition for $\mathcal{S}$ implies the α-condition for $\mathcal{R}$.

EXAMPLE 3.2 (extension of scalars). Let $B \to A$ be a ringmorphism, and $Q \in {}_B\mathcal{M}$. Then $P = A \otimes_B Q \in {}_A\mathcal{M}$, and we have a morphism of right B-modules

$$\gamma : {}^*Q = {}_B\mathrm{Hom}(Q, B) \to {}^*P = {}_A\mathrm{Hom}(P, A), \ \gamma(f)(a \otimes_B q) = af(q)$$

RATIONALITY FOR CORINGS

If Q satisfies the α-condition for ${}^*Q$, then P also satisfies the α-condition for ${}^*Q$ (since $M \otimes_A P \cong M \otimes_B Q$ for every right A-module M), and therefore P satisfies the α-condition for ${}^*P$ as well.

Recall that the *finite topology* on ${}^*P$ is the topology generated by the basis of open sets

$$\mathcal{O}(f, p_1, \ldots, p_n) = \{g \in {}^*P \mid g(p_i) = f(p_i), 1 \leq i \leq n\}$$

where $f \in {}^*P$ and $p_1, \ldots, p_n \in P$.

A subset $\mathcal{R} \subset {}^*P$ is dense with respect to this topology if and only if for every $f \in {}^*P$ and $p_1, \ldots, p_n \in P$, we can find a $g \in \mathcal{R}$ such that $g(p_i) = f(p_i)$, for $1 \leq i \leq n$.

Let N be a subset of P. $\{(e_i, f_i) \mid i \in I\} \subset P \times {}^*P$ is called a dual basis of N if

$$\#\{i \mid e_i^*(n) \neq 0\} < \infty \text{ and } n = \sum_i f_i(n)e_i$$

for every $n \in N$. P is called locally projective if every finite subset (or every finitely generated submodule) of P has a dual basis, and it is well-known that this is equivalent to P having the α-condition for ${}^*P$. If P is finitely generated as an A-module, or if A is left perfect, then local projectivity of P is equivalent to projectivity. We refer to [21] for a survey of known results, and to [14] and [23] for the proofs. In the sequel we will need the following generalization of some results of [14] and [23]; we recover the results of [14] and [23] if we take $\mathcal{R} = {}^*P$.

PROPOSITION 3.3. *Let P be a left A-module, $\mathcal{R} \subset {}^*P$ an additive subgroup and $S = \mathcal{R}A$ the right A-submodule of ${}^*P$ generated by $\mathcal{R}$. Then the following statements are equivalent :*

(i) *P satisfies the α-condition for $\mathcal{R}$;*
(i)' *P satisfies the α-condition for S;*
(ii) *$\xi_{M,\mathcal{R}}$ is injective for every cyclic right A-module M;*
(ii)' *$\xi_{M,S}$ is injective for every cyclic right A-module M;*
(iii) *For every $p \in P$, we have $p \in \mathcal{R}(p)P$;*
(iii)' *For every $p \in P$, we have $p \in S(p)P$;*
(iv) *Every finitely generated submodule N of P has a dual basis contained in $P \times \mathcal{R}$;*
(iv)' *Every finitely generated submodule N of P has a dual basis contained in $P \times S$;*
(v) *P satisfies the α-condition and S is dense with respect to the finite topology.*

If P satisfies any of these equivalent conditions, then we say that P is $\mathcal{R}$-locally projective as a left A-module.

Proof. $\underline{(i) \Rightarrow (i)'}$, $\underline{(i)' \Rightarrow (ii)'}$, $\underline{(i) \Rightarrow (ii)}$ and $\underline{(ii) \Rightarrow (ii)'}$ are trivial. $\underline{(ii)' \Rightarrow (iii)'}$.

$$S(p) = \{f(p) \mid f \in S\} = \left\{\sum_i f_i(p)a_i \mid f_i \in \mathcal{R}, a_i \in A\right\}$$

is a right ideal of A, and $A/S(p)$ is a cyclic right A-module. For all $f \in S$, $1f(p) = 0$ in $A/S(p)$, hence $1 \otimes_A p = p = 0$ in $A/S(p) \otimes_A P \cong P/S(p)P$, and $p \in S(p)P$.

$\underline{(iii)' \Rightarrow (iii)}$ We know that $p = \sum_i s_i(p)e_i$, with $s_i \in S$ and $e_i \in P$. Since S is generated by $\mathcal{R}$, we can write $s_i = \sum_{j_i} r_{j_i}a_{j_i}$ and we find $p = \sum_i \sum_{j_i} (r_{j_i}a_{j_i})(p)e_i = \sum_i \sum_{j_i} r_{j_i}(p)a_{j_i}e_i$ and so $p \in \mathcal{R}(p)P$.

$(iii) \Rightarrow (iv)$ We prove by induction on k that every finite set $\{n_1, \cdots, n_k\}$ has a finite dual basis. If $k = 1$, this follows immediately form (iii). Now suppose we have a dual basis $\{(e_i, r_i) \mid i \in I\}$ for $n_1, \ldots, n_{k-1}$, and consider then $n_k - \sum_i r_i(n_k)e_i \in P$ and take a dual basis $\{(e'_j, r'_j) \mid j \in J\}$ for this element. An easy calculation shows that $\{(e'_j, r'_j) \mid j \in J\} \cup \{(e_i - \sum_j r'_j(e_i)e'_j, r_i) \mid i \in I\}$ is a dual basis for $\{n_1, \cdots, n_k\}$.
$(iv) \Rightarrow (iv)'$ is trivial.
$(iv)' \Rightarrow (iv)$ is similar to $(iii)' \Rightarrow (iii)$.
$(iv) \Rightarrow (i)$ Let M be a right A-module, take $\sum_i m_i \otimes_A p_i \in M \otimes_A P$ and suppose that $\sum_i m_i f(p_i) = 0$ for all $f \in \mathcal{R}$. From (iv) we know we kan find a dual basis $\{e_j, r_j\}$ for the elements p_i.

$$\sum_i m_i \otimes_A p_i = \sum_{i,j} m_i \otimes_A r_j(p_i)e_j = \sum_{i,j} m_i r_j(p_i) \otimes_A e_j = 0$$

$(i) \Rightarrow (v)$. It follows from the comments following Definition 3.1 that P satisfies the α-condition. For every $f \in {}^*P$ and $p_1, \ldots, p_n \in P$, we have to find $g \in S$ such that $f(p_i) = g(p_i)$. From (iv)', we know that $\{p_1, \cdots, p_n\}$ has a dual basis $\{(e_j, s_j) \mid j \in J\} \subset P \times S$. We find

$$f(p_i) = f(\sum_j s_j(p_i)e_j) = \sum_j s_j(p_i)f(e_j) = \sum_j (s_j f(e_j))(p_i)$$

and our statement follows since $\sum_j s_j f(e_j) \in S$.
$(v) \Rightarrow (iv)'$ The α-condition for P, gives us a dual basis $\{(e_i, f_i) \mid i \in I\} \subset P \times {}^*P$ for N. Since S is dense in ${}^*P$, we can find elements $s_i \in S$, such that f_i and s_i have the same action on N. Then $\{(e_i, s_i) \mid i \in I\} \subset P \times S$ is the dual basis that we are looking for. $\qquad \square$

Remark 3.4. Note that the equivalent conditions of Proposition 3.3 do not imply that $\mathcal{R}$ is dense in the finite topology. If one takes $P = A \otimes_B Q$ as in Example 3.2, then ${}^*Q$ is never dense in the finite topology on ${}^*P$ if $B \to A$ is a proper ring extension.

Now let $P = \mathcal{C}$ be an A-coring, and fix $x \in G(\mathcal{C})$.

DEFINITION 3.5. For $M \in \mathcal{M}_{\bullet \mathcal{C}}$ and $\mathcal{R} \subset {}^*\mathcal{C}$, we define

$$M^{\mathcal{R}} = \{m \in M \mid m \cdot f = mf(x), \text{for all } f \in \mathcal{R}\}$$

PROPOSITION 3.6. *If $\mathcal{C}$ is $\mathcal{R}$-locally projective over A, then $M^{\mathcal{R}} = M^{co\mathcal{C}}$ for every $M \in \mathcal{M}^{\mathcal{C}}$, and $({}^*\mathcal{C})^{\mathcal{R}} = Q$.*

Proof. We have already seen that $M^{co\mathcal{C}} \subset M^{\bullet \mathcal{C}}$. The same argument shows that $M^{co\mathcal{C}} \subset M^{\mathcal{R}}$. Conversely, take $m \in M^{\mathcal{R}}$, and write $\rho(m) = \sum_{j=1}^n m_j \otimes_A c_j$. Take a dual basis $\{(e_i, f_i) \mid i \in I\} \subset \mathcal{C} \times \mathcal{R}$ of $\{c_1, \cdots, c_n, x\}$. Then

$$\begin{aligned}
\rho(m) &= \sum_{i,j} m_j \otimes_A f_i(c_j)e_i = \sum_{i,j} m_j f_i(c_j) \otimes_A e_i \\
&= \sum_i mf_i(x) \otimes_A e_i = \sum_i m \otimes_A f_i(x)e_i = m \otimes_A x
\end{aligned}$$

and $m \in M^{coC}$.

We still have to show that $(^*C)^R \subset Q$. If $q \in (^*C)^R$, then $q\#f = qf(x)$, for all $f \in R$. Take $c \in C$ and a dual basis $\{(e_i, f_i) \mid i \in I\} \subset C \times R$ for $c_{(1)}q(c_{(2)})$ and x. Then we find

$$
\begin{aligned}
c_{(1)}q(c_{(2)}) &= f_i(c_{(1)}q(c_{(2)}))e_i \\
&= q(c)f_i(x)e_i = q(c)x
\end{aligned}
$$

and it follows that $q \in Q$. $\qquad\square$

COROLLARY 3.7. *Let C be an A-coring. If C is locally projective as a left A-module, then $M^{*C} = M^{coC}$ for every $M \in \mathcal{M}^C$. In particular, $B = B'$. We also have $Q = Q'$, and the two Morita contexts of Section 2 coincide.*

Local units and local projectivity

DEFINITION 3.8. Let A be a ring without unit, $M \in \mathcal{M}_A$, and consider subsets $R \subset A$ and $N \subset M$. We say that $a \in A$ can be multiplicatively approximated from the right by R on N, if for every finite subset $\{n_1, \cdots, n_k\} \subset N$, there exists $r \in R$ such that $n_i \cdot a = n_i \cdot r$, for all $i \in \{1, \cdots, k\}$.
If every $a \in A$ can be multiplicatively approximated by R on N, then we say R is a right multiplicative approximation of A on N.

If A has a unit, then R has right local units if and only if 1_A can be multiplicatively approximated from the right by R on R. Furthermore, we have the following result.

PROPOSITION 3.9. *Let A be a ring with unit, and consider an additive subset $R \subseteq A$. Then the following assertions are equivalent:*

  (i) *R is a multiplicative approximation from the right of A on R;*
  (ii) *R is a right ideal of A and has right local units.*

If these conditions hold, then R is a multiplicative approximation of A on every right A-module M satisfying $MR = M$.

Proof. (i) $\Rightarrow$ (ii) For every finite set of elements $\{r_1, \cdots, r_n\} \subset R \subset A$ and $a \in A$, we have an element $r \in R$ such that $r_i a = r_i r \in R$. This means R is a right ideal, and taking $a = 1$ we find that R has right local units.
(ii) $\Rightarrow$ (i) R has right local units, so for all $r_1, \cdots, r_n \in R$, we can find $e \in R$ such that $r_i e = r_i$. For every $a \in A$, we then have $r_i ea = r_i a$, and $ea \in R$, since R is a right ideal. The proof of the final statement is similar. $\qquad\square$

These approximation properties have the same flavour as the density properties that we encountered when we discussed R-relative local projectivity. So the question arises wether they are related. To be able to define a finite topology, first remark that the rings we used before now need to be duals of modules. So it is very natural to look at duals of corings, since they have both a finite topology and a ring structure.

PROPOSITION 3.10. *Let C be an A-coring, and $R \subset {}^*C$ a subring.*

  1. *If R is dense in the finite topology on $^*C$, then R is a multiplicative approximation of $^*C$ from the right on every C-comodule (regarded as a $^*C$-module). In particular, $MR = M$ for every $M \in \mathcal{M}^C$.*

2. *If $\mathcal{R}$ is a right ideal in $^*\mathcal{C}$ and has right local units, and $\mathcal{C}\mathcal{R} = \mathcal{C}$, then $\mathcal{R}$ is dense in the finite topology.*

Proof. 1) Take $M \in \mathcal{M}^\mathcal{C}$. For every $m \in M$ and $f \in {}^*\mathcal{C}$, we have $m \cdot f = m_{[0]} f(m_{[1]})$. Now, by the denseness of $\mathcal{R}$, there exists a $g \in \mathcal{R}$ such that $f(m_{[1]}) = g(m_{[1]})$, and so $m \cdot f = m \cdot g$.

2) By Proposition 3.9, $\mathcal{R}$ is a multiplicative approximation of $^*\mathcal{C}$ on $\mathcal{C}$. Hence for every finite $\{c_1, \cdots, c_n\} \subset \mathcal{C}$, and $f \in {}^*\mathcal{C}$, there exists $g \in \mathcal{R}$ such that $c_{i(1)} f(c_{i(2)}) = c_i \cdot f = c_i \cdot g = c_{i(1)} g(c_{i(2)})$, for all i. Applying $\varepsilon_\mathcal{C}$ to both sides, we find that $f(c_i) = g(c_i)$ for all i, which means exactly that $\mathcal{R}$ is dense in the finite topology on $^*\mathcal{C}$. $\square$

COROLLARY 3.11. *If $\mathcal{R} \subset {}^*\mathcal{C}$ is a $\mathcal{C}$-comodule and an ideal, then $\mathcal{R}$ is dense in the finite topology if and only if $\mathcal{R}$ has local units and $\mathcal{C}$ is unitary as an $\mathcal{R}$-module.*

For later use, we give the following generalization of a well-known property of Morita contexts, illustrating the connection between local units and local projectivity.

PROPOSITION 3.12. *Let A be a ring with left local units, and (A, B, P, Q, f, g) a Morita context. If $f : P \otimes_B Q \to A$ is surjective, then P is locally projective as a left B-module. If A has right local units, then $Q \in \mathcal{M}_A$ is locally projective.*

Proof. Take $e \in A$ such that $p = ep$, and $\sum_i p_i \otimes_B q_i \in f^{-1}\{e\}$. Then

$$p = ep = \sum_i (p_i \otimes_B q_i)p = \sum_i p_i g(q_i \otimes_A p)$$

and we have a dual basis $\{(p_i, g(q_i \otimes_A -)) \mid i \in I\}$. $\square$

4. $\mathcal{R}$-RATIONAL MODULES

Let $\mathcal{C}$ be an A-coring, and $\mathcal{R}$ a subring (without unit) of $^*\mathcal{C}$. $\mathcal{C}$ is $\mathcal{R}$-locally projective if and only if

$$\xi_{M,\mathcal{R}} : M \otimes_A \mathcal{C} \to \mathrm{Hom}_\mathbb{Z}(\mathcal{R}, M), \ \xi_{M,\mathcal{R}}(m \otimes_A c)(r) = m \cdot r(c)$$

is injective for all $M \in \mathcal{M}_A$.

Let T be the subring of $^*\mathcal{C}$ generated by A and $\mathcal{R}$. Then T is the $\mathbb{Z}$-module generated by elements of the form $a_1 \# r_1 \# a_2 \# r_2 \# \cdots \# r_n \# a_{n+1}$ with $a_i \in A$ and $r_i \in R$. Note that T has a unit, and that $S = \mathcal{R}A \subset T$. For every $M \in \mathcal{M}_T$, we define

$$\delta_{M,\mathcal{R}} : M \to \mathrm{Hom}_\mathbb{Z}(\mathcal{R}, M), \ \delta_{M,\mathcal{R}}(m)(f) = m \cdot f$$

If $\mathcal{R}$ is also a right A-module, then $\mathrm{Im}\,(\delta_{M,\mathcal{R}}) \subset \mathrm{Hom}_A(\mathcal{R}, M)$.

DEFINITION 4.1. $M \in \mathcal{M}_T$ *is called $\mathcal{R}$-rational if $\delta_{M,\mathcal{R}}(M) \subset \xi_{M,\mathcal{R}}(M \otimes_A \mathcal{C})$, or, equivalently, if for every $m \in M$, there exist finitely many $m_i \in M$ and $c_i \in \mathcal{C}$ such that $m \cdot f = \sum_i m_i f(c_i)$, for all $f \in \mathcal{R}$. $\mathcal{R}\mathcal{M}_\mathcal{R}$ will be the full subcategory of $\mathcal{M}_T$ consisting of $\mathcal{R}$-faithful $\mathcal{R}$-rational T-modules.*

PROPOSITION 4.2. *Assume that $\mathcal{R}$ is a subring of $^*\mathcal{C}$, such that $\mathcal{C}$ is $\mathcal{R}$-locally projective. Then we have the following properties:*

1. *every cyclic submodule of an $\mathcal{R}$-rational module M is finitely generated as an A-module;*

RATIONALITY FOR CORINGS

2. *the direct sum of a family of $\mathcal{R}$-rational modules is again rational;*
3. *any quotient of a rational $\mathcal{R}$-module is rational;*
4. *any submodule of a rational $\mathcal{R}$-module is rational.*

Proof. The proof of the first three statements is similar to the proof of [13, Theorem 2.2.6]. For the proof of part 4, take $n \in N$ and $f \in \mathcal{R}$. Then we know that $n \cdot f = n_i f(c_i)$, with $n_i \in M$ and $c_i \in C$. We have to show that we can find $n'_j \in N$ and $c'_j \in C$ with the same property. By Proposition 3.3, there exists a dual basis $\{(e_j, r_j) \mid j \in J\} \subset C \times \mathcal{R}$ for the c_i. We easily compute that

$$n \cdot f = n_i f(c_i) = n_i f(r_j(c_i)e_j) = n_i r_j(c_i)f(e_j) = n \cdot r_j f(e_j)$$

using the fact that M is $\mathcal{R}$-rational. Since N is an $\mathcal{R}$-module, we know that $n \cdot r_j \in N$. Now just take $n'_j = n \cdot r_j$ and $c'_j = e_j$. $\square$

Let R be a ring and S and T two subrings of R. Observe that $ST \subset TS$ if and only if for all $s \in S$ and $t \in T$, we can find $s_i \in S$ and $t_i \in T$ such that $st = \sum_i t_i s_i$ in R. We present several examples below.

EXAMPLES 4.3. 1. Let (H, A, C) be a right-right Doi-Hopf structure over a commutative ring k. As we have seen, $\mathcal{C} = A \otimes C$ is an A-coring. C^* is a subring of $^*\mathcal{C} \cong \mathrm{Hom}(\mathcal{C}, A)$, and $AC^* \subset C^*A$, since

$$(i(a)\#f)(b \otimes c) = f((b \otimes c)a) = ba_{[0]}f(ca_{[1]})$$
$$= \quad bf(ca_{[1]})a_{[0]} = ((a_{[1]} \rightharpoonup f)\#i(a_{[0]}))(b \otimes c)$$

Recall that the left H-action on C^* is given by $(h \rightharpoonup f)(b \otimes c) = bf(ch)$. It follows that $i(a)\#f = (a_{[1]} \rightharpoonup f)\#i(a_{[0]})$, and $A\#C^* \subset C^*\#A$, as needed. We will see in Section 5 that the same property does not hold for entwining structures, leading to the introduction of factorizable entwining structures.

2. Let S and T be two k-algebras, k is a commutative ring, and (T, S, R) a factorization structure.

$$i_T : T \to T\#S, \ i_T(t) = t\#1_S \text{ and } i_S : S \to T\#S, \ i_S(s) = 1_T\#s$$

are algebra maps, and $ST \subset TS$ since

$$s \cdot t = (1\#s)(t\#1) = t_R\#s_R = (t_R\#1)(1\#s_R) = t_R \cdot s_R \in TS$$

for all $s \in S$ and $t \in T$.

Remark 4.4. Example 1 has been our motivation to work with subrings $\mathcal{R} \subset {}^*\mathcal{C}$ that are not necessarily right A-modules, as is usually done in the literature. In the setting of Example 1, we can consider $\#(C, A)$-rational modules, as well as C^*-modules (see [12]), and C^* is not a right A-module.

Let $M \in \mathcal{M}_A$, and recall that an A-module is subgenerated by M if it is isomorphic to a subobject of a quotient of a direct sum of copies of M. $\sigma[M]$, the full subcategory of $\mathcal{M}_A$ consisting of modules subgenerated by M is the smallest full Grothendieck subcategory of $\mathcal{M}_A$ containing M (see [20]).

LEMMA 4.5. *Let $\mathcal{R}$ be a subring (without unit) of $^*\mathcal{C}$, T the subring of $^*\mathcal{C}$ generated by $\mathcal{R}$ and A, and $S = \mathcal{R}A$. Then the following assertions are equivalent:*

1. $A\mathcal{R} \subset \mathcal{R}A;$

S. CAENEPEEL, J. VERCRUYSSE AND SHUANHONG WANG

2. S is a ring and a left A-module, and $T = A + S$;
3. S is T-bimodule;
4. S is a left A-module.

In this case:

1. $\mathcal{R}\mathcal{M}_S = \mathcal{R}\mathcal{M}_\mathcal{R}$;
2. $\mathrm{Hom}_\mathbb{Z}(S, M) \in \mathcal{M}_T$, for all $M \in \mathcal{M}_T$;
3. $\xi_{M,S} \in \mathcal{M}_T$ for all $M \in \mathcal{M}_T$.

Proof. 1) $\Rightarrow$ 2). Take a generator $g = a_1 \# r_2 \# \cdots \# r_n \# a_{n+1}$ of T as a $\mathbb{Z}$-module. If $n = 0$, then $g = a \in A$. If $n > 1$, then it follows from $A\mathcal{R} \subset \mathcal{R}A$ that $g \in S$, hence $T = A + S$. S is closed under multiplication since $S^2 = \mathcal{R}A\mathcal{R}A \subset \mathcal{R}^2 A^2 \subset \mathcal{R}A = S$. S is a left A-module since $AS = A\mathcal{R}A \subset \mathcal{R}A^2 = S$.

2) $\Rightarrow$ 3). S is a ring, so it is an S-bimodule. It is a right A-module, and, by 2), a left A-module. Since $T = A + S$, S is a T-bimodule.

3) $\Rightarrow$ 4) is trivial.

4) $\Rightarrow$ 1). If S is a left A-module, then $A\mathcal{R} \subset AS \subset S = \mathcal{R}A$.

1) $\mathcal{R} \subset S$, so S-rationality implies $\mathcal{R}$-rationality. Conversely, if M is $\mathcal{R}$-rational, then for all $m \in M$ and $\sum_i f_i \# a_i \in S$, we have

$$m \cdot \sum_i f_i \# a_i = \sum_i (m \cdot f_i) a_i = \sum_i (m_{[0]} f_i(m_{[1]})) a_i = m_{[0]} \left(\sum_i f_i \# a_i \right) (m_{[1]})$$

and it follows that M is S-rational. It is easy to see that M is $\mathcal{R}$-faithful if and M is S-faithful.

2) Take $\varphi \in \mathrm{Hom}_\mathbb{Z}(S, M)$, $f \in S$ and $g \in T$. Then $g \# f \in S$, and we define $\varphi \cdot g$ by $(\varphi \cdot g)(f) = \varphi(g \# f)$.

3) $\xi_{M,S} : M \otimes_A \mathcal{C} \to \mathrm{Hom}_\mathbb{Z}(S, M)$ is right T-linear since

$$\xi_{M,S}((m \otimes_A c) \cdot g)(s) = \xi_{M,S}(m \otimes_A c_{(1)} g(c_{(2)}))(s) = m \cdot s(c_{(1)} g(c_{(2)}))$$

$$= m \cdot (g \# s)(c) = \xi_{M,S}(m \otimes_A c)(g \# s) = \Big(\big(\xi_{M,S}(m \otimes_A c) \big) \cdot g \Big)(s)$$

$\square$

COROLLARY 4.6. *Let $\mathcal{C}$ be an A-coring, and $\mathcal{R}$ a subring of $^*\mathcal{C}$. If $A\mathcal{R} \subset \mathcal{R}A$ and $\mathcal{C}$ is $\mathcal{R}$-locally projective, then $\mathcal{M}^\mathcal{C}$ is a subcategory of $\sigma[\mathcal{C}_S]$ and $\sigma[\mathcal{C}_S]$ is full subcategory of $\mathcal{R}\mathcal{M}_\mathcal{R}$.*

Proof. Take $M \in \mathcal{M}^\mathcal{C}$. Then there exists a nonempty set I such that M is isomorphic as a $\mathcal{C}$-comodule (and a fortiori as an S-module) to a subobject of a direct sum $\mathcal{C}^{(I)}$ of copies of $\mathcal{C}$, and therefore $M \in \sigma[\mathcal{C}_S]$.

It is clear that $\mathcal{C}$ is an $\mathcal{R}$-rational S-module, and it follows from Proposition 4.2 that every $M \in \sigma[\mathcal{C}_S]$ is an $\mathcal{R}$-rational S-module. $\square$

PROPOSITION 4.7. *Let $\mathcal{C}$ be an A-coring, $\mathcal{R}$ a subring of $^*\mathcal{C}$, and $M \in \mathcal{R}\mathcal{M}_\mathcal{R}$. If $\mathcal{C}$ is $\mathcal{R}$-locally projective and $A\mathcal{R} \subset \mathcal{R}A$, then $\delta_{M,\mathcal{R}}$ defines a right $\mathcal{C}$-comodule structure on M.*

Proof. From the $\mathcal{R}$-rationality of M and the fact that $\mathcal{C}$ is $\mathcal{R}$-locally projective, it follows that, for any $m \in M$, there exists a unique $\sum_i m_i \otimes_A c_i \in M \otimes_A \mathcal{C}$ such that $m \cdot f = \sum_i m_i f(c_i)$, for every $f \in \mathcal{R}$. So we have a well-defined map

RATIONALITY FOR CORINGS

$\delta_M : M \to M \otimes_A C$, $\delta_M(m) = \sum_i m_i \otimes_A c_i$, which is equal to $\delta_{M,\mathcal{R}}$ if we regard the injection $\xi_{M,\mathcal{R}}$ as an inclusion. Let us use the notation

$$\delta_M(m) = m_{[0]} \otimes_A m_{[1]}$$

We will show that δ_M defines a C-comodule structure on M. First, δ_M is right A-linear. Since $A\mathcal{R} \subset \mathcal{R}A$, there exist a_k and f_k such that $i(a)\#f = \sum_k f_k\#i(a_k)$, hence

$$
\begin{aligned}
(ma) \cdot f &= m \cdot (i(a)\#f) = m \cdot \left(\sum_k f_k\#i(a_k)\right) \\
&= \sum_k (m \cdot f_k)a_k = \sum_k (m_{[0]} \cdot f_k(m_{[1]}))a_k \\
&= \sum_k m_{[0]} \cdot (f_k(m_{[1]})a_k) = \sum_k m_{[0]} \cdot (f_k\#i(a_k))(m_{[1]}) \\
&= m_{[0]} \cdot (i(a)\#f)(m_{[1]}) = m_{[0]} \cdot f(m_{[1]}a)
\end{aligned}
$$

for all $f \in \mathcal{R}$, $m \in M$ and $a \in A$, so $\delta_M(ma) = m_{[0]} \otimes_A m_{[1]}a = \delta_M(m)a$.

Let us next show the mixed coassociativity. For all $m \in M$, we have to show that

$$y_m = (I \otimes_A \Delta_C)\delta_M(m) - (\delta_M \otimes_A I)\delta_M(m) = 0$$

For all $f, g \in \mathcal{R}$, we have

$$
\begin{aligned}
m \cdot (f\#g) &= m_{[0]}(f\#g)(m_{[1]}) \\
&= m_{[0]}g((m_{[1]})_{(1)}f((m_{[1]})_{(2)})) \\
&= (I \otimes_A g) \circ (I \otimes_A I \otimes_A f)(m_{[0]} \otimes_A (m_{[1]})_{(1)} \otimes_A (m_{[1]})_{(2)}) \\
&= (I \otimes_A g) \circ (I \otimes_A I \otimes_A f)(I \otimes_A \Delta_C)\delta_M(m) \\
= (m \cdot f) \cdot g &= (m_{[0]} \cdot f(m_{[1]})) \cdot g \\
&= (m_{[0]})_{[0]}g((m_{[0]})_{[1]}f(m_{[1]})) \\
&= (I \otimes_A g) \circ (I \otimes_A I \otimes_A f)((m_{[0]})_{[0]} \otimes_A (m_{[0]})_{[1]} \otimes_A m_{[1]}) \\
&= (I \otimes_A g) \circ (I \otimes_A I \otimes_A f)(\delta_M \otimes_A I)\delta_M(m)
\end{aligned}
$$

where we used the right A-linearity of $\delta_{M,\mathcal{R}}$. It follows that

$$(I \otimes_A g)((I \otimes_A I \otimes_A f)(y_m)) = 0$$

for all $f, g \in \mathcal{R}$. Using the α-property for M, we find

$$(I \otimes_A I \otimes_A f)(y_m) = 0$$

and, using the α-property for $M \otimes_A C$, we find that $y_m = 0$, as needed.

Finally, for every $f \in \mathcal{R}$, we have, using the mixed coassociativity,

$$
\begin{aligned}
(m - m_{[0]}\varepsilon_C(m_{[1]})) \cdot f &= m \cdot f - m_{[0]}f(m_{[1]}\varepsilon_C(m_{[2]})) \\
&= m \cdot f - m_{[0]}f(m_{[1]}) = m \cdot f - m \cdot f = 0
\end{aligned}
$$

From the fact that M is faithful as a right $\mathcal{R}$-module, we then deduce that $m = m_{[0]}\varepsilon_C(m_{[1]})$. $\qquad \square$

It was proved in [21] that $\mathcal{M}^C = \sigma[C_{\bullet}C]$ if and only if $^*C$ is locally projective as an A-module. We will now generalize this result.

PROPOSITION 4.8. *Let C be an A-coring and $\mathcal{R}$ a subring of $^*C$. If C is $\mathcal{R}$-locally projective and $A\mathcal{R} \subset \mathcal{R}A$, then the categories $\mathcal{R}\mathcal{M}_\mathcal{R}$, $\mathcal{R}\mathcal{M}_S$, $\sigma[C_S]$ and $\mathcal{M}^C$ are isomorphic full subcategories of $\mathcal{M}_S$.*

Proof. Suppose first C is $\mathcal{R}$-locally projective. We define a functor $F : \mathcal{R}\mathcal{M}_\mathcal{R} \to \mathcal{M}^C$ as follows: $F(M) = M$ as a right A-module, with C-comodule structure as in Proposition 4.7; for $f : M \to N$ in $\mathcal{R}\mathcal{M}_\mathcal{R}$, we put $F(f) = f$.

Let us first prove that f is right C-colinear, as needed. Take $m \in M$, then $\delta_M(f(m)) = f(m)_{[0]} \otimes_A f(m)_{[1]}$, if and only if $f(m) \cdot g = f(m)_{[0]} g(f(m)_{[1]})$ for every $g \in \mathcal{R}$. But $f(m) \cdot g = f(m \cdot g) = f(m_{[0]} g(m_{[1]})) = f(m_{[0]}) g(m_{[1]})$, since f is right $\mathcal{R}$-linear and right A-linear. Using the α-condition, we find $f(m)_{[0]} \otimes_A f(m)_{[1]} = f(m_{[0]}) \otimes_A m_{[1]}$.

Finally, it is easy to see that $F(M) = M$ if $M \in \mathcal{M}^C$. $\qquad\square$

DEFINITION 4.9. Assume that C is $\mathcal{R}$-locally projective, and let M be a right $\mathcal{R}$-module. We define the $\mathcal{R}$-rational part of M as

$$M^{\mathcal{R}\text{-rat}} = \delta_{M,\mathcal{R}}^{-1}(\xi_{M,\mathcal{R}}(M \otimes_A C))$$

The rational part of M is by definition the $^*C$-rational part: $M^{\text{rat}} = M^{^*C\text{-rat}}$.

Observe that $m \in M^{\mathcal{R}\text{-rat}}$ if and only if there exist $m_i \in M$ and $c_i \in C$ such that $m \cdot f = \sum_i m_i \cdot f(c_i)$ for all $f \in \mathcal{R}$. M is $\mathcal{R}$-rational if and only if $M^{\mathcal{R}\text{-rat}} = M$. Obviously $M^\mathcal{R} \subset M^{\mathcal{R}\text{-rat}}$. If $\mathcal{R} \subset \mathcal{R}' \subset {^*C}$, then $M^{\mathcal{R}'} \subset M^\mathcal{R}$. If $M \in \mathcal{M}^C$ and C is $\mathcal{R}$-locally projective, then $M^{\mathcal{R}'} = M^\mathcal{R} = M^{\text{co}C}$.

PROPOSITION 4.10. *Let $\mathcal{R} \subset {^*C}$ be a subring, assume that C is $\mathcal{R}$-locally projective, and take $M \in \mathcal{M}_{{}^*C}$.*

1. *Let $\mathcal{R}'$ be another subring of $^*C$. If $\mathcal{R} \subset \mathcal{R}'$, then C is also $\mathcal{R}'$-locally projective, and*

$$M^{\mathcal{R}'\text{-rat}} \subset M^{\mathcal{R}\text{-rat}} \text{ and } (^*C)^{\mathcal{R}'\text{-rat}} = (^*C)^{\mathcal{R}\text{-rat}}$$

   If M is $\mathcal{R}'$-rational, then M is also $\mathcal{R}$-rational. $^*C$ is $\mathcal{R}'$-rational if and only if $^*C$ is $\mathcal{R}$-rational.

2. *Fix a grouplike element $x \in C$, and assume that*

$$\varepsilon_{M,\mathcal{R}} : M^\mathcal{R} \otimes_B A \to M, \ \varepsilon_{M,\mathcal{R}}(m \otimes a) = ma$$

   is surjective. If $A\mathcal{R} \subset \mathcal{R}A$, then M is $\mathcal{R}$-rational.

3. *$M^{\mathcal{R}\text{-rat}}$ is a right $\mathcal{R}$-submodule of M. If $A\mathcal{R} \subset \mathcal{R}A$, then it is a right T-module, and consequently it is the biggest T-submodule of M that is $\mathcal{R}$-rational. $M^{\mathcal{R}\text{-rat}}$ is then also a C-comodule.*

Proof. 1) The first statement is obvious. Take $g \in (^*C)^{\mathcal{R}\text{-rat}}$. Then there exist $g_j \in {^*C}$ and $c_j \in C$ such that $g\#h = \sum_j g_j \# i(h(c_j))$, for all $h \in \mathcal{R}$. Now $(g\#h)(c) = h(c_{(1)} g(c_{(2)}))$ and $\sum_j (g_j \# i(h(c_j)))(c) = \sum_j g_j(c) h(c_j) = \sum_j h(g_j(c)c_j)$, so it follows that

$$h\Big(c_{(1)} g(c_{(2)}) - \sum_j g_j(c)c_j\Big) = 0$$

RATIONALITY FOR CORINGS

for all $h \in \mathcal{R}$. $\mathcal{C}$ satisfies the α-property for $\mathcal{R}$, so we have a dual basis $\{(e_i, r_i) \mid i \in I\} \subset \mathcal{C} \times \mathcal{R}$, and we find that

$$c_{(1)}g(c_{(2)}) - \sum_j g_j(c)c_j = \sum_{i \in I} r_i \Big(c_{(1)}g(c_{(2)}) - \sum_j h(g_j(c)c_j)\Big)e_i = 0$$

Applying $f \in \mathcal{R}'$, it follows that $g\#f = \sum_j g_j\#f(c_j)$ hence $g \in (^*C)^{\mathcal{R}'\text{-rat}}$.

2) Take $m \in M$ and $f \in \mathcal{R}$. There exist $m_k \in M^{\mathcal{R}}$ and $a_k \in A$ such that $m = \sum_k m_k a_k$. For any k, we can find $a_{kl} \in A$ and $f_{kl} \in \mathcal{R}$ such that $a_k\#f = \sum_l f_{kl}\#a_{kl}$. Now

$$
\begin{aligned}
m \cdot f &= \sum_k (m_k a_k) \cdot f = \sum_k m_k \cdot (a_k\#f) = \sum_{k,l} m_k \cdot (f_{kl}\#a_{kl}) \\
&= \sum_{k,l} \Big(m_k \cdot f_{kl}\Big)a_{kl} = \sum_{k,l} \Big(m_k f_{kl}(x)\Big)a_{kl} = \sum_{k,l} m_k(f_{kl}(x)a_{kl}) \\
&= \sum_{k,l} m_k((f_{kl}\#a_{kl})(x)) = \sum_k m_k((a_k\#f)(x))
\end{aligned}
$$

and it follows that $m \in M^{\mathcal{R}\text{-rat}}$.

3) For $m \in M^{\mathcal{R}\text{-rat}}$, there exists a unique $m_{[0]} \otimes_A m_{[1]} \in M \otimes_A \mathcal{C}$ such that $m \cdot f = m_{[0]}f(m_{[1]})$, for all $f \in \mathcal{R}$. For all $f, g \in \mathcal{R}$, we then have

$$(m \cdot f) \cdot g = m \cdot (f\#g) = m_{[0]}(f\#g)(m_{[1]}) = m_{[0]}g(m_{[1](1)}f((m_{[1](2)}))$$

and this means that $m \cdot f \in M^{\mathcal{R}\text{-rat}}$, as needed. If $A\mathcal{R} \subset \mathcal{R}A$, then the same argument as in the first part of the proof of Proposition 4.7 shows that $(ma) \cdot f = m_{[0]}f(m_{[1]}a)$, for all $f \in \mathcal{R}$, hence $ma \in M^{\mathcal{R}\text{-rat}}$ if $m \in M^{\mathcal{R}\text{-rat}}$ and $a \in A$. It follows from Proposition 4.8 that $M^{\mathcal{R}\text{-rat}}$ is a right $\mathcal{C}$-comodule. $\square$

5. THE IMAGE OF μ AND μ'

Let $(\mathcal{C}, x)$ be a coring with a fixed grouplike element. We have seen in Section 2 how we can associate Morita contexts $(B, {}^*C, A, Q, \tau, \mu)$ and $(B', {}^*C, A, Q', \tau', \mu')$ to $(\mathcal{C}, x)$. We will now apply the rationality results obtained above, to discuss the image of μ and μ'.

PROPOSITION 5.1. *Let $(\mathcal{C}, x)$ be a coring with a fixed grouplike element, and assume that $\mathcal{C}$ is locally projective as a left A-module. Then we have the following properties:*

1. $(^*C)^{\text{rat}}$ *is a two-sided ideal of* $^*C$;
2. $(^*C)^{\text{rat}}$ *has local units if and only if* $(^*C)^{\text{rat}}$ *is dense in* $^*C$, *with respect to the finite topology;*
3. $\operatorname{Im}\mu' \subseteq (^*C)^{\text{rat}}$ *and* $((^*C)^{\text{rat}})^{\text{co}\mathcal{C}} = Q$.

Proof. 1) We have already seen in Proposition 4.10 that $(^*C)^{\text{rat}}$ is a right ideal. Let us show that it is also a left ideal. For $f \in (^*C)^{\text{rat}}$ and $g, h \in {}^*C$, we have

$$(g\#f)\#h = g\#(f\#h) = g\#(f_{[0]}\#h(f_{[1]})) = (g\#f_{[0]})\#h(f_{[1]})$$

It follows that $(g\#f)_{[0]} \otimes_A (g\#f)_{[1]} = g\#f_{[0]} \otimes_A f_{[1]}$, and $g\#f \in (^*C)^{\text{rat}}$.

2) follows from Corollary 3.11.

3) We have to show that $q\#a \in (^*C)^{\mathrm{rat}}$, for all $q \in Q'$ and $a \in A$. For every $f \in {}^*C$, we compute

$$(q\#a\#f)(c) = f(c_{(1)}q(c_{(2)})a) = f(q(c)xa) = q(c)f(xa)$$

This proves that $(q\#a)_{[0]} \otimes_A (q\#a)_{[1]} = q \otimes_A xa$ and $q\#a \in (^*C)^{\mathrm{rat}}$. $((^*C)^{\mathrm{rat}})^{\mathrm{co}C} = Q$, because $f \in ((^*C)^{\mathrm{rat}})^{\mathrm{co}C}$ if and only if

$$f\#g = f_{[0]}\#g(f_{[1]}) = f\#g(x)$$

for all $g \in {}^*C$, if and only if $f \in Q$. $\qquad\square$

COROLLARY 5.2. *Assume that C is locally projective as a left A-module. If μ' : $Q' \otimes_B A \to {}^*C$ is surjective, then C is finitely generated and projective as a left A-module.*

Proof. If μ' is surjective, then it follows from Proposition 5.1 that ${}^*C = (^*C)^{\mathrm{rat}}$. Put $\rho(\varepsilon) = \sum_i f_i \otimes_A c_i \in {}^*C \otimes_A C$. This means that $\varepsilon\#f = f = \sum_i f_i\#f(c_i)$, for all $f \in {}^*C$. Every right ${}^*C$-module M is rational: for all $m \in M$, we have that $m \cdot f = \sum_i (m \cdot f_i)f(c_i)$, and this means that $\rho(m) = \sum_i m \cdot f_i \otimes_A c_i$. In particular, for $M = C$, we find, for all $c \in C$:

$$\rho(c) = c_{(1)} \otimes_A c_{(2)} = \sum_i c \cdot f_i \otimes_A c_i = \sum_i c_{(1)}f_i(c_{(2)}) \otimes_A c_i$$

Applying ε to the first factor, we find

$$c = \varepsilon(c_{(1)})c_{(2)} = \sum_i \varepsilon(c_{(1)}f_i(c_{(2)}))c_i$$

$$= \sum_i \varepsilon(c_{(1)})f_i(c_{(2)})c_i = \sum_i f_i(\varepsilon(c_{(1)})c_{(2)})c_i = \sum_i f_i(c)c_i$$

and it follows that $\{(c_i, f_i) \mid i = 1, \cdots, n\}$ is a finite dual basis of C. $\qquad\square$

We can now prove the main result of this paper.

THEOREM 5.3. *Let C be locally projective as a left A-module, and assume that τ' : $A \otimes_{\bullet C} Q' \to B'$ is surjective. For any $M \in \mathcal{M}^C$, we consider the map.*

$$\Omega_M : \ M \otimes_{\bullet C} (^*C)^{\mathrm{rat}} \to M, \ \Omega_M(m \otimes_{\bullet C} f) = m_{[0]}f(m_{[1]})$$

The following statements are equivalent:

1. (C, x) *satisfies the Strong Structure Theorem;*
2. (C, x) *satisfies the Weak Structure Theorem;*
3. $\mu : Q \otimes_B A \to (^*C)^{\mathrm{rat}}$ *and Ω_M are bijective, for all $M \in \mathcal{M}^C$;*
4. $\mu : Q \otimes_B A \to (^*C)^{\mathrm{rat}}$ *and Ω_M are surjective, for all $M \in \mathcal{M}^C$.*

Proof. We know from Corollary 3.7 that $B = B'$, $Q = Q'$, $\mu = \mu'$ and $\tau = \tau'$. For every $M \in \mathcal{M}^C$, we have $M^{\mathrm{co}C} = M^{*C}$, and we have the following commutative diagram

$$M \otimes_{\bullet C} Q \otimes_B A \xrightarrow{\ \omega_M \otimes_B I_A\ } M^{coC} \otimes_B A \qquad (11)$$

with vertical maps $I_M \otimes_{\bullet C} \mu$ on the left and ε_M on the right, and bottom row

$$M \otimes_{\bullet C} (^*C)^{\mathrm{rat}} \xrightarrow{\ \Omega_M\ } M$$

ω_M is defined as in Theorem 2.1, and is an isomorphism because τ is surjective.

1) $\Rightarrow$ 2) is trivial.

2) $\Rightarrow$ 3). Take $M = (^*C)^{\mathrm{rat}}$. Then $M^{coC} = Q$, and

$$\varepsilon_M = \mu : \ M^{coC} \otimes_B A \cong Q \otimes_B A \to M = (^*C)^{\mathrm{rat}}$$

is an isomorphism. Ω_M is an isomorphism because all the other maps in the commutative diagram (11) are isomorphisms.

3) $\Rightarrow$ 4) is trivial.

4) $\Rightarrow$ 1). It follows immediately from the commutative diagram (11) that every ε_M is surjective. For every right C-comodule P, we have a commutative diagram

$$P^{coC} \otimes_B A \otimes_{\bullet C} \xrightarrow{\ \varepsilon_P \otimes I_Q\ } P \otimes_{\bullet C} Q$$

with vertical maps $I \otimes_B \tau$ on the left and ω_P on the right, and bottom row

$$P^{coC} \otimes_B B \xrightarrow{\ \cong\ } P^{coC}$$

τ and ω_P are isomorphisms, so $\varepsilon_P \otimes I_Q$ is an isomorphism.

Now take $P = \operatorname{Ker}\varepsilon_M \in \mathcal{M}^C$. Then

$$P^{coC} \cong P \otimes_{\bullet C} Q = \operatorname{Ker}(\varepsilon_M) \otimes_{\bullet C} Q \cong \operatorname{Ker}(\varepsilon_M \otimes_{\bullet C} I_Q) = 0$$

Here we used the fact that Q is finitely generated and projective as a left $^*C$-module, and this follows from the fact that τ is surjective, and using the Morita Theorems. Now $\varepsilon_P : 0 = P^{coC} \otimes_B A \to P$ is surjective, so $P = \operatorname{Ker}\varepsilon_M = 0$, and ε_M is injective. Finally it follows from [8, Prop. 2.5] that the unit maps η_N are isomorphisms, for all $N \in \mathcal{M}_B$. $\qquad\square$

PROPOSITION 5.4. *If A is flat as left B-module, then (C, x) is Galois if and only if it satisfies the Weak Structure Theorem.*

Proof. The Weak Structure Theorem means that $\varepsilon_M : \ M^{coC} \otimes_B A \to M$ is an isomorphism for all $M \in \mathcal{M}^C$. If $M = C$, then $M^{coC} \cong A$ and ε_M is the cannonical map. Conversely, if (C, x) is Galois, we have to show that ε_M is an isomorphism for all $M \in \mathcal{M}^C$. We have isomorphisms

$$M \cong M \square_C C \cong M \square_C (A \otimes_B A) \cong (M \square_C A) \otimes_B A = M^{coC} \otimes_B A$$

The flatness of A implies that the cotensor product is associative. The composition of these isomorphisms is the inverse of ε_M. $\qquad\square$

COROLLARY 5.5. *With notation and assumptions as in Theorem 5.3, assume that* $(^*C)^{\mathrm{rat}}$ *has local units (which is the case if* $(^*C)^{\mathrm{rat}}$ *is dense in the finite topology). Then the four equivalent statements of Theorem 5.3 are equivalent to* (C, x) *being Galois. In this situation,* A *and* Q *are locally projective as a left, resp. a right* B-*module.*

Proof. The local projectivity of A and Q follows from Proposition 3.12. Since locally projectivity implies flatness (see [14]), the other statement follows from Proposition 5.4. $\qquad\square$

Remark 5.6. If $(^*C)^{\mathrm{rat}}$ has local units, then Ω_M is surjective for every $M \in \mathcal{M}^C$, and this condition can then be dropped in Corollary 5.5. In this case, we can restrict the Morita context to $(^*C)^{\mathrm{rat}}$. The unital $(^*C)^{\mathrm{rat}}$-modules are precisely the C-comodules: if $M = M \otimes_{(^*C)^{\mathrm{rat}}} (^*C)^{\mathrm{rat}}$, then M is a C-comodule since $(^*C)^{\mathrm{rat}}$ is a C-comodule, and since $(^*C)^{\mathrm{rat}}$ has local units, it is dense in the finite topology, so we can approximate ε_C and find that all C-comodules are unital as $(^*C)^{\mathrm{rat}}$-modules. Furthermore, the tensor product over $^*C$ coincides with the tensor product over $(^*C)^{\mathrm{rat}}$, which is easy to see if one uses Proposition 3.10. We will use this restricted context in Section 7.

6. FACTORIZABLE ENTWINING STRUCTURES

Let k be a commutative ring, and consider a right-right entwining structure (A, C, ψ). We call (A, C, ψ) factorizable if there exists a map $\alpha : A \to A \otimes \mathrm{End}(C)$ such that ψ factorizes as follows:

$$\psi = (I_A \otimes \theta) \circ (\alpha \otimes I_C) \circ \tau :$$
$$C \otimes A \xrightarrow{\ \tau\ } A \otimes C \xrightarrow{\alpha \otimes I_C} A \otimes \mathrm{End}(C) \otimes C \xrightarrow{I_A \otimes \theta} A \otimes C$$

where $\tau : C \otimes A \to A \otimes C$ is the switch map, and $\theta : \mathrm{End}(C) \otimes C \to C$ is the evaluation map. Using the notation $\alpha(a) = a_\alpha \otimes \lambda^\alpha$ (summation implicitly understood), this means that $\psi(c \otimes a) = a_\alpha \otimes \lambda^\alpha(c)$. We will say that (A, C, ψ) is completely factorizable if ψ has an inverse φ, and (A, C, ψ) and (A, C, φ) are both factorizable.

EXAMPLES 6.1. 1) Let H be a bialgebra, (H, A, C) a left-right Doi-Hopf structure. The corresponding entwining structure (A, C, ψ) is factorizable: Take $\alpha(a) = a_{[0]} \otimes m_{a_{[1]}}$, where $m_h : C \to C$, $m_h(c) = ch$, for all $c \in C$.

2) Assume that A is finitely generated projective. We know from [17] that (A, C, ψ) is induced by a Doi-Hopf structure over a bialgebra, so (A, C, ψ) is factorizable. This can also be seen directly: let $\{(e_i, e_i^*) \mid i = 1, \cdots, n\}$ be a finite dual basis of A, and define

$$\alpha(a) = \sum_i e_i \otimes \langle e_i^*, a_\psi \rangle \bullet^\psi$$

3) Now assume that C is finitely generated projective, and let $\{(c_i, c_i^*) \mid i = 1, \cdots, n\}$ be a finite dual basis of C. Then (A, C, ψ) is factorizable: if we take $\alpha(a) = a_\psi \otimes c_i^\psi c_i^*$, then we easily compute that

$$a_\alpha \otimes \lambda^\alpha(c) = a_\psi \otimes c_i^\psi \langle c_i^*, c \rangle = a_\psi \otimes c^\psi$$

RATIONALITY FOR CORINGS   131

Factorizable entwining structures are close to Doi-Hopf structures: the philosophy is that the bialgebra H is replaced by the algebra $\text{End}(C)$. Of course this is just philosophy, since there are no bialgebra structures on $\text{End}(C)$ with the composition as multiplication. We will see that factorizable entwining structures are more general then Doi-Hopf structures, and also that there exist non-factorizable entwining structures. Our examples are inspired by the examples given in [17]. Let (A, C, ψ) be an entwining structure over a field k. Recall from [17] that we can construct the following endomorphisms of A, for every $c \in C$ and $c^* \in C^*$:

$$T_{c,c^*} : A \to A, \ T_{c,c^*}(a) = \langle c^*, c^\psi \rangle a_\psi$$

If (A, C, ψ) originates from a Doi-Koppinen structure (A, C, H), then $T_{c,c^*}(a) = c^*(a_{[1]}c)a_{[0]}$, and we see that every H-subcomodule of A is T_{c,c^*}-invariant. Since we are working over a field, every $a \in A$ is contained in a finite dimensional H-subcomodule of A, so every $a \in A$ is contained in a finite dimensional T_{c,c^*}-invariant subspace of A. This property will be used in the examples in the sequel.

In a similar way, if (A, C, ψ) originates from an alternative Doi-Hopf datum, then every $c \in C$ lies in a finite dimensional T_{a,a^*}-invariant subspace of C. Here T_{a,a^*} is defined by $T_{a,a^*}(c) = a^*(a_\psi)c^\psi$, for every $a \in A$, $a^* \in A^*$ and $c \in C$. Remark that $\alpha(a) = \sum_i e_i \otimes T_{a,e_i^*}$ in Example 6.1 2).

We will now give an example of a factorizable entwining structure that does not originate from a Doi-Hopf datum.

EXAMPLE 6.2. Let $A = k\langle (X_i)_{i \in I_1} \rangle$ be the free algebra with a family of generators indexed by $I_1 = \mathbb{N}$ or $I_1 = \mathbb{Z}$.

Let C be the k-module with free basis $\{1, t\} \cup \{t_i \mid i \in I_2\}$, where $I_2 = \mathbb{N}$ or $I_2 = \mathbb{Z}$. We put a coalgebra structure on k by making 1 grouplike and t and t_i primitive.

We now define the entwining map ψ. For every $a \in A$ and $c \in C$ we define $\psi(a \otimes 1) = a \otimes 1$ and $\psi(1 \otimes c) = 1 \otimes c$. Furthermore we define

$$\psi(X_{i_1} \cdots X_{i_n} \otimes t) = X_{i_1+1} \cdots X_{i_n+1} \otimes t$$

$$\psi(X_{i_1} \cdots X_{i_n} \otimes t_j) = X_{i_1+1} \cdots X_{i_n+1} \otimes t_{j+n}$$

and extend ψ linearly. A straightforward computation shows that (A, C, ψ) is an entwining structure. Let us show that (A, C, ψ) is factorizable. By linearity, it suffices to define α on elements of the form $a = X_{i_1} \cdots X_{i_n} \otimes c$. Write $c = \widetilde{c} + \bar{c}t + \sum_i c_i t_i$ with $\widetilde{c}, \bar{c}, c_i \in k$. Then we have

$$\psi(a \otimes c) = a + \widetilde{c} + X_{i_1+1} \cdots X_{i_n+1} \otimes (\bar{c}t + \sum_j c_j t_{j+n})$$

$$= a \otimes \lambda_1^a(c) + X_{i_1+1} \cdots X_{i_n+1} \otimes \lambda_2^a(c)$$

where $\lambda_1^a, \lambda_2^a : C \to C$ are defined by

$$\lambda_1^a(1) = 1, \ \lambda_1^a(t) = 0, \ \lambda_1^a(t_i) = 0$$

$$\lambda_2^a(1) = 0, \ \lambda_1^a(t) = t, \ \lambda_1^a(t_i) = t_{i+1}$$

so we find that $\alpha(a) = a \otimes \lambda_1^a + X_{i_1+1} \cdots X_{i_n+1} \otimes \lambda_2^a$ and (A, C, ψ) is factorizable. Let us show that there is no Doi-Hopf structure inducing (A, C, ψ). Take $c^* \in C^*$ such that $\langle c^*, t \rangle = 1$ (this is possible since we work over a field). Then $T_{c^*,t}(X_i) = X_{i+1}$, so every $T_{c^*,t}$-invariant subspace that contains X_0 is infinite dimensional.

In a similar way, we find that (A, C, ψ) does not originate from an alternative Doi-Hopf datum: take $a^* \in A^*$ such that $a^*(X_1) = 1$; then we find that $T_{a^*, X_0}(t_i) = t_{i+1}$, so every T_{a^*, X_0}-invariant subspace containing t_0 is infinite dimensional.

If $I_1 = I_2 = \mathbb{Z}$, then ψ is bijective, and $\varphi = \psi^{-1}$ has the same properties as ψ, so we find a completely factorizable entwining structure that cannot be derived from an (alternative) Doi-Hopf structure.

Remark that we could also have taken

$$\psi(X_{i_1} \cdots X_{i_n} \otimes t_j) = X_{i_1+1} \cdots X_{i_n+1} \otimes t_{j+\sum_k i_k}$$

Adapting Example 6.2, we can give an example of an entwining structure that is not factorizable.

EXAMPLE 6.3. Example 6.2 Let A and C be as in Example 6.2, and let $\psi : A \otimes C \to A \otimes C$ be defined by

$$\psi(X_{i_1} \cdots X_{i_n} \otimes t_j) = X_{i_1+j} \cdots X_{i_n+j} \otimes t_{j+n}$$

Consider the k-linear map $p : C \to k$, given by $p(1) = p(t) = p(t_i) = 1$. If (A, C, ψ) is factorizable, then for all $a \in A$, the set

$$A_a = \{(I_A \otimes p)\psi(a \otimes c) \mid c \in C\}$$

is contained in a finite dimensional subspace of A. This is not the case, since A_{X_1} contains $X_2, X_3, \cdots$. Hence (A, C, ψ) is not factorizable.

Let (A, C, ψ) be a right-right entwining structure, and $\mathcal{C} = A \otimes C$ the associated A-coring. The left dual ring is $^*\mathcal{C} = {}_A\mathrm{Hom}(A \otimes C, A) \cong \#(C, A)$, with multiplication

$$(f \# g)(c) = f(c_{(2)})_\psi g(c_{(1)}^\psi)$$

A and $(C^*)^{op}$ are then subalgebras of $\#(C, A)$, via the algebra monomorphisms $i : A \to \#(C, A)$ and $j : (C^*)^{op} \to \#(C, A)$ given by

$$i(a)(c) = \varepsilon(c)a \text{ and } j(c^*)(a) = \langle c^*, c \rangle 1_A$$

and we easily compute that

$$(j(c^*) \# i(a))(c) = \langle c^*, c \rangle a$$

for all $a \in A$, $c \in C$ and $c^* \in C^*$.

PROPOSITION 6.4. *If (A, C, ψ) is a factorizable entwining structure, then*

$$i(A)j((C^*)^{op}) \subset j((C^*)^{op})i(A)$$

and consequently $i(A)j((C^*)^{op})$ is a subalgebra of $\#(C, A)$.

Proof. For all $a \in A$, $c \in C$ and $c^* \in C^*$, we have

$$(i(a) \# j(c^*))(c) = i(a)(c_{(2)})_\psi j(c^*)(c_{(1)}^\psi) = \varepsilon(c_{(2)})a_\psi \langle c^*, c_{(1)}^\psi \rangle 1_A$$

$$= \langle c^*, c^\psi \rangle a_\psi = \langle c^* \circ \lambda^\alpha, c \rangle a_\alpha = \left(j(c^* \circ \lambda^\alpha) \# j(a_\alpha)\right)(c)$$

hence

$$i(a) \# j(c^*) = j(c^* \circ \lambda^\alpha) \# j(a_\alpha) \in j((C^*)^{op}) \# i(A)$$

as needed. $\square$

RATIONALITY FOR CORINGS 133

If C^* is C-locally projective as a k-module (e.g. C^* is projective and C is dense in C^{**}, this is for example the case when k is a field), then $j \otimes i : (C^*)^{op} \otimes A \to \#(C, A)$ is injective, and we obtain

COROLLARY 6.5. *Let (A, C, ψ) be a factorizable entwining structure, and assume that C^* is C-locally projective as a k-module. Then we have a factorization structure $((C^*)^{op}, A, R)$, with $R : A \otimes (C^*)^{op} \to (C^*)^{op} \otimes A$ given by $R(a \otimes c^*) = c^* \circ \lambda^\alpha \otimes a_\alpha$, and $j \otimes i : (C^*)^{op} \#_R A \to \#(C, A)$ is an algebra monomorphism.*
If (A, C, ψ) is a completely factorizable entwining structure, then R is bijective.

7. CO-FROBENIUS PROPERTIES

Let C be an A-coring with a fixed grouplike element, and consider the Morita contexts indtroduced in Section 2. It was shown in [8] that the connecting modules A and Q are isomorphic if the ring morphism $A \to {}^*C$ is Frobenius. In this Section, we will present a weaker property under more general assumptions.
We call an A-coring C right co-Frobenius if there exists an injective right ${}^*C$-linear map $j : C \to {}^*C$.
If C is co-Frobenius and locally projective as a left A-module, then $I = \text{Im} \, j \subset ({}^*C)^{\text{rat}}$.

PROPOSITION 7.1. *If C is co-Frobenius and locally projective as a left A-module, then there exists an injective right B-linear map $J : A \to Q$.*

Proof. By Proposition 5.1 we have to prove that the injection $j : C \to ({}^*C)^{\text{rat}}$ induces an injection $J : A \cong C^{\text{co}C} = \text{Hom}^C(A, C) \to \text{Hom}^C(A, ({}^*C)^{\text{rat}}) \cong Q$. For every $\phi \in \text{Hom}^C(A, C)$, let $J(\phi) = j \circ \phi \in \text{Hom}^C(A, ({}^*C)^{\text{rat}})$. J is injective: if $J(\phi) = j \circ \phi = 0$, then $\phi = 0$, since J is injective. If we view J as a map $A \to Q$, then J is given by $J(a) = j(ax)$, which is obviously right B-linear. $\square$

Remark 7.2. If A is a (commutative) field and C is left and right co-Frobenius. It is shown in [13, Prop. 5.5.3], that there exists a surjective (and a fortiori a bijective) B-linear map J as above.

Let H be a co-Frobenius Hopf algebra, and A an H-comodule algebra. Then $A \otimes H$ is an A-coring, with grouplike element $1_A \otimes 1_H$. In this case ${}^*C = \#(H, A)$. In [4] it is shown that $Q \cong A$, if we restrict the Morita context from $\#(H, A)$ to $A \# H^{*\text{rat}}$. In Proposition 7.8, we will generalize this to completely factorizable entwining structures.

LEMMA 7.3. *Let A and B be two algebras over a commutative ring k and $R : B \otimes A \to A \otimes B$ a factorization map. Then R is a (B, A)-bimodule morphism from $B \otimes A$ to $A \#_R B$.*

Proof. We show that R is a left B-linear:

$$R(b'b \otimes a) = a_R \#(b'b)_R = a_{Rr} \# b'_r b_R = (1 \# b')(a_R \# b_R) = b' R(a \otimes b)$$

The right A-linearity of R can be handled in a similar way. $\square$

Let (A, C, ψ) be a factorizable (right, right) entwining structure over k. In Section 6, we have seen that we have a smash product $C^{*op} \#_R A$ and a ring morphism

$C^{*op}\#_R A \to \#(C, A)$. If C is locally projective as a k-module (e.g. k is a field), then, by extension of scalars, $A \otimes C$ is a C^*-locally projective left A-module. Consider $\#(C, A)$ as a right $\#(C, A)$-module. By Proposition 4.10, we have that

$$\#(C, A)^{\#(C,A)\text{-rat}} = \#(C, A)^{C^*\text{-rat}}$$

and we will denote this module by $\#(C, A)^{\text{rat}}$. $(C^{*op}\#_R A)^{\text{rat}}$ will be the C^{*op}-rational part of $C^{*op}\#_R A$ as a right C^{*op}-module.

LEMMA 7.4. *With notation as above, $(C^{*op}\#_R A)^{\text{rat}}$ is a twosided ideal of $C^{*op}\#_R A$ and a subring of $\#(C, A)^{\text{rat}}$.*

Proof. It is obvious that $(C^{*op}\#_R A)^{\text{rat}} \subset \#(C, A)^{\text{rat}}$.
Furthermore, since $C^{*op}\#_R A$ is a right C^{*op}-module and a right A-module, and since $A\#C^{*op} \subset C^{*op}\#A$, $(C^{*op}\#_R A)^{\text{rat}}$ is also a right C^{*op} and a right A-module, by part 3) of Proposition 4.10, and we conclude that $(C^{*op}\#_R A)^{\text{rat}}$ is a right ideal in $C^{*op}\#_R A$.
In order to prove that $(C^{*op}\#_R A)^{\text{rat}}$ is also a left ideal, take $b\#g \in C^{*op}\#_R A$, $a\#f \in (C^{*op}\#_R A)^{\text{rat}}$ and $h \in C^*$. We compute

$$((b\#g)(a\#f))h = (b\#g)((a\#f)h) = (b\#g)((a_i\#f_i)h(c_i)) = ((b\#g)(a_i\#f_i))h(c_i)$$

and we find that $(b\#g)(a\#f) \in (C^{*op}\#_R A)^{\text{rat}}$, as needed. $\square$

COROLLARY 7.5. *The Morita context associated to the A-coring $\mathcal{C} = A \otimes C$ can be restricted to a Morita context*

$$(B, (C^{*op}\#_R A)^{\text{rat}}, A, \widetilde{Q}, \widetilde{\tau}, \widetilde{\mu})$$

with $\widetilde{Q} = Q \cap (C^{*op}\#_R A)^{\text{rat}}$ and $\widetilde{\tau}, \widetilde{\mu}$ the ristricted maps.

LEMMA 7.6. *Let (A, C, ψ) be a completely factorizable entwining structure, and assume that A is a free k-module (e.g. k is a field). Then*

$$(C^{*op}\#_R A)^{\text{rat}} \cong (A \otimes C^{*op})^{\text{rat}} = A \otimes (C^{*op})^{\text{rat}}$$

and

$$(A \otimes (C^{*op})^{\text{rat}})^{C^*} = A \otimes ((C^{*op})^{\text{rat}})^{C^*}$$

as (A, C^{*op})-bimodules.

Proof. It follows from Lemma 7.3 that R is a morphism of (A, C^{*op})-bimodules and from Corollary 6.5, that R is bijective, hence $(C^{*op}\#_R A)^{\text{rat}} \cong (A \otimes C^{*op})^{\text{rat}}$. Obviously $A\otimes(C^{*op})^{\text{rat}} \subset (A\otimes C^{*op})^{\text{rat}}$. Choose a basis $\{e_i\}_{i\in I}$ of A. Then for every $\sum_i a_i\otimes f_i \in (A\otimes C^{*op})^{\text{rat}}$, we can find $f_{ij} \in C^*$ such that $\sum_i a_i\otimes f_i = \sum_{i,j\in I} e_j\otimes f_{ij}$, with only finitely many of the f_{ij} different from zero.
For every $g \in C^*$, we find

$$\left(\sum_i a_i \otimes f_i\right)g = \left(\sum_i a_i' \otimes f_i'\right)g(c_i) = \left(\sum_{i,j} e_j \otimes f_{ij}'\right)g(c_i) = \sum_{i,j} e_j \otimes g(c_i)f_{ij}'$$

We also have that $(\sum_i a_i \otimes f_i)g = (\sum_{i,j} e_j \otimes f_{ij})g = \sum_{i,j} e_j \otimes gf_{ij}$, and we find that $gf_{ij} = \sum_i g(c_i)f_{ij}'$, since $\{e_i\}_{i\in I}$ is a basis of A. It follows that $f_{ij} \in (C^{*op})^{\text{rat}}$ and $(A \otimes C^{*op})^{\text{rat}} \subset A \otimes (C^{*op})^{\text{rat}}$. The final statement can be proved in a similar way. $\square$

RATIONALITY FOR CORINGS

LEMMA 7.7. *If (A, C, ψ) is a factorizable entwining structure over a field k, then $(ab)_\alpha \otimes \lambda^\alpha = a_\alpha b_\beta \otimes \lambda^\beta \circ \lambda^\alpha$ and $1_\alpha \otimes \lambda^\alpha = 1 \otimes I_C$.*

Let $\widetilde{B} = \{b \in A \mid \alpha(b) = b \otimes I_C\}$. Then $\widetilde{B} \subseteq B$ and in case of a factorizable entwining structure arising form a H-comodule algebra A (i.e. a Doi-Koppinen structure (A, H, H)), we have $\widetilde{B} = B = \{a \in A \mid a_{[0]} \otimes a_{[1]} = a \otimes 1_H\}$.

PROPOSITION 7.8. *Let (A, C, ψ) be a completely factorizable entwining structure, where C is a left and right co-Frobenius coalgebra over a field k. With notation as above, $\widetilde{Q} \cong A$ as right $\widetilde{B}$-modules.*

Proof. It follows from Propositions 3.6 and 5.1 that $Q = (\#(C, A)^{\mathrm{rat}})^{\mathrm{co}C} = (\#(C, A)^{\mathrm{rat}})^{C^*}$. Applying Lemma 7.4, we find that

$$\begin{aligned}
\widetilde{Q} &= (\#(C, A)^{\mathrm{rat}})^{C^*} \cap (C^{*\mathrm{op}}\#_R A)^{\mathrm{rat}} = ((C^{*\mathrm{op}}\#_R A)^{\mathrm{rat}})^{C^*} \\
&\cong (A \otimes (C^{*\mathrm{op}})^{\mathrm{rat}})^{C^*} \cong A \otimes ((C^{*\mathrm{op}})^{\mathrm{rat}})^{C^*}
\end{aligned}$$

$f \in ((C^{*\mathrm{op}})^{\mathrm{rat}})^{C^*}$ if and only if $gf = g(x)f$ for every $g \in C^*$. If C is left and right co-Frobenius, then by (the dual version of) Remark 7.2, we find an isomorphism $k \to ((C^{*\mathrm{op}})^{\mathrm{rat}})^{C^*}$.

$t = j(x)$ is a generator for $((C^{*\mathrm{op}})^{\mathrm{rat}})^{C^*}$, so we find a bijective map $\varphi : A \to \widetilde{Q}$ given by $\varphi(a) = t \circ \lambda^\alpha \#a_\alpha$. We finally have to show that φ is right $\widetilde{B}$-linear:

$$\begin{aligned}
\varphi(ab) &= t \circ \lambda^\alpha \#(ab)_\alpha = t \circ \lambda^\beta \circ \lambda^\alpha \#a_\alpha b_\beta \\
&= t \circ I_C \circ \lambda^\alpha \#a_\alpha b = t \circ \lambda^\alpha \#a_\alpha b \\
&= (t \circ \lambda^\alpha \#a_\alpha)b = \varphi(a)b
\end{aligned}$$

$\square$

Remark 7.9. We can now transport the C^*-module structure from $\widetilde{Q}$ to A. In the case where $C = H$, this is done in [4] using the distinguished grouplike element. Factorizability implies that $(C^{*\mathrm{op}})^{\mathrm{rat}}\#_R A$ is an A-bimodule. This is a generalization of the well-known fact that, for a Hopf algebra H, $H^{*\mathrm{rat}}$ is an H-bimodule.

REFERENCES

[1] J. Abuhlail, Morita contexts for corings and equivalences, in "Hopf algebras in non-commutative geometry and physics", S. Caenepeel and F. Van Oystaeyen, eds., *Lecture Notes Pure Appl. Math.*, Dekker, New York, to appear.

[2] J.Y. Abuhlail, Rational modules for corings, *Comm. Algebra* **31** (2003), 5793–5840.

[3] P.N. Ánh and L. Márki, Morita equivalence for rings without identity, *Tsukuba J. Math.* **11** (1987), 1–16.

[4] M. Beattie, S. Dăscălescu and Ş. Raianu, Galois extensions for co-Frobenius Hopf algebras, *J. Algebra* **198** (1997), 164–183.

[5] T. Brzeziński, The structure of corings. Induction functors, Maschke-type Theorem, and Frobenius and Galois properties, *Algebr. Representat. Theory* **5** (2002), 389–410.

[6] T. Brzeziński and S. Majid, Coalgebra bundles, *Comm. Math. Phys.* **191** (1998), 467–492.

[7] S. Caenepeel, G. Militaru, and Shenglin Zhu, "Frobenius and separable functors for generalized module categories and nonlinear equations", *Lecture Notes in Math.* **1787**, Springer Verlag, Berlin, 2002.

[8] S. Caenepeel, J. Vercruysse and Shuanhong Wang, Morita Theory for corings and cleft entwining structures, *J. Algebra*, to appear.

[9] M. Cohen, D. Fischman and S. Montgomery, Hopf Galois extensions, smash products, and Morita equivalence, *J. Algebra* **133** (1990), 351–372.

[10] S. Chase and M. E. Sweedler, "Hopf algebras and Galois theory", *Lect. Notes in Math.* **97**, Springer Verlag, Berlin, 1969.

[11] Y. Doi, Unifying Hopf modules, *J. Algebra* **153** (1992), 373–385.

[12] Y. Doi, Generalized smash products and Morita contexts for arbitrary Hopf algebras, in "Advances in Hopf algebras", J. Bergen and S. Montgomery, eds., *Lect. Notes Pure Appl. Math.* **158**, Dekker, New York, 1994.

[13] S. Dăscălescu, C. Năstăsescu and Ş. Raianu, "Hopf algebras: an Introduction", *Monographs Textbooks Pure Appl. Math.* **235**, Marcel Dekker, New York, 2001.

[14] G. S. Garfinkel, Universally torsionless and trace modules, *Trans. Amer. Math. Soc.* **215** (1976), 119–144.

[15] J. Gomez Torrecillas, Separable functors in corings, *Int. J. Math. Math. Sci.* **30** (2002), 203–225.

[16] M. Koppinen, Variations on the smash product with applications to group-graded rings, *J. Pure Appl. Algebra* **104** (1995), 61–80.

[17] P. Schauenburg, Doi-Koppinen modules versus entwined modules, *New York J. Math.*, **6** (2000), 325–329.

[18] M. E. Sweedler, The predual Theorem to the Jacobson-Bourbaki Theorem, *Trans. Amer. Math. Soc.* **213** (1975), 391–406.

[19] M. Takeuchi, as referred to in MR 200c 16047, by A. Masuoka.

[20] R. Wisbauer, "Foundations of Module and Ring Theory", Gordon and Breach, Reading, 1991

[21] R. Wisbauer, On the category of comodules over corings, in "Mathematics and mathematics education (Bethlehem, 2000)", World Sci. Publishing, River Edge, NJ, 2002, 325–336.

[22] R. Wisbauer, On Galois corings, in "Hopf algebras in non-commutative geometry and physics", S. Caenepeel and F. Van Oystaeyen, eds., *Lecture Notes Pure Appl. Math.*, Dekker, New York, to appear..

[23] B. Zimmermann-Huisgen, Pure submodules of direct products of free modules, *Math. Ann.* **224** (1976), 233–245.

Morita Duality for Corings over Quasi-Frobenius Rings

L. EL KAOUTIT AND J. GOMEZ-TORRECILLAS   Departamento de Álgebra, Universidad de Granada, E-18071 Granada, Spain
e-mail address: kaoutit@fedro.ugr.es, torrecil@ugr.es

Pentru Profesor Constantin Năstăsescu cu ocazia aniversării a 60 de ani.

ABSTRACT. Given two corings $\mathfrak{C}$ and $\mathfrak{D}$ projective over their Quasi-Frobenius ground rings, then there is a Morita duality between their categories of comodules $\mathcal{M}^{\mathfrak{C}}$ and $^{\mathfrak{D}}\mathcal{M}$ if and only if $\mathfrak{C}$ and $\mathfrak{D}$ are Morita-Takeuchi equivalent and semiperfect.

1. INTRODUCTION

Morita duality for coalgebras over fields is tightly linked to the notion of a semiperfect coalgebra. This was shown in [15, 16], with the help of Lin's fundamental paper [20]. Although the proofs (and hence, the results) from these papers can be transferred to coalgebras over Quasi-Frobenius commutative rings, one soon realizes that this is not the case for corings over Quasi-Frobenius rings. This is, apart from the own difficulties of the use of bimodules, ultimately due to the fact that every coring has two different convolution rings. Here, we show how to overcome these technical difficulties. Our approach is based essentially on three pillars: an appropriate way to extend the dual functors from modules over the ground ring to comodules, some ideas coming from the study of semiperfect coalgebras developed in [20, 15, 16, 25] and [21], and the role played in duality by linearly compact subcategories accordingly to Gómez Pardo and Guil approach [11, 12, 17].

2. THE DUAL FUNCTORS

In this paper, all rings have a unit. The opposite ring of a ring R will be denoted by R^{op}. All undefined notions can be found in [2] and [22]. The category of all right (resp. left) modules over a ring R is denoted by $\mathcal{M}_R$ (resp. $_R\mathcal{M}$). A *coring* over a ring A (an A–coring, for short) is an A–bimodule $\mathfrak{C}$ together with two A–bimodule homomorphisms $\Delta : \mathfrak{C} \rightarrow \mathfrak{C} \otimes_A \mathfrak{C}$ (comultiplication) and $\epsilon : \mathfrak{C} \rightarrow A$ (counit) sastisfying the usual coassociative and counitary properties [23]. A right $\mathfrak{C}$–comodule is a right A–module M with a homomorphism of A–modules $\rho_M :$

2000 *Mathematics Subject Classification.* 16W30.

Key words and phrases. semiperfect coring, Morita duality, Morita-Takeuchi equivalence for corings.

Investigación parcialmente financiada por el Proyecto BFM2001-3141 del Ministerio de Ciencia y Tecnología de España y FEDER.

$M \to M \otimes_A \mathfrak{C}$ verifying natural axioms. These are the objects of the additive category $\mathcal{M}^{\mathfrak{C}}$ of all right $\mathfrak{C}$-comodules, whose morphisms are defined in a natural way. For right $\mathfrak{C}$-comodules M and N, the notation $\mathrm{Hom}_{\mathfrak{C}}(M, N)$ stands for the abelian group of all morphisms in $\mathcal{M}^{\mathfrak{C}}$ from M to N. The category $\mathcal{M}^{\mathfrak{C}}$ is easily shown to have arbitrary coproducts and co-kernels, which can be already computed in the category $\mathcal{M}_A$ of all right modules over A. If $_A\mathfrak{C}$, the left A-module underlying $\mathfrak{C}$, is flat, then $\mathcal{M}^{\mathfrak{C}}$ is an abelian category. The converse is not true (see [8, Example 1.1]). The category $^{\mathfrak{C}}\mathcal{M}$ of left $\mathfrak{C}$-comodules is understood analogously.

Given any right A-module M, we refer to $M^* = \mathrm{Hom}_A(M, A_A)$ as its *dual* left A-module, where the left A-action is given, as usual, by $(af)(m) = af(m)$, for every $a \in A, m \in M, f \in M^*$. This leads to the well known contravariant functor $(-)^* : \mathcal{M}_A \to {_A}\mathcal{M}$ which acts on a morphism $f : M \to N$ in $\mathcal{M}_A$ as $f^*(\alpha) = \alpha \circ f$, for every $\alpha \in M^*$. Of course, there is an analogous contravariant functor $^*(-) : {_A}\mathcal{M} \to \mathcal{M}_A$. Our first objective is to extend these dual functors to comodules; some few preliminaries on rational modules are needed.

The coring itself can be seen as a right $\mathfrak{C}$-comodule with coaction Δ. Given a right $\mathfrak{C}$-comodule M we have a natural isomorphism of left A-modules $\mathrm{Hom}_{\mathfrak{C}}(M, \mathfrak{C}) \cong M^*$ that maps a homomorphism f onto $\epsilon \circ f$. For a left $\mathfrak{C}$-comodule M we have an isomorphism of right A-modules $\mathrm{Hom}_{\mathfrak{C}}(M, \mathfrak{C}) \cong {^*M}$. The product (composition) in the ring $\mathrm{End}(_{\mathfrak{C}}\mathfrak{C})$ is then transferred to the convolution product on $^*\mathfrak{C}$, which reads $\sigma\tau = \sigma \circ (\mathfrak{C} \otimes_A \tau) \circ \Delta$ for $\sigma, \tau \in {^*\mathfrak{C}}$ [23]. Analogously, we have an isomorphism of rings $\mathrm{End}(\mathfrak{C}_{\mathfrak{C}})^{\mathrm{op}} \cong \mathfrak{C}^*$, where the convolution product in $\mathfrak{C}^*$ is now given by $\sigma\tau = \tau \circ (\sigma \otimes_A \mathfrak{C}) \circ \Delta$. Every right comodule $\rho_M : M \to M \otimes_A \mathfrak{C}$ leads in this way to a left $^*\mathfrak{C}$-module structure on M given by the ring homomorphism $^*\mathfrak{C} \to \mathrm{End}(M_{\mathbb{Z}})$ that maps $\sigma \in {^*\mathfrak{C}}$ onto $(M \otimes_A \sigma) \circ \rho_M$. We have in this way a faithful functor $\omega^r : \mathcal{M}^{\mathfrak{C}} \to {_{*\mathfrak{C}}}\mathcal{M}$. This functor becomes an inclusion of $\mathcal{M}^{\mathfrak{C}}$ as a full subcategory of $_{*\mathfrak{C}}\mathcal{M}$ when $_A\mathfrak{C}$ is locally projective (see [26, 3.5]). Locally projective modules are flat [27, Theorem 2.1]. Such an inclusion can be obtained by recognizing the image of ω^r in $_{*\mathfrak{C}}\mathcal{M}$ as the full subcategory of all rational left $^*\mathfrak{C}$-modules. This can be done as follows (see [8] or [1]): define the rational submodule $\mathrm{Rat}^l_{\mathfrak{C}}(M)$ of a left $^*\mathfrak{C}$-module M as the subset of those elements $m \in M$ for which there are finitely many $(m_i, c_i) \in M \times \mathfrak{C}$ (called a system of rational parameters) such that $\sigma m = \sum_i m_i \sigma(c_i)$ for every $\sigma \in {^*\mathfrak{C}}$. That this construction is mathematically sound is guaranteed here by the locally projective condition on $_A\mathfrak{C}$. By [8, Theorem 2.6], ω^r becomes an isomorphism of categories between $\mathcal{M}^{\mathfrak{C}}$ and $\mathrm{Rat}^l_{\mathfrak{C}}(_{*\mathfrak{C}}\mathcal{M})$, where this last is the full subcategory of $_{*\mathfrak{C}}\mathcal{M}$ whose objects are the modules M satisfying $\mathrm{Rat}^l_{\mathfrak{C}}(M) = M$.

PROPOSITION 2.1. *Let $\mathfrak{C}$ be an A-coring.*

a) *The dual M^* of every right $\mathfrak{C}$-comodule M has a structure of right $\mathfrak{C}^*$-module given by*

$$x.\sigma = \sigma \circ (x \otimes_A \mathfrak{C}) \circ \rho_M, \quad \text{for any } x \in M^* \text{ and } \sigma \in \mathfrak{C}^*. \tag{1}$$

In this way, the dual functor $(-)^* : \mathcal{M}_A \to {_A}\mathcal{M}$ extends to a functor $(-)^* : \mathcal{M}^{\mathfrak{C}} \to \mathcal{M}_{\mathfrak{C}^*}$ making the diagram (2) commutative, where $U^r : \mathcal{M}_{\mathfrak{C}^*} \to {_A}\mathcal{M}$ is the restriction of scalars functor associated to the canonical ring homomorphism $\rho : A^{\mathrm{op}} \to \mathfrak{C}^*$ which sends $a \in A$ to $\rho(a) : c \mapsto a\epsilon(c)$.

$$\begin{CD} \mathcal{M}^{\mathcal{C}} @>(-)^*>> \mathcal{M}_{\mathcal{C}^*} \\ @VU_AVV @VVU^rV \\ \mathcal{M}_A @>>(-)^*> {}_A\mathcal{M} \end{CD} \qquad (2)$$

b) *The dual* $^*N$ *of every left* $\mathcal{C}$*-comodule* N *has a structure of left* $^*\mathcal{C}$*-module given by*

$$\delta.y = \delta \circ (\mathcal{C} \otimes_A y) \circ \lambda_N, \quad \text{for any } y \in {}^*N \text{ and } \delta \in {}^*\mathcal{C}.$$

In this way, the dual functor $^*(-) : {}_A\mathcal{M} \to \mathcal{M}_A$ *extends to a functor* $^*(-) : {}^{\mathcal{C}}\mathcal{M} \to {}_{*\mathcal{C}}\mathcal{M}$ *making diagram (3) commutative, where* $U^l : {}_{*\mathcal{C}}\mathcal{M} \to \mathcal{M}_A$ *is the restriction of scalars functor associated to the canonical ring homomorphism* $\lambda : A^{op} \to {}^*\mathcal{C}$ *which sends* $a \in A$ *to* $\lambda(a) : c \mapsto \epsilon(c)a$.

$$\begin{CD} {}^{\mathcal{C}}\mathcal{M} @>{}^*(-)>> {}_{*\mathcal{C}}\mathcal{M} \\ @V{}_AUVV @VVU^lV \\ {}_A\mathcal{M} @>>{}^*(-)> \mathcal{M}_A \end{CD} \qquad (3)$$

Proof. We will only show (a), because (b) is similarly proved. For every right $\mathcal{C}$-comodule M, the abelian group $\text{Hom}_{\mathcal{C}}(M, \mathcal{C})$ becomes a left module over $\text{End}(\mathcal{C}_{\mathcal{C}})$ in a canonical way. Now, use the ring isomorphism $\text{End}(\mathcal{C}_{\mathcal{C}})^{op} \cong \mathcal{C}^*$ and the isomorphism $\text{Hom}_{\mathcal{C}}(M, \mathcal{C}) \cong M^*$ to transfer this left action to a structure of right $\mathcal{C}^*$-module over M^*. Some straightforward computations show that this right action is precisely the given in (1). Since the isomorphism $\text{Hom}_{\mathcal{C}}(-, \mathcal{C}) \cong (-)^*$ is natural, we have that every morphism f in $\mathcal{M}^{\mathcal{C}}$ gives a morphism f^* in $\mathcal{M}_{\mathcal{C}^*}$. Therefore, we have a contravariant functor $(-)^* : \mathcal{M}^{\mathcal{C}} \to \mathcal{M}_{\mathcal{C}^*}$. The commutativity of the displayed functorial diagram comes from the fact that the ring isomorphism $\text{End}(\mathcal{C}_{\mathcal{C}})^{op} \cong \mathcal{C}^*$ acts as the identity over A^{op}. □

The following (non-commutative) diagram of functors summarizes the situation, where the dotted arrows make sense under the locally projective condition on the suitable side

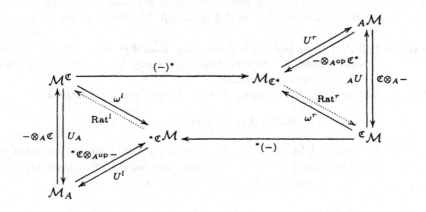

The sideways pairs of arrows represent the canonical adjunctions: $- \otimes_A \mathfrak{C}$ is left adjoint to U_A (see [18, Proposition 3.1], [4, Lemma 3.1]), the adjunction between $^*\mathfrak{C} \otimes_{A^{\mathrm{op}}} -$ and U^l is the associated to the ring homomorphism $A^{\mathrm{op}} \to {}^*\mathfrak{C}$, and the embedding functor $\omega^l : \mathcal{M}^{\mathfrak{C}} = \mathrm{Rat}^l(\cdot_{\mathfrak{C}}\mathcal{M}) \to \cdot_{\mathfrak{C}}\mathcal{M}$ is, as usual, left adjoint to the left exact preradical $\mathrm{Rat}^l : \cdot_{\mathfrak{C}}\mathcal{M} \to \cdot_{\mathfrak{C}}\mathcal{M}$. Using these adjunctions, one can easily deduce the following.

LEMMA 2.2. *Assume $_A\mathfrak{C}$ to be locally projective, and let M be a right $\mathfrak{C}$-comodule. Then*

   a) *$M_{\mathfrak{C}}$ is finitely generated (in $\mathcal{M}^{\mathfrak{C}}$) if and only if M_A is finitely generated if and only if $\cdot_{\mathfrak{C}}M$ is finitely generated.*

   b) *If $M_{\mathfrak{C}}$ is finitely presented (in $\mathcal{M}^{\mathfrak{C}}$), then M_A is finitely presented.*

Proof. For the proof of (a), use that an object X in a Grothendieck category $\mathcal{A}$ is finitely generated if and only if $\mathrm{Hom}_{\mathcal{A}}(X, -)$ preserves direct unions. If, in addition, $\mathcal{A}$ is locally finitely generated, then X is finitely presented if and only if $\mathrm{Hom}_{\mathcal{A}}(X, -)$ preserves direct limits [22, Chapter V, Prop. 3.4]. Since the cyclic rational left $^*\mathfrak{C}$-modules form a generating set for the category $\mathcal{M}^{\mathfrak{C}}$, we easily deduce (b). $\qquad\square$

There is a natural transformation $\Phi : 1 \to {}^*(-) \circ (-)^*$ defined, for every right A-module M, by the evaluation map

$$\Phi_M : M \to {}^*(M^*), \; m \mapsto [x \mapsto x(m)]$$

Analogously, there is a natural transformation $\Phi' : 1 \to (-)^* \circ {}^*(-)$ for left A-modules. Assume that $_A\mathfrak{C}$ and $\mathfrak{C}_A$ are locally projective. For every right $\mathfrak{C}$-comodule M, let $j : \mathrm{Rat}^r_{\mathfrak{C}}(M^*) \to M^*$ denote the inclusion map and define $\sigma_M : M \to {}^*(\mathrm{Rat}^r_{\mathfrak{C}}(M^*))$ as $^*j \circ \Phi_M$. Similarly, we can define $\sigma'_N : N \to (\mathrm{Rat}^l_{\mathfrak{C}}({}^*N))^*$ for every left $\mathfrak{C}$-comodule N.

PROPOSITION 2.3. *Let $\mathfrak{C}$ be an A-coring such that $_A\mathfrak{C}$ and $\mathfrak{C}_A$ are locally projective modules. Let M (resp. N) be a right (resp. left) $\mathfrak{C}$-comodule.*

   a) *σ_M is left $^*\mathfrak{C}$-linear and σ'_N is right $\mathfrak{C}^*$-linear.*

   b) *If M_A (resp. $_AN$) is finitely presented, then the dual M^* (resp. $^*N$) is a left (resp. right) $\mathfrak{C}$-comodule and Φ_M is left $^*\mathfrak{C}$-linear (resp. Φ'_N is right $\mathfrak{C}^*$-linear).*

   c) *Assume that A is right and left self-injective, then Φ_M (resp. Φ'_N) is a natural $^*\mathfrak{C}$-isomorphism (resp. $\mathfrak{C}^*$-isomorphism) for every $M_{\mathfrak{C}}$ (resp. $_{\mathfrak{C}}N$) such that M_A (resp. $_AN$) is finitely presented.*

   d) *If A is right and left Noetherian, then we have a pair of functors*

$$(-)^* : \mathcal{M}^{\mathfrak{C}}_f \leftrightarrows {}^{\mathfrak{C}}_f\mathcal{M} : {}^*(-),$$

where $\mathcal{M}^{\mathfrak{C}}_f$ (resp. $^{\mathfrak{C}}_f\mathcal{M}$) denotes the category of all finitely generated right (resp. left) $\mathfrak{C}$-comodules. Moreover, we have natural transformations $\Phi : 1_{\mathcal{M}^{\mathfrak{C}}_f} \to {}^*(-) \circ (-)^*$ and $\Phi' : 1_{{}^{\mathfrak{C}}_f\mathcal{M}} \to (-)^* \circ {}^*(-)$.

Proof. (a) A straightforward computation shows that $\mathfrak{C}$ is a $^*\mathfrak{C}$-$\mathfrak{C}^*$-bimodule. Consider the commutative diagram

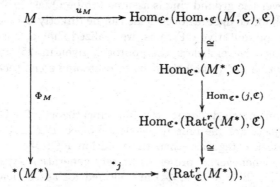

where u_M is the canonical evaluation map, which is left $^*\mathfrak{C}$-linear. The left $^*\mathfrak{C}$-linear isomorphisms in the diagram are obtained from the isomorphisms $M^* \cong \text{Hom}_{\mathfrak{C}}(M, \mathfrak{C}) = \text{Hom}_{^*\mathfrak{C}}(M, \mathfrak{C})$ and $\text{Hom}_{\mathfrak{C}^*}(\text{Rat}^r_{\mathfrak{C}}(M^*), \mathfrak{C}) = \text{Hom}_{\mathfrak{C}}(\text{Rat}^r_{\mathfrak{C}}(M^*), \mathfrak{C}) \cong {}^*(\text{Rat}^r_{\mathfrak{C}}(M^*))$. Thus, $\sigma_M = {}^*j \circ \Phi_M$ becomes a composition of left $^*\mathfrak{C}$-linear maps. The right $\mathfrak{C}^*$-linearity of σ'_N is similarly showed.

(b) If M_A is finitely presented, as $\mathfrak{C}_A$ is flat, then we have a natural isomorphism

$$\eta_M : \mathfrak{C} \otimes_A \text{Hom}_A(M_A, A) \longrightarrow \text{Hom}_A(M_A, \mathfrak{C})$$
$$c \otimes_A x \longmapsto [m \mapsto cx(m)].$$

Fix $x \in M^*$, and consider an arbitrary $\sigma \in \mathfrak{C}^*$, so $(x \otimes_A \mathfrak{C}) \circ \rho_M \in \text{Hom}_A(M_A, \mathfrak{C})$. Therefore, there exists a finite subset $\{(c_i, x_i)\}$ of $\mathfrak{C} \times M^*$ such that $\eta_M(\sum_i c_i \otimes_A x_i)(m) = (x \otimes_A \mathfrak{C}) \circ \rho_M(m) = \sum_i c_i x_i(m)$, for any $m \in M$. Hence $x.\sigma = \sigma \circ (x \otimes_A \mathfrak{C}) \circ \rho_M = \sum_i \sigma \circ (c_i \otimes_A x_i)$, thus $x.\sigma(m) = \sum_i \sigma(c_i) x_i(m)$, for any $m \in M$; that is $x.\sigma = \sum_i \sigma(c_i).x_i$. This implies that $\{(c_i, x_i)\}$ is a set of rational parameters for x. By [8, Theorem 2.6], $M^* \in \text{Rat}^r_{\mathfrak{C}}(M_{\mathfrak{C}^*}) \cong {}^{\mathfrak{C}}M$.

(c) This is a consequence of (a) and (b) in conjunction with [22, p. 47].

(d) Let $M \in \mathcal{M}^{\mathfrak{C}}_f$ then, by Lemma 2.2, M_A is finitely generated and, since A is noetherian, M_A is finitely presented. By (b), M^* is a left $\mathfrak{C}$-comodule which is finitely presented as a left A-module because A is noetherian. Therefore, $M^* \in {}^{\mathfrak{C}}_f\mathcal{M}$, and we have thus defined, with the help of Proposition 2.1, the functor $(-)^* : \mathcal{M}^{\mathfrak{C}}_f \to {}^{\mathfrak{C}}_f\mathcal{M}$. The functor $^*(-)$ is similarly obtained. □

REMARK 2.4. If $M_{\mathfrak{C}}$ is a comodule such that M_A is finitely presented and $\mathfrak{C}_A$ is flat, then the composite map $M^* \cong \text{Hom}_{\mathfrak{C}}(M, \mathfrak{C}) \subseteq \text{Hom}_A(M, \mathfrak{C}) \cong \mathfrak{C} \otimes_A M^*$ gives a structure of left $\mathfrak{C}$-comodule over M^* as in the case of coalgebras [6], [25, 5.4]. This structure coincides with the given in Proposition 2.3.(b).

3. SEMIPERFECT CORINGS OVER QUASI-FROBENIUS RINGS

A coring is said to be *right semiperfect* if its category of right comodules is a Grothendieck category, and each finitely generated right comodule has a projective cover. A well understood class of semiperfect corings is the given by the cosemisimple corings [8], [7], [14].

Here we consider semiperfect corings as a generalization of semiperfect coalgebras over a field. Most of the results on semiperfect coalgebras can be extended to the coring case, whenever the ground ring is assumed to be Quasi-Frobenius. However, although the proofs for coalgebras over fields can be directly transferred to the case of coalgebras over commutative QF rings, we realized that this is not the case for corings over QF rings. Nevertheless, the approach given in [25] and [21] to coalgebras over commutative QF rings helped us to overcome some technical difficulties in this section.

We use the notation $\mathcal{A}_f$ to designate the full subcategory of a Grothendieck category $\mathcal{A}$ whose objects are the finitely generated ones. If $F : \mathcal{A} \to \mathcal{B}$ is a functor between Grothendieck categories, then the notation $F_f : \mathcal{A}_f \to \mathcal{B}_f$ refers to the restriction functor, whenever F preserves finitely generated objects. Some variations of this notation (like $\mathcal{A}^f, {}^f\mathcal{A}, {}_f\mathcal{A}$) are allowed for aesthetical reasons. Thus, the category of all finitely generated right modules over a ring A is $\mathcal{M}_A^f$, while the category of finitely generated left A–modules is ${}_A^f\mathcal{M}$. Recall that a ring A is said to be Quasi-Frobenius (QF, for short) if the functors $(-)^* : \mathcal{M}_A^f \leftrightarrows {}_A^f\mathcal{M} : {}^*(-)$ give a contravariant equivalence of categories. Quasi-Frobenius rings are characterized in several ways, for example, they are just the artinian selfinjective rings. In particular, every flat module is projective over a QF ring. Given a comodule M over a coring, we write $E(M)$ to denote its injective envelope in and $\mathrm{Soc}(M)$ for its socle. Notice that $E(M)$ do exist whenever the corresponding category of comodules is a Grothendieck category.

Recall from [9, page 356] that a Grothendieck category $\mathcal{A}$ is said to be *locally finite* if $\mathcal{A}$ has a generating set consisting of objects of finite length.

PROPOSITION 3.1. *Let A be a QF ring, and $\mathfrak{C}$ be an A–coring such that ${}_A\mathfrak{C}$ is a projective module. Then*

    a) *$\mathfrak{C}_\mathfrak{C}$ is an injective comodule and every right $\mathfrak{C}$–comodule embeds in a coproduct of copies of $\mathfrak{C}_\mathfrak{C}$.*

    b) *$\mathcal{M}^\mathfrak{C}$ is a locally finite category, in particular $\mathfrak{C}_\mathfrak{C} = \oplus_{\omega\in\Omega} E(S_\omega)^{(n_\omega)}$, where $\{S_\omega\}_{\omega\in\Omega}$ is the set of all representatives of simple right $\mathfrak{C}$–comodules, and the n_ω's are cardinal numbers.*

Proof. (a) The ring A is right selfinjective and all right modules embed in free modules (by the Faith-Walker characterization of QF rings). These properties are transferred to $\mathfrak{C}$ because the exact functor $- \otimes_A \mathfrak{C} : \mathcal{M}_A \to \mathcal{M}^\mathfrak{C}$ preserves direct sums and has an exact left adjoint.

(b) This is a consequence of (a) and Lemma 2.2. $\qquad\qquad\square$

The following consequence of Proposition 2.3 is crucial for the development of the theory.

THEOREM 3.2. *Let $\mathfrak{C}$ be a coring over a QF ring A such that ${}_A\mathfrak{C}$ and $\mathfrak{C}_A$ are projective modules. Then we have a contravariant equivalence of categories*

$$(-)^* : \mathcal{M}_f^\mathfrak{C} \leftrightarrows {}_f^\mathfrak{C}\mathcal{M} : {}^*(-)$$

The following proposition illustrates the use of Theorem 3.2 in conjunction with the locally finite property for comodules given in Proposition 3.1.

PROPOSITION 3.3. *Let $\mathfrak{C}$ be a coring over a QF ring A such that $_A\mathfrak{C}$ and $\mathfrak{C}_A$ are projective modules. Consider $M \in \mathcal{M}_f^{\mathfrak{C}}$, a finitely generated comodule. Then*

   a) *If $M_{\mathfrak{C}}$ is an injective comodule, then $M_{\mathfrak{C}^*}^*$ is a projective module.*
   b) *$M_{\mathfrak{C}}$ is projective if and only if $_{\mathfrak{C}}M^*$ is injective.*
   c) *$M_{\mathfrak{C}}$ is injective if and only if $_{\mathfrak{C}}M^*$ is projective.*
   d) *$M_{\mathfrak{C}}$ is a projective comodule if and only if $\cdot_{\mathfrak{C}}M$ is a projective module.*

Proof. (a) We use an argument taken from the proof of [6, Proposition 4]. If M is an injective right $\mathfrak{C}$–comodule then, by Proposition 3.1, there is a splitting monomorphism of right comodules $M \hookrightarrow \mathfrak{C}^{(n)}$, for some finite index n. By applying the functor $(-)^* : \mathcal{M}^{\mathfrak{C}} \to \mathcal{M}_{\mathfrak{C}^*}$ given in Proposition 2.1, we obtain a splitting epimorphism of right $\mathfrak{C}^*$–modules $\mathfrak{C}^{*(n)} \twoheadrightarrow M^*$ which, of course, implies that M^* is a projective right $\mathfrak{C}^*$–module.

(b) If $M_{\mathfrak{C}}$ is a projective comodule, then, by Theorem 3.2, M^* is L–injective for every finitely generated left $\mathfrak{C}$–comodule L. Consequently, by [2, Proposition 16.13.(2)], M^* is $(\oplus L_i)$–injective for every family $\{L_i\}_i$ of finitely generated left $\mathfrak{C}$–comodules. Therefore, by [2, 16.13.(1)], and Proposition 3.1.(b), M^* is an injective left $\mathfrak{C}$–comodule. Conversely, suppose that M^* is an injective left comodule. By the version of part (a) for left comodules, it follows that $M \cong {}^*(M^*)$ is a projective left $^*\mathfrak{C}$–module, and, by the isomorphism of categories $\mathrm{Rat}_{\mathfrak{C}}^l(\cdot_{\mathfrak{C}}M) \cong \mathcal{M}^{\mathfrak{C}}$, a projective right $\mathfrak{C}$–comodule.

(c) Write $N = {}_{\mathfrak{C}}M^*$, and apply the version of part (b) to left comodules to obtain that $_{\mathfrak{C}}N$ is projective if and only if $^*N_{\mathfrak{C}}$ is injective. Finally, apply the duality stated in Theorem 3.2.

(d) The proof of (b) runs here to prove that if $M_{\mathfrak{C}}$ is projective then $\cdot_{\mathfrak{C}}M$ is projective. $\square$

A *projective cover* of an object X of a Grothendieck category $\mathcal{A}$ is an epimorphism $p : P \to X$ in $\mathcal{A}$ such that P is a projective object and $Ker(p)$ is a superfluous (or small) subobject of P (see, e.g. [22, Chapter V, page 120]).

PROPOSITION 3.4. *Let $\mathfrak{C}$ be a coring over a QF ring A such that $_A\mathfrak{C}$ and $\mathfrak{C}_A$ are projective modules. A finitely generated right $\mathfrak{C}$–comodule M has a projective cover in $\mathcal{M}^{\mathfrak{C}}$ if and only if $_AE(_{\mathfrak{C}}M^*)$ is finitely generated.*

Proof. Assume that $P \twoheadrightarrow M$ is a projective cover of M in $\mathcal{M}^{\mathfrak{C}}$. By Proposition 3.3.(d), $P \twoheadrightarrow M$ is a projective cover of M in $\cdot_{\mathfrak{C}}M$, as the lattices of subobjects of $\cdot_{\mathfrak{C}}P$ and $P_{\mathfrak{C}}$ are equal. Then P is a finitely generated left $^*\mathfrak{C}$–module and, by Corollary 2.2, it is a finitely generated right A–module. By Theorem 3.2 and Proposition 3.3.(b) we have that the dual map $M^* \hookrightarrow P^*$ gives an injective hull of M^* in $^{\mathfrak{C}}\mathcal{M}$. Thus, we can write $E(M^*) = P^*$, which is finitely generated as a left A–module.

If $_AE(M^*)$ is finitely generated, then, by Theorem 3.2 and Proposition 3.3.(d), the dual map $^*(E(M^*)) \twoheadrightarrow {}^*(M^*) \cong M$ is a projective cover. $\square$

The following theorem was first proved for coalgebras by Lin in [20].

144   L. EL KAOUTIT AND J. GÓMEZ-TORRECILLAS

THEOREM 3.5. *Let A, $\mathfrak{C}$ be as in Proposition 3.4. The following statements are equivalent.*

  (i) *Every simple right $\mathfrak{C}$–comodule has a projective cover;*
  (ii) *every finitely generated right $\mathfrak{C}$–comodule has a projective cover (i.e. $\mathfrak{C}$ is right semiperfect);*
  (iii) $_AE(N)$ *is finitely generated for every finitely generated left $\mathfrak{C}$–comodule N;*
  (iv) $_AE(S)$ *is finitely generated for every simple left $\mathfrak{C}$–comodule S.*

Proof. $(i) \Rightarrow (iv)$ Let S be a simple left $\mathfrak{C}$–comodule. By Theorem 3.2 we know that $^*S$ is a simple right $\mathfrak{C}$–comodule. By hypothesis, $^*S$ has a projective cover which implies, by Proposition 3.4, that $_AE((^*S)^*)$ is finitely generated. Thus, $_AE(S)$ is finitely generated, as $S \cong (^*S)^*$.

$(iv) \Rightarrow (iii)$ By Proposition 3.1, $E(N) = E(\mathrm{Soc}(N))$ and, since N is finitely generated, $\mathrm{Soc}(N) = S_1 \oplus \cdots \oplus S_r$, for some simple left comodules $S_1, \ldots, S_r$. Therefore, $E(N) = E(S_1) \oplus \cdots \oplus E(S_r)$ and, in particular, it is finitely generated as a left A–module.

$(iii) \Rightarrow (ii)$ If M is a finitely generated right $\mathfrak{C}$–comodule, then M^* is a finitely generated left $\mathfrak{C}$–comodule which, by hypothesis, has a finitely generated injective hull. By Proposition 3.4, M possesses a projective cover.

$(ii) \Rightarrow (i)$ This is obvious. □

The next step is to characterize right semiperfect corings in terms of density. The notion of density we will use is analogue to the usually considered for coalgebras over fields. Let M be a right A–module over a QF ring A, and consider the pairing

$$\langle -, - \rangle : M \times M^* \to A \qquad (\langle m, \varphi \rangle = \varphi(m)).$$

For every right A–submodule U of M, define $U^\perp = \{\varphi \in M^* : \langle U, \varphi \rangle = 0\}$, which is a left A–submodule of M^*. In fact, the mapping $(M/U)^* \to U^\perp$ that sends $\alpha \in (M/U)^*$ to $\alpha \circ \pi$, where $\pi : M \to M/U$ is the canonical projection, is an isomorphism of left A–modules. An analogous definition of $V^\perp$ can be made for left A–submodules V of M^*. Such a submodule V of M^* is said to be *dense* if $V^\perp = 0$. The fact that A_A is an injective cogenerator for $\mathcal{M}_A$ guarantees that V is dense in M^* if and only if $V^{\perp\perp} = M^*$.

The version for coalgebras of the following Lemma 3.6 is [20, Lemma 7]. The first statement of Proposition 3.7 was proved for coalgebras in [25, 6.1], while the second appears in [21, Lemma 1.8].

LEMMA 3.6. *Let $\mathfrak{C}$ be an A–coring over a QF ring such that $\mathfrak{C}_A$ is projective. Let $f : M \hookrightarrow N$ be a monomorphism of right $\mathfrak{C}$–comodules. If $\mathrm{Rat}^r_{\mathfrak{C}}(N^*)$ is dense in N^*, then $\mathrm{Rat}^r_{\mathfrak{C}}(M^*)$ is dense in M^*.*

Proof. This is a consequence of the fact that $f^*(\mathrm{Rat}^r_{\mathfrak{C}}(N^*)) \subseteq \mathrm{Rat}^r_{\mathfrak{C}}(M^*)$. □

PROPOSITION 3.7. *Let $\mathfrak{C}$ be a coring over a QF ring A such that $_A\mathfrak{C}$ and $\mathfrak{C}_A$ are projectives. Then*

  (a) $\mathrm{Soc}(\mathfrak{C}_{\mathfrak{C}}) = \mathrm{Soc}(_{\bullet_{\mathfrak{C}}}\mathfrak{C})$ *is essential in $\mathfrak{C}_{\mathfrak{C}}$, and the Jacobson radical of $\mathfrak{C}^*$ is*

$$\mathrm{Jac}(\mathfrak{C}^*) = (\mathrm{Soc}(_{\bullet_{\mathfrak{C}}}\mathfrak{C}))^\perp$$

(b) *If S is a right simple comodule, then $E(S)^*$ is a cyclic local right $\mathfrak{C}^*$-module with $S^\perp$ the unique maximal right $\mathfrak{C}^*$-submodule.*

Proof. (a) That $\mathrm{Soc}(\mathfrak{C}_\mathfrak{C}) = \mathrm{Soc}(\cdot_\mathfrak{C}\mathfrak{C})$ is essential in $\mathfrak{C}_\mathfrak{C}$ follows from Proposition 3.1. By Proposition 3.1.(a), $\cdot_\mathfrak{C}\mathfrak{C}$ is injective in $\mathcal{M}^\mathfrak{C}$. This implies that the Jacboson radical of $\mathrm{End}(\mathfrak{C}_\mathfrak{C})$ consists of those $f : \mathfrak{C} \to \mathfrak{C}$ with essential kernel, which, in the present case, is equivalent to $f(\mathrm{Soc}(\mathfrak{C}_\mathfrak{C})) = 0$. The ring isomorphism $\mathfrak{C}^* \cong \mathrm{End}(\mathfrak{C}_\mathfrak{C})^{\mathrm{op}}$ gives the description of $\mathrm{Jac}(\mathfrak{C}^*)$.

(b) Consider $S_\mathfrak{C}$ a simple comodule, so we have the following commutative diagram with exact rows

$$
\begin{array}{ccccc}
S & \hookrightarrow & E(S) & \longrightarrow & E(S)/S \\
\downarrow & & \downarrow & & \downarrow \\
\mathrm{Soc}(\cdot_\mathfrak{C}\mathfrak{C}) & \hookrightarrow & \mathfrak{C} & \longrightarrow & \mathfrak{C}/\mathrm{Soc}(\cdot_\mathfrak{C}\mathfrak{C})
\end{array}
$$

Dualising we get

$$
\begin{array}{ccccc}
(\mathrm{Soc}(\cdot_\mathfrak{C}\mathfrak{C}))^\perp \cong (\mathfrak{C}/\mathrm{Soc}(\cdot_\mathfrak{C}\mathfrak{C}))^* & \hookrightarrow & \mathfrak{C}^* & \longrightarrow & \mathrm{Soc}(\cdot_\mathfrak{C}\mathfrak{C})^* \\
\downarrow & & \downarrow & & \downarrow \\
S^\perp \cong (E(S)/S)^* & \hookrightarrow & E(S)^* & \longrightarrow & S^*,
\end{array}
$$

a commutative diagram in $\mathcal{M}_{\mathfrak{C}^*}$. Clearly, $E(S)^*$ is cyclic, and the exactness of the diagram easily gives $S^\perp = E(S)^* \mathrm{Jac}(\mathfrak{C}^*)$. Therefore, $S^\perp$ is superfluous in $E(S)^*$, whence $S^\perp \subseteq \mathrm{Jac}(E(S)^*)$ (the radical of $E(S)^*_{\mathfrak{C}^*}$). By Theorem 3.2, S^* is simple which implies, in view of the second exact row of our diagram, that $S^\perp$ is a maximal right $\mathfrak{C}^*$-submodule of $E(S)^*$, whence $\mathrm{Jac}(E(S)^*) = S^\perp$. $\square$

We are ready to prove the main result in this section. It generalizes, in conjunction with Theorem 4.2 in the next section, [20, Theorem 10].

THEOREM 3.8. *Let $\mathfrak{C}$ be a coring over a QF ring A such that $_A\mathfrak{C}$ and $\mathfrak{C}_A$ are projective modules. The following statements are equivalent.*

(i) *The coring $\mathfrak{C}$ is right semiperfect;*

(ii) $\mathrm{Rat}^l_\mathfrak{C}(\cdot_\mathfrak{C}{}^*\mathfrak{C})$ *is dense in $^*\mathfrak{C}$;*

(iii) $\mathrm{Rat}^l_\mathfrak{C}(^*E(S))$ *is dense in $^*E(S)$ for each simple left $\mathfrak{C}$-comodule S.*

Proof. $(i) \Rightarrow (iii)$ If S is a simple left $\mathfrak{C}$-comodule, then, by Theorem 3.5, $E(S)$ is a finitely generated left A-module. By Proposition 2.3, $^*E(S)$ is a right $\mathfrak{C}$-comodule. Thus, $\mathrm{Rat}^l_\mathfrak{C}(^*E(S)) = {}^*E(S)$ and, obviously, $\mathrm{Rat}^l_\mathfrak{C}(^*E(S))$ is dense in $^*E(S)$.

$(iii) \Rightarrow (ii)$ By Proposition 3.1, $_\mathfrak{C}\mathfrak{C} = \bigoplus_{i \in I} E(S_i)$ for a suitable set S_i of simple left $\mathfrak{C}$-comodules. Clearly, $\bigoplus_{i \in I} {}^*E(S_i)$ is dense in $\prod_{i \in I} {}^*E(S_i) = {}^*\mathfrak{C}$. Therefore $\bigoplus_{i \in I} \mathrm{Rat}^l_\mathfrak{C}(^*E(S_i))$ is dense in $^*\mathfrak{C}$, and so is $\mathrm{Rat}^l_\mathfrak{C}(\cdot_\mathfrak{C}{}^*\mathfrak{C})$.

$(ii) \Rightarrow (iii)$ This is a consequence of (the symmetric version of) Lemma 3.6, as every left comodule of the form $E(S)$ for a simple S embeds in $\mathfrak{C}$.

$(iii) \Rightarrow (i)$ Let S be a left $\mathfrak{C}$-comodule, and consider its injective envelope $E(S) \in {}^\mathfrak{C}\mathcal{M}$. By the version of Proposition 3.7.(b) for left comodules, $^*E(S)$ is a cyclic local left $^*\mathfrak{C}$-module with maximal submodule $S^\perp$. Since $\mathrm{Rat}^l_\mathfrak{C}(^*E(S))$ is dense in $^*E(S)$, we get that $S^\perp \subsetneq S^\perp + \mathrm{Rat}^l_\mathfrak{C}(^*E(S))$; hence $S^\perp + \mathrm{Rat}^l_\mathfrak{C}(^*E(S)) = {}^*E(S)$, as

$S^\perp$ is maximal. Finally, $S^\perp$ is superfluous in $^*E(S)$, whence $\mathrm{Rat}^l_{\mathfrak{C}}(^*E(S)) = {}^*E(S)$. Therefore, $^*E(S)$ is rational and finitely generated $^*\mathfrak{C}$–module. By Theorem 3.2, $E(S) \cong (^*E(S))^*$ is finitely generated (as an A–module, if desired). $\qquad\square$

4. DUALITY FOR SEMIPERFECT CORINGS OVER QF RINGS

This section contains the extension of the duality theory developed in [15, 16] for semiperfect coalgebras over fields to semiperfect corings over QF ground rings. Thus, we assume in this section that the corings are over QF rings.

A *duality* is a contravariant equivalence between two categories. Theorem 3.2 says that the functors $(-)^*$ and $^*(-)$ give a duality between $\mathcal{M}^{\mathfrak{C}}_f$ and $^{\mathfrak{C}}_f\mathcal{M}$. This notion of duality is too restrictive even from the point of view of module theory, where the concept of a Morita duality has been proved to be fundamental. In what follows, let us recall a generalization of Morita duality to Grothendieck categories due to R. R. Colby and K. R. Fuller [5]. Consider contravariant functors between Grothendieck categories

$$H : \mathcal{A} \leftrightarrows \mathcal{A}' : H',$$

together with natural transformations $\tau : 1_{\mathcal{A}} \to H' \circ H$ and $\tau' : 1_{\mathcal{A}'} \to H \circ H'$, satisfying the condition $H(\tau_X) \circ \tau'_{H(X)} = 1_{H(X)}$ and $H'(\tau'_{X'}) \circ \tau_{H'(X')} = 1_{H'(X')}$ for $X \in \mathcal{A}$ and $X' \in \mathcal{A}'$. This situation is called a *right adjoint pair*. Moreover, any pair of natural transformations τ, τ' satisfying these conditions determine a natural isomorphism

$$\eta_{X,X'} : \mathrm{Hom}_{\mathcal{A}}(X, H'(X')) \longrightarrow \mathrm{Hom}_{\mathcal{A}'}(X', H(X)) .$$
$$\alpha \longmapsto H(\alpha) \circ \tau'_{X'}$$
$$H'(\beta) \circ \tau_X \longleftarrow\!\!\!\longmapsto \beta$$

Conversely, given any natural isomorphism

$$\eta_{X,X'} : \mathrm{Hom}_{\mathcal{A}}(X, H'(X')) \to \mathrm{Hom}_{\mathcal{A}'}(X', H(X)),$$

then $\tau_X = \eta^{-1}_{X,H(X)}(1_{H(X)})$ and $\tau'_{X'} = \eta_{H'(X'),X'}(1_{H'(X')})$ satisfy the above conditions. We call an object X of $\mathcal{A}$ (resp. X' of $\mathcal{A}'$) *reflexive* in case τ_X (resp. $\tau'_{X'}$) is an isomorphism. If we denote by $\mathcal{A}_0$ and $\mathcal{A}'_0$ the full subcategories of $\mathcal{A}$ and $\mathcal{A}'$ of the reflexive objects, then H and H' form duality between them. Using the terminology of [11] we will say that the pair of right adjoint functors is a *Colby-Fuller* duality between $\mathcal{A}$ and $\mathcal{A}'$ if and only if the functors H and H' are exact and $\mathcal{A}_0$ and $\mathcal{A}'_0$ are closed under subobjects, quotient objects, and finite direct sums (i.e., they are *finitely closed*) and contain sets of generators for $\mathcal{A}$ and $\mathcal{A}'$ (i.e., they are *generating*).

Now, we return to categories of comodules. Let $\mathfrak{C}$ be a coring over a QF ring A such that $_A\mathfrak{C}$ and $\mathfrak{C}_A$ are projective modules. The duality given in Theorem 3.2 is a Morita duality between $\mathcal{M}^{\mathfrak{C}}$ and $^{\mathfrak{C}}\mathcal{M}$ in the sense of [3], as the subcategories $\mathcal{M}^{\mathfrak{C}}_f$ and $^{\mathfrak{C}}_f\mathcal{M}$ are generating. According to [12, Theorem 2.1], the functors $(-)^*$: $\mathcal{M}^{\mathfrak{C}}_f \rightleftarrows {}^{\mathfrak{C}}_f\mathcal{M} : {}^*(-)$ can be extended to a right adjoint pair of contravariant functors $D : \mathcal{M}^{\mathfrak{C}} \rightleftarrows {}^{\mathfrak{C}}\mathcal{M} : D'$ in a unique way. For our purposes we will need the following description of this extension. For every $M \in \mathcal{M}^{\mathfrak{C}}$ we get, by Proposition 2.2, a left

MORITA DUALITY FOR CORINGS

$^*\mathfrak{C}$–linear map $\sigma_M : M \to {}^*(\mathrm{Rat}^r_{\mathfrak{C}}(M^*))$. Since M is rational, we deduce that the image of σ_M is included in $\mathrm{Rat}^l_{\mathfrak{C}}({}^*(\mathrm{Rat}^r_{\mathfrak{C}}(M^*)))$, so that we have obtained a natural transformation

$$\sigma : 1_{\mathcal{M}^{\mathfrak{C}}} \to \mathrm{Rat}^l_{\mathfrak{C}} \circ {}^*(-) \circ \mathrm{Rat}^r_{\mathfrak{C}} \circ (-)^*.$$

Analogously, we obtain a natural transformation

$$\sigma' : 1_{\mathfrak{C}\mathcal{M}} \to \mathrm{Rat}^r_{\mathfrak{C}} \circ (-)^* \circ \mathrm{Rat}^l_{\mathfrak{C}} \circ {}^*(-).$$

The functors

$$\mathrm{Rat}^r_{\mathfrak{C}} \circ (-)^* : \mathcal{M}^{\mathfrak{C}} \rightleftarrows {}^{\mathfrak{C}}\mathcal{M} : \mathrm{Rat}^l_{\mathfrak{C}} \circ {}^*(-) \tag{4}$$

and the natural transformations σ and σ' give the aforementioned right adjoint pair extending the duality $(-)^* : \mathcal{M}^{\mathfrak{C}}_f \rightleftarrows {}^{\mathfrak{C}}_f\mathcal{M} : {}^*(-)$.

In [16], the authors characterize right semiperfect coalgebras by the locally compactness of the full subcategory of finite-dimensional right comodules. We shall extend this characterization to corings over a QF ring. Recall, from [10], that an object X of a Grothendieck category $\mathcal{A}$, is called *linearly compact* when for each inverse system of epimorphisms $\{p_i : X \twoheadrightarrow X_i\}_{i \in I}$ in $\mathcal{A}$, the induced morphism $\underleftarrow{\lim} p_i : X \to \underleftarrow{\lim} X_i$ is also an epimorphism. A full subcategory $\mathcal{S}$ of $\mathcal{A}$ is said to be *linearly compact* if for every inverse system $\{S_i \to A_i\}$ of epimorphisms, with $S_i \in \mathcal{S}$, the projective limit $\underleftarrow{\lim} S_i \to \underleftarrow{\lim} A_i$ is an epimorphism. Note that if $\mathcal{S}$ is linearly compact, then each of its objects is also linearly compact; but the converse is not true (see [11, Example 4]).

LEMMA 4.1. *Let A, $\mathfrak{C}$ be as in Theorem 3.8. Suppose that $\mathfrak{C}$ is right semiperfect, and let $P \in \mathcal{M}^{\mathfrak{C}}$ be a projective object. Then $\cdot_{\mathfrak{C}}P$ is a projective module.*

Proof. By the locally finite property for right $\mathfrak{C}$–comodules there is an epimorphism of right comodules

$$\bigoplus_{i \in I} M_i \twoheadrightarrow P$$

or, equivalently, of left $^*\mathfrak{C}$–modules, where $\{M_i | i \in I\}$ is a family of finitely generated left $^*\mathfrak{C}$–modules. Now consider the following epimorphism

$$\bigoplus_{i \in I} P_i \twoheadrightarrow P, \tag{5}$$

where each $P_i \to M_i$ is a projective cover of M_i, and P_i is finitely generated right $\mathfrak{C}$–comodule. Since $P_{\mathfrak{C}}$ is projective, (5) is splitting so $\oplus_{i \in I} P_i = P \oplus P'$, for some $P'_{\mathfrak{C}}$. By Proposition 3.3, each P_i is a projective left $^*\mathfrak{C}$–module, thus $\cdot_{\mathfrak{C}}P$ is a projective module. $\square$

The following result extends characterizations of semiperfect coalgebras over fields and commutative QF rings [20, Theorem 10], [15, Theorem 3.3], [16, Theorem 1.5], [25, 6.3] to corings over QF rings. The category $\mathcal{M}^{\mathfrak{C}}$ is said *to have enough projectives* if it has a projective generator.

THEOREM 4.2. *Let $\mathfrak{C}$ be a coring over a QF ring A such that $_A\mathfrak{C}$ and $\mathfrak{C}_A$ are projective modules. The following conditions are equivalent*

  (i) *$\mathfrak{C}$ is right semiperfect;*
  (ii) *the category $\mathcal{M}^{\mathfrak{C}}$ has enough projectives;*
  (iii) *$\mathcal{M}^{\mathfrak{C}}_f$ is a linearly compact subcategory of $\mathcal{M}^{\mathfrak{C}}$;*

148   L. EL KAOUTIT AND J. GÓMEZ-TORRECILLAS

  (iv) *the functor* $\mathrm{Rat}^l_{\mathfrak{C}} \circ {}^*(-) : {}^{\mathfrak{C}}\mathcal{M} \to {}_{\cdot\mathfrak{C}}\mathcal{M}$ *is exact;*

  (v) $\sigma'_M : M \to (\mathrm{Rat}^l_{\mathfrak{C}}({}^*M))^*$ *is a monomorphism for every left* $\mathfrak{C}$-*comodule* M;

 (vi) $\mathrm{Rat}^l_{\mathfrak{C}}({}^*M)$ *is dense in* ${}^*M$ *for every right* $\mathfrak{C}$-*comodule* M;

(vii) *the functor* $\mathrm{Rat}^l_{\mathfrak{C}} : {}_{\cdot\mathfrak{C}}\mathcal{M} \to {}_{\cdot\mathfrak{C}}\mathcal{M}$ *is exact.*

Therefore, if $\mathfrak{C}$ *is right semiperfect, then* $\mathrm{Rat}^l_{\mathfrak{C}}({}_{\cdot\mathfrak{C}}\mathcal{M})$ *is a localizing subcategory of* ${}_{\cdot\mathfrak{C}}\mathcal{M}$.

Proof. $(i) \Rightarrow (ii)$ Since $\mathcal{M}^{\mathfrak{C}}$ is locally finite, then the coproduct of the projective covers of a complete set of simple right $\mathfrak{C}$-comodules is a projective generator.

$(ii) \Rightarrow (iii)$ Let U be a projective generator of $\mathcal{M}^{\mathfrak{C}}$, and $T = \mathrm{End}(U_{\mathfrak{C}})$ its endomorphism ring. The exact faithful and full functor $F = \mathrm{Hom}_{\mathfrak{C}}(U, -) : \mathcal{M}^{\mathfrak{C}} \to \mathcal{M}_T$ has a left adjoint $G = - \otimes_T U : \mathcal{M}_T \to \mathcal{M}^{\mathfrak{C}}$, let $\eta : 1 \to FG$ be its unity and $\nu : GF \to 1$ be its counity. Let us first prove that F preserves locally compact objects. Given a locally compact comodule M and an inverse system of epimorphisms $\{F(M) \to N_i\}$ in $\mathcal{M}_T$, we get an inverse system of epimorphisms $\{GF(M) \to G(N_i)\}$ in $\mathcal{M}^{\mathfrak{C}}$, and a new inverse system of epimorphisms of modules $\{FGF(M) \to FG(N_i)\}$. Therefore, we have commuting diagrams

$$
\begin{array}{ccc}
FGF(M) & \longrightarrow & FG(N_i) \\
\big\uparrow{\scriptstyle \eta_{F(M)}} & & \big\uparrow{\scriptstyle \eta_{N_i}} \\
F(M) & \longrightarrow & N_i,
\end{array}
$$

which give a commutative diagram

$$
\begin{array}{ccc}
FGF(M) & \longrightarrow & \varprojlim FG(N_i) \cong F(\varprojlim G(N_i)) \\
\big\uparrow{\scriptstyle \eta_{F(M)}} & & \big\uparrow \\
F(M) & \longrightarrow & \varprojlim N_i
\end{array} \tag{6}
$$

Now, ν_M gives an isomorphism $GF(M) \cong M$ and $\eta_{F(M)} = F(\nu_M^{-1})$ is an isomorphism, too. Therefore, the left vertical arrow in (6) is an isomorphism and the top arrow is an epimorphism (since $GF(M) \cong M$ is linearly compact and F is exact). We get thus that $F(M) \to \varprojlim N_i$ is an epimorphism and $F(M)$ is then linearly compact. Let $\mathcal{C}$ be the image under F of $\mathcal{M}^{\mathfrak{C}}_f$, which is a full subcategory of $\mathcal{M}_T$ consisting of linearly compact modules. By [19, Theorem 7.1] (see also [11, Lemma 6]), $\mathcal{C}$ is a linearly compact subcategory of $\mathcal{M}_T$. Now, given any inverse system of epimorphisms $\{M_i \to L_i\}$ with $M_i \in \mathcal{M}^f_{\mathfrak{C}}$ we get an inverse system of epimorphisms of modules $\{F(M_i) \to F(L_i)\}$. By [12, Proposition 3.1] and Theorem 3.2, the M_i's are linearly compact and, hence, the $F(M_i)$'s are in $\mathcal{C}$. We thus get that $F(\varprojlim M_i) \cong \varprojlim F(M_i) \to \varprojlim F(N_i) \cong F(\varprojlim N_i)$ is an epimorphism. Since F is faithful, we deduce that $\varprojlim M_i \to \varprojlim N_i$ is an epimorphism. Therefore, $\mathcal{M}^{\mathfrak{C}}_f$ is a linearly compact subcategory of $\mathcal{M}^{\mathfrak{C}}$.

$(iii) \Rightarrow (iv)$ The proof of [16, Theorem 1.5] given for coalgebras runs in the framework of corings.

MORITA DUALITY FOR CORINGS

$(iv) \Rightarrow (v)$ Every $m \in M$ is contained in a $\mathfrak{C}$–subcomodule N of M such that $_A N$ is finitely presented. We have the following commutative diagram

$$
\begin{array}{ccc}
N & \xrightarrow{\;\;\Phi'_N\;\;} & (^*N)^* \\[2mm]
\downarrow{\scriptstyle i} & & \downarrow{\scriptstyle (\mathrm{Rat}^l_{\mathfrak{C}}(^*i))^*} \\[2mm]
M & \xrightarrow{\;\;\sigma'_M\;\;} & (\mathrm{Rat}^l_{\mathfrak{C}}(^*M))^*,
\end{array}
\tag{7}
$$

where i is the inclusion map. The injectivity of the map σ'_M is easily deduced from this diagram, as $(\mathrm{Rat}^l_{\mathfrak{C}}(^*i))^*$ and Φ'_N are monomorphisms.

$(v) \Rightarrow (vi)$ It follows because $Ker\sigma'_M = \mathrm{Rat}^l_{\mathfrak{C}}(^*M)^\perp$.

$(vi) \Rightarrow (i)$ Apply Theorem 3.8.

$(i) \Rightarrow (vii)$ It follows from Lemma 4.1, since $\omega^l : \mathrm{Rat}^l_{\mathfrak{C}}(\cdot_{\mathfrak{C}}\mathcal{M}) \to \cdot_{\mathfrak{C}}\mathcal{M}$ is left adjoint to $\mathrm{Rat}^l_{\mathfrak{C}}$.

$(vii) \Rightarrow (iv)$ This is clear, as $^*(-)$ is an exact functor. $\qquad\square$

REMARK 4.3. Examples of semiperfect corings can be constructed as follows: let U_A be a finitely generated and projective module with dual basis $\{e_i, e_i^*\}$ and let $T \subseteq \mathrm{End}(U_A)$ be a semiperfect subring such that $_T U$ is faithfully flat. Consider the *comatrix A–coring* $U^* \otimes_T U$ with comultiplication given by $\Delta(\varphi \otimes_T u) = \sum_i \varphi \otimes_T e_i \otimes_A e_i^* \otimes_T u$ and counit given by $\epsilon(\varphi \otimes_T u) = \varphi(u)$, for $\varphi \otimes_T u \in U^* \otimes_T U$ (see [7]). By [7, Theorem 2], the category $\mathcal{M}^{U^* \otimes_T U}$ is equivalent to $\mathcal{M}_T$ and, hence, $U^* \otimes_T U$ is a right semiperfect A–coring with finitely many non isomorphic simple right comodules (no assumption is made here on the ground ring A). In fact, if A is a QF ring, every right semiperfect A–coring $\mathfrak{C}$ with finitely many simples such that $_A\mathfrak{C}$ is projective is a comatrix coring as described before. To see this, consider a set $S_1, \ldots, S_n$ of representatives of all simple right $\mathfrak{C}$–comodules. Then a finitely generated projective generator of $\mathcal{M}^{\mathfrak{C}}$ is $U = P_1 \oplus \cdots \oplus P_n$, where P_i is a projective cover of S_i. Let $T = \mathrm{End}(U_{\mathfrak{C}})$, which is a semiperfect subring of $\mathrm{End}(U_A)$. By [7, Theorem 1], U_A is finitely generated projective and $_T U$ is faithfully flat. Moreover, there is a canonical isomorphism of A–corings $can : U^* \otimes_T U \cong \mathfrak{C}$ and the category $\mathcal{M}^{\mathfrak{C}}$ is equivalent to $\mathcal{M}_T$.

We are now ready to give a characterization of left and right semiperfect corings in terms of duality, which generalizes the obtained for semiperfect coalgebras in [15, Theorem 3.5].

THEOREM 4.4. *Let A be a QF ring and $\mathfrak{C}$ an A–coring with $_A\mathfrak{C}$ and $\mathfrak{C}_A$ projective modules. Then $\mathfrak{C}$ is left and right semiperfect if and only if the functors $\mathrm{Rat}^r_{\mathfrak{C}} \circ (-)^* : \mathcal{M}^{\mathfrak{C}} \leftrightarrows {}^{\mathfrak{C}}\mathcal{M} : \mathrm{Rat}^l_{\mathfrak{C}} \circ {}^*(-)$ give a Colby-Fuller duality.*

Proof. Assume $\mathfrak{C}$ to be left and right semiperfect, so the subcategories $\mathcal{M}^{\mathfrak{C}}_f$ and $^{\mathfrak{C}}_f\mathcal{M}$ are linearly compact. By Theorem 3.2 and [11, Theorem 3], we get the Colby-Fuller duality. Conversely, apply [11, Theorem 3] to see that the subcategories of linearly compact objects $\mathcal{M}^{\mathfrak{C}}_f$ and $^{\mathfrak{C}}_f\mathcal{M}$ are linearly compact subcategories and then apply Theorem 4.2. $\qquad\square$

Let us now develop the Morita duality theory for corings which extends the given in [16] in the coalgebra case. We will work over two ground rings A and B. We will

150   L. EL KAOUTIT AND J. GÓMEZ-TORRECILLAS

use the same notation for the duals of modules over A or B. Which dual is acting each time will be clarified by the context.

THEOREM 4.5. *Let $\mathfrak{C}$ and $\mathfrak{D}$ be corings over QF rings A and B, respectively, such that $_A\mathfrak{C}$, $_B\mathfrak{D}$, $\mathfrak{C}_A$ and $\mathfrak{D}_B$ are projective modules. Assume that there exists a Colby-Fuller duality $H : \mathcal{M}^{\mathfrak{C}} \leftrightarrows {}^{\mathfrak{D}}\mathcal{M} : H'$. Then*

  (a) $\mathfrak{C}$, $\mathfrak{D}$ *are left and right semiperfect corings.*
  (b) *There exists a Colby-Fuller duality between ${}^{\mathfrak{C}}\mathcal{M}$ and $\mathcal{M}^{\mathfrak{D}}$.*

Proof. (a) By [16, Lemma 1.4] $\mathcal{M}_f^{\mathfrak{C}}$ and $_f^{\mathfrak{D}}\mathcal{M}$ are reflexive subcategories of $\mathcal{M}^{\mathfrak{C}}$ and ${}^{\mathfrak{D}}\mathcal{M}$, respectively; hence [11, Lemma 2, Theorem 3] imply that $\mathcal{M}_f^{\mathfrak{C}}$ and $_f^{\mathfrak{D}}\mathcal{M}$ are linearly compact subcategories. Therefore, by Theorem 4.2, $\mathfrak{C}$ is right semiperfect and $\mathfrak{D}$ is left semiperfect. Let us show that $\mathfrak{D}$ is right semiperfect; for this consider an arbitrary simple left $\mathfrak{D}$-comodule $_{\mathfrak{D}}S$. So $H'(S) \in \mathcal{M}_f^{\mathfrak{C}}$ is simple, hence $H'(S)^* \in {}_f^{\mathfrak{C}}\mathcal{M}$ is simple too. Since $\mathfrak{C}$ is a right semiperfect, $_A E(H'(S)^*)$ is finitely generated. Using Proposition 3.3, $^*E(H'(S)^*)_{\mathfrak{C}}$ is a finitely generated projective comodule, and $^*E(H'(S)^*) \twoheadrightarrow {}^*(H'(S)^*) \cong H'(S)$. If we apply the functor H to this last sequence, we get

$$S \cong HH'(S) \hookrightarrow H(^*E(H'(S)^*)). \tag{8}$$

By [17, Proposición 3.3.10] or [16, Lemma 1.10], $H(^*E(H'(S)^*))$ is a finitely generated injective left $\mathfrak{D}$-comodule. So $_A E(S)$ is a finitely generated module, by (8). This implies, by Theorem 3.5, that $\mathfrak{D}$ is right semiperfect. Analogously, $\mathfrak{C}$ is shown to be left semiperfect.
(b) It follows from part (a) and Theorem 4.4 that the following are Colby-Fuller dualities

$$F = \mathrm{Rat}_{\mathfrak{C}}^r \circ (-)^* : \mathcal{M}^{\mathfrak{C}} \leftrightarrows {}^{\mathfrak{C}}\mathcal{M} : \mathrm{Rat}_{\mathfrak{C}}^l \circ {}^*(-) = F',$$

$$G = \mathrm{Rat}_{\mathfrak{D}}^r \circ (-)^* : \mathcal{M}^{\mathfrak{D}} \leftrightarrows {}^{\mathfrak{D}}\mathcal{M} : \mathrm{Rat}_{\mathfrak{D}}^l \circ {}^*(-) = G'.$$

So, by [16, Lemma 1.4],

$$L = G'_f \circ H_f \circ F'_f : {}_f^{\mathfrak{C}}\mathcal{M} \leftrightarrows \mathcal{M}_f^{\mathfrak{D}} : L' = F_f \circ H'_f \circ G_f$$

is a duality, where $(-)_f$ denotes the restriction of a functor $(-)$ to the full subcategory of finitely generated comodules. Since $_f^{\mathfrak{C}}\mathcal{M}$ and $\mathcal{M}_f^{\mathfrak{D}}$ are finitely closed generating, linearly compact full subcategories of ${}^{\mathfrak{C}}\mathcal{M}$ and $\mathcal{M}^{\mathfrak{D}}$, respectively, the assertion (b) is deduced by [11, Theorem 5].   □

Let $\mathfrak{C}$ and $\mathfrak{D}$ be as in Theorem 4.5. If $L : \mathcal{M}^{\mathfrak{C}} \to \mathcal{M}^{\mathfrak{D}}$ is an equivalence of categories, then $L_f : \mathcal{M}_f^{\mathfrak{C}} \to \mathcal{M}_f^{\mathfrak{D}}$ is also an equivalence of categories. Moreover, the assignment $L \mapsto L_f$ defines a bijective correspondence (up to natural isomorphisms) between equivalences $\mathcal{M}^{\mathfrak{C}} \sim \mathcal{M}^{\mathfrak{D}}$ and equivalences $\mathcal{M}_f^{\mathfrak{C}} \sim \mathcal{M}_f^{\mathfrak{D}}$, see [16, Proposition 1.1]. On the other hand, if $\mathfrak{C}$ is right semiperfect and $\mathfrak{D}$ is a left semiperfect, then the assignment

$$H : \mathcal{M}^{\mathfrak{C}} \leftrightarrows {}^{\mathfrak{D}}\mathcal{M} : H' \longmapsto H_f : \mathcal{M}_f^{\mathfrak{C}} \leftrightarrows {}_f^{\mathfrak{D}}\mathcal{M} : H'_f$$

is also a bijective correspondence (up to natural isomorphism) between Colby-Fuller dualities and dualities, because each duality between $\mathcal{M}_f^{\mathfrak{C}}$ and $_f^{\mathfrak{D}}\mathcal{M}$ is uniquely extended to a Colby-Fuller duality, see [12, Theorem 2.1].

MORITA DUALITY FOR CORINGS

THEOREM 4.6. *Let $\mathfrak{C}$ and $\mathfrak{D}$ be corings over Quasi-Frobenius rings A and B, respectively, such that $_A\mathfrak{C}$, $_B\mathfrak{D}$, $\mathfrak{C}_A$, and $\mathfrak{D}_B$ are projective modules. Assume that $\mathfrak{C}$ is right semiperfect, and that $\mathfrak{D}$ is left semiperfect. There is a bijective correspondence (up to natural isomorphism) between equivalences $\mathcal{M}^{\mathfrak{C}} \sim \mathcal{M}^{\mathfrak{D}}$ and Colby-Fuller dualities $H : \mathcal{M}^{\mathfrak{C}} \leftrightarrows {}^{\mathfrak{D}}\mathcal{M} : H'$.*

Proof. Let $L : \mathcal{M}^{\mathfrak{C}} \to \mathcal{M}^{\mathfrak{D}}$ be an equivalence of categories, and let $L_f : \mathcal{M}^{\mathfrak{C}}_f \to \mathcal{M}^{\mathfrak{D}}_f$ the induced equivalence. Then $(-)^* \circ L_f : \mathcal{M}^{\mathfrak{C}}_f \to {}^{\mathfrak{D}}_f\mathcal{M}$ is a duality, which can be uniquely extended, by [11, Theorem 5] or [12, Theorem 2.1], to a Colby-Fuller duality between $\mathcal{M}^{\mathfrak{C}}$ and $^{\mathfrak{D}}\mathcal{M}$, as $\mathcal{M}^{\mathfrak{C}}_f$ and $^{\mathfrak{D}}_f\mathcal{M}$ are linearly compact subcategories by Theorem 4.2. Conversely, given a Colby-Fuller duality $H : \mathcal{M}^{\mathfrak{C}} \leftrightarrows {}^{\mathfrak{D}}\mathcal{M} : H'$, then $^*(-) \circ H_f : \mathcal{M}^{\mathfrak{C}}_f \to \mathcal{M}^{\mathfrak{D}}_f$ is an equivalence of categories which extends uniquely to an equivalence of categories between $\mathcal{M}^{\mathfrak{C}}$ and $\mathcal{M}^{\mathfrak{D}}$. $\square$

It seems to be unknown for general corings $\mathfrak{C}$ and $\mathfrak{D}$ whether an equivalence of categories $\mathcal{M}^{\mathfrak{C}} \sim \mathcal{M}^{\mathfrak{D}}$ does imply an equivalence $^{\mathfrak{C}}\mathcal{M} \sim {}^{\mathfrak{D}}\mathcal{M}$. Up to our knowledge, this is far from being a trivial problem. Our next result gives a positive answer in a (restrictive) particular case.

THEOREM 4.7. *Let $\mathfrak{C}$ and $\mathfrak{D}$ be corings over Quasi-Frobenius rings A and B, respectively, such that $_A\mathfrak{C}$, $_B\mathfrak{D}$, $\mathfrak{C}_A$, and $\mathfrak{D}_B$ are projective modules. The following statements are equivalent.*

   (i) *There is a Colby-Fuller duality $\mathcal{M}^{\mathfrak{C}} \leftrightarrows {}^{\mathfrak{D}}\mathcal{M}$;*
   (ii) *there is an equivalence $\mathcal{M}^{\mathfrak{C}} \sim \mathcal{M}^{\mathfrak{D}}$, $\mathfrak{C}$ is right semiperfect, and $\mathfrak{D}$ is left semiperfect;*
   (iii) *there is an equivalence $^{\mathfrak{C}}\mathcal{M} \sim {}^{\mathfrak{D}}\mathcal{M}$, $\mathfrak{C}$ is left semiperfect and $\mathfrak{D}$ is right semiperfect;*
   (iv) *there is a Colby-Fuller duality $^{\mathfrak{C}}\mathcal{M} \leftrightarrows \mathcal{M}^{\mathfrak{D}}$.*

Proof. It follows easily from Theorems 4.5 and 4.6. $\square$

To close this section we give a characterization of Colby-Fuller dualities between comodules by means of the existence of a quasi–finite injective cogenerator comodule; such characterization was given in [16, Corollary 1.8] for a coalgebras over a fixed base field. It is convenient, first, to recall the notion of quasi–finite comodule.

Let $\mathfrak{C}$ (resp. $\mathfrak{D}$) be a coring over A (resp. over B). Here, A and B are not assumed to be QF-rings. Let N be an $A - B$–bimodule with a right $\mathfrak{D}$–structure map $\rho_N : N \to N \otimes_B \mathfrak{D}$ which is assumed to be left A–linear. Assume that $N_{\mathfrak{D}}$ is quasi-finite, that is, the functor $- \otimes_A N : \mathcal{M}_A \to \mathcal{M}^{\mathfrak{D}}$ has a left adjoint $F : \mathcal{M}^{\mathfrak{D}} \to \mathcal{M}_A$, see [13, Section 4]. This functor is called the *cohom* functor by analogy with the case of coalgebras over fields (see [24]); notation $F = h_{\mathfrak{D}}(N, -)$. Let $\eta_{-,-} : \mathrm{Hom}_{\mathfrak{D}}(-, - \otimes_A N) \to \mathrm{Hom}_A(F(-), -)$ denote the natural isomorphism of the adjunction, and $\theta : 1_{\mathcal{M}^{\mathfrak{D}}} \to F(-) \otimes_A N$ the unity of the adjunction. The canonical map $A_A \to \mathrm{Hom}_{\mathfrak{D}}(N, N) \to \mathrm{Hom}_A(F(N), F(N))$ gives a structure of left A–module on $F(N)$ such that $F(N)$ becomes an A–bimodule, which is endowed with a structure of A–coring as follows. Define a comultiplication $\Delta : F(N) \to F(N) \otimes_A F(N)$ by $\Delta = \eta_{N, F(N) \otimes_A F(N)}((F(N) \otimes_A \theta_N)\theta_N)$, that is, Δ is determined

L. EL KAOUTIT AND J. GÓMEZ-TORRECILLAS

by the condition $(F(N) \otimes_A \theta_N)\theta_N = (\Delta \otimes_A N)\theta_N$. The counit is given by $\epsilon = \eta_{N,A}(\iota)$, where $\iota : N \to A \otimes_A N$ is the canonical isomorphism. This A–coring will be denoted by $e_{\mathfrak{D}}(N)$.

Now, assume that N is a $\mathfrak{C} - \mathfrak{D}$–bicomodule and that $_B\mathfrak{D}$ is a flat module. By [13, Proposition 4.2], $F = h_{\mathfrak{D}}(N, -)$ factors through the category $\mathcal{M}^{\mathfrak{C}}$, and $h_{\mathfrak{D}}(N, -) : \mathcal{M}^{\mathfrak{D}} \to \mathcal{M}^{\mathfrak{C}}$ becomes a left adjoint to the cotensor product functor $-\square_{\mathfrak{C}} N : \mathcal{M}^{\mathfrak{C}} \to \mathcal{M}^{\mathfrak{D}}$ with unity $\theta : 1_{\mathcal{M}^{\mathfrak{D}}} \to F(-)\square_{\mathfrak{C}} N$ and counity $\chi : F(-\square_{\mathfrak{C}} N) \to 1_{\mathcal{M}^{\mathfrak{C}}}$.

PROPOSITION 4.8. [7] *Let N be a $\mathfrak{C} - \mathfrak{D}$–bicomodule. Assume that N is quasi-finite as a right $\mathfrak{D}$–comodule, and that $_B\mathfrak{D}$ is flat. The map $f : e_{\mathfrak{D}}(N) \to \mathfrak{C}$ defined by $f = \chi_{\mathfrak{C}} \circ h_{\mathfrak{D}}(N, \lambda_N)$, where $\lambda_N : N \to \mathfrak{C} \otimes_A N$ is the left comodule structure map, is a homomorphism of A–corings.*

THEOREM 4.9. *Let $\mathfrak{C}$ and $\mathfrak{D}$ be corings over Quasi-Frobenius rings A and B, respectively, such that $_A\mathfrak{C}$, $_B\mathfrak{D}$, $\mathfrak{C}_A$, and $\mathfrak{D}_B$ are projective modules. If $H : \mathcal{M}^{\mathfrak{C}} \leftrightarrows {}^{\mathfrak{D}}\mathcal{M} : H'$ is a Colby-Fuller duality, then there exists a $\mathfrak{C} - \mathfrak{D}$–bicomodule P such that $P_{\mathfrak{D}}$ is a quasi-finite injective cogenerator of $\mathcal{M}^{\mathfrak{D}}$ with an isomorphism of A–corings $e_{\mathfrak{D}}(P) \cong \mathfrak{C}$, and a natural isomorphism*

$$H(M) \cong \mathrm{Rat}^r_{\mathfrak{D}}\left((M\square_{\mathfrak{C}}P)^*\right)$$

for every right $\mathfrak{C}$-comodule M.

Proof. Let $F : \mathcal{M}^{\mathfrak{C}} \to \mathcal{M}^{\mathfrak{D}}$ be the equivalence of categories corresponding by Theorem 4.6 to the given Colby–Fuller duality $H : \mathcal{M}^{\mathfrak{C}} \to {}^{\mathfrak{D}}\mathcal{M}$. Given $M \in \mathcal{M}^{\mathfrak{C}}$, write $M = \varinjlim M_i$, where the M_i's are finitely generated right $\mathfrak{C}$-comodules. Then

$$
\begin{aligned}
H(M) &= \mathrm{Rat}^r_{\mathfrak{D}}\left(\varprojlim(F_f(M_i))^*\right) \\
&\cong \mathrm{Rat}^r_{\mathfrak{D}}\left((\varinjlim F_f(M_i))^*\right) \\
&\cong \mathrm{Rat}^r_{\mathfrak{D}}\left((F(M))^*\right).
\end{aligned}
$$

By [13, Theorem 3.5] there is a $\mathfrak{C} - \mathfrak{D}$-bicomodule $P = F(\mathfrak{C})$ such that $F \cong -\square_{\mathfrak{C}} F(\mathfrak{C})$. Using [13, Proposition 4.2.(2)], we deduce that $P_{\mathfrak{D}}$ is a quasi-finite right comodule which is clearly an injective cogenerator. Now, $e_{\mathfrak{D}}(P) = F(\mathfrak{C})\square_{\mathfrak{D}} G(\mathfrak{D}) \cong GF(\mathfrak{C}) \cong \mathfrak{C}$, where G is the inverse functor of F; which is an isomorphism of A–corings, by Proposition 4.8. $\qquad\square$

Acknowledgements

The authors thank the referee for his interesting remarks resulting in an improvement of the presentation of this paper. L. El Kaoutit quiere agradecer al Ministerio de Educación, Cultura y Deporte de España la concesión de una beca de movilidad para estudiar en el programa de doctorado FISYMAT de la Universidad de Granada.

REFERENCES

[1] J.Y. Abuhlail, Rational modules for corings, *Comm. Algebra.* **31** (2003), 5793–5840.
[2] F. W. Anderson and K. R. Fuller, "Rings and categories of modules", Springer-Verlag, Berlin, 1974.

MORITA DUALITY FOR CORINGS

[3] P. N. Ành and R. Wiegandt, Morita duality for Grothendieck categories, *J. Algebra* **168** (1994), 273–293.

[4] T. Brzeziński, The structure of corings. Induction functors, Maschke-type Theorem, and Frobenius and Galois-type properties, *Algebr. Represent. Theory* **5** (2002), 389–410

[5] R. R. Colby and K. R. Fuller, Exactness of the double dual and Morita duality for Grothendieck categories, *J. Algebra*, **82** (1983), 546–558.

[6] Y. Doi, Homological coalgebra, *J. Math. Soc. Japan*, **33** (1981), 31–50.

[7] L. EL Kaoutit and J. Gómez-Torrecillas, Comatrix coring: Galois coring, Descent theory, and a structure theorem for cosemisimple corings, *Math. Z.* **244** (2003), 887-906.

[8] L. ELKaoutit, J. Gómez-Torrecillas, and F. J. Lobillo, Semisimple corings, *Algebra Colloquium*, to appear, preprint math.RA/0201070.

[9] P. Gabriel, Des catégories abéliennes, *Bull. Soc. Math. France* **90** (1962), 323–448.

[10] J.L. Gómez Pardo, Counterinjective modules and duality, *J. Pure. Appl. Algebra* **61** (1989), 165–170.

[11] J.L. Gómez Pardo and P.A. Guil Asensio, Linear compactness and Morita duality for Grothendieck categories, *J. Algebra* **148** (1992), 53–67.

[12] J.L. Gómez Pardo and P.A. Guil Asensio, Morita duality for Grothendieck categories, *Publ. Mat.* **36** (1992), 625–635.

[13] J. Gómez-Torrecillas, Separable functors in corings, *Int. J. Math. Math. Sci.* **30** (2002), 203–225.

[14] J. Gómez-Torrecillas and A. Louly, Coseparable corings, *Comm. Algebra* **31** (2003), 4455–4471.

[15] J. Gómez-Torrecillas and C. Năstăsescu, Quasi-co-Fronenius coalgebras, *J. Algebra* **174** (1995), 909–923.

[16] J. Gómez-Torrecillas and C. Năstăsescu, Colby-Fuller duality between coalgebras, *J. Algebra* **185** (1996), 527–543.

[17] P.A. Guil Asensio, Dualidades de Morita entre categorías de Grothendieck y anillos de endomorfismos, Ph.D. thesis, Universidad de Murcia, 1990.

[18] F. Guzman, Cointegration, Relative Cohomology for Comodules, and Coseparable Coring, *J. Algebra* **126** (1989), 211–224.

[19] C.U. Jensen, "Les foncteurs dérivés de lim et leurs Applications en Théorie des Modules", *Lect. Notes Math.* **254**, Springer-Verlag, Berlin, 1972.

[20] I.-P. Lin, Semiperfect coalgebras, *J. Algebra* **49** (1977), 357–373.

[21] C. Menini, B. Torrecillas, and R. Wisbauer, Strongly rational comodules and semiperfect Hopf algebras over QF rings, *J. Pure. Appl. Algebra* **155** (2001), 237–255.

[22] B. Stenström, "Rings of Quotients", Springer-Verlag, Berlin, 1975.

[23] M. Sweedler, The predual to the Jacobson-Bourbaki Theorem, *Trans. Amer. Math. Soc.* **213** (1975), 391–406.

[24] M. Takeuchi, Morita Theorems for categories of comodules, *J. Fac. Sci. Univ. Tokyo* **24** (1977), 629–644.

[25] R. Wisbauer, Semiperfect Coalgebras over Rings, in ' Algebras and Combinatorics, An International Congress, ICAC'97", Hong Kong, Springer-Verlag, Singapore, 1999, 487–512.

[26] R. Wisbauer, On the category of comodules over corings, preprint 2001.

[27] B. Zimmermann-Huigsen, Pure submodules of direct products of free modules, *Math. Ann.* **224** (1976), 233-245.

Quantized Coinvariants at Transcendental q

K. R. GOODEARL   Department of Mathematics,
University of California, Santa Barbara, CA 93106, USA
e-mail: *goodearl@math.ucsb.edu*

T. H. LENAGAN   School of Mathematics, James Clerk Maxwell Building,
Kings Buildings, Mayfield Road, Edinburgh EH9 3JZ, Scotland
e-mail: *tom@maths.ed.ac.uk*

> ABSTRACT. A general method is developed for deriving Quantum First and
> Second Fundamental Theorems of Coinvariant Theory from classical analogs
> in Invariant Theory, in the case that the quantization parameter q is tran-
> scendental over a base field. Several examples are given illustrating the utility
> of the method; these recover earlier results of various researchers including
> Domokos, Fioresi, Hacon, Rigal, Strickland, and the present authors.

1. INTRODUCTION

In the classic terminology of Hermann Weyl [18], a full solution to any invariant
theory problem should incorporate a *First Fundamental Theorem*, giving a set of
generators (finite, where possible) for the ring of invariants, and a *Second Fundamental Theorem*, giving generators for the ideal of relations among the generators
of the ring of invariants. Many of the classical settings of Invariant Theory have
quantized analogs, and one seeks corresponding analogs of the classical First and
Second Fundamental Theorems. However, the setting must be dualized before potential quantized analogs can be framed, since there are no quantum analogs of
the original objects, only quantum analogs of their coordinate rings. Hence, one
rephrases the classical results in terms of rings of coinvariants (see below), and then
seeks quantized versions of these. A first stumbling block is that in general these
coactions are not algebra homomorphisms, and so at the outset it is not even obvious that the coinvariants form a subalgebra. However, this can often be established;
cf. [9, Proposition 1.1]; [4, Proposition 1.3].

Typically, a classical invariant theoretical setting is quantized uniformly with respect to a parameter q, that is, there is a family of rings of coinvariants to be
determined, parametrized by nonzero scalars in a base field, such that the case
$q = 1$ is the classical one. (Many authors, however, restrict attention to special
values of q, such as those transcendental over the rational number field, and do

---

2000 *Mathematics Subject Classification.* 16W35, 16W30, 20G42, 17B37, 81R50.
Key words and phrases. coinvariants, First Fundamental Theorem, Second Fundamental The-
orem, quantum group, quantized coordinate ring.
This research was partially supported by NSF research grant DMS-9622876 and by NATO
Collaborative Research Grant CRG.960250.

not address the general case.) As in the classical setting, one usually has identified natural candidates to generate the ring of coinvariants. Effectively, then, one has a parametrized inclusion of algebras (candidate subalgebras inside the algebras of coinvariants), which is an equality at $q = 1$, and one seeks equality for other values of q. In the best of all worlds, the equality at $q = 1$ could be "lifted" by some general process to equality at arbitrary q. Lifting to transcendental q has been done succesfully in some cases, by ad hoc methods – see, for example, [6], [17]. Also, an early version of [9] for the transcendental case was obtained in this way. Quantum Second Fundamental Theorems can be approached in a similar manner. We develop here a general method for lifting equalities of inclusions from $q = 1$ to transcendental q, which applies to many analyses of quantized coinvariants.

In order to apply classical results as indicated above, we must be able to transform invariants to coinvariants. For morphic actions of algebraic groups, the setting of interest to us, invariants and coinvariants are related as follows. Suppose that $\gamma : G \times V \to V$ is a morphic action of an algebraic group G on a variety V. This action induces an action of G on $\mathcal{O}(V)$, where $(x.f)(v) = f(x^{-1}.v)$ for $x \in G$, $f \in \mathcal{O}(V)$, and $v \in V$. The invariants for this action are, of course, those functions in $\mathcal{O}(V)$ which are constant on G-orbits. The comorphism of γ is an algebra homomorphism $\gamma^* : \mathcal{O}(V) \to \mathcal{O}(G) \otimes \mathcal{O}(V)$, with respect to which $\mathcal{O}(V)$ becomes a left $\mathcal{O}(G)$-comodule. Now a function $f \in \mathcal{O}(V)$ is a coinvariant in this comodule when $\gamma^*(f) = 1 \otimes f$. Since $1 \otimes f$ corresponds to the function $(x, v) \mapsto f(v)$ on $G \times V$, we see that $\gamma^*(f) = 1 \otimes f$ if and only if $f(x.v) = f(v)$ for all $x \in G$ and $v \in V$, that is, if and only if f is an invariant function. To summarize:

$$\mathcal{O}(V)^{\mathrm{co}\,\mathcal{O}(G)} = \mathcal{O}(V)^G.$$

Quantized coordinate rings have been constructed for all complex semisimple algebraic groups G. These quantized coordinate rings are Hopf algebras, which we denote $\mathcal{O}_q(G)$, since we are concentrating on the single parameter versions. In those cases where a morphic action of G on a variety V has been quantized, we have a quantized coordinate ring $\mathcal{O}_q(V)$ which supports an $\mathcal{O}_q(G)$-coaction. The coaction is often not an algebra homomorphism, but nonetheless – as mentioned above – the set of $\mathcal{O}_q(G)$-coinvariants in $\mathcal{O}_q(V)$ is typically a subalgebra. The goal of Quantum First and Second Fundamental Theorems for this setting is to give generators and relations for the algebra $\mathcal{O}_q(V)^{\mathrm{co}\,\mathcal{O}_q(G)}$ of $\mathcal{O}_q(G)$-coinvariants in $\mathcal{O}_q(V)$. We discuss several standard settings in later sections of the paper, and outline how our general method applies. We recover known Quantum First and Second Fundamental Theorems at transcendental q in these settings, with some simplifications to the original proofs, and in some cases extending the range of the theorems.

Throughout, k will denote a field, which may be of arbitrary characteristic and need not be algebraically closed.

2. REDUCTION MODULO $q - 1$

Throughout this section, we work with a field extension $k^\circ \subset k$ and a scalar $q \in k \backslash k^\circ$ which is transcendental over k°. Thus, the k°-subalgebra $R = k^\circ[q, q^{-1}] \subset k$ is a

QUANTIZED COINVARIANTS AT TRANSCENDENTAL q

Laurent polynomial ring. Let us denote reduction modulo $q - 1$ by overbars, that is, given any R-module homomorphism $\phi : A \to B$, we write $\overline{\phi} : \overline{A} \to \overline{B}$ for the induced map $A/(q-1)A \to B/(q-1)B$.

PROPOSITION 2.1. *Let* $A \xrightarrow{\phi} B \xrightarrow{\psi} C$ *be a complex of R-modules, such that C is torsionfree. Suppose that there are R-module decompositions*

$$A = \bigoplus_{j \in J} A_j, \qquad B = \bigoplus_{j \in J} B_j, \qquad C = \bigoplus_{j \in J} C_j$$

such that B_j is finitely generated, $\phi(A_j) \subseteq B_j$, and $\psi(B_j) \subseteq C_j$ for all $j \in J$. If the reduced complex $\overline{A} \xrightarrow{\overline{\phi}} \overline{B} \xrightarrow{\overline{\psi}} \overline{C}$ is exact, then so is

$$k \otimes_R A \xrightarrow{\ \mathrm{id} \otimes \phi\ } k \otimes_R B \xrightarrow{\ \mathrm{id} \otimes \psi\ } k \otimes_R C.$$

Proof. The hypotheses and the conclusions all reduce to the direct sum components of the given decompositions, so it is enough to work in one component. Hence, there is no loss of generality in assuming that B is finitely generated.

Let S denote the localization of R at the maximal ideal $(q-1)R$. Set

$$\widetilde{A} = S \otimes_R A, \qquad \widetilde{B} = S \otimes_R B, \qquad \widetilde{C} = S \otimes_R C,$$

and let $\widetilde{A} \xrightarrow{\widetilde{\phi}} \widetilde{B} \xrightarrow{\widetilde{\psi}} \widetilde{C}$ denote the induced complex of S-modules. Since $\widetilde{A}/(q-1)\widetilde{A}$ is naturally isomorphic to $A/(q-1)A = \overline{A}$, and similarly for B and C, there is a commutative diagram

$$
\begin{array}{ccccc}
\widetilde{A} & \xrightarrow{\ \widetilde{\phi}\ } & \widetilde{B} & \xrightarrow{\ \widetilde{\psi}\ } & \widetilde{C} \\
{\scriptstyle \alpha}\downarrow & & {\scriptstyle \beta}\downarrow & & \downarrow{\scriptstyle \gamma} \\
\overline{A} & \xrightarrow{\ \overline{\phi}\ } & \overline{B} & \xrightarrow{\ \overline{\psi}\ } & \overline{C}
\end{array}
$$

where α, β, γ are epimorphisms with kernels $(q-1)\widetilde{A}$ etc.

The bottom row of the diagram is exact by hypothesis, and we claim that the top row is exact. Consider an element $x \in \ker \widetilde{\psi}$. Chasing x around the diagram, we see that $x - \widetilde{\phi}(y) = (q-1)z$ for some $y \in \widetilde{A}$ and $z \in \widetilde{B}$. Note that $(q-1)z \in \ker \widetilde{\psi}$. Since $\widetilde{C}$ is torsionfree, it follows that $z \in \ker \widetilde{\psi}$. Thus $\ker \widetilde{\psi} \subseteq \widetilde{\phi}(\widetilde{A}) + (q-1) \ker \widetilde{\psi}$, whence

$$(\ker \widetilde{\psi})/\widetilde{\phi}(\widetilde{A}) = (q-1)\left[(\ker \widetilde{\psi})/\widetilde{\phi}(\widetilde{A})\right].$$

Since $\ker \widetilde{\psi}$ is a finitely generated S-module, it follows from Nakayama's Lemma that $(\ker \widetilde{\psi})/\widetilde{\phi}(\widetilde{A}) = 0$, establishing the claim.

Since k is a flat S-module, the sequence $k \otimes_S \widetilde{A} \xrightarrow{\ \mathrm{id} \otimes \widetilde{\phi}\ } k \otimes_S \widetilde{B} \xrightarrow{\ \mathrm{id} \otimes \widetilde{\psi}\ } k \otimes_S \widetilde{C}$ is exact. This is isomorphic to the sequence

$$k \otimes_R A \xrightarrow{\ \mathrm{id} \otimes \phi\ } k \otimes_R B \xrightarrow{\ \mathrm{id} \otimes \psi\ } k \otimes_R C$$

and therefore the latter is exact. $\qquad\square$

Proposition 2.1 is useful in obtaining quantized versions of both First and Second Fundamental Theorems. For ease of application, we write out general versions of both situations with appropriate notation, as follows.

THEOREM 2.2. *Let H be an R-bialgebra and B a left H-comodule with structure map $\lambda : B \to H \otimes_R B$, and let $\phi : A \to B$ be an R-module homomorphism whose image is contained in $B^{co\,H}$. Assume that B and H are torsionfree R-modules, and suppose that there exist R-module decompositions $A = \bigoplus_{j \in J} A_j$ and $B = \bigoplus_{j \in J} B_j$ such that B_j is finitely generated, $\phi(A_j) \subseteq B_j$, and $\lambda(B_j) \subseteq H \otimes_R B_j$ for all $j \in J$. Now $\overline{B}$ is a left comodule over the k°-bialgebra $\overline{H}$, and $k \otimes_R B$ is a left comodule over the k-bialgebra $k \otimes_R H$. If $\overline{B}^{co\,\overline{H}}$ equals the image of the induced map $\overline{\phi} : \overline{A} \to \overline{B}$, then $[k \otimes_R B]^{co\,k \otimes_R H}$ equals the image of the induced map $\mathrm{id} \otimes \phi : k \otimes_R A \to k \otimes_R B$.*

Proof. Since R is a principal ideal domain, $H \otimes_R B$ is a torsionfree R-module. We may identify $\overline{H \otimes_R B}$ with $\overline{H} \otimes_{k^\circ} \overline{B}$ and $k \otimes_R (H \otimes_R B)$ with $(k \otimes_R H) \otimes_k (k \otimes_R B)$. Now apply Proposition 2.1 to the complex $A \xrightarrow{\phi} B \xrightarrow{\lambda - \gamma} H \otimes_R B$, where $\gamma : B \to H \otimes_R B$ is the map $b \mapsto 1 \otimes b$. $\square$

THEOREM 2.3. *Let $\phi : A \to B$ be a homomorphism of R-algebras, and I an ideal of A contained in $\ker \phi$. Assume that B is a torsionfree R-module, and suppose that there exist R-module decompositions*

$$I = \bigoplus_{j \in J} I_j, \qquad A = \bigoplus_{j \in J} A_j, \qquad B = \bigoplus_{j \in J} B_j$$

such that A_j is finitely generated, $I_j \subseteq A_j$, and $\phi(A_j) \subseteq B_j$ for all $j \in J$.
If $\ker \overline{\phi}$ equals the image of $\overline{I}$, then the kernel of the k-algebra homomorphism $\mathrm{id} \otimes \phi : k \otimes_R A \to k \otimes_R B$ equals the image of $k \otimes_R I$.

Proof. Apply Proposition 2.1 to the complex $I \xrightarrow{\eta} A \xrightarrow{\phi} B$, where η is the inclusion map. $\square$

Proposition 2.1 can easily be adapted to yield other exactness conclusions. In particular, suppose that, in addition to the given hypotheses, the cokernels of ϕ and ψ are torsionfree R-modules. Then one can show that for all nonzero scalars $\lambda \in k^\circ$, the induced complex

$$A/(q - \lambda)A \longrightarrow B/(q - \lambda)B \longrightarrow C/(q - \lambda)C$$

is exact. Unfortunately, in the applications to quantized coinvariants, it appears to be a difficult task to verify the above additional hypotheses.

3. THE INTERIOR ACTION OF GL_t ON $M_{m,t} \times M_{t,n}$

Let m, n, t be positive integers with $t \leq \min\{m, n\}$. The group $GL_t(k)$ acts on the variety $V := M_{m,t}(k) \times M_{t,n}(k)$ via the rule

$$g \cdot (A, B) = (Ag^{-1}, gB),$$

and consequently on the coordinate ring $\mathcal{O}(V) \cong \mathcal{O}(M_{m,t}(k)) \otimes \mathcal{O}(M_{t,n}(k))$. The invariant theory of this action is closely related to the matrix multiplication map

$$\mu : V \longrightarrow M_{m,n}(k)$$

QUANTIZED COINVARIANTS AT TRANSCENDENTAL q 159

and its comorphism μ^*. The classical Fundamental Theorems for this situation ([2, Theorems 3.1, 3.4]) are

(1) The ring of invariants $\mathcal{O}(V)^{GL_t(k)}$ equals the image of μ^*.
(2) The kernel of μ^* is the ideal of $\mathcal{O}(M_{m,n}(k))$ generated by all $(t+1) \times (t+1)$ minors.

As discussed earlier, the comorphism of the action map $\gamma : GL_t(k) \times V \to V$ is an algebra homomorphism $\gamma^* : \mathcal{O}(V) \to \mathcal{O}(GL_t(k)) \otimes \mathcal{O}(V)$ which makes $\mathcal{O}(V)$ into a left comodule over $\mathcal{O}(GL_t(k))$, and the coinvariants for this coaction equal the invariants for the action above: $\mathcal{O}(V)^{GL_t(k)} = \mathcal{O}(V)^{\mathrm{co}\,\mathcal{O}(GL_t(k))}$.

We may consider quantum versions of this situation, relative to a parameter $q \in k^\times$, using the standard quantized coordinate rings $\mathcal{O}_q(M_{m,t}(k))$, $\mathcal{O}_q(M_{t,n}(k))$, and $\mathcal{O}_q(GL_t(k))$. We write X_{ij} for the standard generators in each of these algebras; see, e.g., [8, (1.1)], for the standard relations.

To quantize the classical coaction above, one combines quantized versions of the actions of $GL_t(k)$ on $M_{m,t}(k)$ and $M_{t,n}(k)$ by right and left multiplication, respectively. The latter quantizations are algebra morphisms

$$\rho_q^* : \mathcal{O}_q(M_{m,t}(k)) \longrightarrow \mathcal{O}_q(M_{m,t}(k)) \otimes \mathcal{O}_q(GL_t(k))$$
$$\lambda_q^* : \mathcal{O}_q(M_{t,n}(k)) \longrightarrow \mathcal{O}_q(GL_t(k)) \otimes \mathcal{O}_q(M_{t,n}(k))$$

which are both determined by $X_{ij} \mapsto \sum_{l=1}^{t} X_{il} \otimes X_{lj}$ for all i, j. The right comodule structure on $\mathcal{O}_q(M_{m,t}(k))$ determines a left comodule structure in the standard way, via the map $\tau \circ (\mathrm{id} \otimes S) \circ \rho_q^*$, where τ is the flip and S is the antipode in $\mathcal{O}_q(GL_t(k))$. We set $\mathcal{O}_q(V) = \mathcal{O}_q(M_{m,t}(k)) \otimes \mathcal{O}_q(M_{t,n}(k))$, which is a left comodule over $\mathcal{O}_q(GL_t(k))$ in the standard way. The resulting structure map $\gamma_q^* : \mathcal{O}_q(V) \to \mathcal{O}_q(GL_t(k)) \otimes \mathcal{O}_q(V)$ is determined by the rule

$$\gamma_q^*(a \otimes b) = \sum_{(a),(b)} S(a_1) b_{-1} \otimes a_0 \otimes b_0$$

for $a \in \mathcal{O}_q(M_{m,t}(k))$ and $b \in \mathcal{O}_q(M_{t,n}(k))$, where $\rho_q^*(a) = \sum_{(a)} a_0 \otimes a_1$ and $\lambda_q^*(b) = \sum_{(b)} b_{-1} \otimes b_0$. It is, unfortunately, more involved to work with γ_q^* than with its classical predecessor γ^* because γ_q^* is not an algebra homomorphism.

The analog of μ^* is the algebra homomorphism $\mu_q^* : \mathcal{O}_q(M_{m,n}(k)) \to \mathcal{O}_q(V)$ such that $X_{ij} \mapsto \sum_{l=1}^{t} X_{il} \otimes X_{lj}$ for all i, j. We can now state the Fundamental Theorems for the quantized coinvariants in the current situation:

(1_q) The set of coinvariants $\mathcal{O}_q(V)^{\mathrm{co}\,\mathcal{O}_q(GL_t(k))}$ equals the image of μ_q^* (and so is a subalgebra of $\mathcal{O}_q(V)$).
(2_q) The kernel of μ_q^* is the ideal of $\mathcal{O}_q(M_{m,n}(k))$ generated by all $(t+1) \times (t+1)$ quantum minors.

These theorems were proved for arbitrary q in [9, Theorem 4.5], and [8, Proposition 2.4], respectively.

To illustrate the use of Theorems 2.2 and 2.3, we specialize to the case that q is transcendental over a subfield $k^\circ \subset k$. Set $R = k^\circ[q, q^{-1}]$ as in the previous section. Since the construction of quantum matrix algebras requires only a commutative base ring and an invertible element in that ring, we can form $\mathcal{O}_q(M_{m,t}(R))$ and $\mathcal{O}_q(M_{t,n}(R))$. These R-algebras are iterated skew polynomial extensions of R, and thus are free R-modules, as is the algebra $\mathcal{O}_q(V_R) := \mathcal{O}_q(M_{m,t}(R)) \otimes_R \mathcal{O}_q(M_{t,n}(R))$. Similarly, $\mathcal{O}_q(M_t(R))$ is a free R-module and an integral domain, and so the localization $\mathcal{O}_q(GL_t(R))$, obtained by inverting the quantum determinant, is a torsionfree R-module, as well as a Hopf R-algebra. Since the restrictions of ρ_q^* and λ_q^* make $\mathcal{O}_q(M_{m,t}(R))$ and $\mathcal{O}_q(M_{t,n}(R))$ into right and left comodules over $\mathcal{O}_q(GL_t(R))$, respectively, the restriction γ_R^* of γ_q^* makes $\mathcal{O}_q(V_R)$ into a left $\mathcal{O}_q(GL_t(R))$-comodule. Finally, μ_q^* restricts to an R-algebra homomorphism $\mu_R^* : \mathcal{O}_q(M_{m,n}(R)) \to \mathcal{O}_q(V_R)$. It is easily checked that the image of μ_R^* is contained in $\mathcal{O}_q(V_R)^{\mathrm{co}\,\mathcal{O}_q(GL_t(R))}$, and that the kernel of μ_R^* contains the ideal I generated by all $(t+1) \times (t+1)$ quantum minors (cf. [9, Prop. 2.3], and [8, (2.1)]).

All of the quantum matrix algebras $\mathcal{O}_q(M_{\bullet,\bullet}(R))$ are positively graded R-algebras, with the generators X_{ij} having degree 1, and $\mathcal{O}_q(V_R)$ inherits a positive grading from its two factors. In each of these algebras, the homogeneous components are finitely generated free R-modules. It is easily checked that

$$\mu_R^*(\mathcal{O}_q(M_{m,n}(R))_j) \subseteq \mathcal{O}_q(V_R)_{2j} \text{ and } \gamma_R^*(\mathcal{O}_q(V_R)_j) \subseteq \mathcal{O}_q(GL_t(R)) \otimes_R \mathcal{O}_q(V_R)_j$$

for all $j \geq 0$. Moreover, the ideal I of $\mathcal{O}_q(M_{m,n}(R))$ is homogeneous with respect to this grading. Note that when we come to apply Theorem 2.3, we should replace the grading on $\mathcal{O}_q(V_R)$ by the decomposition $\bigoplus_{j=0}^{\infty} (\mathcal{O}_q(V_R)_{2j} \oplus \mathcal{O}_q(V_R)_{2j+1})$, for instance.

The classical First and Second Fundamental Theorems (1) and (2) say that $\overline{\mu_R^*}$ maps $\overline{\mathcal{O}_q(M_{m,n}(R))}$ onto the coinvariants of $\overline{\mathcal{O}_q(V_R)}$, and that the kernel of $\overline{\mu_R^*}$ equals the image of $\overline{I}$. Therefore Theorems 2.2 and 2.3 yield the quantized Fundamental Theorems (1_q) and (2_q) with no further work in the transcendental case.

4. THE RIGHT ACTION OF SL_r ON $M_{n,r}$

Fix positive integers $r < n$. In this section, we consider the right action of $SL_r(k)$ on $M_{n,r}(k)$ by multiplication: $A.g = Ag$ for $A \in M_{n,r}(k)$ and $g \in SL_r(k)$. The First Fundamental Theorem for this case (cf. [7, Prop. 2, p. 138]) says that

(1) The ring of invariants $\mathcal{O}(M_{n,r}(k))^{SL_r(k)}$ equals the subalgebra of $\mathcal{O}(M_{n,r}(k))$ generated by all $r \times r$ minors.

To state the Second Fundamental Theorem for this case, let $k[X]$ be the polynomial ring in a set of variables X_I, where I runs over all r-element subsets of $\{1, \ldots, n\}$. In view of the theorem above, there is a natural homomorphism of $\phi : k[X] \to \mathcal{O}(M_{n,r}(k))^{SL_r(k)}$.

(2) The kernel of ϕ is generated by the Plücker relations ([7, Prop. 2, p. 138]).

To quantize this situation, we make $\mathcal{O}_q(M_{n,r}(k))$ into a right $\mathcal{O}_q(SL_r(k))$-comodule via the algebra homomorphism $\rho : \mathcal{O}_q(M_{n,r}(k)) \to \mathcal{O}_q(M_{n,r}(k)) \otimes \mathcal{O}_q(SL_r(k))$ such that $\rho(X_{ij}) = \sum_{l=1}^{r} X_{il} \otimes X_{lr}$ for all i, j. The First Fundamental Theorem for the quantized coinvariants is the statement

(1_q) The set of coinvariants $\mathcal{O}_q(M_{n,r}(k))^{\mathrm{co}\,\mathcal{O}_q(SL_r(k))}$ is equal to the subalgebra of $\mathcal{O}_q(M_{n,r}(k))$ generated by all $r \times r$ quantum minors.

This follows from work of Fioresi and Hacon under the assumptions that k is algebraically closed of characteristic 0 and q is transcendental over some subfield of k ([6, Theorem 3.12]) – in fact, they prove (in our notation) that (1_q) holds with

QUANTIZED COINVARIANTS AT TRANSCENDENTAL q 161

k replaced by a Laurent polynomial ring $k^\circ[q, q^{-1}]$. The result is obtained for any nonzero q in an arbitrary field k in [11, Theorem 5.2]. The Fioresi-Hacon version of the Second Fundamental Theorem yields a presentation of $\mathcal{O}_q(M_{n,r}(k))^{co\,\mathcal{O}_q(SL_r(k))}$ in terms of generators $\lambda_{i_1\cdots i_r}$ where $i_1\cdots i_r$ runs over all unordered sequences of r distinct integers from $\{1,\ldots,n\}$. Each $\lambda_{i_1\cdots i_r}$ is mapped to $(-q)^{\ell(\pi)}D_{\{i_1,\ldots,i_r\}}$ where $\ell(\pi)$ denotes the length of the permutation π sending $l \mapsto i_l$ for $l = 1,\ldots,r$, and $D_{\{i_1,\ldots,i_r\}}$ denotes the $r \times r$ quantum minor with row index set $\{i_1,\ldots,i_r\}$.

(2_q) The algebra $\mathcal{O}_q(M_{n,r}(k))^{co\,\mathcal{O}_q(SL_r(k))}$ is isomorphic to the quotient of the free algebra $k\langle\lambda_{i_1\cdots i_r}\rangle$ modulo the ideal generated by the relations (c) and (y) given in [6, Theorem 3.14].

The hypotheses on k and q are as above.

Now let k°, k, q, R be as in Section 3, and define everything over R. In particular, $\mathcal{O}_q(M_{n,r}(R))$ is a right comodule algebra over $\mathcal{O}_q(SL_r(R))$. As already mentioned, Fioresi and Hacon have proved the R-algebra versions of (1_q) and (2_q), assuming that k is algebraically closed of characteristic zero. The easier parts of their work – proving inclusions rather than equalities in these R-algebra forms – lead directly to the k-algebra forms of these theorems, as follows.

Let $R\langle\lambda_\bullet\rangle$ denote the free R-algebra on generators $\lambda_{i_1\cdots i_r}$, and let $\phi_q : R\langle\lambda_\bullet\rangle \to \mathcal{O}_q(M_{n,r}(R))$ be the R-algebra homomorphism sending each $\lambda_{i_1\cdots i_r}$ to the weighted quantum minor $(-q)^{\ell(\pi)}D_{\{i_1,\ldots,i_r\}}$. The image of ϕ_q is the R-subalgebra generated by the $D_{\{i_1,\ldots,i_r\}}$, which is contained in $\mathcal{O}_q(M_{n,r}(R))^{co\,\mathcal{O}_q(SL_r(R))}$ by [6, Lemma 3.5]. Let I_q be the ideal of $R\langle\lambda_\bullet\rangle$ generated by the relations (c) and (y) of [6, Theorem 3.14]; that $I_q \subseteq \ker\phi_q$ was checked by Fioresi ([5, Prop. 2.21 and Theorem 3.6]). Now $\mathcal{O}_q(M_{n,r}(R))$ and $R\langle\lambda_\bullet\rangle$ are positively graded R-algebras, with the X_{ij} having degree 1 and the $\lambda_\bullet$ having degree r. The homogeneous components with respect to these gradings are finitely generated free R-modules, I_q is a homogeneous ideal of $R\langle\lambda_\bullet\rangle$, and the map ϕ_q is homogeneous of degree 0.

The classical Fundamental Theorems (1) and (2) say that $\overline{\phi_q}$ maps $\overline{R\langle\lambda_\bullet\rangle}$ onto the coinvariants of $\overline{\mathcal{O}_q(M_{n,r}(R))}$, and the kernel of $\overline{\phi_q}$ equals $\overline{I_q}$. (The latter statement requires a change of relations using the results of [7, Chapter 8], as observed in [6, p. 435].) Therefore Theorems 2.2 and 2.3 yield the quantized Fundamental Theorems (1_q) and (2_q). Note that the hypotheses of algebraic closure and characteristic zero on k are not needed, although q is still assumed to be transcendental over a subfield of k.

5. THE RIGHT ACTION OF Sp_{2n} ON $M_{m,2n}$

Fix positive integers m and n, and consider the right action of $Sp_{2n}(k)$ on $M_{m,2n}(k)$ by multiplication. Here we take $Sp_{2n}(k)$ to preserve the standard alternating bilinear form on k^{2n}, which we denote $\langle -, - \rangle$, and we view $M_{m,2n}(k)$ as the variety of m vectors of length $2n$. Thus, each row $x_i = (X_{i1},\ldots,X_{i,2n})$ of generators in $\mathcal{O}(M_{m,2n}(k))$ corresponds to the i-th of m generic $2n$-vectors. We shall need the functions describing the values of $\langle -, - \rangle$ on two of these generic vectors:

$$z_{ij} := \langle x_i, x_j \rangle = \sum_{l=1}^{n}(X_{i,2l-1}X_{j,2l} - X_{i,2l}X_{j,2l-1})$$

for $i, j = 1, \ldots, m$.

Next, consider the variety $\text{Alt}_m(k)$ of alternating $m \times m$ matrices over k; its coordinate ring is the algebra

$$\mathcal{O}(\text{Alt}_m(k)) = \mathcal{O}(M_m(k))/\langle X_{ii}, X_{ij} + X_{ji} \mid i, j = 1, \ldots, m \rangle.$$

Let us write Y_{ij} for the coset of X_{ij} in $\mathcal{O}(\text{Alt}_m(k))$. There is a morphism $\nu : M_{m,2n}(k) \to \text{Alt}_m(k)$ given by the rule $\nu(A) = ABA^{\text{tr}}$, where B is the matrix of the symplectic form $\langle -, - \rangle$, and the comorphism $\nu^* : \mathcal{O}(\text{Alt}_m(k)) \to \mathcal{O}(M_m(k))$ sends $Y_{ij} \mapsto z_{ij}$ for all i, j. For any even number h of distinct indices $i_1, \ldots, i_h$ from $\{1, \ldots, m\}$, let $[i_1, \ldots, i_h] \in \mathcal{O}(\text{Alt}_m(k))$ be the Pfaffian of the submatrix of (Y_{ij}) obtained by taking the rows and columns with indices $i_1, \ldots, i_h$.

The First and Second Fundamental Theorems for the present situation ([2, Theorems 6.6, 6.7]) state that

(1) The ring of invariants $\mathcal{O}(M_{m,2n}(k))^{Sp_{2n}(k)}$ coincides with the subalgebra of $\mathcal{O}(M_{m,2n}(k))$ generated by the z_{ij} for $i < j$, that is, the image of ν^*.

(2) If $m \leq 2n + 1$, the kernel of ν^* is zero, while if $m \geq 2n + 2$, the kernel of ν^* is the ideal of $\mathcal{O}(\text{Alt}_m(k))$ generated by all the Pfaffians $[i_1, \ldots, i_{2n+2}]$.

A quantized version of this situation was studied by Strickland [17], who quantized the $U(\mathfrak{sp}_{2n}(k))$-module action on $\mathcal{O}(M_{2n,m}(k))$ rather than its dual, the $\mathcal{O}(Sp_{2n}(k))$ coaction. We transpose the matrices and indices from her paper in order to match the notation for the classical situation used above. Since the approaches to quantized enveloping algebras for $\mathfrak{sp}_{2n}(k)$ and quantized coordinate rings for $Sp_n(k)$ become more complicated at roots of unity, let us restrict our discussion to a quantum parameter $q \in k^\times$ which is not a root of unity.

Recall that the quantized coordinate rings of the different classical groups coact on different quantized coordinate rings of affine spaces. In particular, $\mathcal{O}_q(Sp_{2n}(k))$ coacts on an algebra that we denote $\mathcal{O}_q(\mathfrak{sp}\, k^{2n})$, the *quantized coordinate ring of symplectic $2n$-space* ([16, Def. 14]; see [15, §1.1] for a simpler set of relations). Thus, $\mathcal{O}(M_{m,2n}(k))$ needs to be quantized by a suitable algebra with $m \cdot 2n$ generators, each row of which generates a copy of $\mathcal{O}_q(\mathfrak{sp}\, k^{2n})$. Let us call this algebra $\mathcal{O}_q((\mathfrak{sp}\, k^{2n})^m)$ and take it to be the k-algebra with generators $X_{ij} = x_{j,i}$ (for $i = 1, \ldots, m$ and $j = 1, \ldots, 2n$) satisfying the relations given by Strickland in [17, Equations (2.1), (2.2), (2.3)]. (The $x_{j,i}$ are Strickland's generators for the algebra she denotes B; we transpose the indices for the reasons indicated above.) Similarly, let us write $\mathcal{O}_q(\text{Alt}_m(k))$ for the k-algebra with generators $Y_{ij} = a_{j,i}$ for $1 \leq j < i \leq m$ satisfying the relations given in [17, Equations (1.1)]. As in [17, Theorem 2.5(2)], there is a k-algebra homomorphism $\phi : \mathcal{O}_q(\text{Alt}_m(k)) \to \mathcal{O}_q((\mathfrak{sp}\, k^{2n})^m)$ such that

$$\phi(Y_{ts}) = \sum_{l=1}^{n} q^{l-1-n} X_{sl} X_{t,2n-l+1} - \sum_{l=1}^{n} q^{n-l+1} X_{s,2n-l+1} X_{tl}$$

for $s < t$.

Set $B_q = \mathcal{O}_q((\mathfrak{sp}\, k^{2n})^m)$ for the moment. Strickland defines an action of $U_q(\mathfrak{sp}_{2n}(k))$ on B_q ([17, p. 87]). It is clear from the definition of this action that B_q is a locally finite dimensional left $U_q(\mathfrak{sp}_{2n}(k))$-module. Hence, B_q becomes a right comodule over the Hopf dual $U_q(\mathfrak{sp}_{2n}(k))^\circ$ in the standard way ([14, Lemma 1.6.4(2)]), and the $U_q(\mathfrak{sp}_{2n}(k))$-invariants coincide with the $U_q(\mathfrak{sp}_{2n}(k))^\circ$-coinvariants ([14, Lemma 1.7.2(2)]). The standard quantized coordinate ring of $Sp_{2n}(k)$, which we denote $\mathcal{O}_q(Sp_{2n}(k))$, is a sub-Hopf-algebra of $U_q(\mathfrak{sp}_{2n}(k))^\circ$, and one can check that the coaction $B_q \to B_q \otimes U_q(\mathfrak{sp}_{2n}(k))^\circ$ actually maps B_q into $B_q \otimes \mathcal{O}_q(Sp_{2n}(k))$. Consequently, B_q is a right comodule over $\mathcal{O}_q(Sp_{2n}(k))$, and its $\mathcal{O}_q(Sp_{2n}(k))$-coinvariants coincide with its $U_q(\mathfrak{sp}_{2n}(k))^\circ$-coinvariants.

Statements of First and Second Fundamental Theorems for quantized coinvariants in the symplectic situation above can be given as follows, where the q-*Pfaffians* $[i_1, \ldots, i_h]$ are defined as in [17, p. 82]:

(1_q) The set of coinvariants $\mathcal{O}_q((\mathfrak{sp}\, k^{2n})^m)^{\operatorname{co} \mathcal{O}_q(Sp_{2n}(k))}$ equals the image of ϕ.

(2_q) If $m \le 2n + 1$, the kernel of ϕ is zero, while if $m \ge 2n + 2$, the kernel of ϕ is the ideal of $\mathcal{O}_q(\operatorname{Alt}_m(k))$ generated by all the q-Pfaffians $[i_1, \ldots, i_{2n+2}]$.

Both statements have been proved by Strickland (modulo the changes of notation discussed above) under the assumptions that char$(k) = 0$ and q is transcendental over a subfield of k ([17, Theorem 2.5]). Via Theorems 2.2 and 2.3, we obtain the transcendental cases of (1_q) and (2_q) from the classical results (1) and (2) in arbitrary characteristic. (We note that part of Strickland's development also involves reduction modulo $q - 1$. See the proof of [17, Theorem 1.5].) As far as we are aware, it is an open question whether (1_q) and (2_q) hold when q is algebraic over the prime subfield of k.

6. THE CONJUGATION OF GL_n ON M_n

Fix a positive integer n, and assume that k has characteristic zero. In this section, we consider the quantum analogue of the classical conjugation action of $GL_n(k)$ on $M_n(k)$. We shall need the trace functions tr_i for $i = 1, \ldots, n$, where tr_i is the sum of the $i \times i$ principal minors. Note that tr_1 is the usual trace function, and that tr_n is the determinant function. The First and Second Fundamental Theorems for this situation ([13, Satz 3.1]) can be stated as follows:

(1) The ring of invariants $\mathcal{O}(M_n(k))^{GL_n(k)}$ equals the subalgebra of $\mathcal{O}(M_n(k))$ generated by $\operatorname{tr}_1, \ldots, \operatorname{tr}_n$.

(2) There are no relations among the tr_i, that is, $\operatorname{tr}_1, \ldots, \operatorname{tr}_n$ are algebraically independent over k.

The coinvariants of a quantum analogue of the conjugation action have been studied in [3]. The *right conjugation coaction* of $\mathcal{O}_q(GL_n(k))$ on $\mathcal{O}_q(M_n(k))$ is the right coaction $\beta : \mathcal{O}_q(M_n(k)) \longrightarrow \mathcal{O}_q(M_n(k)) \otimes \mathcal{O}_q(GL_n(k))$ given by $\beta(u) := \sum_{(u)} u_2 \otimes S(u_1)u_3$, where we are using the Sweedler notation. In particular, $\beta(X_{ij}) = \sum_{l,m} X_{lm} \otimes S(X_{il})X_{mj}$ for all i, j. However, as with the interior coaction studied in Section 3, the map β is not an algebra homomorphism.

Recall that if I and J are subsets of $\{1, \ldots, n\}$ of the same size, then the quantum determinant of the quantum matrix subalgebra generated by the X_{ij} with $i \in I$ and $j \in J$ is denoted by $[I|J]$ and called a *quantum minor* of the relevant quantum matrix algebra. For each $i = 1, \ldots, n$, define the weighted sums of principal quantum minors by $\tau_i := \sum_I q^{-2w(I)}[I|I]$, where I runs through all i element subsets of $\{1, \ldots, n\}$ and $w(I)$ is the sum of the entries in the index set I. These weighted sums of principal quantum minors provide quantized coinvariants for β

by [3, Proposition 7.2]. One can state First and Second Fundamental Theorems for quantized coinvariants in this situation as follows:

(1_q) The set of coinvariants $\mathcal{O}_q(M_n(k))^{co\,\mathcal{O}_q(GL_n(k))}$ is equal to the subalgebra of $\mathcal{O}_q(M_n(k))$ generated by $\tau_1, \ldots, \tau_n$.

(2_q) The subalgebra $k[\tau_1, \ldots, \tau_n]$ is a commutative polynomial algebra of degree n.

These statements were proved for $k = \mathbb{C}$ and q not a root of unity in [3, Theorem 7.3], by using the corepresentation theory of the cosemisimple Hopf algebra $\mathcal{O}_q(GL_n(\mathbb{C}))$. We give a partial extension below.

Now let $k^\circ = \mathbb{Q}$. Suppose that $q \in k^\times$ is transcendental over $\mathbb{Q}$, set $R = \mathbb{Q}[q, q^{-1}]$ and define everything over R. Then $\mathcal{O}_q(M_n(R))$ is a right comodule over $\mathcal{O}_q(GL_n(R))$ by using β as above. A straightforward calculation shows that the R-subalgebra generated by $\tau_1, \ldots, \tau_n$ is contained in $\mathcal{O}_q(M_n(R))^{co\,\mathcal{O}_q(GL_n(R))}$; see, for example, [3, Proposition 7.2]. Also, note that when $q = 1$ the coinvariants $\tau_1, \ldots, \tau_n$ coincide with the classical traces $\mathrm{tr}_1, \ldots, \mathrm{tr}_n$. Let $F(R)$ denote the free R-algebra on generators $\gamma_1, \ldots, \gamma_n$, and consider $F(R)$ to be graded by setting $\deg(\gamma_i) = i$. Let ϕ_q be the R-algebra homomorphism sending γ_i to τ_i. Then ϕ_q is homogeneous of degree 0 and the image of ϕ_q is contained in $\mathcal{O}_q(M_n(R))^{co\,\mathcal{O}_q(GL_n(R))}$. The homogeneous components of $\mathcal{O}_q(M_n(R))$ and $F(R)$ are finitely generated free R-modules with $\phi_q(F(R)_j) \subseteq \mathcal{O}_q(M_n(R))_j$ and $\beta(\mathcal{O}_q(M_n(R))_j) \subseteq \mathcal{O}_q(M_n(R))_j \otimes \mathcal{O}_q(GL_n(R))$. The classical First Fundamental Theorem (1) says that $\overline{\phi_q}$ maps $\overline{F(R)}$ onto the coinvariants of $\overline{\mathcal{O}_q(M_n(R))}$. Theorem 2.2 now yields the quantized First Fundamental Theorem (1_q) for the case that char(k) = 0 and q is transcendental over $\mathbb{Q}$.

In [1, Corollary 2.3], Cohen and Westreich show that the τ_i commute, by exploiting the coquasitriangular structure of $\mathcal{O}_q(GL_n(k))$ (see, e.g., [10, Theorem 3.1 and Prop. 4.1]; a more detailed proof for the case $k = \mathbb{C}$ is given in [12, Theorem 10.9]). Thus, if we set I to be the ideal of $F(R)$ generated by the commutators $\gamma_i\gamma_j - \gamma_j\gamma_i$, we see that $I \subseteq \ker(\phi_q)$. It is obvious that I is a homogeneous ideal. The classical theory shows that $\ker(\overline{\phi_q}) = \overline{I}$. Thus, Theorem 2.3 yields the Second Fundamental Theorem (2_q) in the case under discussion.

REFERENCES

[1] M. Cohen and S. Westreich, Some interrelations between Hopf algebras and their duals, preprint Ben Gurion University, 2002.

[2] C. De Concini and C. Procesi, A characteristic free approach to invariant theory, *Adv. Math.* **21** (1976), 330–354.

[3] M. Domokos and T. H. Lenagan, Conjugation coinvariants of quantum matrices, *Bull. London Math. Soc.* **35** (2003), 117–127.

[4] M. Domokos and T. H. Lenagan, Weakly multiplicative coactions of quantized function algebras, *J. Pure Appl. Algebra* **183** (2003), 45–60.

[5] R. Fioresi, Quantum deformation of the Grassmannian manifold, *J. Algebra* **214** (1999), 418–447.

[6] R. Fioresi and C. Hacon, Quantum coinvariant theory for the quantum special linear group and quantum Schubert varieties, *J. Algebra* **242** (2001), 433–446.

[7] W. Fulton, "Young Tableaux", *London Math. Soc. Student Texts* **35**, Cambridge Univ. Press, Cambridge, 1997.

[8] K. R. Goodearl and T. H. Lenagan, Quantum determinantal ideals, *Duke Math J.* **103** (2000) 165–190.

QUANTIZED COINVARIANTS AT TRANSCENDENTAL q

[9] K. R. Goodearl, T. H. Lenagan and L. Rigal, The first fundamental theorem of coinvariant theory for the quantum general linear group, *Publ. RIMS (Kyoto)* **36** (2000), 269–296.

[10] T Hayashi, Quantum groups and quantum determinants, *J. Algebra* **152** (1992), 146–165.

[11] A. C. Kelly, T. H. Lenagan and L. Rigal, Ring theoretic properties of quantum grassmannians, *J. Algebra Appl.*, to appear; posted at http://arXiv.org/math.QA/0208152.

[12] A. Klimyk and K. Schmüdgen, Quantum Groups and Their Representations, Springer-Verlag, Berlin, 1997.

[13] H. Kraft, Klassische Invariantentheorie: Eine Einführung, in "Algebraische Transformationsgruppen und Invariantentheorie" (H. Kraft, P. Slodowy, and T. A. Springer, Eds.), Birkhäuser, Basel, 1989, pp 41–62.

[14] S. Montgomery, "Hopf Algebras and Their Actions on Rings", *CBMS Regional Conf. Series in Math.* **82**, Amer. Math. Soc., Providence, RI, 1993.

[15] I. M. Musson, Ring-theoretic properties of the coordinate rings of quantum symplectic and Euclidean space, in "Ring Theory, Proc. Biennial Ohio State–Denison Conf., 1992" (S. K. Jain and S. T. Rizvi, Eds.), World Scientific, River Edge, NJ, 1993, pp. 248–258.

[16] N. Yu. Reshetikhin, L. A. Takhtadzhyan, and L. D. Faddeev, Quantization of Lie groups and Lie algebras, *Leningrad Math. J.* **1** (1990), 193–225.

[17] E. Strickland, Classical invariant theory for the quantum symplectic group, *Adv. Math.* **123** (1996), 78–90.

[18] H. Weyl, The Classical Groups, Princeton Univ. Press, Princeton, 1939.

Classification of Differentials on Quantum Doubles and Finite Noncommutative Geometry

SHAHN MAJID   School of Mathematical Sciences
Queen Mary, University of London
327 Mile End Rd, London E1 4NS, UK
e-mail address: s.majid@qmul.ac.uk

> ABSTRACT. We discuss the construction of finite noncommutative geometries on Hopf algebras and finite groups in the 'quantum groups approach'. We apply the author's previous classification theorem, implying that calculi in the factorisable case correspond to blocks in the dual, to classify differential calculi on the quantum codouble $D^*(G) = kG{\blacktriangleright\!\!\triangleleft}k(G)$ of a finite group G. We give $D^*(S_3)$ as an example including its exterior algebra and lower cohomology. We also study the calculus on $D^*(\mathcal{A})$ induced from one on a general Hopf algebra $\mathcal{A}$ and specialise to $D^*(G) = U(\mathfrak{g}){\blacktriangleright\!\!\triangleleft}k[G]$ as a noncommutative isometry group of an enveloping algebra $U(\mathfrak{g})$ as a noncommutative space.

1. INTRODUCTION

Coming out of the theory of quantum groups has emerged an approach to noncommutative geometry somewhat different from the operator algebras and K-theory one of Connes and others but with some points of contact with that as well as considerable contact with abstract Hopf algebra theory. The approach is based on building up the different layers of geometry: the differential structure, line bundles, frame bundles, etc. eventually arriving at spinors and a Dirac operator as naturally constructed and not axiomatically imposed as a definition of the geometry (as in Connes' spectral triple theory).

This article has three goals. The first, covered in Section 2, is an exposition of the overall dictionary as well as the differences between the Hopf algebraic quantum groups approach and the more well-known operator algebras and K-theory one. We also discuss issues that appear to have led to confusion in the literature, notably the role of the Dirac operator. The quantum groups methods are very algebraically computable and hence should be interesting even to readers coming from the Connes spectral triple side of noncommutative geometry. In fact, the convergence and interaction between the two approaches is a very important recent development and we aim to bridge between them with our overview.

The second aim, in Sections 3, 4, is a self-contained demonstration of the starting point of the quantum groups approach, namely the classification of bicovariant

---

2000 *Mathematics Subject Classification.* 58B32, 58B34, 20C05.
Key words and phrases. Quantum groups, quantum double, noncommutative geometry.
This paper is in final form and no version of it will be submitted for publication elsewhere.

168 S. MAJID

differential calculi on Hopf algebras. We start with the seminal work of Woronowicz [33] for any Hopf algebra, which is mostly known to Hopf algebraists and is
included mainly for completeness. We then explain our classification [22] for factorisable quantum groups and apply it to finite-dimensional Hopf algebras such
as $\mathbb{C}_q[SL_2]$ at a cube root of unity. We also discuss several subtleties not so well
known in the literature. Let us mention that in addition there are the Beggs–Majid
classification theorem [4] for bicrossproduct quantum groups, see also [30], and the
Majid–Oeckl twisting theorem [27] (which covers most triangular quantum groups),
which altogether pretty much cover all classes of interest. In this way the first layer
of geometry, choosing the differential structure, is more or less well understood in
the quantum groups approach. The theory is interesting even for finite-dimensional
Hopf algebras as 'finite geometries' as our example of $\mathbb{C}_q[SL_2]$ demonstrates.

Our third aim is to present some specific new results about differential calculi
on quantum doubles. Section 5 applies the classification theorem of Section 4
to differential calculi on $D^*(\mathcal{A})$ where $\mathcal{A}$ is a finite-dimensional Hopf algebra and
$D^*(\mathcal{A})$ is the dual of its Drinfeld double [11]. We demonstrate this on the nice
example of the dual of the Drinfeld quantum double of a finite group, which is in
fact a cross coproduct $kG{>}{\blacktriangleleft}k(G)$. Even this case is interesting as we demonstrate
for $G = S_3$, when we compute also its exterior algebra and cohomology. Section 6
looks at the general case of $D^*(\mathcal{A})$, where we show that a calculus on $\mathcal{A}$ extends in
a natural way to one on the double, generalising the finite group case in [30]. The
results also make sense for infinite-dimensional quantum doubles such as $D^*(G)$ in
the form $U(g){>}{\blacktriangleleft}k[G]$ for an algebraic group of Lie type. We propose the general
construction as a step towards a duality for quantum differentials inspired by T-
duality in physics.

2. COMPARING THE K-THEORETIC AND QUANTUM GROUPS APPROACHES

A dictionary comparing the two approaches appears in Table 1, and let me say
right away that we are not comparing like with like. The Connes K-theoretic
approach is mathematically purer and has deeper theorems, while the quantum
groups one is more organic and experience-led. Both approaches of course start
with an algebra M, say with unit (to keep things simple), thought of as 'functions'
but not necessarily commutative. The algebra plays the role of a topological space in
view of the Gelfand-Naimark theorem. I suppose this idea goes back many decades
to the birth of quantum mechanics. Integration of course is some kind of linear
functional. In either approach it has to be made precise using analysis (and we do
not discuss this here) but at a conceptual level there is also the question: which
functional? The cyclic cohomology approach gives good motivation to take here an
element of $HC^0(M)$, i.e. a trace functional as an axiom. We don't have a good
axiom in the quantum groups approach except when M is itself a quantum group,
when translation invariance implies a unique one. In examples, it isn't usually a
trace. For other spaces we would hope that the geometry of the situation, such as a
(quantum group) symmetry would give the natural choice. So here we already see
a difference in scope and style of the approaches. Also, since we usually proceed
algebraically in the quantum groups approach, we work over k a general field.

DIFFERENTIALS ON QUANTUM DOUBLES

Classical	Connes approach	Quantum groups approach
topol. space	algebra M	algebra M
integration	$\int : M \to \mathbb{C}$, trace	$\int : M \to k$, symmetry
differential calc.	DGA $\Omega^{\cdot}$, d	bimodule $\Omega^1 \Rightarrow$ DGA $\Omega^{\cdot}$
contruction of :	spe. triple $(\rho(M), \mathcal{H}, D)$	classify all by symmetry
inner calculus	$-$	typ. $\exists \theta :$ d $= [\theta, \}$
top form	impose cycle	typ. $\exists$ Top
principal bundle	$-$	$(P, \Omega^1(P), A, \Omega^1(A), \Delta_R)$
		A quantum group fiber
vector bundle	projective module $(\mathcal{E}, e)$	$\mathcal{E} = (P \otimes V)^A$, typ. $\exists\ e$
connection	∇	$\Omega^1(P) = \Omega^1_{hor} \oplus \Omega^1_{ver}$
		$\Leftrightarrow \omega : \Lambda^1(A) \to \Omega^1(P) \Rightarrow D_\omega$
Chern classes	Chern $-$ Connes pairing	$-$
frame bundle	$-$	princ. $P +$ soldering $\Rightarrow \mathcal{E} \cong \Omega^1(M)$
metric	$\|[D, M]\| \Rightarrow d(\psi, \phi)$	$g \in \Omega^1 \otimes_M \Omega^1$; coframing
Levi $-$ Civta	contained in D	$D_\omega +$ soldering $\Rightarrow \nabla_\omega$
Ricci tensor		$d\omega + \omega \wedge \omega \Rightarrow$ Ricci
spin bundle	assumed $(D, \mathcal{H})$	assoc. to frame bundle
Dirac operator		$(d, \gamma_a, \omega) \Rightarrow \not{D}$
Hodge star	in $\| \ \|$ + orientation	g, Top $\Rightarrow *$

TABLE 1. Two approaches to noncommutative geometry

Next, also common, is the notion of 'differential structure' specified as an exterior algebra of differential forms Ω over M. An example is the universal differential calculus Ω_{univ} going back to algebraic topology, Hochschild cohomology etc. in the works of Quillen, Loday, Connes, Karoubi and others. Here

$$\Omega^1_{univ} = \ker(\cdot), \quad df = f \otimes 1 - 1 \otimes f, \quad \forall f \in M, \tag{1}$$

where $\cdot : M \otimes M \to M$ is the product. Similarly for higher degrees. All other exterior algebras are quotients, so to classify the abstract possibilities for Ω as a differential graded algebra (DGA) we have to construct suitable differential graded ideals. This is the line taken in [8] and an idea explained there is to start with a spectral triple $(\rho(M), \mathcal{H}, D)$ as a representation of Ω_{univ}, where d is represented by commutator with an operator D called 'Dirac', and then to divide by the kernel. This does *not* however, usually, impose enough constraints for a reasonable exterior algebra $\Omega(M)$, e.g. it is often not finite-dimensional over M, so one forces higher degrees to be zero, as well as additional relations with the aid of a cyclic cocycle of the desired top degree d corresponding to a 'cycle' $\int : \Omega \to \mathbb{C}$. For an orientation there is also a $\mathbb{Z}_2$ grading operator (trivial in the odd case) and a charge conjugation operator, and Connes requires a certain Hochschild cycle of degree d whose image in the spectral triple representation is the grading operator. This data together with the operator norm and a K-theoretic condition for Poincaré duality subsumes the role played in geometry by the usual Hodge * operator and the volume form.

Here again the quantum groups approach is less ambitious and we don't have a general construction. Instead we *classify all possible* $\Omega^1(M)$, typically restricted by some symmetry to make the problem manageable. We postpone $\Omega^2(M)$ and higher till later. Thus a first order calculus over M is a nice notion all by itself:

- An $M - M$-bimodule Ω^1
- A linear map d $: M \to \Omega^1$ such that $\mathrm{d}(fg) = (\mathrm{d}f)g + f\mathrm{d}g$ for all $f, g \in M$ and such that the map $M \otimes M \to \Omega^1$ given by $f\mathrm{d}g$ is surjective.

It means classifying sub-bimodules of Ω^1_{univ} to quotient by. In the last 5 years it was achieved on a case-by-case bass for all main classes of quantum groups and bicovariant calculi (see section 3). Then homogeneous spaces will likewise be constrained by smoothness of desired actions and hence inherit natural choices for their calculi, and so on for an entire noncommutative universe of objects. Then we look at natural choices for $\Omega^2(M)$, $\Omega^3(M)$ etc. layer by layer and making choices only when needed. There is a key lemma [6] that if $\Omega^m(M)$ for $m \leq n$ are specified with d obeying $\mathrm{d}^2 = 0$ then these have a maximal prolongation to higher degree, where we add in only those relations at higher degree implied by $\mathrm{d}^2 = 0$ and Leibniz. And when M is a quantum group there is a canonical extension to all of $\Omega(M)$ due to [33] and the true meaning of which is Poincaré duality in the sense of a nondegenerate pairing between forms and skew-tensor fields. We will say more about this in Section 3.

Apart from these differences in style the two approaches are again broadly similar at the level of the differential forms and there is much scope for convergence. Let us note one difference in terminology. In the quantum groups approach we call a calculus 'inner' if there is a 1-form θ which generates d as graded commutator. This looks a lot like Connes 'Dirac operator' D in the spectral triple representation

$$\rho_D(\mathrm{d}f) = [D, \rho(f)] \tag{2}$$

but beware: our θ is a 1-form and nothing to do with the geometric Dirac operator $\slashed{D}$ that we construct only much later via gamma-matrices, spin connections etc. Moreover D in a spectral triple does *not* need to be in the image under the representation ρ_D of the space of 1-forms and should not be thought of as a one form.

The next layers of geometry require bundles. In the traditional approach to vector bundles coming from the theorems of Serre and Swan one thinks of these as finitely generated projective modules over M. However, in the quantum groups approach, as in differential geometry, we think that important vector bundles should really come as associated to principal ones. Principal bundles should surely have a quantum group fiber so this is the first place where the quantum group approach comes into its own. The Brzeziński-Majid theory of these is based on an algebra P (the 'total space'), a quantum group coordinate algebra A, *and differential structures on them*, a coaction $\Delta_R : P \to P \otimes A$ such that $M = P^A$ and an exact sequence [7]

$$0 \to P\Omega^1(M)P \to \Omega^1(P) \xrightarrow{\mathrm{ver}} P \otimes \Lambda^1(A) \to 0 \tag{3}$$

which expresses 'local triviality'. Here $\Lambda^1(A)$ are the left-invariant differentials in $\Omega^1(A)$ and the map ver is the generator of the vertical vector fields. We require its kernel to be exactly the 'horizontal' forms pulled up from the base. In the

DIFFERENTIALS ON QUANTUM DOUBLES

geometrically less interesting case of the universal differential calculus the exactness of (3) is equivalent to a Hopf-Galois extension. A connection is a splitting of this sequence and characterised by a connection form ω. Of course, we have associated vector bundles $\mathcal{E} = (P \otimes V)^A$ for every A-comodule V, analogous to the usual geometric construction. A connection ω induces a covariant derivative D_ω with the 'derivation-like' property that one expects for an abstract covariant derivative ∇, so the two approaches are compatible.

Just as a main example in the K-theory approach used to be the 'noncommutative torus' T_θ^2[9], with vector bundles classified in [10], the main example for noncommutative bundles in the quantum groups approach was the q-monopole bundle [7] over the Podleś 'noncommutative sphere' S_q^2. Probably the first nontrivial convergence between these approaches was in 1997 with the Hajac-Majid projector e where [16]:

$$1 - e = \begin{pmatrix} q^2(1 - b_3) & -qb_+ \\ b_- & b_3 \end{pmatrix} \in M_2(S_q^2), \quad ede \cong D_\omega, \quad \langle [\tau], [e] \rangle = \tau(\text{Tr}(e)) = 1 \quad (4)$$

where ω is the q-monopole connection and τ is the Masuda et al. trace on S_q^2 which had been found in [29]. The pairing is the Chern-Connes one between HC^0 and the class of e as an element of the K-group K_0. We use the standard coordinates on S_q^2 namely with generators $b_\pm, b_3$ and relations

$$b_\pm b_3 = q^{\pm 2} b_3 b_\pm + (1 - q^{\pm 2}) b_\pm, \quad q^2 b_- b_+ = q^{-2} b_+ b_- + (q - q^{-1})(b_3 - 1)$$

$$b_3^2 = b_3 + q b_- b_+. \quad (5)$$

In this way one may exhibit the lowest charge monopole bundle $\mathcal{E}_1$ as a projective module. At the same time, since the Chern-Connes pairing is nonzero, it means that the q-monopole bundle is indeed nontrivial as expected. The computations in [16] are algebraic and, moreover, only for the universal differential calculus, but we see that matching up the two approaches is useful even at this level. This projector has sparked quite a bit of interest in recent years. Dabrowski and Landi looked for similar constructions and projectors for q-instantons, replacing complex numbers by quaternions. An inspired variant of that became the Connes-Landi projectors for twisted spheres S_θ^4. In these cases the new idea is (as far as I understand) to work backwards from an ansatz for the form of projector to the forced commutation relations for those matrix entries as required by $e^2 = e$ and a desired pairing with cyclic cohomology. Note that, as twists, the noncommutative differential geometry of such examples is governed at the algebraic level by the Majid-Oeckl twisting theorem [27].

Finally, we come to the 'top layer' which is Riemannian geometry. In the approach of Connes this is all contained in the 'Dirac operator' D which was assumed at the outset as an axiom. In the quantum groups approach we have been building up the different layers and hope to construct a particular family of $\not{D}$ reflecting all the choices of differential structure, spin connections etc. that we have made at lower levels. So the two approaches are going in opposite directions. In the table we show the quantum groups formulation of Riemannian geometry introduced in [23] and studied recently for finite sets [24], based on the notion of a quantum frame bundle. The main idea is a principal bundle (P, A, Δ_R) and an A-comodule V along with a 'soldering form' $\theta_V : V \to \Omega^1(P)$ that ensures that

$$\Omega^1(M) \cong \mathcal{E} = (P \otimes V)^A \tag{6}$$

i.e. that the cotangent bundle really can be identified as an associated bundle to the principal 'frame' one. It turns out that usual notions proceed independently of the choice of A, though what ω are possible and what ∇_ω they induce depends on the choice of A. Here $\nabla_\omega : \Omega^1(M) \to \Omega^1(M) \otimes_M \Omega^1(M)$ is D_ω mapped over under the soldering isomorphism. A certain $\bar{D}_\omega \wedge \theta_V$ corresponds similarly to its torsion tensor. In terms of ∇_ω the Riemann and torsion tensor are given by

$$\text{Riemann} = (\text{id} \wedge \nabla_\omega - \text{d} \otimes \text{id})\nabla_\omega, \quad \text{Tor} = \text{d} - \nabla_\omega \tag{7}$$

as a 2-form valued operator an 1-forms and a map from 1-forms to 2-forms, respectively. For the Ricci tensor we need to lift the 2-form values of Riemann to $\Omega^1(M) \otimes_M \Omega^1(M)$ in a way that splits the projection afforded by the wedge product, after which we can take a trace over Ω^1. This is possible at least in some cases. This much is a 'framed manifold' with connection.

A *framed Riemannian manifold* needs in addition a nondegenerate tensor $g \in \Omega^1(M) \otimes \Omega^1(M)$, which is equivalent to a coframing $\theta_{V^*} : V^* \to \Omega^1(P)$ via

$$g = \langle \theta_V \underset{M}{\otimes} \theta_{V^*} \rangle, \tag{8}$$

where we pair by evaluating on the canonical element of $V \otimes V^*$, and θ_{V^*} is such that $\mathcal{E}^* \cong \Omega^1(M)$. Since $\mathcal{E}^*$ under the original framing is isomorphic to $\Omega^{-1}(M)$, such a coframing is the same as an isomorphism $\Omega^{-1}(M) \cong \Omega^1(M)$. The coframing point of view is nice because it points to a natural self-dual generalization of Riemannian geometry: instead of demanding zero torsion and $\nabla_\omega g = 0$ as one might usually do for a Levi-Civita connection, one can demand zero torsion and zero cotorsion (i.e. torsion with respect to θ_{V^*} as framing). The latter is a skew-version

$$(\nabla_\omega \wedge \text{id} - \text{id} \wedge \nabla_\omega)g = 0 \tag{9}$$

of metric compatibility as explained in [23]. We can also demand symmetry, if we want, in the form

$$\wedge(g) = 0 \tag{10}$$

and we can limit ourselves to θ_{V^*} built from θ_V induced by an A-invariant local metric $\eta \in V \otimes V$. Let us also note that if M is parallelizable we can frame with a trivial tensor product bundle and θ_V, θ_{V^*} reduce to a vielbein $e_V : V \hookrightarrow \Omega^1(M)$ and covielbein $e_{V^*} : V^* \hookrightarrow \Omega^1(M)$, i.e. subspaces forming left and right bases respectively over M and dual as A-comodules. A connection ω now reduces to a 'Lie algebra-valued' 1-form

$$\alpha : \Lambda^1(A) \to \Omega^1(M), \tag{11}$$

etc., in keeping with the local picture favoured by physicists. One has to solve for zero torsion and zero cotorsion in the form $\bar{D} \wedge e$ and $D \wedge e^* = 0$. At the time of writing the main noncommutative examples are when M is itself a quantum group. In the coquasitriangular case the dual of the space of invariant 1-forms forms a braided-Lie algebra [15], which comes with a braided-Killing form η. This provides a natural metric and in several examples one finds for it (by hand; a general theorem is lacking) that there is a unique associated generalised Levi-Civita connection in the sense above [24, 25].

We are then able to take a different A-comodule W, say, for spinors. The associated bundle $S = (P \otimes W)^A$ gets its induced covariant derivative from the spin connection ω on the principal bundle, and in many cases there is a reasonable choice of 'gamma matrices' appropriate to the local metric η. We then define the Dirac operator from these objects much as usual. By now the approach is somewhat different from the Connes one and we do *not* typically obtain something obeying the axioms for D. This seems the case even for finite groups [24] as well as for q-quantum groups [25]. The fundamental reason is perhaps buried in the very notion of vector field: in the parallelizable case an M-basis $\{e_a\}$ of $\Omega^1(M)$ implies 'partial derivatives' ∂^a defined by $df = \sum_a e_a \partial^a(f)$. These are not usually derivations but more typically 'braided derivations' (e.g. on a quantum group this is shown in [22]). In cases such as the noncommutative torus one has in fact ordinary derivations around. The noncommutative differential calculus is a twist so that the constructions look close to classical. But for q-examples and even finite group examples, this is not at all the case. Perhaps this is at the root of the mismatch and may stimulate a way to fix the problem.

Also in the presence of a metric we obtain a Hodge $*$ operator $\Omega^m \to \Omega^{d-m}$ where d is the 'volume dimension' or degree of the top (volume) form, assuming of course that it exists. Once we have this we can write down actions such as $-\frac{1}{4} F \wedge *F$ etc where $F = d\alpha$ is the curvature of $\alpha \in \Omega^1$ modulo exact forms (Maxwell theory) or $F = d\alpha + \alpha \wedge \alpha$ is the curvature of α viewed modulo gauge transformation by invertible functions (this is $U(1)$-Yang-Mills theory). Integration and the Hodge $*$ thus play the role of the operator norm $\| \ \|$ in the Connes approach. The two approaches were compared in a simple model in [28]. In principle one should be able to extend these ideas to the non-Abelian gauge theory on bundles as well, to construct a variety of Lagrangian-based models.

3. CLASSIFYING CALCULI ON GENERAL HOPF ALGEBRAS

We now focus for the rest of the paper on a small part of the quantum groups approach discussed above, namely just the differential calculus, and for the most part just Ω^1. In this section we let $M = A$ be a Hopf algebra over a field k. Following Woronowicz [33], a differential structure is bicovariant if:

- Ω^1 is a bicomodule with $\Delta_L : \Omega^1 \to A \otimes \Omega^1$ and $\Delta_R : \Omega^1 \to \Omega^1 \otimes A$ bimodule maps.
- d is a bicomodule map.

Here a Hopf algebra means a coproduct $\Delta : A \to A \otimes A$, a counit $\epsilon : A \to k$ and an antipode $S : A \to A$ such that A is a coalgebra, Δ an algebra map etc. [20, 26]. Coalgebras and (bi)comodules are defined in the same way as algebras and (bi)modules but with the directions of structure maps reversed. In the bicovariance condition A is itself a bi(co)module via the (co)product and $\Omega^1 \otimes A, A \otimes \Omega^1$ have the tensor product (bi)module structure. The second condition in particular fully determines Δ_L, Δ_R by compatibility with Δ, so a bicovariant calculus means precisely one where left and right translation expressed by Δ extend consistently to

Ω^1. The universal calculus $\Omega^1_{univ} \subset A \otimes A$ is bicovariant with coactions the tensor product of the regular coactions defined by the coproduct on each copy of A.

The result in [33] is that Ω^1 in the bicovariant case is fully determined by the subspace Λ^1 of (say) Δ_L-invariant 1-forms. Indeed, there is a standard bi(co)module isomorphism

$$A \otimes A \cong A \otimes A, \quad a \otimes b \mapsto a\Delta b \tag{12}$$

under which $\Omega^1_{univ} \cong A \otimes A^+$, where $A^+ = \ker \epsilon$ is the augmentation ideal or kernel of the counit (classically it would be the functions vanishing at the group identity). The bimodule structure on the right hand side of (12) is left multiplication in the first A from the left and the tensor product of two right multiplications from the right. The bicomodule structure is the left coproduct on the first A from the left and the tensor product of right comultiplication and the right quantum adjoint coaction from the right. Hence we arrive at the classic result:

PROPOSITION 3.1. *(Woronowicz) Bicovariant Ω^1 are in 1-1 correspondence with quotient objects Λ^1 of A^+ as an A-crossed module under right multiplication and the right quantum adjoint coaction.*

We recall that an A-crossed module means a vector space which is both an A-module and a compatible A-comodule; the compatibility conditions are due to Radford and correspond in the finite-dimensional case to a module of the Drinfeld double $D(A) = A^{*\mathrm{op}} \bowtie A$ when we view a right coaction of A as a right action of $A^{*\mathrm{op}}$ by evaluation. Given the crossed module Λ^1 we define $\Omega^1 = A \otimes \Lambda^1$ with the regular left(co)modules and the tensor product (co)actions from the right. Because the category is prebraided, there is a Yang-Baxter operator $\Psi = \Psi_{\Lambda^1,\Lambda^1} : \Lambda^1 \otimes \Lambda^1 \to \Lambda^1 \otimes \Lambda^1$ which is invertible when A has bijective antipode. One has the same results for right-invariant 1-forms $\bar\Lambda^1$ with $\Omega = \bar\Lambda^1 \otimes A$.

Finally we recall that in the quantum groups approach we only need to classify Ω^1 because, at least for a bicovariant calculus, there is a natural extension to an entire exterior algebra $\Omega = A \otimes \Lambda$ (where Λ is the algebra of left-invariant differential forms). There is a d operation obeying $\mathrm{d}^2 = 0$ i.e. we have an entire DGA. The construction in [33] is a quotient of the tensor algebra over A,

$$\Omega = T_A(\Omega^1) / \oplus_n \ker A_n, \quad A_n = \sum_{\sigma \in S_n} (-1)^{l(\sigma)} \Psi_{i_1} \cdots \Psi_{i_{l(\sigma)}} \tag{13}$$

where $\Psi_i \equiv \Psi_{i,i+1}$ denotes a certain braiding $\Psi : \Omega^1 \otimes_A \Omega^1 \to \Omega^1 \otimes_A \Omega^1$ acting in the $i, i+1$ place of $(\Omega^1)^{\otimes^n_A}$ and $\sigma = s_{i_1} \cdots s_{i_{l(\sigma)}}$ is a reduced expression in terms of simple reflections. This turns out to be equivalent to defining Λ directly as the tensor algebra of Λ^1 over k with A_n defined similarly by $\Psi_{\Lambda^1,\Lambda^1}$ above. As such, Λ is manifestly a braided group or Hopf algebra in a braided category of the linear braided space type [20]. Similarly Λ^* starting from the tensor algebra of Λ^{1*} is a braided group, dually paired with Λ. The relations in (13) can be interpreted as the minimal such that this pairing is nondegenerate, which is a version of Poincaré duality. Also, Λ being a braided group means among other things that if there is a top form (an integral element) then it is unique, which defines an 'epsilon tensor' by the coefficient of the top form. This combines with a metric to form a Hodge * operator as mentioned in Section 2. More details and an application to Schubert calculus on flag varieties are in our companion paper in this volume.

4. CLASSIFICATION ON COQUASITRIANGULAR HOPF ALGEBRAS

After Woronowicz's 1989 work for any Hopf algebra, the next general classification result is my result for factorisable coquasitriangular Hopf algebras such as appropriate versions of the coordinate algebras $\mathbb{C}_q[G]$ of the Drinfeld-Jimbo quantum groups, presented in Goslar, July 1996 [21, 22]. We start with the finite-dimensional theory which is our main interest for later sections, e.g. the above quantum groups at q a primitive odd root of unity. We recall that a Hopf algebra H is quasitriangular if there is a 'universal R-matrix' $\mathcal{R} \in H \otimes H$ obeying certain axioms (due to Drinfeld). It is factorisable if $Q = \mathcal{R}_{21}\mathcal{R}$ is nondegenerate when viewed as a linear map from the dual. The dual notion to quasitriangular is coquasitriangular in the sense of a functional $\mathcal{R}$ on the tensor square, a notion which works well in the infinite-dimensional case also [20].

THEOREM 4.1. *(Majid) Let A be a finite-dimensional factorisable coquasitriangular Hopf algebra with dual H. Bicovariant $\Omega^1(A)$ are in 1-1 correspondence with two-sided ideals of H^+.*

Proof. This is [22, Prop. 4.2] which states that the required action of the quantum double on $\Lambda^{1*} \subset H^+$ under the quantum Killing form isomorphism $Q : A^+ \to H^+$ is the left and right coregular representation of two copies of H on A^+ (the coregular action of H on A is obviously adjoint to the regular action of H on H if we wish to phrase it explicitly in terms of H, i.e. Λ^1 is isomorphic to a quotient of H^+ by a 2-sided ideal). $\square$

As far as the actual proof in [22] is concerned, there were two ideas. First of all, instead of classifying quotients Λ^1 of A^+ it is convenient to classify their duals or 'quantum tangent spaces' as subcrossed modules $\mathcal{L} \subset H^+$. For coirreducible calculi (ones with no proper quotients) we want irreducible $\mathcal{L}$, but we can also classify indecomposable ones, etc. Of course an A-crossed module can be formulated as an H-crossed module (the roles of action and coaction are swapped) or $D(H)$-module, so actually we arrive at a self-contained classification for quantum tangent spaces for any Hopf algebra H. One has to dualise back (as above) to get back to the left-invariant 1-forms. The second idea was that when H is factorisable the quantum Killing form is a nondegenerate map $Q : A^+ \to H^+$ and, moreover, $D(H) \cong H \blacktriangleright\!\blacktriangleleft H$, which as an algebra is a tensor product $H \otimes H$. We refer to [20] where the forward direction of the isomorphism was proven for the first time. So the crossed module structures $\mathcal{L} \subset H^+$ that we must classify become submodules of A^+ under the action of H from the left and the right (viewed as left via S), which we computed as the left and right coregular ones. Converting back to Λ^1 means of course 2-sided ideals as stated.

At present we are interested in the finite-dimensional case where we need only the algebraic theorem as above. Note that every Artinian algebra has a unique block decomposition, which includes all finite-dimensional algebras H over a field

k. The decomposition is equivalent to finding a set of orthogonal centrally primitive idempotents e_i with

$$1 = \sum_i e_i. \tag{14}$$

These generate ideals $e_i H$. Note that $e_i^2 = e_i$ implies that $\epsilon(e_i) = 0, 1$ and the above implies that exactly one is nonzero. Similarly $H^+ = \oplus_i e_i H^+$ is a decomposition of H^+. Hence

$$\Omega^1_{univ} \cong \oplus_i \Omega^1_{e_i} \tag{15}$$

where, for any central projector e we have a calculus

$$\Omega^1_e = A \otimes \Lambda^1_e, \quad \Lambda^1_e = eH^+. \tag{16}$$

We build the left-invariant 1-forms directly on the block as isomorphic to H^+ modulo the kernel of multiplication by e. In these terms (tracing through the details of the classification theorem [22]) we have explicitly:

$$(eh).a = \sum a_{(1)} e \mathcal{R}_2(a_{(2)}) h \mathcal{R}_1(a_{(3)}), \quad \mathrm{d}a = \sum a_{(1)} e \mathcal{Q}_1(a_{(2)}) - ae; \quad \theta = e \tag{17}$$

(the calculus here is inner). We use the Sweedler notation $\Delta a = \sum a_{(1)} \otimes a_{(2)}$ and the notation from [20] where $\mathcal{R}_1(a) = (a \otimes \mathrm{id})(\mathcal{R})$ etc. Since $\mathcal{R}_1$ is an algebra map and $\mathcal{R}_2$ an antialgebra map, one may easily verify that this defines a bimodule and that the Leibniz rule is obeyed. These formulae generalise ones given usually in terms of R-matrices. One has, cf. [22]:

COROLLARY 4.2. *(Majid) In the finite-dimensional semisimple case, coirreducible bicovariant calculi on A in the setting of Theorem 4.1 are in 1-1 correspondence with nontrivial irreducible representations of the dual Hopf algebra H. The dimension of the calculus is the square of the dimension of the corresponding representation.*

Proof. Indeed, in the case of H finite-dimensional semisimple, each block will be a matrix block corresponding to an irreducible representation. In this case among the projectors there will be exactly one where $eh = e\epsilon(h)$ for all $h \in H$ (the normalized unimodular integral). It has counit 1 and corresponds to the trivial representation; we exclude it in view of $eH^+ = 0$. $\square$

This semsimple case is the situation covered in [22] in the form of an assumed Peter-Weyl decomposition. We will apply it to the quantum double example in the next section. Moreover, for each irredicble representation, one may write (17) in terms of the matrix R given by $\mathcal{R}$ in the representation. In this case one has formulae first used by Jurco [18] for the construction of bicovariant calculi on the standard quantum groups such as $\mathbb{C}_q[SL_n]$. That is not our context at the moment but we make some remarks about it at the end of the section. Let us note only that at the time they appeared such results in [22] were the first of any kind to identify the full moduli of all coirreducible bicovariant calculi in some setting with irreducible representations.

In the nonsemisimple case the algebra H has a nontrivial Jacobson radical J defined as the intersection of all its maximal 2-sided ideals. It lies in H^+. Hence by Theorem 4.1 there is a calculus

$$\Omega^1_{ss} = A \otimes H^+/J \tag{18}$$

DIFFERENTIALS ON QUANTUM DOUBLES 177

which we call the 'semisimple quotient' of the universal calculus. H/J has a decomposition into matrix blocks giving a decomposition of Ω_{ss}^1 along the lines of the semisimple case.

EXAMPLE 4.3. *Let $\mathbb{C}_q^{\mathrm{red}}[SL_2]$ be the 27-dimensional reduced quantum group at q a primitive cube root of unity. Here*

$$a^3 = d^3 = 1, \quad b^3 = c^3 = 0$$

in terms of the usual generators. The enveloping Hopf algebra $u_q(sl_2)$ is known to have the block decomposition

$$u_q(sl_2) = M_3(\mathbb{C}) \oplus B$$

where B is an 18-dimensional non-matrix block (the algebra is not semisimple) with central projection of counit 1. Hence the universal calculus decomposes into nontrivial calculi of dimensions $9, 17$. Moreover, $J \subset B$ is 13-dimensional and the quotient

$$B/J = \mathbb{C} \oplus M_2(\mathbb{C})$$

implies a 4-dimensional calculus as a quotient of the 17-dimensional one (the other summand $\mathbb{C}$ gives zero). Here Ω_{ss}^1 is the direct sum of the 4 and 9 dimensional matrix calculi.

The natural choice of calculus here is 4-dimensional and has the same form as the lowest dimension calculus for generic q, the 4D one first found by hand by Woronowicz [33]. For roots of unity the cohomology and entire geometry are, however, completely different from the generic or real q case as shown in [14]. This takes $\mathbb{C}_q^{\mathrm{red}}[SL_2]$ at 3,5,7-th roots as a finite geometry where all computations can be done and all ideas explored completely. We find, for example, the Hodge * operator for the natural q-metric, and show that the moduli space of solutions of Maxwell's equations without sources or 'self-propagating electromagnetic modes' decomposes into a direct sum

$$\{\text{Maxwell zero modes}\} = \{\text{zero} - \text{curvature}\} \oplus \{\text{self} - \text{dual}\} \oplus \{\text{anti} - \text{selfdual}\}$$

of topological (cohomology) modes, self-dual and anti-selfdual ones. The noncommutative de Rham cohomology here in each degree has the same dimension as the space of left-invariant forms, for reasons that are mysterious. We also find that the number of self-dual plus zero curvature modes appears to coincide with the number of harmonic 1-form modes. We find in general that the reduced $\mathbb{C}_q^{\mathrm{red}}[SL_2]$, although finite dimensional and totally algebraic, behaves geometrically like a 'noncompact' manifold apparently linked to the nonsemisimplicity of $u_q(sl_2)$. Also interesting is the Riemannian geometry of the reduced $\mathbb{C}_q^{\mathrm{red}}[SL_2]$ in [25]. The Ricci tensor for the same q-metric turns out to be essentially proportional to the metric itself, i.e. an 'Einstein space'. It would also be interesting to look at the differential geometry induced by taking the other blocks according to our classification theorem, particularly the non-matrix block.

Finally, let us go back and comment on the situation for the usual (not reduced) q-deformation quantum groups $\mathbb{C}_q[G]$ associated to simple Lie algebras $\mathfrak{g}$. There are two versions of interest, Drinfeld's deformation-theoretic formal powerseries setting over $\mathbb{C}[[\hbar]]$ where $q = e^{\frac{\hbar}{2}}$, and the algebraic fixed q setting. The first is

178   S. MAJID

easier and we can adapt the proof of Theorem 4.1 immediately to it [22]. Here $H = U_q(\mathfrak{g})$ and $A = U_q(\mathfrak{g})'$ is its topological dual Hopf algebra rather than the more usual coordinate algebra (the main difference is that some of its generators have logarithms).

PROPOSITION 4.4. *(Majid) In the formal powerseries setting over $\mathbb{C}[[\hbar]]$, bicovariant Ω^1 on the dual of $U_q(\mathfrak{g})$ are in 1-1 correspondence with 2-sided ideals in $U(\mathfrak{g})[[\hbar]]^+$*

Proof. In fact the key point of Drinfeld's formulation [11] is that one has a category of deformation Hopf algebras with all the axioms of usual Hopf algebra including duality holding over the ring $\mathbb{C}[[\hbar]]$ and the additional axioms of a quasitriangular structure holding in particular for his version of $U_q(\mathfrak{g})$. The arguments in the proof of [22, Prop. 4.2] and Theorem 4.1, as outlined above, require only these general axiomatic properties and take the same form line by line in the Drinfeld setting. Thus, the map Q has the same intertwiner properties used to convert an ad-stable right ideal in the dual to a 2-sided ideal in $U_q(\mathfrak{g})$. That $U_q(\mathfrak{g})$ are indeed factorizable (so that Q is an isomorphism) when we use the topological dual is the Reshetikhin-Semenov-Tian-Shanksy theorem [31]. Its underlying reason is visible at the level of the subquantum groups $U_q(b_\pm)$ where $U_q(b_\pm)' \cong U_q(b_\mp)$ was shown by Drinfeld in [11], which combines with the triangular decomposition of $U_q(\mathfrak{g})$ to yield factorisibility. Finally, also due to Drinfeld, is that $U_q(\mathfrak{g}) \cong U(\mathfrak{g})[[\hbar]]$ as algebras (since given by a coproduct-twist of the latter as a quasiHopf algebra [12]). Hence two sided ideals in $U_q(\mathfrak{g})^+$ are in correspondence with 2-sided ideals in the undeformed $U(\mathfrak{g})[[\hbar]]^+$. $\square$

This reduces the classification of calculi to a classical question about the undeformed algebra $U(\mathfrak{g})[[\hbar]]$. The natural two-sided ideals of interest here are those given by two-ideals from $U(\mathfrak{g})$ as a Hopf algebra over $\mathbb{C}$. Let us call these ideals of 'classical type' and likewise the corresponding calculi 'of classical type'. We are interested in those contained in $U(\mathfrak{g})^+$ and which are cofinite so that the corresponding calculi are finite-dimensional (f.d.). Maximal such ideals correspond to coirreducible calculi. Maximal cofinite ideals of $U(\mathfrak{g})$, as for any algebra over a field, correspond (via the kernel) to its finite-dimensional irreducible representations and hence to such representations of $\mathfrak{g}$ as a Lie algebra. Intersection with $U(\mathfrak{g})^+$ gives corresponding ideals which are maximal there, and for a proper ideal we drop the trivial representation. The correspondence here uses the central character to separate maximal ideals in one direction, and the annihilator of the quotient of $U(\mathfrak{g})^+$ by the ideal in the other direction. Hence Proposition 4.4 has the corollary

$$\left\{ \begin{array}{c} \text{coirreducible classical} - \text{type} \\ \text{f.d. bicovariant calculi on } U_q(\mathfrak{g})' \end{array} \right\} \leftrightarrow \left\{ \begin{array}{c} \text{nontrivial f.d. irreducible} \\ \text{representations of } \mathfrak{g} \end{array} \right\} \tag{19}$$

which is in the same spirit as Corollary 4.2. Let us note that when one speaks of representations of $U_q(\mathfrak{g})$ in the deformation setting one usually has in mind ones similarly of classical-type deforming ones of $U(\mathfrak{g})$ (so that one speaks of an equivalence of categories). On the other hand, without a detailed study of the Peter-Weyl decomposition for $U_q(\mathfrak{g})$ in the $\mathbb{C}[[\hbar]]$ setting, we do not claim that the universal calculus decomposes into a direct sum of these calculi, which would be the full-strength version of the theory in [22] (as explained there).

DIFFERENTIALS ON QUANTUM DOUBLES 179

The classification for the $q \in \mathbb{C}$ case with $\mathbb{C}_q[G]$ algebraic is more complicated and shows a different 'orthogonal' kind of phenomenon. Here one finds, as well as one calculus for each finite-dimensional irreducible representation (in the spirit of (19)), further 'twisted variants' of the same square dimension. This possibility was mentioned in [22] and attributed to the fact that in this case the Hopf algebra is not quite factorisable. Points of view differ on the significance of these additional twists but our own is the following. First of all, one can guess that the calculi that fit with the $\mathbb{C}[[\hbar]]$ analysis above should correspond in the algebraic setting to calculi which are commutative as $q \to 1$, while the 'twisted variants' should not have this property and hence could be considered as pathological from a deformation-theoretic point of view. As evidence, we demonstrate this below for $\mathbb{C}_q[SL_n]$. Moreover, this is a pathology that exists for most R-matrix constructions, not only calculi. Indeed, we already explained in [19, 20] that to fit with Hopf algebra theory R-matrices must be normalised in the 'quantum group normalisation'. For $\mathbb{C}_q[SL_n]$ let R_{hecke} be in the usual Hecke normalisation where the braiding has eigenvalues $q, -q^{-1}$. Then the correct quantum group normalisation is

$$R = q^{-\frac{1}{n}} R_{hecke} \tag{20}$$

and we explained that one has the ambiguity of the choice of n-th root. The Jurco construction for the n^2-squared calculus was given entirely in terms of R-matrices (we give some explicit formulae in Section 5) so one has an n-fold ambiguity for the choice of $q^{-\frac{1}{n}}$. The unique choice compatible with the $\mathbb{C}[[\hbar]]$ point of view is the principal root that tends to 1 as $q \to 1$. The other choices differ by n-th root of unity factors in the normalisation and thereby in all formulae and these are the 'additional twists'. Similarly in the contragradient and other representations. This reproduces exactly what was found for the specific analysis of calculi on $\mathbb{C}_q[SL_n]$ of dimension n^2 in [32] (note, however, that this was not a classification of all calculi in the sense above since the dimension was fixed at n^2 or less). It was found that for $n > 2$ there were $2n$ calculi labelled by $\pm$ and z a primitive nth root of unity (or just the parameter z if $n = 2$). As remarked already in [32], only the $z = 1$ cases have a commutative limit as $q \to 1$ and we identify them now as the canonical choices for the fundamental and conjugate fundamental representations (these being identified if $n = 2$). One can also view this at the level of the universal R-matrix for all representations as a variation of $\mathcal{R}$.

Following [22] (their preprint appeared several months after [22] was archived), Bauman and Schmidt [3] studied the classification for $\mathbb{C}_q[G]$ using the same factorizablity method as above. This is a more formal treatment but not a complete analysis of the possible twists any more than [22] was. Meanwhile, about a year after [22], Heckenberger and Schmüdgen [17] gave a full classification for certain but not all $\mathbb{C}_q[G]$ using a different method, which appears to be the current state of play.

5. DIFFERENTIALS ON THE QUANTUM CODOUBLE OF A FINITE GROUP

In this section we are going to demonstrate the classification theorem in the previous section for the most famous factorisable coquasitriangular Hopf algebra of

180 S. MAJID

all, namely the coordinate algebra of the Drinfeld quantum double itself. So $A = D^*(\mathcal{A}) = \mathcal{H}^{cop} \blacktriangleright\!\!\blacktriangleleft \mathcal{A}$ where $\mathcal{A}$ is a finite dimensional Hopf algebra and $\mathcal{H}$ is its dual. As an algebra the codouble is a tensor product but we are interested in bicovariant calculi on it, which depends on the doubly-twisted coproduct. By our theorem, these are classified by two-sided ideals in $D(\mathcal{H})$. Actually, we compute only the case where $\mathcal{A} = k(G)$, the functions on a finite group G, but the methods apply more generally. Then $A = kG \!\!>\!\!\blacktriangleleft k(G)$. We assume k is of characteristic zero.

THEOREM 5.1. *Differential calculi on* $D^*(G) = kG \!\!>\!\!\blacktriangleleft k(G)$ *are classified by pairs* $(\mathcal{C}, V)$ *where* $\mathcal{C} \subset G$ *is a conjugacy class and* V *is an irreducible representation of the centralizer, and at least one of* $\mathcal{C}, V$ *are nontrivial. The calculus has dimension* $|\mathcal{C}|^2 \dim(V)^2$.

Proof. Here $H = D(G) = k(G) \!\!>\!\!\triangleleft kG$ is a semidirect product by the adjoint action. Taking basis $\{\delta_s \otimes u| \ s, u \in G\}$, the product is $(\delta_s \otimes u)(\delta_t \otimes v) = \delta_{s, utu^{-1}}(\delta_s \otimes uv)$. Consider an element of the form $e = \sum_s \delta_s \otimes e_s$ where $e_s \in kG$. To be central, we consider

$$\delta_t . e = \delta_t \otimes e_t, \quad e.\delta_t = \sum_s \delta_s \delta_{Ad_{e_s}(t)} \otimes e_s.$$

Writing $e_s = \sum_u e_{s,u} u$, say, equality requires

$$e_{t,u} \delta_t = e_{utu^{-1}, u} \delta_{utu^{-1}}, \quad \forall t, u \in G.$$

This implies that $e_s \in kG_s$ the group algebra of the centralizer of s in G. Next, $u.e = e.u$ requires that $e_{usu^{-1}} = ue_s u^{-1}$ for all $u, s \in G$. Thus central elements e are of the form

$$e = \sum_{s \in \mathcal{C}} \delta_s \otimes e_s, \quad e_s \in kG_s, \quad ue_s u^{-1} = e_{usu^{-1}}, \quad \forall u \in G \qquad (21)$$

for some Ad-stable subset $\mathcal{C} \subset G$. In that case one may compute $e^2 = \sum_s \delta_s \otimes e_s^2$. Hence projectors are precisely of the above form with each e_s a central idempotent of kG_s. For a centrally primitive idempotent we need $\mathcal{C}$ a conjugacy class and a centrally primitive idempotent e_0 on the group algebra of the centralizer G_0 (any one point $s_0 \in \mathcal{C}$ determines the rest). The choice of e_0 comes from the block decomposition of kG_0 and in characteristic zero this is given by irreducible representations V. The relation is

$$e_0 = \frac{\dim(V)}{|G_0|} \sum_{u \in G_0} \mathrm{tr}_V(u^{-1}) u. \qquad (22)$$

This gives the block decomposition of $D(G)$. For $D(G)^+$ we remove the trivial case. $\qquad \square$

Note that this result is exactly in line with Corollary 4.2 since the result is the same data as for the classification of irreducible representations of $D(G)$, which are also the same as irreducible crossed G-modules. This is to be expected as $D(G)$ is semisimple because the square of its antipode is the identity. Let $\{e_i\}$ be a basis of V and $\mathcal{C} = \{a, b, c, \cdots\}$. Note also that $\mathcal{C}$ is just the data for a calculus $\Omega^1(k(G))$ on a finite group function algebra, while V is just the data for a calculus $\Omega^1(kG_0)$ on a finite group algebra viewed 'up side down' as a noncommutative space, a result

DIFFERENTIALS ON QUANTUM DOUBLES

in [22]. The calculus on the double glues together these calculi. We now obtain explicit formulae as follows.

Given the data above, the associated representation W of $D(G)$ has basis $\{e_{ai}\}$ say, with action

$$\delta_s.e_{ai} = \delta_{s,a}e_{ai}, \quad u.e_{ai} = \sum_j e_{uau^{-1}j}\,\zeta_a(u)^j{}_i,$$

where $\zeta : \mathcal{C} \times G \to G_0$ is a cocycle

$$\zeta_a(u) = g_{uau^{-1}}^{-1}ug_a; \quad \zeta_a(uv) = \zeta_{vav^{-1}}(u)\zeta_a(v) \tag{23}$$

defined by any section map $g : \mathcal{C} \to G$ such that $g_a s_0 g_a^{-1} = a$ for all $a \in \mathcal{C}$, and we use its matrix in the representation V. Completely in terms of matrices, we have the representation of $D(G)$ by

$$\rho(\delta_s \otimes u)^{ai}{}_{bj} = \delta_{s,a}\delta_{u^{-1}au,b}\zeta_b(u)^i{}_j. \tag{24}$$

We also need the universal R-matrix and quantum Killing form

$$\mathcal{R} = \sum_{u \in G} \delta_u \otimes 1 \otimes 1 \otimes u, \quad \mathcal{Q} = \sum_{u,v} \delta_{uvu^{-1}} \otimes u \otimes \delta_u \otimes v \tag{25}$$

of $D(G)$. Finally, we need the Hopf algebra structure of $A = kG{>\!\!\blacktriangleleft}k(G) = \{s \otimes \delta_u|\ s, u \in G\}$ with coproduct

$$\Delta\delta_u = \sum_{vw=u} \delta_v \otimes \delta_w, \quad \Delta s = \sum_u s\delta_u \otimes u^{-1}su \tag{26}$$

and s, δ_u commuting.

We insert these formulae into the Jurco-type construction for a representation (ρ, W) and the conventions of [22]. One can do it either for left-invariant forms Λ as elsewhere in the paper, or, which gives nicer results for our present conventions for $D^*(G)$, we can do it for right invariant forms $\bar{\Lambda}^1$. In this case we define $\bar{\Lambda}^1 = \mathrm{End}(W)$ with basis $\{e_\alpha{}^\beta\}$ and use the standard formulae for quasitriangular Hopf algebras, cf. Section 4:

$$\Omega^1 = \mathrm{End}(W) \otimes A, \quad a.e_\alpha{}^\beta = \sum \rho(\mathcal{R}_1(a_{(1)}))^\gamma{}_\alpha\, e_\gamma{}^\delta\, \rho(\mathcal{R}_2(a_{(2)}))^\beta{}_\delta\, a_{(3)}$$

$$\mathrm{d}a = \sum \rho(\mathcal{Q}_2(a_{(1)}))a_{(2)} - \theta a, \quad \forall a \in A; \quad \theta = \sum_\alpha e_\alpha{}^\alpha, \tag{27}$$

where $\rho(h) = \rho(h)^\alpha{}_\beta e_\alpha{}^\beta$ and summations of indices are understood. Putting in the explicit formulae (24)–(26) we have basis $\{e_{ai}{}^{bj}\}$ say, and find

$$\Omega^1 = \mathrm{End}(W).D^*(G), \quad fe_{ai}{}^{bj} = e_{ai}{}^{bj}L_b(f), \quad \mathrm{d}f = \sum_{a,i} e_{ai}{}^{ai}\partial_a(f)$$

$$se_{ai}{}^{bj} = \sum_k e_{sas^{-1}k}{}^{bj}\,\zeta_a(s)^k{}_i b^{-1}sb, \quad \mathrm{d}s = \sum_{a,i,j} e_{sas^{-1}i}{}^{aj}\,\zeta_a(s)^i{}_j a^{-1}sa - \theta s \tag{28}$$

for all $s \in G$, $f \in k(G)$. Here $\partial_a = L_a - \mathrm{id}$, where $L_a(f) = f(a(\ ))$ denotes left-translation in the direction of a and $\theta = \sum_{ai} e_{ai}{}^{ai}$ makes the calculus inner. The special case where we take the trivial representation of the centralizer is a canonical calculus of dimension $|\mathcal{C}|^2$ associated to any conjugacy class, cf. some similar

182   S. MAJID

formulae from a different point of view (viewing the double as a bicrossproduct) in [30].

EXAMPLE 5.2. *For S_3 with $u = (12), v = (23), w = (13)$, the possible conjugacy classes are $\{e\}, \{uv, vu\}, \{u, v, w\}$ of orders 1,2,3. Their centralizers are $S_3, \mathbb{Z}_3, \mathbb{Z}_2$. Hence on $D^*(S_3)$ we have calculi of dimension 1,4 for the first class (the nontrivial irreducibles of S_3), three calculi of dimension 4 (the three irreducibles of $\mathbb{Z}_3$) for the second class and two of dimension 9 (the two irreducibles of $\mathbb{Z}_2$) for the third class.*

Of these, we now focus on the 9-dimensional calculus since the associated conjugacy class of order 3 defines the usual differential calculus on S_3 and the $D^*(S_3)$ calculus is an extension of that. As basepoint we take $s_0 = u$ and as section we take $g_u = e$, $g_v = w$, $g_w = v$. We let $q = \pm 1$ according to the trivial or nontrivial representation of $\mathbb{Z}_2$. The resulting cocycle as a function on S_3 is

$$
\begin{array}{c|cccccc}
S_3 & 1 & u & v & w & uv & vu \\
\hline
\zeta_u & 1 & q & 1 & 1 & q & q \\
\zeta_v & 1 & q & q & 1 & 1 & q \\
\zeta_w & 1 & q & 1 & q & q & 1
\end{array}
\tag{29}
$$

For the calculus, we obtain from the above that $df = \sum_a e_a \partial_a(f)$ is the usual calculus on S_3 with $e_a \equiv e_a{}^a$, and

$$
du = q(e_u u + e_w{}^v w + e_v{}^w v) - \theta u, \quad dv = e_u{}^w u + q e_v v + e_w{}^u w - \theta v \tag{30}
$$

and the same with v, w interchanged in the second expression. Here $\theta = e_u + e_v + e_w$. We also have

$$
d(uv) = (q e_u{}^w + q e_v{}^u + e_w{}^v) vu - \theta uv
$$
$$
d(vu) = (q e_u{}^v + e_v{}^w + q e_w{}^u) uv - \theta vu.
$$

Of course, it is enough to work with generators u, v say of S_3 using the commutation relations from (28), namely

$$
u e_a{}^b = q e_{uau}{}^b u b u, \quad v e_u{}^b = e_w{}^b v b v, \quad v e_v{}^b = q e_v{}^b v b v, \quad v e_w{}^b = e_u{}^b v b v. \tag{31}
$$

Finally, for physics, we also need the higher exterior algebra as explained in Section 3. For coquasitriangular Hopf algebras and in the present right-invariant conventions we have the standard braiding

$$
\Psi(e_\alpha{}^\beta \otimes e_\gamma{}^\delta) = e_\mu{}^\nu \otimes e_\sigma{}^\tau (R^{-1})^{\alpha_1}{}_\alpha{}^\mu{}_{\alpha_2} R^\beta{}_{\alpha_3}{}^{\alpha_2}{}_\gamma R^\delta{}_{\alpha_4}{}^{\alpha_3}{}_\tau \tilde{R}^{\alpha_4}{}_\nu{}^\sigma{}_{\alpha_1} \tag{32}
$$

where $R = (\rho \otimes \rho)(\mathcal{R})$ and $\tilde{R} = (\rho \otimes \rho \circ S)(\mathcal{R})$. This is adjoint to the braided-matrix relations of the braided matrices $B(R)$, i.e. $\bar{\Lambda}^{1*}$ is a standard matrix braided Lie algebra [15]. We can compute this braiding explicitly for $D^*(G)$ above to find:

$$
\Psi(e_{ai}{}^{bj} \otimes e_{ck}{}^{dl})
$$
$$
= e_{a^{-1}bcb^{-1}am}{}^{dl} \zeta_c(a^{-1}b)^m{}_k \otimes \zeta_b^{-1}(d^{-1})^j{}_p e_{d^{-1}adn}{}^{d^{-1}bdp} \zeta_a(d^{-1})^n{}_i \tag{33}
$$

where we sum over the matrix indices m, n, p of the representation of the centralizer group. The relations of the exterior algebra are then computed using the braided-factorial matrices [20].

Then for our example $D^*(S_3)$ with the standard order 3 conjugacy class and either representation $q = \pm 1$ of the isotropy group $\mathbb{Z}_2$, we have dimensions and cohomology in low degree:

$$\dim(\Omega(D^*(S_3))) = 1 : 9 : 48 : 198 : \cdots,$$

$$H^0(D^*(S_3)) = k.1, \quad H^1(D^*(S_3)) = k.\theta \tag{34}$$

which is the same as for an isomorphic bicrossproduct example computed in [30]. We expect the exterior algebra to be finite-dimensional and the Hilbert series to have a symmetric form (our computer did not have enough memory for further degrees to check this). The exterior algebra appears to be quadratic with the relations in degree 2 from (33) as follows. For simplicity we state them only for $q = 1$:

$$e_a{}^b \wedge e_{aba^{-1}}{}^b = 0, \quad (e_{aba^{-1}}{}^a)^2 + \{e_a{}^a, e_b{}^a\} = 0, \quad \forall a, b$$

$$e_u \wedge e_v + e_v \wedge e_w + e_w \wedge e_u = 0 \tag{35}$$

and the conjugate (product-reversal) of this. Here $e_a \equiv e_a{}^a$ obey $e_a^2 = 0$ and we recover precisely the usual $\Omega(S_3)$ as a subalgebra generated by them. In addition, we have

$$e_u{}^v \wedge e_v{}^u + e_v{}^w \wedge e_w{}^v + e_w{}^u \wedge e_u{}^w = 0$$

$$e_u{}^v \wedge e_u{}^w + e_v{}^w \wedge e_v{}^u + e_w{}^u \wedge e_w{}^v = 0 \tag{36}$$

and their opposites, and

$$e_u \wedge e_u{}^w + e_u{}^w \wedge e_v + e_u{}^v \wedge e_w{}^u = 0 \tag{37}$$

and all permutations of u, v, w in this equation, plus all their conjugates.

6. CALCULUS ON GENERAL $D^*(\mathcal{A})$ AND T-DUALITY

Here we describe the differential calculus on a general quantum codouble Hopf algebra $D^*(\mathcal{A})$ associated to any $\mathcal{A}$-crossed module. These are not in general all calculi (they are the block decomposition of the semisimple calculus Ω_{ss}) and we do not attempt a classification as we did for $D^*(G)$ above. Let $\mathcal{A}$ for the moment be a finite-dimensional Hopf algebra with dual $\mathcal{H}$ and $A = D^*(\mathcal{A}) = \mathcal{H}^{cop} \blacktriangleright\!\!\blacktriangleleft \mathcal{A}$ as in Section 5. Its coproduct and coquasitriangular structure are, in the conventions of [20],

$$\Delta(h \otimes a) = \sum h_{(1)} \otimes f^a a_{(1)} f^b \otimes (Se_a) h_{(1)} e_b \otimes a_{(2)}$$

$$\mathcal{R}(h \otimes a, g \otimes b) = \epsilon(a)\epsilon(g)\langle h, b \rangle, \quad \forall h, g \in \mathcal{H}, \quad a, b \in \mathcal{A}$$

where $\{e_a\}$ is a basis of $\mathcal{H}$ and $\{f^a\}$ a dual basis. Meanwhile, $H = D(\mathcal{H}) = \mathcal{A}^{op} \bowtie \mathcal{H}$ has the product

$$(a \otimes h)(b \otimes g) = \sum b_{(1)} a \otimes h_{(2)} g \langle Sh_{(1)}, b_{(1)} \rangle \langle h_{(3)}, b_{(3)} \rangle$$

as in [20]. According to Corollary 4.2 we get a matrix block calculus for any representation of $D(\mathcal{H})$, which means a (left) $\mathcal{H}$-crossed module W. We let $\{e_\alpha\}$ be a basis for W, $\{f^\alpha\}$ a dual basis and $e_\alpha{}^\beta = e_\alpha \otimes f^\beta$ as in the previous section. We write the left crossed module action as $\triangleright$ and coaction as $\sum w^{(\bar{1})} \otimes w^{(\bar{\infty})}$.

184 S. MAJID

PROPOSITION 6.1. *Let W be an $\mathcal{H}$-crossed module. The corresponding calculus* $\Omega^1(D^*(\mathcal{A})) = \text{End}(W) \otimes D^*(\mathcal{A})$ *has the relations*

$$dh = \sum (h_{(2)} \triangleright e_\beta{}^{(\tilde{\infty})} \otimes f^\beta) \text{Ad}_{e_\beta{}^{(\tilde{1})}}(h_{(1)}) - \theta h$$

$$da = \sum \langle \text{Ad}_{f^a}(a_{(1)}), e_\beta{}^{(\tilde{1})} \rangle (e_\beta{}^{(\tilde{\infty})} \otimes f^\beta) e_a \, a_{(2)} - \theta a$$

$$h.e_\alpha{}^\beta = \sum h_{(2)} \triangleright e_\alpha \otimes f^\gamma \text{Ad}_{e_\gamma{}^{(\tilde{1})}}(h_{(1)}) \langle e_\gamma{}^{(\tilde{\infty})}, f^\beta \rangle$$

$$a.e_\alpha{}^\beta = \sum e_\alpha \otimes f^\gamma \langle \text{Ad}_{f^a}(a_{(1)}), e_\gamma{}^{(\tilde{1})} \rangle e_a \langle e_\gamma{}^{(\tilde{\infty})}, f^\beta \rangle a_{(2)}$$

for all $a \in \mathcal{A}$ and $h \in \mathcal{H}$, where Ad is the right adjoint action and $\theta = \sum e_\alpha{}^\alpha$.

Proof. From the definition of the matrices ρ we have

$$\rho(a) = \sum \langle a, e_\alpha{}^{(\tilde{1})} \rangle e_\alpha{}^{(\tilde{\infty})} \otimes f^\alpha, \quad \rho(h) = \sum h \triangleright e_\alpha \otimes f^\alpha.$$

We then use the above formulae for the quantum double and find in particular the required map

$$\mathcal{Q}_2(h \otimes a) = (1 \otimes h)(a \otimes 1) = \sum a_{(2)} \otimes h_{(2)} \langle Sh_{(1)}, a_{(1)} \rangle \langle h_{(3)}, a_{(3)} \rangle.$$

We then use (27), making routine Hopf algebra computations, including (for dh) the crossed-module compatibility conditions

$$\sum (h_{(1)} \triangleright g)^{(\tilde{1})} h_{(2)} \otimes (h_{(1)} \triangleright g)^{(\tilde{\infty})} = \sum h_{(1)} g^{(\tilde{1})} \otimes h_{(2)} \triangleright g^{(\tilde{\infty})}$$

to obtain the result. We use $\text{Ad}_h(g) = \sum (Sh_{(1)}) g h_{(2)}$. Note also that $\sum f^a \otimes \text{Ad}_{e_a}(h)$ is the left $\mathcal{A}$-coadjoint coaction on $\mathcal{H}$ adjoint to the right adjoint coaction $\text{Ad}(a) = \sum a_{(2)} \otimes (Sa_{(1)}) a_{(3)}$ on $\mathcal{A}$. $\square$

Next, since $\mathcal{A}$ is the geometric quantity for us, it is useful to recast these results in terms of an $\mathcal{A}$-crossed module. First, we can replace the left $\mathcal{H}$ coaction on W by a right action of $\mathcal{A}$ on W, and this by a left action of $\mathcal{A}$ on W^*. The result (after some computations) is

$$dh = \sum (h_{(2)} \triangleright e_\beta \otimes f^a \triangleright f^\beta) \text{Ad}_{e_a}(h_{(1)}) - \theta h \tag{38}$$

$$da = \sum (e_\beta \otimes \text{Ad}_{f^a}(a_{(1)}) \triangleright f^\beta) e_a a_{(2)} - \theta a \tag{39}$$

$$h.e_\alpha{}^\beta = \sum (h_{(2)} \triangleright e_\alpha \otimes f^a \triangleright f^\beta) \text{Ad}_{e_a}(h_{(1)}) \tag{40}$$

$$a.e_\alpha{}^\beta = \sum (e_\alpha \otimes \text{Ad}_{f^a}(a_{(1)}) \triangleright f^\beta) e_a a_{(2)}. \tag{41}$$

We can also write the left action of $\mathcal{H}$ as a right $\mathcal{A}$-coaction, or if possible a left $\mathcal{A}$-coaction on W^*. Using the same notation for left coactions and $\sum w^{(\tilde{0})} \otimes w^{(\tilde{1})}$ for right coactions, the affected formulae become

$$dh = \sum \langle h_{(2)}, f^{\beta (\tilde{1})} \rangle (e_\beta \otimes f^a \triangleright f^{\beta (\tilde{\infty})}) \text{Ad}_{e_a}(h_{(1)})$$

$$h.e_\alpha{}^\beta = \sum \langle h_{(2)}, e_\alpha{}^{(\tilde{1})} \rangle (e_\alpha{}^{(\tilde{0})} \otimes f^a \triangleright f^\beta) \text{Ad}_{e_a}(h_{(1)}).$$

Now suppose $\mathcal{A}$ has a bicovariant calculus. Then $\bar{\Lambda}^1$ is a left $\mathcal{A}$-crossed module as explained in Section 3 (a quotient of $\mathcal{A}^+$ under left multiplication and left adjoint coaction). We therefore set $W = \bar{\Lambda}^{1*}$ in the above and obtain an *induced calculus*

DIFFERENTIALS ON QUANTUM DOUBLES 185

on $D^*(\mathcal{A})$. One may check easily that if the initial $\Omega^1(\mathcal{A})$ is inner with a generator $\bar{\theta}$ say, then

$$\pi(e_\alpha{}^\beta) = \langle e_\alpha, \bar{\theta}\rangle f^\beta, \quad \pi(h \otimes a) = \epsilon(h)a \tag{42}$$

defines a surjection $\Omega^1(D^*(\mathcal{A})) \to \Omega^1(\mathcal{A})$ extending the canonical Hopf algebra surjection $D^*(\mathcal{A}) \to \mathcal{A}$ as stated.

Of course not all crossed modules are of the type which comes from an initial differential structure on $\mathcal{A}$. Let us give two examples, one which is and one (the first) which is not. They both show a different phenomenon whereby a calculus on $\mathcal{A}$ (and on $\mathcal{H}$) can instead be sometimes included in the calculus on the double. Note also that while the theory used is for finite-dimensional Hopf algebras, all formulae can be used also with appropriate care for infinite-dimensional Hopf algebras $\mathcal{A}, \mathcal{H}$ dually paired (or skew-paired). For example, if $\mathcal{A} = k[G]$ is an algebraic group of Lie type dually paired with an enveloping algebra $\mathcal{H} = U(\mathfrak{g})$ then its double $D^*(G) = U(\mathfrak{g}) \blacktriangleright\!\!\triangleleft k[G]$ is the tensor product algebra and as a coalgebra crossed by the coadjoint coaction, but the latter does not play a role in the differential structure that results. We let $\{e_i\}$ be a basis of $\mathfrak{g}$. We note that

$$\partial_i f \equiv \sum \langle e_i, f_{(1)}\rangle f_{(2)}, \quad \forall f \in k[G] \tag{43}$$

is the action of the classical right-invariant vector field generated by $e_i \in \mathfrak{g}$. We assume the summation convention for repeated such indices. We use the 2-action formulation (38)-(41).

EXAMPLE 6.2. *We take* $W = \bar{\mathfrak{g}} = 1 \oplus \mathfrak{g} \subset U(\mathfrak{g})$ *as a subcrossed* $U(\mathfrak{g})$-*module under the left adjoint action and the regular coaction given by the coproduct. We write* $e_0 = 1$ *as the additional basis element and let* $\{f^i, f^0\}$ *be the dual basis. The left* $U(\mathfrak{g})$-*crossed module structure and left action of* $k[G]$ *on* W^* *are*

$$\xi\triangleright e_i = [\xi, e_i], \quad \xi\triangleright e_0 = 0, \quad \Delta_L e_i = e_i \otimes e_0 + 1 \otimes e_i, \quad \Delta_L e_0 = 1 \otimes e_0$$

$$f\triangleright f^i = f^i f(1), \quad f\triangleright f^0 = f^0 f(1) + f^i\langle e_i, f\rangle$$

for $\xi \in \mathfrak{g}$ *and* $f \in k[G]$. *Then the resulting calculus* $\Omega^1(D^*(G))$ *has structure*

$$\mathrm{d}\xi = [\xi, e_i] \otimes f^i + e_0{}^i[\xi, e_i], \quad \mathrm{d}f = e_0{}^i\partial_i f$$

$$[\xi, e_0{}^0] = e_0{}^i[\xi, e_i], \quad [\xi, e_0{}^i] = 0, \quad [\xi, e_i{}^0] = [\xi, e_i] \otimes f^0 + e_i{}^j[\xi, e_j]$$

$$[\xi, e_i{}^j] = [\xi, e_i] \otimes f^j, \quad [f, e_\alpha{}^i] = 0, \quad [f, e_\alpha{}^0] = e_\alpha{}^i\partial_i f; \quad \alpha = 0, j.$$

Note that it is possible to restrict the calculus to basic forms of the type $\{e^i{}_j, e_0{}^i\}$ as a subcalculus $\Omega^1_{res}(D^*(G))$. When this is restricted to $k[G]$ we have $\Omega^1_{res}|_{k[G]} = \Omega^1(G)$ its usual classical calculus. When restricted to $U(\mathfrak{g})$ we have $\Omega^1_{res}|_{U(\mathfrak{g})}$ the calculus with $\mathrm{d}\xi = \rho(\xi) + e_0{}^i[\xi, e_i]$ where the first term is a standard type for $\Omega^1(U(\mathfrak{g}))$ as a noncommutative space. Thus Ω^1_{res} 'factorizes' into something close to these standard constructions. In fact we need not all $\mathrm{span}\{e_i{}^j\}$ here but only the image $\rho(U(\mathfrak{g})^+)$ which could be different unless $\mathfrak{g}$ is simple. Also note that while the classical calculus on $k[G]$ is not inner, when viewed inside Ω^1_{res} the element $\theta_0 = \sum e_i{}^i$ generates $\mathrm{d}f$. The full calculus above is necessarily inner, by construction.

As far as applications to physics are concerned let us note that if one (perversely) regards $U(su_2)$ as a noncommutative $\mathbb{R}^3$, i.e. as quantising the Kirillov-Kostant bracket then $D(SU_2)$ can be viewed as the appropriate deformation of the isometry group of $\mathbb{R}^3$. We refer to [1] for some recent work in this area. Hence if one wants to construct an affine frame bundle etc., as in Section 2 then one will need a calculus on $D^*(SU_2)$ such as the above. Details will appear elsewhere [2].

Let us now give a different example on the same quantum double, this time of the type induced by a calculus on $k[G]$. If one tries to start with the classical calculus on $k[G]$, where $\bar{\Lambda}^1 = \mathfrak{g}^*$ the dual of the Lie algebra, one will get zero for differentials df in the induced calculus. That is why we had to work with an extension $\bar{\mathfrak{g}}$ above. Similarly, at least in characteristic zero (and assuming an invertible Killing form):

EXAMPLE 6.3. *Let $\mathfrak{g}$ be a semisimple Lie algebra over k and $c = \frac{1}{2}K^{ij}e_ie_j$ the quadratic Casimir defined by the inverse Killing form. We let $W = kc \oplus \mathfrak{g} = \bar{\mathfrak{g}} \subset U(\mathfrak{g})^+$ as a subcrossed module under* Ad *and $\Delta_L = \Delta - \mathrm{id} \otimes 1$. We write $e_0 = c$ to complete the basis of $\mathfrak{g}$. The left $U(\mathfrak{g})$-crossed module structure and left action of $k[G]$ on W^* are*

$$\xi \triangleright e_i = [\xi, e_i], \quad \xi \triangleright e_0 = 0, \quad \Delta_L e_i = 1 \otimes e_i, \quad \Delta_L e_0 = 1 \otimes e_0 + K^{ij}e_i \otimes e_j$$

$$f \triangleright f^i = f^i f(1) + f^0 K^{mi}\langle e_m, f \rangle, \quad f \triangleright f^0 = f^0 f(1)$$

for $\xi \in \mathfrak{g}$ and $f \in k[G]$. Then the resulting calculus $\Omega^1(D^*(G))$ has structure

$$d\xi = [\xi, e_i] \otimes f^i + e_i{}^0 K^{mi}[\xi, e_m], \quad df = e_i{}^0 K^{mi}\partial_m f$$

$$[\xi, e_0{}^0] = 0, \quad [\xi, e_0{}^i] = e_0{}^0 K^{mi}[\xi, e_m], \quad [\xi, e_i{}^0] = [\xi, e_i] \otimes f^0$$

$$[\xi, e_i{}^j] = [\xi, e_i] \otimes f^j + e_i{}^0 K^{mj}[\xi, e_m], \quad [f, e_\alpha{}^0] = 0, \quad [f, e_\alpha{}^i] = e_\alpha{}^0 K^{mi}\partial_m f.$$

We use the same conventions as the previous example. Again we have a restricted subcalculus spanned by $\{e_i{}^j, e_i{}^0\}$ and indeed the restriction of this to $k[G]$ is again the classical calculus in terms of a new basis $e^m \equiv K^{mi}e_i$. The restriction to $U(\mathfrak{g})$ is different, however. Unlike the previous example, the input crossed module $\bar{\mathfrak{g}}$ here is of the form such that its dual is $\bar{\mathfrak{g}}^* = \bar{\Lambda}^1$ for an initial differential structure $\Omega^1_{init}(\mathcal{A})$ on $\mathcal{A}$. Namely

$$df = f^i \partial_i f + f^0 \frac{1}{2} K^{ij} \partial_j \partial_i f, \quad [f, f^i] = f^0 K^{mi} \partial_m f, \quad [f, f^0] = 0 \tag{44}$$

for all $f \in k[G]$. This is a standard 1-dimensional noncommutative extension of the usual classical calculus in which the second-order Laplacian evident here is viewed as a 'first order' partial derivative ∂_0 in the extra f^0 direction (the other ∂_i are the usual classical differentials as above). Thus a reasonable but noncommutative calculus can induce a reasonable one on the double.

Finally, we consider the above results as a step towards a 'T-duality' theory for differential calculi whereby Hopf algebra duality is extended to noncommutative geometry. In physics, Poisson-Lie T-duality refers to an equivalence between a σ-model on Poisson Lie group G and one on its Drinfeld dual G^* with dual Lie bialgebra, see [13, 5] and elsewhere. The transfer of solutions is via the Lie bialgebra double $D(\mathfrak{g})$. We could hope for a similar theory in the quantum group case. For this one would need first of all to extend the Hopf algebra duality functor to differentials. Our construction indicates a way to do this: given a calculus on $\mathcal{A}$ we can induce one on the codouble and then project that down to one on $\mathcal{H}$. Probably some modifications of this idea will be needed for nontrivial results (as the examples

DIFFERENTIALS ON QUANTUM DOUBLES 187

above already indicated) but this is a general idea for 'transference of calculi' that we propose.

Acknowledgements

I would like to thank Xavier Gomez for discussions. The author is a Royal Society University Research Fellow.

REFERENCES

[1] E. Batista and S. Majid, Noncommutative geometry of angular momentum space $U(su_2)$, J. Math. Phys. **44** (2003), 107–137.

[2] E. Batista and S. Majid, in preparation.

[3] P. Bauman and F. Schmidt, Classification of bicovariant differential calculi over quantum groups (a representation-theoretic approach), Comm. Math. Phys., **194** (1998),71–86.

[4] E.Beggs and S.Majid, Quasitriangular and differential structures on bicrossproduct Hopf algebras, J. Algebra **219** (1999), 682–727.

[5] E. Beggs and S. Majid, Poisson-Lie T-Duality for Quasitriangular Lie Bialgebras, Comm. Math. Phys. **220** (2001), 455–488.

[6] T. Brzeziński and S. Majid, Quantum differentials and the q-monopole revisited, Acta Appl. Math. **54** (1998), 185–232.

[7] T. Brzeziński and S. Majid, Quantum group gauge theory on quantum spaces. Commun. Math. Phys. **157** (1993), 591–638. Erratum **167** (1995), 235.

[8] A. Connes, "Noncommutative Geometry", Academic Press, 1994.

[9] A. Connes, C^* algebres et géométrie différentielle, C.R. Acad. Sc. Paris **290** (1980), 599–604.

[10] A. Connes and M. Rieffel, Yang-Mills theory over quantum tori, Contemp. Math. **62** (1987), 237.

[11] V. G. Drinfeld, Quantum groups, in "Proc. ICM at Berkeley", Amer. Math. Soc., Providence, 1987, 798–820.

[12] V.G. Drinfeld, QuasiHopf algebras, Leningrad Math. J. **1** (1990), 1419–1457.

[13] C. Klimcik and P. Severa, Dual non-Abelian duality and the Drinfeld double, Phys. Lett. B **351** (1995), 455-462.

[14] X. Gomez and S. Majid, Noncommutative cohomology and electromagnetism on $\mathbb{C}_q[SL_2]$ at roots of unity, Lett. Math. Phys. **60** (2002), 221–237.

[15] X. Gomez and S. Majid, Braided Lie algebras and bicovariant differential calculi over coquasitriangular Hopf algebras, J. Algebra **261** (2003), 334–388.

[16] P. Hajac and S. Majid, Projective module description of the q-monopole, Comm. Math. Phys. **206** (1999), 246–464.

[17] I. Heckenberger and K. Schmüdgen, Classification of bicovariant differential calculi on the quanutm groups $SL_q(n+1)$ and $Sp_q(2n)$, J. Reine Angew. Math. **502** (1998), 141–162.

[18] B. Jurco, Differential calculi on quantized Lie groups, Lett. Math. Phys. **22** (1991), 177–186.

[19] S. Majid, Quasitriangular Hopf algebras and Yang-Baxter equations, Int. J. Mod. Phys. A**5** (1990), 1–91.

[20] S. Majid, "Foundations of Quantum Group Theory", Cambridge Univeristy Press, Cambridge, 1995.

[21] S. Majid, Advances in quantum and braided geometry, in "Quantum Group Symposium at Group XXI" (H.-D. Doebner and V.K. Dobrev, eds.), Heron Press, Sofia, 1997, 11–26.

[22] S. Majid, Classification of bicovariant differential calculi, J. Geom. Phys. **25** (1998), 119–140.

[23] S. Majid, Quantum and braided group Riemannian geometry, J. Geom. Phys. **30** (1999), 113–146.

[24] S. Majid, Riemannian geometry of quantum groups and finite groups with nonuniversal differentials, Comm. Math. Phys. **225** (2002), 131–170.

[25] S. Majid, Ricci tensor and Dirac operator on $\mathbb{C}_q[SL_2]$ at roots of unity, Lett. Math. Phys. **63** (2003), 39–54.

188   S. MAJID

[26] S. Majid, "A Quantum Groups Primer", *Lect. Notes London Math. Soc.* **292**, Cambridge University Press, Cambridge, 2002.

[27] S. Majid and R. Oeckl, Twisting of quantum differentials and the Planck scale Hopf algebra, *Comm. Math. Phys.* **205** (1999), 617-655.

[28] S. Majid and T. Schucker, $\mathbb{Z}_2 \times \mathbb{Z}_2$ lattice as a Connes-Lott-quantum group model, *J. Geom. Phys.* **43** (2002), 1-26.

[29] T. Masuda, K. Mimachi, Y. Nakagami, M. Noumi, and K. Ueno, Representations of the quantum group $SU_q(2)$ and the little q-Jacobi polynomials, *J. Funct. Anal.* **99** (1991), 357-387.

[30] F. Ngakeu, S. Majid, and J-P. Ezin, Cartan calculus for quantum differentials on bicrossproducts, preprint math. QA/0205194.

[31] N.Yu. Reshetikhin and M.A. Semenov-Tian-Shansky, Quantum R-matrices and factorization problems, *J. Geom. Phys.*, **5** (1990), 533.

[32] K. Schmüdgen and A. Schüler, Classification of bicovariant differential calculi on quantum groups, *Comm. Math. Phys.* **170** (1995), 315-335.

[33] S.L. Woronowicz, Differential calculus on compact matrix pseudogroups (quantum groups), *Comm. Math. Phys.* **122** (1999), 125-170.

Noncommutative Differentials and Yang-Mills on Permutation Groups S_n

SHAHN MAJID   School of Mathematical Sciences
Queen Mary, University of London
327 Mile End Rd, London E1 4NS, UK
e-mail address: s.majid@qmul.ac.uk

> ABSTRACT. We study noncommutative differential structures on the group of permutations S_N, defined by conjugacy classes. The 2-cycles class defines an exterior algebra Λ_N which is a super analogue of the Fomin-Kirillov algebra $\mathcal{E}_N$ for Schubert calculus on the cohomology of the GL_N flag variety. Noncommutative de Rahm cohomology and moduli of flat connections are computed for $N < 6$. We find that flat connections of submaximal cardinality form a natural representation associated to each conjugacy class, often irreducible, and are analogues of the Dunkl elements in $\mathcal{E}_N$. We also construct Λ_N and $\mathcal{E}_N$ as braided groups in the category of S_N-crossed modules, giving a new approach to the latter that makes sense for all flag varieties.

1. INTRODUCTION

In recent years there has been developed a fully systematic approach to the noncommutative differential geometry on (possibly noncommutative) algebras, starting with differential forms on quantum groups [21] and including principal bundles with Hopf algebra fiber, connections and Riemannian structures, etc, see [17] or our companion paper in the present volume for a review. These constructions successfully extend conventional concepts of differential geometry to the q-deformed case such as q-spheres and q-coordinate rings of quantum groups.

However, this constructive noncommutative geometry can also be usefully specialised to finite-dimensional Hopf algebras and from there to finite groups, where differentials and functions noncommute (even though the functions themselves commute). Indeed, one has then a rich 'Lie theory of finite groups' complete with differentials, Yang-Mills theory, metrics and Riemannian structures. If $k(G)$ denotes the functions on the finite group, then the differential structures are defined by exterior algebras of the form $\Omega = k(G).\Lambda$ where Λ is the algebra of left-invariant differential forms. These in turn are determined by conjugacy classes. The case of the symmetric group S_3 of permutations of 3 elements, with its 2-cycle conjugacy class, was fully studied in [17] and [18]. Among other results, it was shown that S_3 has the same noncommutative de Rham cohomology as the quantum group $SL_q(2)$

2000 *Mathematics Subject Classification.* 58B32, 58B34, 14N15.
Key words and phrases. noncommutative geometry, braided categories, finite groups, quantum groups, flag variety, orbit method.
This paper is in final form and no version of it will be submitted for publication elsewhere.

190   S. MAJID

(or 3-sphere S_q^3 in a unitary setting). The moduli space of flat $U(1)$ connections on S_3 is likewise nontrivial and was computed. The goal of the present article is to extend some of these results to higher S_N, with some results for all N and others by explicit computation for $N < 6$.

In particular, we make a thorough study of the invariant exterior algebra $\Lambda = \Lambda_N$ with the 2-cycles calculus, and explain its close connection with other algebras in mainstream representation theory (in Schubert calculus) and algebraic topology. Our first result is a description of the algebra Λ_N as generated by $\{e_{(ij)}\}$ labelled by 2-cycles with relations

$$e_{(ij)} \wedge e_{(ij)} = 0, \quad e_{(ij)} \wedge e_{(km)} + e_{(km)} \wedge e_{(ij)} = 0$$

$$e_{(ij)} \wedge e_{(jk)} + e_{(jk)} \wedge e_{(ki)} + e_{(ki)} \wedge e_{(ij)} = 0$$

where i, j, k, m are distinct. We consider $e_{(ij)} = e_{(ji)}$ since they are labelled by the same 2-cycle. Our first observation is that Λ_N has identical form to the noncommutative algebra $\mathcal{E}_N$ introduced in [6] with generators $[ij]$ and relations

$$[ij] = -[ji], \quad [ij]^2 = 0, \quad [ij][km] = [km][ij], \quad [ij][jk] + [jk][ki] + [ki][ij] = 0$$

for distinct i, j, k, m. The main difference is that our $e_{(ij)}$ are symmetric and partially anticommute whereas the $[ij]$ are antisymmetric and partially commute. We will show that many of the problems posed in [6] and some of the results there have a direct noncommutative-geometrical meaning in our super version. For example, the algebra $\mathcal{E}_N$ has a subalgebra isomorphic to the cohomology of the flag variety associated to GL_N and among our analogous results we have a subalgebra of flat connections with constant coefficients. These results are in Section 3, with some further metric aspect on Section 5. Moreover, using our methods we obtain several new results about the algebras $\mathcal{E}_N$. These are in Section 6. The first and foremost is our result that the $\mathcal{E}_N$ are braided groups or Hopf algebras in braided categories. We conjecture that as such they are self-dual and show that this unifies and implies several disparate conjectures in [6]. We show that the extended divided-difference operators Δ_{ij} in that paper are indeed the natural braided-differential operators on any braided group, and that the cross product Hopf algebras in [7] are the natural bosonisations. We also prove that if $\mathcal{E}_N$ is finite-dimensional then it has a unique element of top degree. Our approach works for all flag varieties associated to other Lie algebras with Weyl groups beyond S_N.

The reasons for the close relation between Λ_N and $\mathcal{E}_N$ is not known in detail but can be expected to be something like this: the flag variety has a cell decomposition labelled by S_N and its differential geometric invariants should correspond in some sense to the noncommutative discrete geometry of the 'skeleton' of the variety provided by the cell decomposition. One can also consider this novel phenomenon as an extension of Schur-Weyl duality. Let us also note the connnection between flag varieties and the configuration space $C_N(d)$ of ordered N-tuples in $\mathbb{R}^d$ with distinct entries, as emphasized in the recent works of Lehrer, Atiyah and others. Its cohomology ring in the Arnold form can be written as generated by $d - 1$-forms E_{ij} labelled by pairs $i \neq j$ in the range $1, \cdots, N$ with relations [4][9]

$$E_{ij} = (-1)^d E_{ji}, \quad E_{ij} E_{km} = (-1)^{d-1} E_{km} E_{ij}, \quad E_{ij} E_{jk} + E_{jk} E_{ki} + E_{ki} E_{ij} = 0$$

for all i, j, k, m not necessarily distinct (to the extent allowed for the labels to be valid). We have rewritten the third relation in the required suggestive form using the first two relations. We see that $H(C_N(d))$ is precisely a graded-commutative version of the algebra Λ_N if d is even and of $\mathcal{E}_N$ extended by dropping the $[ij]^2 = 0$ relation if d is odd. Thus one can say that the noncommutative geometry of S_N and the extended Fomin-Kirillov algebra $\mathcal{E}_N$ together 'quantize' the cohomology of this configuration space in the sense that some of the graded-commutativity relations are dropped.

In the preliminary Section 2 we recall the basic ingredients of the constructive approach to noncommutative geometry (coming out of quantum groups) that we use. Its relation to other approaches such as [5] is only partly understood, see [17]. In Section 4 of the paper we look at other differential calculi on symmetric groups as defined by other conjugacy classes. To be concrete we look at S_4, S_5 and compute moduli of flat connections with constant coefficients. The result suggests the (incomplete) beginnings of an approach to construct an irreducible representation associated to each conjugacy class by noncommutative-geometrical means and in a manner that would make sense in principle for any finite group G. Since all of the geometry is G-equivariant there is plenty of scope to associate representations; here we explore one such method and tabulate the results.

2. PRELIMINARIES ON NONCOMMUTATIVE DIFFERENTIALS

Noncommutative differential geometry works over a general unital (say) algebra A. The main idea is to define the differential structure by specifying an $A-A$-bimodule Ω^1 of '1-forms' equipped with an exterior derivative $\mathrm{d} : A \to \Omega^1$ obeying the Leibniz rule. When A is a Hopf algebra there is a natural notion of Ω^1 bicovariant [21] and in this case it can be shown that $\Omega^1 = A.\Lambda^1$ (a free left A-module) where Λ^1 is the space of left-invariant 1-forms. This space has the natural structure of a right A-crossed module (in the case of A finite-dimensional it means a right module over the right quantum double of A) and as a result a braiding operator $\Psi : \Lambda^1 \otimes \Lambda^1 \to \Lambda^1 \otimes \Lambda^1$ obeying the Yang-Baxter equations. This can be used to define the wedge product between invariant 1-forms in such a way that they 'skew-commute' with respect to Ψ. The naive prescription is a quadratic algebra Λ_{quad} but there is also a more sophisticated Woronowicz prescription Λ_w; in both cases the exterior algebra Ω is defined as freely generated by these over A. Here

$$\Lambda_{quad} = T\Lambda^1 / \ker(\mathrm{id} - \Psi), \quad \Lambda_w = T\Lambda^1 / \oplus_n \ker A_n \tag{1}$$

as quotients of the tensor algebra, where the Woronowicz antisymmetrizer is

$$A_n = \sum_{\sigma \in S_n} (-1)^{l(\sigma)} \Psi_{i_1} \cdots \Psi_{i_{l(\sigma)}} : (\Lambda^1)^{\otimes n} \to (\Lambda^1)^{\otimes n}. \tag{2}$$

Here $\Psi_i \equiv \Psi_{i,i+1}$ denotes Ψ acting in the $i, i+1$ place and $\sigma = s_{i_1} \cdots s_{i_{l(\sigma)}}$ is a reduced expression in terms of simple reflections. There is also an operator $\mathrm{d} : \Lambda^1 \to \Lambda^2$ which extends to the entire exterior algebra with $\mathrm{d}^2 = 0$, and defines the noncommutative de Rham cohomology as closed forms modulo exact.

PROPOSITION 2.1. [14] $A_n = [n; -\Psi]!$ where

$$[n; -\Psi] = \mathrm{id} - \Psi_{12} + \Psi_{12}\Psi_{23} + \cdots + (-1)^{n-1}\Psi_{12} \cdots \Psi_{n-1,n}$$

192 S. MAJID

are the braided integer matrices and $[n; -\Psi]! = (\mathrm{id} \otimes [n - 1; -\Psi]!)[n; -\Psi]$.

This is a practical method to compute the A_n, which we will use. It comes from the author's theory of braided binomials (or sometimes called braided shuffles) introduced in [11]. See also the later works [20][1]. For example,

$$[3; -\Psi]! = (\mathrm{id} \otimes [2; -\Psi])[3; -\Psi] = (\mathrm{id} - \Psi_{23})(\mathrm{id} - \Psi_{12} + \Psi_{12}\Psi_{23})$$
$$= \mathrm{id} - \Psi_{12} - \Psi_{23} + \Psi_{12}\Psi_{23} + \Psi_{23}\Psi_{12} - \Psi_{23}\Psi_{12}\Psi_{23} = A_3.$$

For other formulae it is enough for our purposes to specialise directly to finite sets and finite groups. We work over a field k of characteristic zero. Let $A = k(\Sigma)$ a finite set. Then the differential structures are easily seen from the axioms to correspond to subsets $E \subset \Sigma \times \Sigma -$ diag of 'allowed directions'. Thus

$$\Omega^1 = \mathrm{span}\{\delta_x \otimes \delta_y | (x, y) \in E\}, \quad \mathrm{d}f = \sum_{(x,y)\in E} (f(y) - f(x))\delta_x \otimes \delta_y \qquad (3)$$

where δ_x is the Kronecker delta-function. Note that $\delta_x \otimes \delta_y = \delta_x \mathrm{d}\delta_y$ for all $(x, y) \in E$. This result for finite sets is common to all approaches to noncommutative geometry, e.g. in [5]. If $\Sigma = G$ is a finite group then a natural choice of E is given by

$$E = \{(x, y) \in G \times G | x^{-1}y \in \mathcal{C}\} \qquad (4)$$

for any subset $\mathcal{C}$ not containing the group identity e. Such a calculus is manifestly invariant under translation by G and all covariant differential calculi are of this form. Bicovariant ones (as above) are given precisely by those $\mathcal{C}$ which are Ad-stable, so that E is invariant from both sides. The 'simple' such differential structures (with no proper quotient) are classified precisely by the nontrivial conjugacy classes. They take the form of a free left $k(G)$-module

$$\Omega^1 = k(G) \cdot \mathrm{span}\{e_a | a \in \mathcal{C}\}, \quad \mathrm{d}f = \sum_{a\in\mathcal{C}}(R_a(f) - f)e_a, \quad e_a f = R_a(f)e_a \qquad (5)$$

where $R_a(f)(g) = f(ga)$ denotes right translation and, explicitly, $e_a = \sum_{g\in G} \delta_g \mathrm{d}\delta_{ga}$. Such formulae follow at once from Woronowicz's paper as a special case. An early study of this case, in the physics literature, is in [3].

Moreover, in the case of a finite group G, a right $k(G)$-crossed module is the same thing (by evaluation) as a left G-crossed module in the sense of Whitehead, i.e. a G-graded G-module with the degree map $|\ |$ from the module to kG being equivariant (where G acts on kG by Ad), see[15]. The particular crossed module structure on $\Lambda^1 = k\mathcal{C}$ and induced braiding are

$$|e_a| = a, \quad g.e_a = e_{gag^{-1}}, \quad \Psi(e_a \otimes e_b) = e_{aba^{-1}} \otimes e_a. \qquad (6)$$

PROPOSITION 2.2. [17] *For each* $g \in G$, *consider the set* $\mathcal{C} \cap g\mathcal{C}^{-1}$. *This has an automorphism* $\sigma(a) = a^{-1}g$ *corresponding to the braiding under the decomposition* $k\mathcal{C} \otimes k\mathcal{C} = \sum_g k(\mathcal{C} \cap g\mathcal{C}^{-1})$. *Hence if* $V_g = (k\mathcal{C} \cap g\mathcal{C}^{-1})^\sigma$ *(the fixed subspace) has basis* $\{\lambda^{(g)\alpha}\}$, *the full set of relations of* Λ_{quad} *are*

$$\forall g \in G: \quad \sum_{a,b\in\mathcal{C},\, ab=g} \lambda_a^{(g)\alpha} e_a e_b = 0.$$

NONCOMMUTATIVE DIFFERENTIALS AND YANG-MILLS ON S_n

These are also the relations of Ω_{quad} over $k(G)$. Meanwhile, the exterior derivative is provided by

$$\mathrm{d}e_a = \theta e_a + e_a \theta, \quad \theta = \sum_{a \in C} e_a. \tag{7}$$

It follows that d is given in all degrees by graded-commutation with the 1-form θ. It is easy to see that it obeys $\theta^2 = 0$ and $\mathrm{d}\theta = 0$ and that θ is never exact (so the noncommutative de Rham cohomology H^1 always contains the class of θ).

3. 2-CYCLE DIFFERENTIAL STRUCTURE ON S_N

It is straightforward to compute the quadratic exterior algebra for $G = S_N$ from the above definitions. We are particularly interested in the invariant differential forms since these generate the full structure over $k(G)$. In this section, we take the differential structure defined by the conjugacy class C consisting of 2-cycles described as unordered pairs (ij) for distinct $i, j \in \{1, \cdots, N\}$.

PROPOSITION 3.1. *The quadratic exterior algebra* $\Lambda_N \equiv \Lambda_{quad}(S_N)$ *for the 2-cycles class is the algebra with generators* $\{e_{(ij)}\}$ *and relations*

$$(i) \quad e_{(ij)}^2 = 0, \quad (ii) \quad e_{(ij)}e_{(km)} + e_{(km)}e_{(ij)} = 0$$

$$(iii) \quad e_{(ij)}e_{(jk)} + e_{(jk)}e_{(ik)} + e_{(ik)}e_{(ij)} = 0$$

where i, j, k, m *are distinct.*

Proof. There are three kinds of elements $g \in G$ for which $C \cap gC^{-1}$ is not empty. These are (i) $g = e$, in which case σ is trivial and $V_e = kC$. This gives the relations (i) stated; (ii) $g = (ij)(km)$ where i, j, k, m are disjoint. In this case $C \cap gC^{-1}$ has two elements (ij) and (km), interchanged by σ. The basis of $V_{(ij)(km)}$ is 1-dimensional, namely $(ij) + (km)$ and this gives the relation (ii) stated; (iii) The element $g = (ij)(jk)$ where i, j, k are disjoint. Here $C \cap gC^{-1}$ has 3 elements $(ij), (jk), (ik)$ cyclically rotated by σ. The invariant subspace is 1-dimensional with basis $(ij) + (jk) + (ik)$ giving the relation (iii). $\qquad\square$

We note that

$$\dim(\Lambda_N^1) = \binom{N}{2}, \quad \dim(\Lambda_N^2) = \frac{N(N-1)(N-2)(3N+7)}{24}$$

which are the same dimensions as for the algebra $\mathcal{E}_N$ in [6]. The first of these is the 'cotangent dimension' of the noncommutative manifold structure on S_N. It is more or less clear from the form of the two algebras that their dimensions coincide in all degrees (and for $N = 3$ they are actually isomorphic). We have computed these dimensions for the exterior algebra for $N < 6$ using the explicit form of the braiding Ψ defining the algebra, and they indeed coincide with the corresponding dimensions for $\mathcal{E}_N$ listed in [6]. These data are listed in Table 1 with the compact form of the Hilbert series taken from [6] (for S_4, S_5 only the low degrees have been explicitly verified by us). Also of interest is the top degree d in the last column. From our noncommutative geometry point of view this is the 'volume dimension' of the noncommutative manifold structure where the top form plays the role of the volume form. Note that the cotangent dimension and volume dimension need not coincide even though they would do so in classical geometry. Also note (thanks

dim	Ω^0	Ω^1	Ω^2	Ω^3	Ω^4	Hilbert polynomial(q)	Top degree
S_2	1	1				$[2]_q$	1
S_3	1	3	4	3	1	$[2]_q^2[3]_q$	4
S_4	1	6	19	42	71	$[2]_q^2[3]_q^2[4]_q^2$	12
S_5	1	10	55	220	711	$[4]_q^4[5]_q^2[6]_q^4$	40

TABLE 1. Dimensions and Hilbert polynomial for the exterior algebras Λ_{quad} for $N < 6$ as for $\mathcal{E}_N$ in [6]. Here $[n]_q = (q^n-1)/(q-1)$.

to a comment by R. Marsh) that these volume dimensions are exactly the number of indecomposable modules of the preprojective algebra of type SL_N. The latter is a quotient of the path algebra of the doubled quiver of the associated oriented Dynkin diagram and a module means an assignment of 'parallel transport' operators to arrows of the quiver, i.e. some kind of 'connection'. A classic theorem of Lusztig-Kashiwara-Saito states that there is a 1-1 correspondence between the irreducible components of its module variety with fixed dimension vector and the canonical basis elements of the same weight. Since the representation theory for the next preprojective algebra in the series is tame but infinite, we therefore expect Λ_6 and higher to be infinite-dimensional, and similarly for $\mathcal{E}_6$ and higher.

The infinite-dimensionality or not of $\mathcal{E}_6$ has been posed in [6], where it was conjectured that if finite dimensional then the top form should be unique (we will prove this in Section 6) and that the Hilbert series should have a symmetric increasing and decreasing form. Without proving this second conjecture here, let us outline a noncommutative-geometric strategy for its proof. Namely, the Woronowicz quotient Λ_w by its very construction will be nondegenerately paired with a similar algebra Λ_w^* of 'skew tensor fields' (see the Appendix). Moreover, if finite dimensional, and in the presence of a nondegenerate metric (see in Section 5) we then expect Hodge $*$ isomorphisms $\Lambda_w^m \to \Lambda_w^{d-m}$ and ultimately an increasing-decreasing symmetric form of the Hilbert series for Λ_w as familiar in differential geometry. All of this was concretely demonstrated for S_3 in [18]. The main ingredient missing then is that Λ_N is the quadratic quotient whereas the Woronowicz one could in principle be a quotient of that. The same strategy and considerations apply to $\mathcal{E}_N$. As a step we have,

THEOREM 3.2. *For the 2-cycle differential calculus on S_N, $\Lambda_w = \Lambda_{quad}$ in degree < 4 (we conjecture this for all degrees).*

Proof. We will decompose the space $k\mathcal{C} \otimes k\mathcal{C} \otimes k\mathcal{C} = V_3 \oplus V_2 \oplus V_1 \oplus V_0$ where each V_i is stable under the braiding operators Ψ_{12}, Ψ_{23}. Since A_3 can be factorised either through $\mathrm{id} - \Psi_{12}$ or $\mathrm{id} - \Psi_{23}$, its kernel contains that of these operators. So it suffices to show on each V_i that the dimension of the kernel of A_3 equals the dimension of the sum of the kernels of $\mathrm{id} - \Psi_{12}, \mathrm{id} - \Psi_{23}$. We say $a \sim b$ if the 2-cycles a, b have exactly one entry in common and $a \perp b$ if disjoint. We then decompose $\mathcal{C} \times \mathcal{C} \times \mathcal{C}$ as follows. For V_0 we take triples (a, b, c) which are pairwise either $\perp$ or equal, but not all three equal. Here the braiding is trivial. For V_1 we take triples where two pairs are mutually $\perp$ and one is $\sim$. It suffices to let the totally disjoint element be fixed, say (45) and the others to have entries taken from a fixed set, say $\{1, 2, 3\}$ (i.e. V_1 is a direct sum of stable subspaces spanned by basis triples with these properties fixed). On such a 9-dimensional space one may compute Ψ_{12}, Ψ_{23} explicitly and verify the required kernel dimensions (for A_3 it is 7). For V_2 we take triples (a, b, c) where two pairs are $\sim$ and one is $\perp$, or where all three pairs are $\sim$ through the same entry occurring in all three 2-cycles. This time it suffices to take entries from $\{1, 2, 3, 4\}$, say, and verify the kernels on such a 16-dimensional subspace (for A_3 it is 11 dimensional). For V_3 we take triples which are pairwise either $\sim$ or $=$, excluding the special subcase of three $\sim$ used in V_2. Here it suffices to take entries from $\{1, 2, 3\}$ and the braidings become as for S_3, where the result is known from [18]. The remaining type of triple, where there is one pair $\perp$, one $=$ and one $\sim$, is not possible. $\qquad\square$

The absence of additional cubic relations strongly suggests that the Woronowicz exterior algebra on S_N coincides with the quadratic one in all degrees (this is known for $N = 2, 3$ by direct computation). In view of the above theorem, we continue to work throughout with the quadratic exterior algebra. On the other hand, it should be stressed that we expect $\Lambda_w = \Lambda_{quad}$ to be a special feature of S_N. The evidence for this is that one may expect a kind of 'Schur-Weyl duality' between the noncommutative geometry of the finite group on one side and that of the classical or quantum group on the other. And on the quantum group side it is known that the Woronowicz exterior algebra of $SL_q(N)$ coincides with the quadratic one for generic q, but not for the other classical families. Therefore for other than the SL_N series we would expect to need to work with the nonquadratic Λ_w and likewise propose a corresponding nonquadratic antisymmetric version generalising the $\mathcal{E}_N$.

Next, for any differential graded algebra we define cohomology as usual, namely closed forms modulo exact. It is easy to see that $H^0(S_N) = k.1$ for all N.

PROPOSITION 3.3. *The first noncommutative de Rham cohomology of S_N at least for $N < 6$ with the 2-cycle differential structure is*

$$H^1(S_N) = k.\theta$$

Proof. This is done by direct computation of the dimension of the kernel of d, along the same lines as in [18], after which the result follows. We expect that in fact $H^1(S_N) = k.\theta$ for all N, but the general proof requires some elaboration. $\quad\square$

It follows from Poincaré duality that $H^2(S_3) = 0$ and $H^3(S_3) = k$, $H^4(S_3) = k$ as computed explicitly in [18], which is the same as for $SL_q(2)$ and gives some small evidence for the Schur-Weyl duality mentioned above (up to a shift or mismatch in the rank). Next, beyond the cohomology H^1 is a nonlinear variant which can be called '$U(1)$ Yang-Mills theory'. Here a connection or gauge field is again a 1-form $\alpha \in \Omega^1$. But rather than modulo exact 1-forms we are interested in working modulo the gauge transformation

$$\alpha \mapsto u\alpha u^{-1} + udu^{-1}$$

for invertible u in our coordinate algebra. The covariant curvature of a connection is

$$F(\alpha) = d\alpha + \alpha^2$$

196   S. MAJID

and transforms by conjugation. This is like nonAbelian gauge theory but is non-linear even for the $U(1)$ case because the differential calculus is noncommutative.

PROPOSITION 3.4. *For the 2-cycle differential calculus on S_N,*

$$\alpha_i = -\theta_i, \quad \theta_i = \sum_{j \neq i} e_{(ij)}$$

are flat connections with constant coefficients. The 1-forms θ_i obey $\theta_i \theta_j + \theta_j \theta_i = 0$ for $i \neq j$.

Proof. We first check the anticommutativity for $i \neq j$. In the sum

$$\theta_i \theta_j + \theta_j \theta_i = \sum_{k \neq i, l \neq j} e_{(ik)} e_{(jl)} + e_{(jl)} e_{(ik)}$$

only the cases where i, j, k, l are not distinct contribute due to relation (ii) in Proposition 3.1. Likewise the terms where $(ik) = (jl)$ do not contribute by (i). There are three remaining and mutually exclusive cases: $k = l$, or $k = j$ or $l = i$. Relabelling the summation variable k in each case we have,

$$\theta_i \theta_j + \theta_j \theta_i = \sum_{k \neq i, j} e_{(ik)} e_{(jk)} + e_{(jk)} e_{(ik)} + e_{(ij)} e_{(jk)} + e_{(jk)} e_{(ij)} + e_{(ik)} e_{(ji)} + e_{(ji)} e_{(ik)} = 0$$

by relation (iii). Next, we note that $\sum_i \theta_i = 2\theta$. Hence, $d\alpha_i = \theta \alpha_i + \alpha_i \theta = -\frac{1}{2} \sum_k \alpha_k \alpha_i + \alpha_i \alpha_k = -\alpha_i^2$ as required. $\square$

In the algebra $\mathcal{E}_N$ the similar elements

$$\theta_i = \sum_{i<j} [ij] - \sum_{j<i} [ji] = \sum_{j \neq i} [ij] \tag{8}$$

form a commutative subalgebra isomorphic to the cohomology of the flag variety[6]. In our case we see that they anticommute rather than commute. Also, while the elementary symmetric polynomials of the θ_i in $\mathcal{E}_N$ vanish, we have in Λ_N

$$\sum_i \theta_i = 2\theta, \quad \sum_i \theta_i^2 = 0$$

as above. On the other hand, in [6] the generators θ_i are motivated from Dunkl operators on the cohomology of the flag variety but in our case they have a direct noncommutative geometrical interpretation as flat connections. We will see in the next section that they are precisely the flat connections with constant coefficients of minimal support.

4. GENERAL DIFFERENTIALS AND FLAT CONNECTIONS UP TO S_5

So far we have studied only one natural conjugacy class. However, our approach associates a similar exterior algebra for any nontrivial conjugacy class in a finite group G. Moreover, since our constructions are G-invariant, we will obtain 'geometrically' plenty of G-modules naturally associated to the conjugacy class. The cohomology does not tend to be a very interesting representation but the moduli of flat connections turns out to be more nontrivial and we will see that for S_N it

NONCOMMUTATIVE DIFFERENTIALS AND YANG-MILLS ON S_n 197

does yield interesting irreducible modules. We begin with some remarks for general finite groups G equipped with a choice of nontrivial conjugacy class.

First of all, the space of connections is an affine space. We take as 'reference' the form $-\theta$. Then one may easily see that the differences $\phi \equiv \alpha + \theta$ transform covariantly as

$$\phi = \sum_a \phi^a e_a \mapsto u\phi u^{-1} = \sum_a \frac{u}{R_a(u)}\phi^a e_a. \tag{9}$$

Moreover, the curvature of α is

$$F(\alpha) = \mathrm{d}\alpha + \alpha^2 = \mathrm{d}(\phi - \theta) + (\phi - \theta)^2 = \phi^2 \tag{10}$$

in view of the properties of θ.

LEMMA 4.1. *Let $\alpha \in \Omega^1(G)$ be a connection. We define its 'cardinality' to be the number of nonzero components of $\alpha + \theta$ in the basis $\{e_a\}$. This is gauge-invariant and stratifies the moduli of connections.*

Proof. A gauge transformation u is invertible hence the support of each component ϕ^a is gauge-invariant under the transformation shown above. In particular, the number of ϕ^a with nontrivial support is invariant. $\square$

We are particularly interested in invariant forms Λ_{quad} and hence connections with constant coefficients ϕ^a (otherwise we have more refined gauge-invariant support data, namely an integer-valued vector whose entries are the cardinality of the support of each ϕ^a). To simplify the problem further we restrict to constant coefficients in $\{0, 1\}$. We can project any connection to such a $\{0, 1\}$-connection by replacing non-zero ϕ^a by 1, so this limited class of connections gives useful information about any connection.

PROPOSITION 4.2. *Flat connections with constant coefficients in $\{0, 1\}$ are in correspondence with subsets $X \subseteq \mathcal{C}$ such that*

$$\mathrm{Ad}_x(X) = X, \quad \forall x \in X.$$

The intersection of such subsets provides a product in the moduli of such flat connections that non-strictly lowers cardinality. The stratum F_n of subsets of a given cardinality n is G-invariant under Ad.

Proof. The correspondence between $\{0, 1\}$-connections and subsets is via the support of the components ϕ^a regarded as a function of $a \in \mathcal{C}$ (so the cardinality of the connection is that of the subset.) We have to solve the equation $\phi^2 = 0$. But the relations in Λ_{quad} are defined by the braiding Ψ and hence this equation is

$$\Psi(\phi \otimes \phi) = \phi \otimes \phi.$$

Using the form of Ψ this is

$$0 = \sum_{a,b \in \mathcal{C}} \phi^a \phi^b (e_{aba^{-1}} \otimes e_a - e_a \otimes e_b)$$

or

$$\phi^a (\phi^{a^{-1}ba} - \phi^b) = 0 \quad \forall a, b \in \mathcal{C}.$$

This translates into the characterisation shown. On the other hand, this characterisation is clearly closed under intersection. Finally, if X is such a subset then

198   S. MAJID

S_3	$|\mathcal{C}|$	Solutions/k	Repn $kF_n/k\theta$	Specht
(12)	3	$F_3 = \{\theta\}$ $F_1 = \{e_{23}, e_{13}, e_{12}\}$	fund	fund
(123)	2	$F_2 = \{\cdot e_{123} + \cdot e_{132}\}$ $F_1 = \{e_{123}, e_{321}\}$	sign	sign

TABLE 2. Flat connections with constant coefficients on S_3 for each conjugacy class, listed by cardinality. $\cdot$ denotes independent nonzero multiples are allowed.

$Y = \mathrm{Ad}_g(X)$ is another such subset because if $y = gxg^{-1}$ and $z = gwg^{-1}$ for $x, w \in X$ then $\mathrm{Ad}_y(z) = (gxg^{-1})gwg^{-1}(gx^{-1}g^{-1}) = g(\mathrm{Ad}_x(w))g^{-1}$ is in Y. $\qquad\square$

Over $\{0, 1\}$ the stratum of top cardinality has one point, $X = \mathcal{C}$, which corresponds to $\alpha = 0$ or $\phi = \theta$. The stratum of zero cardinality likewise has one point, $X = \emptyset$ corresponding to $\alpha = -\theta$ or $\phi = 0$, and the stratum with cardinality 1 can be identified with $\mathcal{C}$, with $\alpha = e_a - \theta$ or $\phi = e_a$ for $a \in \mathcal{C}$. In between these, the spans kF_n are natural sources of permutation G-modules. More precisely one typically has $\theta \in kF_n$ and we look at the module

$$V_n = kF_n/k\theta. \tag{11}$$

For example, the 'submaximal stratum' (the one below the top one) associates a representation to a conjugacy class, i.e. is an example of an 'orbit method' for finite groups. Like the usual orbit method for Lie groups, it does not always yield an irreducible representation, but does sometimes. It should be stressed that this is only one example of the use of our geometrical methods to define representations and we present it only as a first idea towards a more convincing orbit method.

We now use the permutation groups S_N to explore these ideas concretely. For S_3 the full moduli of (unitary) flat connections has already been found in [18] for the 2-cycles class, while the other class is more trivial. The exterior algebra in the second case is $\Lambda_w = \Lambda_{quad} = k\langle e_{123}, e_{132}\rangle$ modulo the relations

$$e_{123}^2 = 0, \quad e_{132}^2 = 0, \quad e_{123}e_{132} + e_{132}e_{123} = 0$$

(a Grassmann 2-plane). For brevity, we suppress the brackets, so $e_{123} \equiv e_{(123)}$, etc.

PROPOSITION 4.3. *The set of flat connections with constant coefficients for S_3, S_4 with their various conjugacy classes are as shown in Tables 2,3. For each stratum F_n of cardinality n, we list the corresponding ϕ up to an overall scale. The associated representations V_n turn out to be irreducible.*

Proof. This is done by direct computation. Note that the entries ϕ of a discrete stratum each define a line of flat connections $\alpha = \lambda\phi - \theta$ for a parameter λ. The entries $\cdot e_{123} + \cdot e_{132}$, etc., define a plane of connections $\alpha = \lambda e_{123} + \mu e_{132}$. As above, we omit the brackets on the cycles labelling the e_a, for example $e_{12,34}$ denotes $e_{(12)(34)}$. For the V_n we enumerate the flat connections in the stratum with coefficients $\{0, 1\}$. The resulting representation is then recognised using character theory. Representations are labelled by dimension and by $^-$ if the character at (12) is negative. The fundamental representation of S_4 means the standard 3-dimensional one. $\qquad\square$

S_4	$|\mathcal{C}|$	Solutions/k	Repn $kF_n/k\theta$	Specht
(12)	6	$F_6 = \{\theta\}$ $F_3 = \{\theta - \theta_i\}$ $F_2 = \{\cdot e_{14} + \cdot e_{23},\, \cdot e_{13} + \cdot e_{24},\, \cdot e_{12} + \cdot e_{34}\}$ $F_1 = \{e_a\}$	fund 2	fund
(12)(34)	3	$F_3 = \{\cdot e_{12,34} + \cdot e_{13,24} + \cdot e_{14,23}\}$ $F_2 = \left\{ \begin{array}{l} \cdot e_{12,34} + \cdot e_{13,24},\ \cdot e_{13,24} + \cdot e_{14,23}, \\ \cdot e_{12,34} + \cdot e_{14,23} \end{array} \right\}$ $F_1 = \{e_a\}$	2	2
(123)	8	$F_8 = \left\{ \begin{array}{l} \cdot(e_{123} + e_{142} + e_{134} + e_{243}) \\ + \cdot (e_{132} + e_{124} + e_{143} + e_{234}) \end{array} \right\}$ $F_4 = \left\{ \begin{array}{l} e_{123} + e_{142} + e_{134} + e_{243}, \\ e_{132} + e_{124} + e_{143} + e_{234} \end{array} \right\}$ $F_2 = \left\{ \begin{array}{l} \cdot e_{123} + \cdot e_{132},\ \cdot e_{142} + \cdot e_{124}, \\ \cdot e_{134} + \cdot e_{143},\ \cdot e_{243} + \cdot e_{234} \end{array} \right\}$ $F_1 = \{e_a\}$	sign fund	$\overline{\text{fund}}$
(1234)	6	$F_6 = \{\theta\}$ $F_2 = \left\{ \begin{array}{l} \cdot e_{1234} + \cdot e_{1432},\ \cdot e_{1243} + \cdot e_{1342}, \\ \cdot e_{1324} + \cdot e_{1423} \end{array} \right\}$ $F_1 = \{e_a\}$	2	sign

TABLE 3. Flat connections with constant coefficients on S_4 for each conjugacy class, listed by cardinality. $\cdot$ denotes independent nonzero multiples are allowed.

For comparison, the tables also list the standard Specht module of the Young tableau of conjugate shape to that of the conjugacy class. We see that our 'orbit method' produces comparable (although different) answers. As was to be expected, we do not obtain all irreducibles from consideration of $\{0, 1\}$ connections alone. Similarly for the S_5 case:

PROPOSITION 4.4. *The set of flat connections with constant coefficients $\{0, 1\}$ for S_5 with its various conjugacy classes are as shown in Table 4, organised by stratum F_n of cardinality n. The submaximal strata are shown in detail as well as their associated representations V_n.*

Proof. These results have been obtained with GAP to compute Ad tables followed by MATHEMATICA running for several days to enumerate the flat connections. In the tables θ_i denotes a sum over the relevant size cycles containing i, extending our previous notation. $\square$

We see that for low N the V_n tend to be irreducible, but they are not always. Notably, the (123)(45) conjugacy class for S_5 has a 9-dimensional representation associated to the submaximal stratum. A possible refinement would be to consider only the 'discrete series' i.e. flat connections where ϕ of a fixed normalisaiton is not deformable. For S_4 this means from Table 3 the strata F_3 for the 2-cycles class and F_4 for 3-cycles. A different problem is that one does not get all irreducibles in

S_5	$|\mathcal{C}|$	Solutions/$\mathbb{Z}_2$	Repn $kF_n/k\theta$
(12)	10	$F_{10} = \{\theta\}$ $F_6 = \{\theta - \theta_i\}$ $\|F_4\| = 10,\ \|F_3\| = 10,\ \|F_2\| = 15$ $F_1 = \{e_a\}$	fund
(12)(34)	15	$F_{15} = \{\theta\}$ $F_5 = \left\{ \begin{array}{l} e_{14,23}+e_{12,35}+e_{13,45}+e_{25,34}+e_{15,24}, \\ e_{14,23}+e_{13,25}+e_{24,35}+e_{15,34}+e_{12,45}, \\ e_{13,24}+e_{12,35}+e_{24,35}+e_{12,45}+e_{15,24}, \\ e_{13,24}+e_{15,23}+e_{25,34}+e_{12,45}+e_{14,35}, \\ e_{12,34}+e_{13,25}+e_{23,45}+e_{15,24}+e_{14,35}, \\ e_{12,34}+e_{15,23}+e_{24,35}+e_{14,25}+e_{13,45} \end{array} \right\}$ $\|F_3\| = 15,\ \|F_2\| = 15$ $F_1 = \{e_a\}$	$\bar{5}$
(123)	20	$F_{20} = \{\theta\}$ $F_8 = \{\theta - \theta_i\}$ $\|F_4\| = 10,\ \|F_2\| = 10$ $F_1 = \{e_a\}$	fund
(123)(45)	20	$F_{20} = \{\theta\}$ $F_2 = \left\{ \begin{array}{l} e_{xyz,12} + e_{xzy,12}, \cdots \\ \text{(all } 2-\text{cycles; } xyz \text{ complementary)} \end{array} \right\}$ $F_1 = \{e_a\}$	fund $\oplus 5$
(1234)	30	$F_{30} = \{\theta\}$ $F_{10} = \left\{ \begin{array}{l} e_{1234}+e_{1523}+e_{2435}+e_{2534}+e_{1245} \\ \quad +e_{1542}+e_{1354}+e_{1453}+e_{1432}+e_{1325}, \\ e_{1234}+e_{1253}+e_{2453}+e_{2354}+e_{1524} \\ \quad +e_{1425}+e_{1345}+e_{1543}+e_{1432}+e_{1352}, \\ e_{1234}+e_{1253}+e_{2453}+e_{2354}+e_{1524} \\ \quad +e_{1425}+e_{1345}+e_{1543}+e_{1432}+e_{1352}, \\ e_{1523}+e_{2453}+e_{2354}+e_{1243}+e_{1452} \\ \quad +e_{1254}+e_{1534}+e_{1435}+e_{1342}+e_{1325}, \\ e_{1532}+e_{2435}+e_{2534}+e_{1452}+e_{1254} \\ \quad +e_{1345}+e_{1324}+e_{1543}+e_{1423}+e_{1235}, \\ e_{1253}+e_{2345}+e_{2543}+e_{1245}+e_{1542} \\ \quad +e_{1534}+e_{1435}+e_{1324}+e_{1423}+e_{1352} \end{array} \right\}$ $F_6 = \{\theta - \theta_i\}$ $\|F_5\| = 12,\ \|F_2\| = 15$ $F_1 = \{e_a\}$	$\bar{5}$
(12345)	24	$F_{24} = \{\theta\}$ $F_{12} = \left\{ \begin{array}{l} e_{12345}+e_{12453}+e_{12534}+\cdots\text{(sum even)}, \\ e_{12354}+e_{12435}+e_{12543}+\cdots\text{(sum odd)} \end{array} \right\}$ $\|F_4\| = 6,\ \|F_3\| = 24,\ \|F_2\| = 36$ $F_1 = \{e_a\}$	sign

TABLE 4. Flat connections with constant coefficients in $\{0,1\}$ on S_5 for each conjugacy class. The submaximal strata are listed in detail, as well as the associated representation.

NONCOMMUTATIVE DIFFERENTIALS AND YANG-MILLS ON S_n 201

this way (since one only gets permutation modules). To go beyond this one could consider the full moduli of flat connections including the non-discrete series but with other constraints. For example, one could consider connections with values in $\{-1, 0, 1\}$, or one could introduce further geometric ideas such as 'polarizations' to our noncommutative setting.

Finally, each of our conjugacy classes on S_N has its associated quadratic algebra Λ_{quad} of interest in its own right. We give just one example.

PROPOSITION 4.5. *For S_4 with its 3-cycle conjugacy class the exterior algebra Λ_{quad} has relations*

$$e_{xyz}^2 = 0, \quad e_{xyz}e_{xzy} + e_{xzy}e_{xyz} = 0,$$

$$e_{123}e_{134} + e_{134}e_{142} + e_{142}e_{123} = 0, \quad e_{123}e_{243} + e_{243}e_{134} + e_{134}e_{123} = 0,$$

$$e_{134}e_{243} + e_{243}e_{142} + e_{142}e_{134} = 0, \quad e_{123}e_{142} + e_{142}e_{243} + e_{243}e_{123} = 0,$$

$$e_{123}e_{124} + e_{124}e_{134} + e_{134}e_{234} + e_{234}e_{123} = 0,$$

$$e_{123}e_{143} + e_{143}e_{243} + e_{243}e_{124} + e_{124}e_{123} = 0,$$

$$e_{123}e_{234} + e_{234}e_{142} + e_{142}e_{143} + e_{143}e_{123} = 0$$

and their seven conjugate-transposes (i.e. replacing e_{xyz} by e_{xzy} and reversing products).

Proof. Direct computation of the kernel of $\mathrm{id} - \Psi$ using GAP and MATHEMATICA. We omit the brackets around the 3-cycle labels (as above). □

We conclude with one general result pertaining to the above ideas.

PROPOSITION 4.6. *For the 2-cycles conjugacy class on S_N, the flat connections with constant coefficients in $\{0, 1\}$ of submaximal cardinality are precisely the $\alpha_i = -\theta_i$ in Proposition 3.4. The associated module is the fundamental representation of S_N.*

Proof. The α_i correspond to $\phi_i = \theta - \theta_i$ and have cardinality $\binom{N-1}{2}$. Consider any flat connection with constant coefficients in $\{0, 1\}$ with corresponding ϕ or corresponding subset X in Proposition 4.2 of cardinality $|X| \geq \binom{N-1}{2}$. Suppose there exists $i \in \{1, \cdots, N\}$ such that for all i', $(ii') \notin X$. But there are only $\binom{N-1}{2}$ such elements of $\mathcal{C}$ (those not containing i in the 2-cycle) so X cannot have cardinality greater than this, hence $|X| = \binom{N-1}{2}$ and $\phi = \phi_i$. Otherwise, we suppose that for all i there exists i' such that $(ii') \in X$. Then for any i, j we have $(ii'), (jj') \in X$ hence by the Ad closure of X we have $(ij) \in X$, i.e. $X = \mathcal{C}$ or $\phi = \theta$. □

5. METRIC STRUCTURE

In this section we look at some more advanced aspects of the differential geometry for S_N, but for the 2-cycle calculus. First of all, just as the dual of the invariant 1-forms on a Lie group can be identified with the Lie algebra, the space $\mathcal{L} = \Lambda^{1*}$ for a bicovariant differential calculus on a coquasitriangular Hopf algebra A is typically a braided-Lie algebra in the sense introduced in [13]. Moreover, every braided-Lie algebra has an enveloping algebra[13] which in our case means

$$U(\mathcal{L}) = T\Lambda^{1*}/\mathrm{image}(\mathrm{id} - \Psi^*) = \Lambda_{quad}^!, \tag{12}$$

202   S. MAJID

where ! is the quadratic algebra duality operation. There is also a canonical algebra homomorphism $U(\mathcal{L}) \to H$ where H is dual to A. We will call a differential structure 'connected' if this is a surjection. This theory applies to the Drinfeld-Jimbo $U_q(g)$ and gives it as generated by a braided-Lie algebra for each connected calculus.

However, the theory also applies to finite groups and in this case the axioms of a braided-Lie algebra reduce to what is called in algebraic topology a rack. Thus, given a conjugacy class on a finite group G, the associated rack or braided-Lie algebra is[17]

$$\mathcal{L} = \{x_a\}_{a \in \mathcal{C}}, \quad [x_a, x_b] = x_{b^{-1}ab}, \quad \Delta x_a = x_a \otimes x_a, \quad \epsilon(x_a) = 1. \tag{13}$$

The analogue of the Jacobi identity is

$$[[x_a, x_c], [x_b, x_c]] = [[x_a, x_b], x_c]. \tag{14}$$

The enveloping algebra is the ordinary bialgebra $U(\mathcal{L}) = k\langle x_a \rangle$ modulo the relations $x_a x_b = x_b x_{b^{-1}ab}$ and its homomorphism to the group algebra of G is $x_a \mapsto a$. This is surjective precisely when any element of G can be expressed as a product of elements of $\mathcal{C}$, i.e. by a path with respect to our differential structure (which determines the allowed steps as elements of $\mathcal{C}$) connecting the element to the group identity. Thus, in our finite group setting, the quadratic algebra Λ_{quad} is the !-dual of a fairly natural quadratic extension of the group algebra as an infinite-dimensional bialgebra. Note also that the flat connections in Proposition 4.2 define braided sub-Lie algebras.

Next, associated to any braided-Lie algebra is an Ad-invariant and braided-symmetric (with respect to Ψ) braided-Killing form, which may or may not be nondegenerate. This is computed in [17] for finite groups and one has

$$\eta^{a,b} \equiv \eta(x_a, x_b) = \#\{c \in \mathcal{C} \mid cab = abc\}. \tag{15}$$

The associated metric tensor in $\Omega^1 \otimes_{k(G)} \Omega^1$ is

$$\eta = \sum_{a,b} \eta^{a,b} e_a \otimes e_b$$

It is easy to see that among Ad-invariant η, 'braided symmetric' under Ψ is equivalent to symmetric in the usual sense. It is also equivalent (by definition of $\wedge$) to $\wedge(\eta) = 0$ under the exterior product.

PROPOSITION 5.1. *For the braided-Lie algebra associated to the 2-cycle calculus on S_N, the braided-Killing form is*

$$\eta^{(ij),(ij)} = \binom{N}{2}, \quad \eta^{(ij),(km)} = \binom{N-4}{2} + 2, \quad \eta^{(ij),(jk)} = \binom{N-3}{2}$$

for distinct i, j, k, m. Moreover, the calculus is 'connected'.

Proof. All of $\mathcal{C}$ commutes with $(ij)^2 = e$. In the second case all elements disjoint from i, j, k, m and $(ij), (km)$ themselves commute with $(ij)(km)$. For the third case all elements disjoint from i, j, k commute with $(ij)(jk)$. The connectedness is the well-known property that the 2-cycles can be taken as generators of S_N. $\square$

NONCOMMUTATIVE DIFFERENTIALS AND YANG-MILLS ON S_n 203

To be a metric, we need η to be invertible, which we have verified explicitly at least up to $N < 30$. Other symmetric and invariant metrics also exist, not least $\eta^{a,b} = \delta_{a,b^{-1}}$, the Kronecker δ-function which is always invertible and works for any conjugacy class on any finite group that is stable under inversion. The general situation for S_N is:

PROPOSITION 5.2. *The most general conjugation-invariant metric for the 2-cycle calculus on S_N has the symmetric form*

$$\eta^{(ij),(ij)} = \alpha, \quad \eta^{(ij),(km)} = \beta, \quad \eta^{(ij),(jk)} = \gamma$$

for distinct i, j, k, m, where α, β, γ are three arbitrary constants. Moreover,

$$\det(\eta) = (\alpha + \beta - 2\gamma)^{\frac{N(N-3)}{2}} (\alpha - (N-3)\beta + (N-4)\gamma)^{N-1}$$
$$\cdot \left(\alpha + \frac{(N-2)(N-3)}{2}\beta + 2(N-2)\gamma \right)$$

at least up to $N \leq 10$.

Proof. Invariance here means $\eta^{gag^{-1}, gbg^{-1}} = \eta^{a,b}$ for all $g \in G$. We use the mutually exclusive notations $a = b$, $a \perp b$ and $a \sim b$ as in the proof of Theorem 3.2, which is clearly an Ad-invariant decomposition of $\mathcal{C} \times \mathcal{C}$ (since the action of S_N is by a permutation of the 2-cycle entries). Clearly all the diagonal cases $a = b$ have the same value since $\mathcal{C}$ is a conjugacy class. Moreover, any $(ij) \perp (km)$ (for $N \geq 4$) is conjugate to $(12) \perp (34)$ by the choice of a suitable permutation (which we use to make the conjugation), so all of these have the same value. Similarly every $(ij) \sim (jk)$ (for $N \geq 3$) is conjugate to $(12) \sim (23)$, so these all have the same value. We then compute the determinants for $N \leq 10$ and find that they factorise in the form stated. The first two factors cancel in the case of $N = 2$. $\qquad \square$

Armed with an invertible metric, one may compute the associated Hodge-* operator, etc. as in [18] for S_3. The computation of this for S_N is beyond our present scope as it would require knowledge of Λ_{quad} in all degrees (we do not even know the dimensions for large N). It is also beyond our scope to recall all the details of noncommutative Riemannian geometry, but along the same lines as for S_3 in [17] we would expect a natural regular Levi-Civita connection with Ricci curvature tensor proportional to the metric modulo $\theta \otimes \theta$. Moreover, the same questions can be examined for the other conjugacy classes or 'Riemannian manifold' structures on S_N.

6. BRAIDED GROUP STRUCTURE ON $\mathcal{E}_N$

In this section we show that that the Fomin-Kirillov algebra $\mathcal{E}_N$ is a Hopf algebra in the braided category of crossed S_N-modules. In fact, we will find that like the exterior algebras Λ_N, it is a 'braided linear space' with additive coproduct on the generators[15]. We recall that a braided group B has a coproduct $\underline{\Delta} : B \to B \underline{\otimes} B$ which is coassociative and an algebra homomorphism provided the algebra $B \underline{\otimes} B$ is the braided-tensor product where

$$(a \otimes b)(c \otimes d) = a \Psi_{B,B}(b \otimes c) d$$

204   S. MAJID

where $a, b, c, d \in B$ and $\Psi_{B,B}$ is the braiding on B. We show how the cross prod-
uct (usual) Hopf algebras $kS_N \triangleright\!\!\!< \mathcal{E}_N$ in [7] and the skew derivations Δ_{ij} related
to divided differences in [6] arise immediately as corollaries of the braided group
structure. While the $\mathcal{E}_N$ are already-well studied by explicit means, we provide
a more conceptual approach that is also more general and applies both to other
conjugacy classes and to other groups beyond S_N.

As in [6] we consider that the algebra $\mathcal{E}_N$ is generated by an $[{N \atop 2}]$-dimensional
vector space E_N (say) with basis $[ij]$ where $i < j$, and we extend the notation to
$i > j$ by $[ij] = -[ji]$.

THEOREM 6.1. *The algebras $\mathcal{E}_N$ are 'braided groups' or Hopf algebras in the category of S_N-crossed modules. Here*

$$g.[ij] = [g(i)\ g(j)] = \begin{cases} [g(i)\ g(j)] & \text{if } g(i) < g(j) \\ -[g(j)\ g(i)] & \text{if } g(i) > g(j) \end{cases}, \ \forall g \in S_N, \quad |[ij]| = (ij)$$

*is the crossed module structure on E_N, where $|\ |$ denotes the S_N-degree. Let Ψ
denote the induced braiding, then*

$$\mathcal{E}_N = TE_N / \ker(\mathrm{id} + \Psi), \quad \underline{\Delta}[ij] = [ij] \otimes 1 + 1 \otimes [ij], \quad \underline{\epsilon}[ij] = 0$$

*is an additive braided group or 'linear braided space' in the category of S_N-crossed
modules.*

Proof. It is easy to verify that this is a crossed module structure. Thus $|g.[ij]| =
\pm(g(i)\ g(j)) = \pm g(ij)g^{-1} = |g[ij]g^{-1}|$ for the two cases (note that we consider the
S_N-degree extended by linearity). The braiding is then

$$\Psi([ij] \otimes [km]) = (ij).[km] \otimes [ij]$$

as defined by the crossed module structure. This is a signed version of the braiding
used in Proposition 3.1 and by a similar analysis to the proof there, one finds that
the kernel of $\mathrm{id} + \Psi$ is precisely spanned by the relations of $\mathcal{E}_N$. In particular, note
that

$$\Psi([ij] \otimes [ij]) = -[ij] \otimes [ij], \quad \Psi([ij] \otimes [km]) = [km] \otimes [ij]$$

if disjoint, which gives the relations $[ij][ij] = 0$ and $[ij][km] = [km][ij]$ when
disjoint. Similarly for the 3-term relations $[ij][jk] + [jk][ki] + [ki][ij] = 0$ when
i, j, k are distinct. Next, we define the coalgebra structure on the generators as
stated and verify that these extend in a well-defined manner to a braided group
structure on $\mathcal{E}_N$. there. This part is the same as for any braided-linear space [15]
and we do not repeat it. The only presentational difference is that we directly
define the relations as $\ker(\mathrm{id} + \Psi) = 0$ rather then seeking some other matrix Ψ'
such that $\mathrm{image}(\mathrm{id} - \Psi') = \ker(\mathrm{id} + \Psi)$. $\quad\square$

COROLLARY 6.2. *If $\mathcal{E}_N$ is finite-dimensional then it has a unique element in top
degree.*

Proof. A top degree element would be an integral in the braided-Hopf algebra. But
as for a usual Hopf algebra, the integral if it exists is unique (a formal proof in the
braided case is in [10][2]). $\quad\square$

NONCOMMUTATIVE DIFFERENTIALS AND YANG-MILLS ON S_n

Also, the biproduct bosonisation of any braided group B in the category of left A-crossed modules is an ordinary Hopf algebra $B{>}{\triangleleft}A$ (where A is an ordinary Hopf algebra with bijective antipode). This is the simultaneous cross product and cross coproduct in the construction of [19], in the braided group formulation [12, Appendix]. In our case A is finite dimensional so B also lives in the category of right A^*-crossed modules. Hence we immediately have two ordinary Hopf algebras, the first of which recovers the cross product observed in [7] and studied further there.

COROLLARY 6.3. *Biproduct bosonisation in the category of left crossed S_N-module structure gives an ordinary Hopf algebra $\mathcal{E}_N{>}{\triangleleft}kS_N$ with*

$$g[ij] = [g(i)\,g(j)]g, \quad \forall g \in S_N, \quad \Delta[ij] = [ij]\otimes 1 + (ij)\otimes[ij], \quad \epsilon[ij] = 0$$

extending that of kS_N, as in [7]. Bosonisation in the equivalent category of right $k(S_N)$-crossed modules gives ordinary Hopf algebra $k(S_N){\triangleright}{\ltimes}\mathcal{E}_N$ with

$$[ij]f = R_{(ij)}(f)[ij], \quad \forall f \in k(S_N), \quad \Delta[ij] = \sum_{g\in S_N}[g(i)\,g(j)]\otimes\delta_g + 1\otimes[ij].$$

Proof. The kS_N-module structure defines the cross product and the kS_N-coaction $\Delta_L[ij] = (ij)\otimes[ij]$ defines the cross coproduct. In the second case the S_N-grading defines an action of $k(S_N)$ and the kS_N-module structure defines the $k(S_N)$-coaction $\Delta_R[ij] = \sum_g[g(i)\,g(j)]\otimes(ij)$ by dualisation. $\qquad\square$

Next, from a geometrical point of view the $\mathcal{E}_N$ are 'linear braided spaces', i.e. the coproduct $\underline{\Delta}$ corresponds to the additive group law on usual affine space in terms of its usual commutative polynomial algebra in several variables, but now in a braided-commutative version. We will use several results from this theory of linear braided spaces. For clarity we explicitly label the generators of $\mathcal{E}_N$ by 2-cycles. Thus $[ij] = e_{(ij)}$ if $i < j$. The products are different from those of Λ_N but we identify the basis of generators. In this notation we have

$$g.e_b = \zeta_{g,b}e_{gbg^{-1}}, \quad |e_a| = a \tag{16}$$

$$\zeta_{(ij),(ij)} = -1, \quad \zeta_{(ij),(km)} = 1, \quad \zeta_{(ij),(jk)} = \begin{cases} 1 & \text{if } i < j < k \\ 1 & \text{if } j < i < k \\ -1 & \text{if } j < k < i \\ -1 & \text{if } i < k < j \\ 1 & \text{if } k < i < j \\ 1 & \text{if } k < j < i \end{cases} \tag{17}$$

for i, j, k, m distinct, where ζ extends to S_N in its first argument by $\zeta_{gh,b} = \zeta_{g,hbh^{-1}}\zeta_{h,b}$ for all $g, h \in S_N$, i.e.,

$$\zeta \in Z^1_{\mathrm{Ad}}(S_N, k(\mathcal{C})); \quad \zeta(g)(b) = \zeta_{g,b},$$

as a multiplicative cocycle (using the multiplication of the algebra $k(\mathcal{C})$ of functions on $\mathcal{C}$ and with Ad the right action on $k(\mathcal{C})$ induced by conjugation). Thus, the algebras $\mathcal{E}_N$ differ from the exterior algebras Λ_N precisely by the introduction of a cocycle. This makes precise how to construct analogues of the $\mathcal{E}_N$ for other finite groups.

In this notation we have for any braided linear space[11][15]

$$\underline{\Delta}(e_{b_1} \cdots e_{b_m}) = \sum_{r=1}^{m} e_{c_1} \cdots e_{c_r} \otimes e_{c_{r+1}} \cdots e_{c_m} \begin{bmatrix} m \\ r \end{bmatrix}, \Psi \end{bmatrix}_{b_1 \cdots b_m}^{c_1 \cdots c_m} \tag{18}$$

on products of generators. We view the braiding Ψ as a matrix (denoted PR in[15]). The braided binomial matrices have been introduced by the author in exactly this context and are not assumed to be invertible. There is also a braided antipode $\underline{S} : \mathcal{E}_N \to \mathcal{E}_N$ defined as -1 on the generators and extended braided-antimultiplicatively in the sense $\underline{S}(fg) = \cdot \Psi(\underline{S}f \otimes \underline{S}g)$ for all $f, g \in \mathcal{E}_N$, see [15]. In our case this comes out inductively as

$$\underline{S}(e_a f) = - \cdot \Psi(e_a \otimes \underline{S}f) = -(a.\underline{S}f)e_a, \quad \forall f \in \mathcal{E}_N. \tag{19}$$

Next we note that the braided group $\mathcal{E}_N$ is certainly finite-dimensional in each degree, so it has a graded-dual braided group $\mathcal{E}_N^*$. However, a linear braided space and its dual can typically be identified in the presence of an invariant metric. In our case we use the Kronecker $\delta_{a,b}$ metric and have:

PROPOSITION 6.4. *$\mathcal{E}_N$ is self-dually paired as a braided group, with pairing*

$$\langle e_{a_n} \cdots e_{a_1}, e_{b_1} \cdots e_{b_m} \rangle = \delta_{n,m}([n, \Psi]!)_{b_1 \cdots b_n}^{a_1 \cdots a_n}$$

Proof. The pairing we take on the generators is $\langle e_a, e_b \rangle = \delta_{a,b}$, which is compatible with the S_N-grading since all a have order 2, and compatible with the action of $g \in S_N$ since $(\pm 1)^2 = 1$, i.e. the pairing is a morphism to the trivial crossed module. We extend this to products via the axioms of a braided group as explained in [15] to obtain the pairing stated, using the above formula for $\underline{\Delta}$ and properties of the braided binomial operators in relation to braided factorial matrices. It follows from the construction that the pairing is well-defined in its second input. This part is the same as in [15]. It is also well-defined in its first input after we observe that $\Psi^* = \Psi$, where Ψ^* is defined as the adjoint on $E_N \otimes E_N$ with respect to the braided-tensor pairing $E_N \otimes E_N \otimes E_N \otimes E_N \to k$ (in which we apply $\langle \, , \, \rangle$ to the inner $E_N \otimes E_N$ first and then the outer two.) $\qquad\square$

This implies in particular that the two Hopf algebras $k(S_N) \bowtie \mathcal{E}_N$ and $\mathcal{E}_N \rtimes k S_N$ in Corollary 6.3 are dually paired. There are many more applications of the braided-linear space structure. As a less obvious one we compute the braided-Fourier theory [8] introduced for q-analysis on braided spaces.

PROPOSITION 6.5. *For $\mathcal{E}_3$ the coevaluation for the pairing in Proposition 6.4 is*

$$\exp = 1 \otimes 1 + [12] \otimes [12] + [23] \otimes [23] + [31] \otimes [31]$$
$$- [12][23] \otimes [12][31] + [23][12] \otimes [12][23] + [23][31] \otimes [31][23] - [31][23] \otimes [31][12]$$
$$+ [31][12][23] \otimes [31][12][23] + [12][23][31] \otimes [12][23][31] + [23][31][12] \otimes [23][31][12]$$
$$+ [12][23][12][31] \otimes [12][23][12][31]$$

and this along with the integration $\int$ defined as the coefficient of the top element $[12][23][12][31]$ (and zero in lower degree) defines braided Fourier transform S :

$$\mathcal{E}_3 \to \mathcal{E}_3$$

$$\mathcal{S}(1) = [12][23][12][31], \quad \mathcal{S}([12]) = [31][12][23], \quad \mathcal{S}([23]) = [12][23][31]$$
$$\mathcal{S}([31]) = [23][31][12], \quad \mathcal{S}([12][23]) = [31][12], \quad \mathcal{S}([23][12]) = [31][23]$$
$$\mathcal{S}([23][31]) = [12][23], \quad \mathcal{S}([31][23]) = [12][31], \quad \mathcal{S}([31][12][23]) = -[12]$$
$$\mathcal{S}([12][23][31]) = -[23], \quad \mathcal{S}([23][31][12]) = -[31], \quad \mathcal{S}([12][23][12][31]) = 1.$$

It obeys $\mathcal{S}^2 = \mathrm{id}$ in degrees 0,4, $\mathcal{S}^2 = -\mathrm{id}$ in degrees 1,3 and $\mathcal{S}^3 = \mathrm{id}$ in degree 2.

Proof. Let $\{e_A^{(r)}\}$ be a basis of $\mathcal{E}_N$ in degree r and $\{f^{(r)A}\}$ the dual basis with respect to the pairing. In the nondegenerate case this is given by the inverse of the quotient operator $[r, \Psi]!$ acting on the degree r component of $\mathcal{E}_N$. The coevaluation for the pairing is

$$\exp = \sum_r \sum_A e_A^{(r)} \otimes f^{(r)A}$$

which computes as stated for $\mathcal{E}_3$. We take basis $\{[12], [23], [31]\}$ for degree 1, which is orthonormal with respect to the duality pairing. For degree 2 we take basis $\{[12][23], [23][12], [23][31], [31][23]\}$ with dual $\{-[12][31], [12][23], [31][23], -[31][12]\}$. The basis in degree 3 is the Fourier transform of the basis in degree 1 and orthonormal. The braided Fourier transform is defined on general braided groups possessing duals and integrals [8]. Here

$$\mathcal{S}(f) = (\int \otimes \,\mathrm{id}) f \exp, \quad \forall f \in \mathcal{E}_N$$

which computes as stated using the relations of $\mathcal{E}_3$. A similar formula including the braided antipode $\underline{S}$ provides the inverse Fourier transform. We compute $\mathcal{S}^2$ as stated. In fact $\mathcal{S}^2 = \mathcal{T}$, where $\mathcal{T}(f) = |f|.f$ on $f \in \mathcal{E}_N$ of homogeneous S_N-degree. One also has $\mathcal{S} = \mathcal{T}^{-1} = -\underline{S}$ in degree 2. $\qquad\square$

Also from the linear braided space structure, $\mathcal{E}_N^*$ and in our case $\mathcal{E}_N$ acts on the algebra $\mathcal{E}_N$ by infinitesimal translation from the left and right, which means respectively partial derivatives $D_a, \bar{D}_a : \mathcal{E}_N \to \mathcal{E}_N$ for each $a \in \mathcal{C}$ (they are denoted $\partial^a, \bar{\partial}^a$ in the general theory of [15]). Thus the left partial derivative is defined as the coefficient of $e_a \otimes$ in the operator $\underline{\Delta}$, which from the above yields

$$D_a(e_{a_1} \cdots e_{a_m}) = e_{b_2} \cdots e_{b_m} [m, \Psi]_{a_1 \cdots a_m}^{a b_2 \cdots b_m}. \tag{20}$$

As for any braided linear space these necessarily represent $\mathcal{E}_N$ on itself and obey

$$D_a(fg) = D_a(f)g + \cdot \Psi^{-1}(D_a \otimes f)g, \quad \forall f, g \in \mathcal{E}_N, \tag{21}$$

making $\mathcal{E}_N$ an opposite braided $\mathcal{E}_N$-module algebra (i.e., in the braided category with inverted braiding). Similarly the right partial derivatives $\bar{D}_a$ are defined via right translations but converted into an action from the left via the braiding. This yields

$$\bar{D}_a(e_{a_1} \cdots e_{a_m}) = e_{b_2} \cdots e_{b_m} [m, \Psi^{-1}]_{a_1 \cdots a_m}^{a b_2 \cdots b_m} \tag{22}$$

and necessarily represent $\mathcal{E}_N$ on itself with

$$\bar{D}_a(fg) = \bar{D}_a(f)g + \cdot \Psi(\bar{D}_a \otimes f)g, \quad \forall f, g \in \mathcal{E}_N, \tag{23}$$

208   S. MAJID

making $\mathcal{E}_N$ a $\mathcal{E}_N$-module algebra in its original braided category. The partial derivatives and their conjugates are related by the braided antipode according to

$$\underline{S}D_a = -\bar{D}_a\underline{S} \tag{24}$$

as shown in [16]. Proofs of all of these facts are by braid-diagram methods as part of our established theory of braided groups.

COROLLARY 6.6. *In the case of $\mathcal{E}_N$ the braided partial derivatives $D_a, \bar{D}_a$ are covariant (morphisms in the category of crossed modules) in the sense*

$$|D_a f| = a|f], \quad g.D_a(f) = \zeta_{g,a} D_{gag^{-1}}(g.f), \quad \forall g \in S_N, \ f \in \mathcal{E}_N$$

(and similarly for $\bar{D}_a$). They obey $D_a(e_b) = \bar{D}_a(e_b) = \delta_{a,b}$ and the braided Leibniz rules

$$D_a(fg) = D_a(f)g + f\zeta_{|f|^{-1},a}D_{|f|^{-1}a|f|}(g), \quad \bar{D}_a(fg) = \bar{D}_a(f)g + (a.f)\bar{D}_a(g)$$

for all $f, g \in \mathcal{E}_N$ and f of homogeneous S_N-degree $|f|$ in the first case.

Proof. Whereas the above review of $D_a, \bar{D}_a$ holds for any additive braided group as part of a general theory, we specialize now to the particular braided category for the case of $\mathcal{E}_N$. First of all, the $D_a, \bar{D}_a$ are defined above as evaluation on e_a of morphisms $D, \bar{D} : E_N \otimes \mathcal{E}_N \to \mathcal{E}_N$. Thus $D(g.e_a \otimes g.f) = D(\zeta_{g,a}e_{gag^{-1}} \otimes g.f) = g.D(e_a \otimes f)$ translates to the condition as shown for $g \in S_N$ and $f \in \mathcal{E}_N$. Likewise, commuting with the total S_N-degree gives the other part of the morphism condition. Next we compute the braiding and its inverse on $E_N \otimes \mathcal{E}_N$ in the category of left crossed modules for the crossed module structure stated. In general $\Psi(f \otimes g) = |f|.g \otimes f$ for f of homogeneous S_N-degree $|f|$. Applying $D, \bar{D}$ yields the Leibniz rules as stated with

$$\Psi^{-1}(D_a \otimes f) = \zeta_{|f|^{-1},a} f \otimes D_{|f|^{-1}a|f|}, \quad \Psi(\bar{D}_a \otimes f) = a.f \otimes \bar{D}_a.$$

The braiding of D_a in these expressions corresponds by definition to the braiding of the element e_a which it represents (similarly for $\bar{D}_a$). $\qquad\square$

These $\bar{D}_{(ij)}$ therefore coincide with the operators denoted Δ_{ij} in the notation of [6]. It is proven there that their restriction to polynomials in the $\{\theta_i\}$ (i.e. to the cohomology of the flag variety) yields the finite difference operators

$$\partial_{ij}f = \frac{f - (ij).f}{\theta_i - \theta_j} \tag{25}$$

where $(ij).f$ interchanges the i, j arguments of $f(\theta_1, \cdots, \theta_N)$. We see that these $\bar{D}_a$ follow directly from the braided group structure as infinitesimal translations, which ensures that they are well-defined and form a representation of $\mathcal{E}_N$ on itself. It also provides computational tools, for example braided-Fourier transform intertwines the braided derivatives with multiplication in $\mathcal{E}_N$, as shown in general in [8]. Another canonically-defined representation of a braided group on itself with a similar braided-Leibniz property to the $\bar{D}_a$ is the braided adjoint action, which comes out for $\mathcal{E}_N$ as

$$\underline{\mathrm{Ad}}_{e_a}(f) \equiv e_a f + \cdot\Psi(\underline{S}e_a \otimes f) = e_a f - (a.f)e_a \tag{26}$$

$$\underline{\mathrm{Ad}}_{e_a}(fg) = \underline{\mathrm{Ad}}_{e_a}(f)g + (a.f)\underline{\mathrm{Ad}}_{e_a}(g). \tag{27}$$

NONCOMMUTATIVE DIFFERENTIALS AND YANG-MILLS ON S_n 209

This has no direct geometrical analogue (the usual polynomial algebra is commutative so that $\underline{\mathrm{Ad}}$ is zero).

Let us also note that the full bosonisation theorem[15, Thm. 9.4.12] of $\mathcal{E}_N$ provides another ordinary Hopf algebra $\mathcal{E}_N \rtimes D(kS_N)$ such that its category of modules is fully equivalent to the category of braided modules of the braided group $\mathcal{E}_N$. In particular, module-algebras of this ordinary Hopf algebra are the same thing as braided $\mathcal{E}_N$-module algebras, such as provided by $\bar{D}_a, \mathrm{Ad}_{e_a}$ above. The Drinfeld double of a finite group is itself a semidirect product $D(kS_N) = k(S_N)_{\mathrm{Ad}} \rtimes kS_N$.

COROLLARY 6.7. *The full bosonisation Hopf algebra* $\mathcal{E}_N \rtimes (k(S_N) \rtimes kS_N)$ *contains* $\mathcal{E}_N \rtimes kS_N$ *in Corollary 6.3 and* $k(S_N)$ *as sub-Hopf algebras with additional relations*

$$f[ij] = [ij] L_{(ij)}(f), \quad gf = \mathrm{Ad}_{g^{-1}}(f)g, \quad \forall g \in S_N, \ f \in k(S_N)$$

where $L_g(f) = f(g(\ ))$ *and* $\mathrm{Ad}_g = L_g R_{g^{-1}}$. *The same algebra has another 'conjugate' coproduct containing* $\mathcal{E}_N \rtimes k(S_N)$ *and* kS_N *as sub-Hopf algebras.*

Proof. The right action of $k(S_N)$ on $\mathcal{E}_N$ given by the grading can also be used as a left action. This action and the action of kS_N is the action of the Drinfeld double corresponding to the crossed module. We make the semidirect product by this. The quasitriangular structure $\mathcal{R} = \sum_{g \in S_N} \delta_g \otimes g$ defines a left coaction $\Delta_L(e_a) = \mathcal{R}_{21}.e_a = \sum_{g \in S_N} g \otimes \delta_g.e_a = a \otimes e_a$ induced from the action, so the same cross coproduct as for $\mathcal{E}_N \rtimes kS_N$. On the other hand every quasitriangular Hopf algebra has a conjugate quasitriangular structure $\bar{\mathcal{R}} = \mathcal{R}_{21}^{-1}$. We regard the same algebra $\mathcal{E}_N$ developed as a braided group in this opposite braided category (the opposite braided coproduct looks the same on the generators E_N.) Using $\bar{\mathcal{R}}$ gives a second induced coaction $\bar{\Delta}_L(e_a) = \sum_{g \in S_N} \delta_{g^{-1}} \otimes g.e_a$. This gives a second ordinary coproduct

$$\bar{\Delta}[ij] = [ij] \otimes 1 + \sum_{g \in S_N} \delta_{g^{-1}} \otimes [g(i), g(j)]$$

which is a left handed version $\mathcal{E}_N \rtimes k(S_N)$ of the second biproduct bosonisation in Corollary 6.3. This is an example of a general theory in [16] where the two coproducts are related by complex conjugation in a $*$-algebra setting over $\mathbb{C}$. $\square$

Having understood the structure of $\mathcal{E}_N$ in a natural way, let us note now that all of the above applies equally well to the full quotient of it

$$\mathcal{E}_w = TE_N / \oplus_n \ker Sym_n, \quad Sym_n = \sum_{\sigma = s_{i_1} \cdots s_{i_{l(\sigma)}} \in S_n} \Psi_{i_1} \cdots \Psi_{i_{l(\sigma)}} \quad (28)$$

where in principle there could be nonquadratic relations. In this case, since $Sym_n = [n, \Psi]!$, it is clear that here the pairings are now nondegenerate (we have divided by the coradicals of the pairing in Proposition 6.4). Therefore $\mathcal{E}_w$ is a self-dual braided group. If finite-dimensional then it would inherit a symmetric Hilbert series as explained in Section 3. Also for the reasons given there, we expect that $\mathcal{E}_N$ and $\mathcal{E}_w$ coincide and the latter if finite dimensional will have a symmetric Hilbert series which will prove the conjecture of a symmetric Hilbert series for $\mathcal{E}_N$ made in [6]. But if they do not coincide, we propose $\mathcal{E}_w$ as the better-behaved version of $\mathcal{E}_N$; it may be that $\mathcal{E}_w$ is finite-dimensional while the $\mathcal{E}_N$ is likely not to be for $N \geq 6$.

S. MAJID

Thus we propose a potential and better behaved quotient of $\mathcal{E}_N$. Moreover, our braided group methods work for general finite groups where we would not expect $\mathcal{E}_w$ to be quadratic and which would probably be needed for flag varieties associated to different Lie algebras beyond SL_N. This is a proposal for further work.

Acknowledgements

I would like to thank S. Fomin and A. Zelevinsky for suggesting to compare with the algebra $\mathcal{E}_N$ after a presentation of [17] at the Erwin Schroedinger Institute in 2000. I also want to thank R. Marsh for the comment about the preprojective algebra after a presentation in Leicester in 2001, and G. Lehrer for the comment about cohomology of configuration spaces on a recent visit to Sydney. The work itself was presented at the Trieste/SISSA conference, March 2001, at the Banach Center quantum groups conference in September 2001 and in part at the present conference. The article was originally submitted to J. Pure and Applied Algebra in August 2001 and archived on math.QA/0105253; since then the introduction was redone and some of the more technical material was moved to an Appendix. The author is a Royal Society University Research Fellow.

Appendix A. BRAIDED GROUP STRUCTURE OF Λ_w

Here we will say a little more about the general theory behind exterior algebras Λ_{quad} or Λ_w than covered in the Preliminaries in Section 2. This is needed for some of the remarks about Hodge * operator mentioned in Sections 3,5 and is also the motivation behind the results given directly for $\mathcal{E}_N$ in Section 6. It was considered too technical to be put in the main text.

First of all, the Woronowicz construction Ω_w on a quantum group A is usually given as a quotient of the tensor algebra on Ω^1 over A. We have instead moved everything over to the left-invariant forms Λ_w which is a 'braided approach' to the exterior algebra in [14][1]. See also [20]. The starting point is that associated to any linear space Λ^1 equipped with a Yang-Baxter or braid operator (in our case $-\Psi$) one has Λ_{quad} (and similarly Λ_w) braided linear spaces with additive coproduct

$$\underline{\Delta} e_a = e_a \otimes 1 + 1 \otimes e_a, \quad \underline{\varepsilon} e_a = 0. \tag{29}$$

In our case these live in the braided category which is a $\mathbb{Z}_2$ extension of the category of A-crossed modules, with Λ^1 odd. Thus one may verify:

$$\underline{\Delta}(e_a e_b) = (e_a \otimes 1 + 1 \otimes e_a)(e_b \otimes 1 + 1 \otimes e_b) = e_a e_b \otimes 1 + 1 \otimes e_a e_b + (\mathrm{id} - \Psi)(e_a \otimes e_b).$$

If $\lambda_{a,b} e_a e_b = 0$ (summation understood) then $\underline{\Delta}$ of it is also zero since the relation in degree 2 is exactly that $(\mathrm{id} - \Psi)(\lambda_{a,b} e_a \otimes e_b) = 0$. This covers Λ_{quad}. For Λ_w one has to similarly look at the higher degrees. Similarly to Section 6 there is then a super-biproduct bosonisation theorem which yields Ω_{quad} and Ω_w as super-Hopf algebras by crossed module constructions. We also have super-braided-partial derivatives $D_a, \bar{D}_a$, which define interior products[14].

Here we would like to say a little more as an explanation of the definition of Λ_w. Let Λ^{1*} be the crossed module with adjoint braiding Ψ^*. It has its own algebra of 'skew invariant tensor fields'

$$\Lambda_{quad}^* = T\Lambda^{1*} / \ker(\mathrm{id} - \Psi^*), \quad \Lambda_w^* = T\Lambda^{1*} / \oplus_n A_n^*. \tag{30}$$

NONCOMMUTATIVE DIFFERENTIALS AND YANG-MILLS ON S_n 211

PROPOSITION A.1. *The tensor algebras $T\Lambda^1$ and $T\Lambda^{1*}$ are dually paired braided groups as induced by the pairing in degree 1, and Λ_w, Λ_w^* are their quotients by the kernel of the pairing.*

Proof. Let $\{f^a\}$ be the dual basis of Λ^{1*}. The pairing between monomials in the tensor algebra is then

$$\langle f^{a_n} \cdots f^{a_1}, e_{b_1} \cdots e_{b_m} \rangle = \delta_{n,m}[n, -\Psi]!^{a_1 \cdots a_n}_{b_1 \cdots b_n}$$

as for any braided linear space [15]. In view of Proposition 2.1 we are therefore defining Λ_w exactly by killing the kernel of the pairing from that side. Similarly from the other side. $\qquad\square$

This the meaning of the Woronowicz construction is that one adds enough relations that the pairing with its similar dual version is non-degenerate. Moreover, as in Section 6, we know that by the theory of integrals on braided groups, if Λ_w is finite dimensional then there is a unique top form Top, of degree d say. In this case there is an approach to a Hodge * pairing in [14] based on braided-differentiation of the top form and related to braided Fourier transform. A similar and more explicit version of this which has been used in [18] to define an 'epsilon tensor' by $e_{a_1} \cdots e_{a_d} = \epsilon_{a_1 \cdots a_d}$Top and then use this to define a map $\Lambda_w^m \to \Lambda_w^{*(d-m)}$. In the presence of an invariant metric we have $\Lambda^1 \cong \Lambda^{*1}$ as crossed modules and hence isomorphisms of their generated braided groups. In this case we have a Hodge * operator $\Lambda_w^m \to \Lambda_w^{d-m}$. Similarly if Λ_{quad} is finite dimensional.

In the case of a finite group G with calculus defined by a conjugacy class $\mathcal{C}$, we compute

$$\Psi^*(f^a \otimes f^b) = f^{a^{-1}ba} \otimes f^a \qquad (31)$$

where the adjoint is taken with respect to the pairing on tensor powers (recall that conventionally this is defined by pairing the inner factors first and moving outwards, to avoid unnecessary braid crossings). We let Λ denote either Λ_{quad} or Λ_w (or an intermediate quotient).

COROLLARY A.2. *If $\mathcal{C}$ is stable under group inversion then Λ is self-dually paired as a braided group. If Λ is finite-dimensional with top degree d we have*

$$*(e_{a_1} \cdots e_{a_m}) == d_m^{-1} \epsilon_{a_1 \cdots a_d} e_{a_d^{-1}} \cdots e_{a_{m+1}^{-1}} \qquad (32)$$

for some normalisations d_m.

Proof. In this case we have an invariant metric $\eta^{a,b} = \delta_{a,b^{-1}}$ whereby we identify $f^a = e_{a^{-1}}$. For Λ_w in the algebraically closed case one would typically chose the d_m so that $*^2 = $ id. The formula as in [18] is arranged to be covariant so that if Top is invariant under the $k(G)$-action (which implies that it commutes with functions) then $*$ will extend to a bimodule map $\Omega^m \to \Omega^{d-m}$. $\qquad\square$

Similarly, the exterior algebra Ω is generated in the finite group case by $k(G)$ and Λ with the cross relations (5), which is manifestly a cross product $k(G){\rtimes}\Lambda$. The super coalgebra explicitly is

$$\Delta e_a = \sum_{g \in G} e_{gag^{-1}} \otimes \delta_g + 1 \otimes e_a, \quad \epsilon e_a = 0 \qquad (33)$$

and extends the group coordinate Hopf algebra. Here δ_g is a delta-function on G. Indeed, the G-grading part of the crossed module structure on Λ^1 extends to all of Λ and defines a right action of $k(G)$ on it (by evaluating against the total G-degree) which is used in the cross product algebra. Meanwhile, the left G-action defines a right coaction of $k(G)$,

$$\Delta_R(e_a) = \sum_g e_{gag^{-1}} \otimes \delta_g \tag{34}$$

which extends as an algebra homomorphism to Λ because Ψ is Ad-covariant. Semidirect coproduct by this defines the coalgebra of $k(G){\rtimes}\Lambda$. The two fit together to form a super-Hopf algebra just because the original structure on Λ^1 was a crossed module. For example, one may check

$$\Delta(e_a e_b) = (\sum_g e_{gag^{-1}} \otimes \delta_g + 1 \otimes e_a)(\sum_h e_{hbh^{-1}} \otimes \delta_h + 1 \otimes e_b)$$

$$= 1 \otimes e_a e_b + \sum_g e_{gag^{-1}} e_{gbg^{-1}} \otimes \delta_g + e_{gag^{-1}} \otimes \delta_g e_b - e_{gbg^{-1}} \otimes e_a \delta_g$$

$$= (1 \otimes \cdot + \Delta_R \circ \cdot + (\Delta_R \otimes \mathrm{id})(\mathrm{id} - \Psi)) \, e_a \otimes e_b$$

since we are extending as a super-Hopf algebra (so Λ^1 is odd). We used the relations in the algebra and changed variables in the last term. From this it is clear that Δ is well defined in the quotient by $\ker(\mathrm{id} - \Psi)$. This covers $\Lambda = \Lambda_{quad}$ but the same holds also for Λ_w.

Similarly, since a right $k(G)$-crossed module is the same thing as a left kG-crossed module, we can make another super-Hopf algebra $\Lambda{\rtimes}kG$. We extend the G-action $g.e_a = e_{gag^{-1}}$ to Λ for the cross product and the grading defines a left coaction

$$\Delta_L e_a = a \otimes e_a \tag{35}$$

which we extend to products (expressing the total G-degree). Semidirect product and coproduct by these gives

$$g e_a = e_{gag^{-1}} g, \quad \Delta e_a = e_a \otimes 1 + a \otimes e_a, \quad \epsilon e_a = 0 \tag{36}$$

extending the Hopf algebra structure of the group algebra kG. This time

$$\Delta(e_a e_b) = e_a e_b \otimes 1 + ab \otimes e_a e_b + e_a b \otimes e_b - a e_b \otimes e_a$$

$$= (\cdot \otimes 1) + \Delta_L \circ \cdot + (\mathrm{id} \otimes \Delta_L)(\mathrm{id} - \Psi)) \, e_a \otimes e_b$$

which is well-defined on the quotient . Geometrically, this is the dual of the super-Hopf algebra $k(G){\rtimes}\Lambda^*$ of skew-vector fields. These are the direct contructions of the cross products analogous to those in Section 6 for $\mathcal{E}_N$. We have similar pairing results.

Finally, let us note that at this level of generality all the same proofs work with $-\Psi$ replaced by Ψ. Thus for any crossed module E with braiding Ψ we have a braided space

$$\mathcal{E}_{quad} = TE/\ker(\mathrm{id} + \Psi) \tag{37}$$

and similarly $\mathcal{E}_w$ defined by Sym_n as in (28), both forming additive braided groups. Moreover, one should be able to construct a suitable crossed module from any conjugacy class on a finite group and possibly a cocycle ζ. This indicates how the analogues of the Fomin-Kirillov algebra could be extended to other types.

REFERENCES

[1] Yu. N. Bespalov and B. Drabant, Differential calculus in braided Abelian categories, preprint q-alg/9703036.

[2] Y. Bespalov, T. Kerler, V. Lyubashenko and V. Turaev, Integrals for braided Hopf algebras, *J. Pure Appl. Algebra* **148** (2000), 113–164.

[3] K. Bresser, F. Mueller-Hoissen, A. Dimakis and A. Sitarz, Noncommutative geometry of finite groups, *J. Phys. A* **29** (1996), 2705–2736.

[4] F.R. Cohen, T.J. Lada and J.P. May, "The homology of iterated loop spaces", *Lect. Notes in Math.* **533**, Springer Verlag, Berlin, 1976.

[5] A. Connes, "Noncommutative Geometry", Academic Press, 1994.

[6] S. Fomin and A.N. Kirillov, Quadratic algebras, Dunkl elements, and Schubert calculus, *Adv. Geom, Progr. Math.* **172** (1989), 147–182.

[7] S. Fomin and C. Procesi, Fibered quadratic Hopf algebras related to Schubert calculus, *J. Algebra* **230** (2000), 174–183.

[8] A. Kempf and S. Majid, Algebraic q-integration and Fourier theory on quantum and braided Spaces, *J. Math. Phys.* **35** (1994), 6802–6837.

[9] G.I. Lehrer, Equivariant cohomology of configurations in $\mathbb{R}^d$, *Algebras and Repn. Theory* **3** (2000), 377–384.

[10] V. Lyubashenko, Modular transformations for tensor categories, *J. Pure Appl. Algebra* **98** (1995), 279–327.

[11] S. Majid, Free braided differential calculus, braided binomial theorem and the braided exponential map, *J. Math. Phys.* **34** (1993), 4843–4856.

[12] S. Majid, Braided matrix structure of the Sklyanin algebra and of the quantum Lorentz group, *Comm. Math. Phys.* **156** (1993), 607–638.

[13] S. Majid, Quantum and braided Lie algebras, *J. Geom. Phys.* **13** (1994), 307–356.

[14] S. Majid, q-Epsilon tensor for quantum and braided spaces, *J. Math. Phys.* **36** (1995), 1991–2007.

[15] S. Majid, "Foundations of Quantum Group Theory", Cambridge Univeristy Press, Cambridge, 1995.

[16] S. Majid, Quasi-* structure on q-Poincaré algebras, *J. Geom. Phys.* **22** (1997), 14–58.

[17] S. Majid, Riemannian geometry of quantum groups and finite groups with nonuniversal differentials, *Comm. Math. Phys.* **225** (2002), 131-170.

[18] S. Majid and E. Raineri, Electromagnetism and gauge theory on the permutation group S_3, *J. Geom. Phys.* **44** (1992) 129-155.

[19] D. Radford, The structure of Hopf algebras with a projection, *J. Algebra* **92** (1985), 322–347.

[20] M. Rosso, Groupes quantiques et qlgebres de battage quantiques, *C.R.A.C.* **320** (1995), 145–148.

[21] S. L. Woronowicz, Differential calculus on compact matrix pseudogroups (quantum groups), *Comm. Math. Phys.* **122** (1989), 125–170.

The Affineness Criterion for Doi-Koppinen Modules

CLAUDIA MENINI   Department of Mathematics, University of Ferrara,
Via Machiavelli 35, I-44100 Ferrara, Italy
e-mail: *men@dns.unife.it*

GIGEL MILITARU   Faculty of Mathematics, University of Bucharest,
Str. Academiei 14, RO-70109 Bucharest 1, Romania
e-mail: *gmilit@al.math.unibuc.ro*

> ABSTRACT. Let (H, A, C) be a threetuple, where H is a Hopf algebra coacting
> on an algebra A and acting on a coalgebra C, and $^C\mathcal{M}(H)_A$ the category of
> representations of (H, A, C). Let $z \in C \otimes A$ be a generalized grouplike element
> of (H, A, C) and B the subalgebra of z-coinvariants of the Verma structure
> $A \in {}^C\mathcal{M}(H)_A$. We prove the following affineness criterion: if there exist a
> total z-normalized integral $\gamma : C \to \operatorname{Hom}(C, A)$ and if the canonical map
> $\beta : A \otimes_B A \to C \otimes A, \beta(a \otimes_B b) = a \cdot z \cdot b$ is surjective, then the induction
> functor $- \otimes_B A : \mathcal{M}_B \to {}^C\mathcal{M}(H)_A$ is an equivalence of categories.

1. INTRODUCTION

The affineness criterion for affine algebraic group schemes was proved by Cline,
Parshall and Scott [4], and independently by Oberst [10]. A purely algebraic proof
was given by Doi [5, Theorem 3.2] in the equivalent context of commutative Hopf
algebras coacting on commutative algebras. The general noncommutative case was
proved by Schneider [12, Theorem 3.4]: in these last two papers, the category of
relative Hopf modules $\mathcal{M}_A^H$ play the key role. Recently, a quantum version of the
affineness criterion was obtained in the context of quantum Yetter-Drinfeld modules
$^H \mathcal{YD}_A$, where now A is an H-bicomodule algebra [8].
In this note we will prove the affineness criterion for a Doi-Koppinen datum (H, A, C),
where H is a Hopf algebra coacting on an algebra A and acting on a coalgebra C. Let $^C\mathcal{M}(H)_A$ be the category of representations of (H, A, C), also called
Doi-Koppinen modules. $^C\mathcal{M}(H)_A$ unifies modules, comodules, Sweedler's Hopf
modules, relative Hopf modules, graded modules, Long dimodules and Yetter-
Drinfeld modules [3]. The general concept of integral of (H, A, C), introduced
in [8], will be the main tool of our approach. We fix a generalized grouplike element $z = \sum_i c_i \otimes a_i \in C \otimes A$. Such an element exists if and only if A has a

---

2000 *Mathematics Subject Classification.* 16W30.

Key words and phrases. Doi-Koppinen modules, Hopf Galois theory.

This paper was written while the first author was a member of G.N.S.A.G.A. with partial
financial support from M.I.U.R. and the second author was a visiting professor at the University
of Ferrara, supported by I.N.D.A.M.

structure of an object in $^C\mathcal{M}(H)_A$. Let B be the subalgebra of z-coinvariants of A. Assume that there exists $\gamma : C \to \mathrm{Hom}(C, A)$, a z-normalized integral of (H, A, C). Then the induction functor $- \otimes_B A : \mathcal{M}_B \to {}^C\mathcal{M}(H)_A$ is a fully faithful functor (Theorem 3.7). If, furthermore, the canonical map $\beta : A \otimes_B A \to C \otimes A$, $\beta(a \otimes_B b) = \sum c_i b_{<-1>} \otimes aa_i b_{<0>}$ is surjective, then the induction functor $- \otimes_B A : \mathcal{M}_B \to {}^C\mathcal{M}(H)_A$ is an equivalence of categories (Theorem 3.9). All the above affineness criteria are special cases of Theorem 3.9. A new application for categories of X-graded representation of an algebra A is given (Theorem 4.3).

2. PRELIMINARIES

Throughout this paper, k will be a commutative ring with unit. Unless specified otherwise, all modules, algebras, coalgebras, bialgebras, tensor products and homomorphisms are over k. For a k-algebra A, $\mathcal{M}_A$ (resp. ${}_A\mathcal{M}$) will be the category of right (resp. left) A-modules and A-linear maps. H will be a Hopf algebra over k, and we will use Sweedler's sigma-notation extensively: $\Delta(c) = \sum c_{(1)} \otimes c_{(2)} \in C \otimes C$ for coproducts and $\rho_M(m) = \sum m_{<-1>} \otimes m_{<0>} \in C \otimes M$ for left coactions. $^C\mathcal{M}$ will be the category of left C-comodules and C-colinear maps.

A left H-comodule algebra A is an algebra in the monoidal category $^H\mathcal{M}$ of left H-comodules. This means that A is an algebra and a left H-comodule such that $\rho_A(ab) = \sum a_{<-1>}b_{<-1>} \otimes a_{<0>}b_{<0>}$ and $\rho_A(1_A) = 1_H \otimes 1_A$, for all $a, b \in A$. In a similar way, a right H-module coalgebra C is a coalgebra in $\mathcal{M}_H$, that is, C is a coalgebra and a right H-module such that $\Delta_C(c \cdot h) = \sum c_{(1)} \cdot h_{(1)} \otimes c_{(2)} \cdot h_{(2)}$ and $\varepsilon_C(c \cdot h) = \varepsilon_C(c)\varepsilon_H(h)$, for all $c \in C$ and $h \in H$. A triple (H, A, C) consisting of a Hopf algebra H, a left H-comodule algebra A and a right H-module coalgebra C will be called a (left-right) Doi-Koppinen datum. A representation of (H, A, C), or a right-left Doi-Koppinen module, is a k-module M that has a structure of right A-module and left C-comodule such that the following compatibility relation holds (cf. [6, 7])

$$\rho_M(ma) = \sum m_{<-1>} \cdot a_{<-1>} \otimes m_{<0>}a_{<0>}, \tag{1}$$

for all $a \in A$, $m \in M$. $^C\mathcal{M}(H)_A$ will be the category of right-left Doi-Koppinen modules and A-linear, C-colinear maps. $^C\mathcal{M}(H)_A$ is a Grothendieck category, if C is flat as a k-module (see e.g. [3]). For a right A-module M, $C \otimes M \in {}^C\mathcal{M}(H)_A$ with the following structures:

$$(c \otimes m) \cdot a = \sum c \cdot a_{<-1>} \otimes ma_{<0>}, \qquad \rho_{C \otimes M}(c \otimes m) = \sum c_{(1)} \otimes c_{(2)} \otimes m \tag{2}$$

for any $c \in C$, $a \in A$ and $m \in M$. In particular, $C \otimes A$ is a Doi-Koppinen module via:

$$(c \otimes b)a = \sum ca_{<-1>} \otimes ba_{<0>}, \qquad \rho^l_{C \otimes A}(c \otimes b) = \sum c_{(1)} \otimes c_{(2)} \otimes b \tag{3}$$

Let V be a k-module. Then $C \otimes A \otimes V \in {}^C\mathcal{M}(H)_A$ via the structures arising from the ones of $C \otimes A$, i.e

$$(c \otimes a \otimes v)b = \sum cb_{<-1>} \otimes ab_{<0>} \otimes v, \qquad \rho_{C \otimes A \otimes V}(c \otimes a \otimes v) = \sum c_{(1)} \otimes c_{(2)} \otimes a \otimes v \tag{4}$$

for all $c \in C$, a, $b \in A$, $v \in V$. If N is a left C-comodule, $N \otimes A \in {}^C\mathcal{M}(H)_A$ with the structures

$$(n \otimes a) \cdot b = n \otimes ab, \quad \rho_{N \otimes A}(n \otimes a) = \sum n_{<-1>}a_{<-1>} \otimes n_{<0>} \otimes a_{<0>} \quad (5)$$

for any $a, b \in A$ and $n \in N$. C is a left C-comodule via Δ; hence $C \otimes A$ can be also viewed as a Doi-Koppinen module via

$$(c \otimes b) \cdot' a = c \otimes ba, \quad \rho'^l_{C \otimes A}(c \otimes b) = \sum c_{(1)} \cdot b_{<-1>} \otimes c_{(2)} \otimes b_{<0>} \quad (6)$$

for all $c \in C$, a, $b \in A$. These two types of structures of Doi-Koppinen module on $C \otimes A$, coming from (6) and (3) are isomorphic; more precisely, the map

$$f : C \otimes A \to C \otimes A, \quad f(c \otimes a) = \sum ca_{<-1>} \otimes a_{<0>} \quad (7)$$

is an isomorphism of Doi-Koppinen modules, with inverse

$$g : C \otimes A \to C \otimes A, \quad g(c \otimes a) = \sum cS(a_{<-1>}) \otimes a_{<0>}.$$

A k-linear map $\gamma : C \to \mathrm{Hom}(C, A)$ is called an *integral* [8] of the Doi-Koppinen datum (H, A, C) if

$$\sum c_{(1)} \otimes \gamma(c_{(2)})(d) = \sum d_{(2)}\gamma(c)(d_{(1)})_{<-1>} \otimes \gamma(c)(d_{(1)})_{<0>}, \quad (8)$$

for all $c, d \in C$. An integral $\gamma : C \to \mathrm{Hom}(C, A)$ is called *total* if

$$\sum \gamma(c_{(1)})(c_{(2)}) = \varepsilon(c)1_A, \quad (9)$$

for all $c \in C$. For examples and the motivation of this general concept of integral, generalizing Doi's total integrals and classical integrals on Hopf algebras, we refer to [8]. In the next Theorem, we have collected some basic properties of integrals, we refer [8, Prop. 2.5, Th. 2.6, Th.2.9].

THEOREM 2.1. *Let (H, A, C) be a Doi-Koppinen datum, $M \in {}^C\mathcal{M}(H)_A$ and suppose that there exists an integral $\gamma : C \to \mathrm{Hom}(C, A)$ of (H, A, C). Then:*

1. *The map*

$$\lambda_M = \lambda_M(\gamma) : C \otimes M \to M, \quad \lambda_M(c \otimes m) = \sum m_{<0>}\gamma(c)(m_{<-1>}) \quad (10)$$

   is left C-colinear; λ_M splits the coaction $\rho_M : M \to C \otimes M$ if γ is a total integral;

2. *if γ is a total integral, then the map $f : C \otimes A \otimes M \to M$ given by*

$$f(c \otimes a \otimes m) = \sum m_{<0>}\gamma(cS(a_{<-1>}))(m_{<-1>})a_{<0>}, \quad (11)$$

   is a k-split epimorphism in ${}^C\mathcal{M}(H)_A$. In particular, $C \otimes A$ is a generator of the category ${}^C\mathcal{M}(H)_A$;

3. *assume that $f : M \to N$ is a morphism in ${}^C\mathcal{M}(H)_A$ which is a k-split injection (resp. a k-split surjection) and γ is a total integral. Then f has a C-colinear retraction (resp. a C-colinear section).*

3. THE AFFINENESS CRITERION FOR DOI-HOPF MODULES

In the next Proposition, we give a necessary and sufficient condition for A to be a Doi-Hopf module. It can be viewed as a special case of [1, Lemma 5.1].

PROPOSITION 3.1. *Let (H, A, C) be a Doi-Koppinen datum. There is a bijection between*

- *the set left C-coactions $\tilde{\rho} : A \to C \otimes A$ on A such that $(A, \cdot, \tilde{\rho}) \in {}^C\mathcal{M}(H)_A$;*
- *the set of elements $z = \sum_i c_i \otimes a_i \in C \otimes A$ such that*

$$\sum_i \varepsilon(c_i)a_i = 1_A, \qquad \sum_i c_{i_{(1)}} \otimes c_{i_{(2)}} \otimes a_i = \sum_{i,j} c_i \otimes c_j a_{i_{<-1>}} \otimes a_j a_{i_{<0>}}. \tag{12}$$

$z \in C \otimes A$ satisfying (12) will be called a grouplike element of (H, A, C).

Proof. Assume that $(A, \cdot, \tilde{\rho}) \in {}^C\mathcal{M}(H)_A$ and define $z := \tilde{\rho}(1_A)$. Conversely, let $z = \sum_i c_i \otimes a_i$ be a grouplike element of (H, A, C). Then $(A, \cdot, \tilde{\rho}) \in {}^C\mathcal{M}(H)_A$ where,

$$\tilde{\rho} : A \to C \otimes A, \quad \tilde{\rho}(a) := z \cdot a = \sum_i c_i a_{<-1>} \otimes a_i a_{<0>}$$

for all $a \in A$. $\qquad\square$

EXAMPLES 3.2. 1. Assume that $A = k$. Then $z \in C \otimes k = C$ is a grouplike element of (H, k, C) if and only if $z \in G(C)$, i.e. z is a grouplike element in the usual sense.
2. Let $x \in G(C)$ be a grouplike element of C. Then $x \otimes 1_A$ is a grouplike of (H, A, C).
3. Let $H = kG$, where G is a group, let X be a right G-set, let A be a G-graded k-algebra and consider the Doi-Koppinen datum (kG, A, kX). Let $z = \sum_{i \in I} x_i \otimes a_i \in kX \otimes A$, where $x_i \in X$ and $a_i \in A$ for every $i \in I$. We can assume, without loss of generality, that $x_i \neq x_j$ for $i \neq j$. Then z is a grouplike element of (kG, A, kX) if and only if for every $i, u \in I$,

$$\sum_{i \in I} a_i = 1_A, \qquad \sum_{x_j g = x_i} a_j a_{i_g} = a_i, \qquad \sum_{x_j g = x_u \neq x_i} a_j a_{i_g} = 0,$$

where a_{i_g} denotes the g-component of a_i, $a_i = \sum_{g \in G} a_{i_g}$.

From now on, we fix a grouplike element $z = \sum_i c_i \otimes a_i \in C \otimes A$ of the Doi-Koppinen datum (H, A, C) and we consider the corresponding Doi-Koppinen module structure $\tilde{\rho}$ on A. We call $\tilde{\rho}$ the generalized *Verma Doi-Koppinen module structure* of A.

DEFINITION 3.3. Let $z = \sum_i c_i \otimes a_i \in C \otimes A$ be a grouplike element of (H, A, C). An integral $\gamma : C \to \mathrm{Hom}(C, A)$ is called z-normalized if

$$\sum a_j \gamma\big(c_i S(a_{i_{<-1>}})\big)(c_j)a_{i_{<0>}} = 1_A. \tag{13}$$

EXAMPLE 3.4. Let $x \in G(C)$ be a grouplike element of C and $z = x \otimes 1_A$. Then any total integral $\gamma : C \to \mathrm{Hom}(C, A)$ is z-normalized, since $1_A \gamma(x)(x)1_A = \varepsilon(x)1_A = 1_A$.

PROPOSITION 3.5. *Let $z = \sum_i c_i \otimes a_i \in C \otimes A$ be a grouplike element of (H, A, C), and assume that there exists a z-normalized integral $\gamma : C \to \mathrm{Hom}(C, A)$ of (H, A, C). Then $\tilde{\rho} : A \to C \otimes A$ splits in ${}^C\mathcal{M}(H)_A$.*

THE AFFINENESS CRITERION FOR DOI-KOPPINEN MODULES 219

Proof. Using Theorem 2.1, we easily find that the map

$$\lambda : C \otimes A \to A, \quad \lambda(c \otimes a) = \sum a_i a_{<0>} \gamma(c)(c_i a_{<-1>})$$

is left C-colinear. Now consider the map $\Lambda : C \otimes A \to A$ given by

$$\Lambda(c \otimes a) = \sum \lambda \Big(cS(a_{<-1>}) \otimes 1_A\Big)a_{<0>} = \sum a_i \gamma\Big(cS(a_{<-1>})\Big)(c_i)a_{<0>}, \quad (14)$$

for all $c \in C$, $a \in A$. Using the fact that γ is z-normalized, we compute that

$$\Lambda(z) = \sum \lambda \Big(c_i S(a_{i<-1>}) \otimes 1_A\Big)a_{i<0>} = \sum a_j \gamma\Big(c_i S(a_{i<-1>})\Big)(c_j)a_{i<0>} = 1_A.$$

Λ is right A-linear since

$$\begin{aligned}
\Lambda((c \otimes a)b) &= \sum \Lambda(cb_{<-1>} \otimes ab_{<0>}) \\
&= \sum \lambda\Big(cb_{<-2>}S(b_{<-1>})S(a_{<-1>}) \otimes 1_A\Big)a_{<0>}b_{<0>} \\
&= \sum \lambda\Big(cS(a_{<-1>}) \otimes 1_A\Big)a_{<0>}b = \Lambda(c \otimes a)b,
\end{aligned}$$

for all $c \in C$ and $a, b \in A$. Λ is a retraction of $\tilde{\rho}$ since

$$(\Lambda \circ \tilde{\rho})(a) = \sum \Lambda(\sum c_i a_{<-1>} \otimes a_i a_{<0>}) = \Lambda(z \cdot a) = \Lambda(z)a = a,$$

for all $a \in A$. Let us finally show that Λ is C-colinear. Indeed, for all $c \in C$ and $a \in A$, we have

$$\begin{aligned}
\tilde{\rho}(\Lambda(c \otimes a)) &= \sum \tilde{\rho}\Big(\lambda(cS(a_{<-1>}) \otimes 1_A)a_{<0>}\Big) \\
&= \sum c_i \Big(\lambda(cS(a_{<-2>}) \otimes 1_A)\Big)_{<-1>} a_{<-1>} \otimes \\
&\qquad a_i \Big(\lambda(cS(a_{<-2>}) \otimes 1_A)\Big)_{<0>} a_{<0>} \\
&= \sum c_{(1)}S(a_{<-2>})_{(1)}a_{<-1>} \otimes \lambda(c_{(2)}S(a_{<-2>})_{(2)} \otimes 1_A)a_{<0>} \\
&= \sum c_{(1)}S(a_{<-2>})a_{<-1>} \otimes \lambda(c_{(2)}S(a_{<-3>}) \otimes 1_A)a_{<0>} \\
&= \sum c_{(1)} \otimes \lambda(c_{(2)}S(a_{<-1>}) \otimes 1_A)a_{<0>} \\
&= (Id \otimes \Lambda)\rho_{C \otimes A}(c \otimes a).
\end{aligned}$$

In the third equality, we used the fact that λ is left C-colinear. $\qquad\square$

For $M \in {}^C\mathcal{M}(H)_A$, let

$$M_z = \{m \in M \mid \rho_M(m) = \sum_i c_i \otimes ma_i\}$$

be the subgroup of *z-coinvariants* of M. In particular,

$$B = A_z = \{a \in A \mid \tilde{\rho}(a) = \sum_i c_i \otimes aa_i\} = \{a \in A \mid \sum_i c_i a_{<-1>} \otimes a_i a_{<0>} = \sum_i c_i \otimes aa_i\}$$

is a subalgebra of A, called the subalgebra of *z-coinvariants* of the Verma structure $(A, \cdot, \tilde{\rho})$. M_z is a right B-module.

C. MENINI AND G. MILITARU

PROPOSITION 3.6. *Let $z = \sum_i c_i \otimes a_i \in C \otimes A$ be a grouplike element of (H, A, C), and assume that there exists a z-normalized integral $\gamma : C \to \mathrm{Hom}(C, A)$ of (H, A, C). Then B is a direct summand of A in $_B\mathcal{M}$ and in $\mathcal{M}_B$.*

Proof. 1. Consider the map $t^l : A \to B$ given by the formula

$$t^l(a) = \sum \lambda\Big(c_i S(a_{i<-1>}) \otimes a\Big) a_{i<0>} = \sum a_j a_{<0>} \gamma\Big(c_i S(a_{i<-1>})\Big)(c_j a_{<-1>}) a_{i<0>}. \tag{15}$$

We will prove that t^l is a left trace: t^l is left B-linear, and a retraction for the inclusion $B \subset A$. From the fact that λ is C-colinear, it follows that

$$
\begin{aligned}
\tilde\rho(t^l(a)) &= \sum \tilde\rho\Big(\lambda(c_i S(a_{i<-1>}) \otimes a) a_{i<0>}\Big) \\
&= \sum c_j \Big(\lambda(c_i S(a_{i<-1>}) \otimes a)\Big)_{<-1>} a_{i<0><-1>} \otimes \\
&\qquad a_j \Big(\lambda(c_i S(a_{i<-1>}) \otimes a)\Big)_{<0>} a_{i<0><0>} \\
&= \sum (c_i S(a_{i<-1>}))_{(1)} a_{i<0><-1>} \otimes \lambda\Big((c_i S(a_{i<-1>}))_{(2)} \otimes a\Big) a_{i<0><0>} \\
&= \sum c_{i(1)} S(a_{i<-2>}) a_{i<-1>} \otimes \lambda\Big(c_{i(2)} S(a_{i<-3>}) \otimes a\Big) a_{i<0>} \\
&= \sum c_{i(1)} \otimes \lambda\Big(c_{i(2)} S(a_{i<-1>}) \otimes a\Big) a_{i<0>} \\
(12) \quad &= \sum c_i \otimes \lambda\Big(c_j a_{i<-1>} S\big((a_j a_{i<0>})_{<-1>}\big) \otimes a\Big)(a_j a_{i<0>})_{<0>} \\
&= \sum c_i \otimes \lambda\Big(c_j a_{i<-2>} S(a_{i<-1>}) S(a_{j<-1>}) \otimes a\Big) a_{j<0>} a_{i<0>} \\
&= \sum c_i \otimes \lambda\Big(c_j S(a_{j<-1>}) \otimes a\Big) a_{j<0>} a_i \\
&= \sum c_i \otimes t^l(a) a_i,
\end{aligned}
$$

for all $a \in A$. This means that $t^l(a) \in B$, for all $a \in A$. For $b \in B$ and $a \in A$, we have

$$\lambda(c \otimes ba) = \sum b a_i a_{<0>} \gamma(c)(c_i a_{<-1>}) = b\lambda(c \otimes a),$$

hence

$$t^l(ba) = \sum \lambda\Big(c_i S(a_{i<-1>}) \otimes ba\Big) a_{i<0>} = b \sum \lambda\Big(c_i S(a_{i<-1>}) \otimes a\Big) a_{i<0>} = bt^l(a),$$

and it follows that t^l is left B-linear. Using (13), we find that

$$
\begin{aligned}
t^l(b) &= bt^l(1_A) = b\lambda\Big(c_i S(a_{i<-1>}) \otimes 1_A\Big) a_{i<0>} \\
&= b \sum a_j \gamma\Big(c_i S(a_{i<-1>})\Big)(c_j) a_{i<0>} = b,
\end{aligned}
$$

for all $b \in B$. We can conclude that t^l is a retraction of the inclusion $B \subset A$.

2. In a similar way, we have a right trace $t^r : A \to B$, given by the formula

$$t^r(a) = \Lambda\Big(\sum c_i \otimes a a_i\Big) = \sum a_j \gamma\Big(c_i S(a_{<-1>} a_{i<-1>})\Big)(c_j) a_{<0>} a_{i<0>}, \tag{16}$$

for all $a \in A$. $\qquad\qquad\square$

THE AFFINENESS CRITERION FOR DOI-KOPPINEN MODULES 221

we will now construct functors connecting $^C\mathcal{M}(H)_A$ and $\mathcal{M}_B$. First, the assignment $M \to M_z$, $M \in {}^C\mathcal{M}(H)_A$, gives a covariant functor,

$$(-)_z : {}^C\mathcal{M}(H)_A \to \mathcal{M}_B.$$

For $N \in \mathcal{M}_B$, $N \otimes_B A \in {}^C\mathcal{M}(H)_A$ via the structures

$$(n \otimes_B a)a' = n \otimes_B aa', \quad \rho_{N \otimes_B A}(n \otimes_B a) = \sum c_i a_{<-1>} \otimes n \otimes_B a_i a_{<0>} \quad (17)$$

for all $n \in N$, a, $a' \in A$ (the map $\rho_{N \otimes_B A}$ is well-defined as the map $n \otimes a \mapsto \sum c_i a_{<-1>} \otimes n \otimes_B a_i a_{<0>}$ is B-balanced). In this way, we have constructed the induction functor

$$- \otimes_B A : \mathcal{M}_B \to {}^C\mathcal{M}(H)_A,$$

which is a left adjoint of the coinvariants functor $(-)_z : {}^C\mathcal{M}(H)_A \to \mathcal{M}_B$. The unit and the counit of the adjunction are given by

$$\eta_N : N \to (N \otimes_B A)_z, \quad \eta_N(n) = n \otimes_B 1_A, \quad (18)$$

for all $N \in \mathcal{M}_B$, $n \in N$ and

$$\beta_M : M_z \otimes_B A \to M, \quad \beta_M(m \otimes_B a) = ma, \quad (19)$$

for all $M \in {}^C\mathcal{M}(H)_A$, $m \in M_z$ and $a \in A$.

THEOREM 3.7. *Let $z = \sum_i c_i \otimes a_i \in C \otimes A$ be a grouplike element of (H, A, C) and assume that there exists a z-normalized integral $\gamma : C \to \mathrm{Hom}(C, A)$ of (H, A, C). Then*

$$\eta_N : N \to (N \otimes_B A)_z, \quad \eta_N(n) = n \otimes_B 1_A$$

is an isomorphism of left B-modules, for all $N \in \mathcal{M}_B$. Therefore the induction functor $- \otimes_B A : \mathcal{M}_B \to {}^C\mathcal{M}(H)_A$ is fully faithful.

Proof. For $N \in \mathcal{M}_B$, we consider the map $\theta_N : (N \otimes_B A)_z \to N$, defined by the formula

$$\theta_N\left(\sum_i n_i \otimes_B \alpha_i\right) = \sum_i n_i t^l(\alpha_i). \quad (20)$$

Then $\theta_N \circ \eta_N = Id_N$, since $t^l(1_A) = 1_A$. If $\sum_i n_i \otimes_B \alpha_i \in (N \otimes_B A)_z$, then

$$\sum c_k \alpha_{i_{<-1>}} \otimes n_i \otimes_B a_k \alpha_{i_{<0>}} = \sum c_k \otimes n_i \otimes_B \alpha_i a_k,$$

and it follows that

$$\sum n_i \otimes_B a_k \alpha_{i_{<0>}} \gamma\Big(c_j S(a_{j_{<-1>}})\Big)(c_k \alpha_{i_{<-1>}})a_{j_{<0>}}$$
$$= \sum n_i \otimes_B \alpha_i a_k \gamma\Big(c_j S(a_{j_{<-1>}})\Big)(c_k)a_{j_{<0>}}, \quad (21)$$

hence

$$\sum n_i \otimes_B t^l(\alpha_i) \overset{(21)}{=} \sum n_i \otimes_B \alpha_i a_k \gamma\Big(c_j S(a_{j_{<-1>}})\Big)(c_k)a_{j_{<0>}}$$
$$= \sum n_i \otimes_B \alpha_i \lambda\Big(c_j S(a_{j_{<-1>}}) \otimes 1_A\Big)a_{j_{<0>}}$$
$$= \sum n_i \otimes_B \alpha_i t^l(1_A) = \sum n_i \otimes_B \alpha_i.$$

It now follows that

$$(\eta_N \circ \theta_N)(\sum_i n_i \otimes_B \alpha_i) = \sum_i n_i t^l(\alpha_i) \otimes_B 1_A = \sum n_i \otimes_B t^l(\alpha_i) = \sum n_i \otimes_B \alpha_i,$$

and we have shown that θ_N is an inverse of η_N. $\qquad\square$

Let V be a k-module. Then $A \otimes V \in {}^C\mathcal{M}(H)_A$ via the structures induced by A, namely

$$(a \otimes v)b = ab \otimes v, \quad \rho_{A \otimes V}(a \otimes v) = \sum c_i a_{<-1>} \otimes a_i a_{<0>} \otimes v, \qquad (22)$$

for all a, $b \in A$ and $v \in V$. In particular, $A \otimes A \in {}^C\mathcal{M}(H)_A$ with

$$(a \otimes a')b = ab \otimes a', \quad \rho_{A \otimes A}(a \otimes a') = \sum c_i a_{<-1>} \otimes a_i a_{<0>} \otimes a', \qquad (23)$$

for all a, a', $b \in A$.

LEMMA 3.8. *Let $z = \sum_i c_i \otimes a_i \in C \otimes A$ be a grouplike element of (H, A, C) and assume that there exists a z-normalized integral $\gamma : C \to \mathrm{Hom}(C, A)$ of (H, A, C). Then the counit map*

$$\beta_{A \otimes V} : (A \otimes V)_z \otimes_B A \to A \otimes V, \quad \beta_{A \otimes V}(\sum \alpha_i \otimes v_i \otimes_B a) = \sum \alpha_i a \otimes v_i$$

is an isomorphism in ${}^C\mathcal{M}(H)_A$, with inverse ψ given by $\psi(a \otimes v) = 1_A \otimes v \otimes_B a$.

Proof. It is clear that $\beta_{A \otimes V}\psi(a \otimes v) = \beta_{A \otimes V}(1_A \otimes v \otimes_B a) = a \otimes v$, for all $a \in A$ and $v \in V$. Take $\sum_i \alpha_i \otimes v_i \in (A \otimes V)_z$ and $a \in A$. A calculation similar to that one that appears at the beginning of the proof of Theorem 3.7 shows that,

$$\sum_i \alpha_i \otimes v_i = \sum_i t^l(\alpha_i) \otimes v_i, \qquad (24)$$

and we compute that

$$\psi(\beta_{A \otimes V}(\sum \alpha_i \otimes v_i \otimes_B a))$$

$$\overset{(24)}{=} \psi\beta_{A \otimes V}\left(\sum t^l(\alpha_i) \otimes v_i \otimes_B a\right) = \psi\left(\sum t^l(\alpha_i)a \otimes v_i\right)$$

$$= \sum 1_A \otimes v_i \otimes_B t^l(\alpha_i)a = \sum (1_A \otimes v_i)t^l(\alpha_i) \otimes_B a$$

$$= \sum t^l(\alpha_i) \otimes v_i \otimes_B a \overset{(24)}{=} \sum \alpha_i \otimes v_i \otimes_B a,$$

and it follows that ψ is an inverse of $\beta_{A \otimes V}$. $\qquad\square$

$C \otimes A \in {}^C\mathcal{M}(H)_A$ with structure given by (3); we have an isomorphism $A \cong (C \otimes A)_z$, sending a to $\sum_i c_i \otimes aa_i$. The counit map $\beta_{C \otimes A}$ can be viewed as a map in ${}^C\mathcal{M}(H)_A$ as follows

$$\beta = \beta_{C \otimes A} : A \otimes_B A \to C \otimes A, \quad \beta(a \otimes_B b) = \sum c_i b_{<-1>} \otimes aa_i b_{<0>}, \qquad (25)$$

for all a, $b \in A$, where $A \otimes_B A \in {}^C\mathcal{M}(H)_A$ via

$$(a \otimes_B b)a' = a \otimes_B ba', \quad a \otimes_B b \to \sum c_i b_{<-1>} \otimes a \otimes_B a_i b_{<0>}$$

for all a, a', $b \in A$.

We can now prove our main result, the affineness criterion for Doi-Koppinen modules.

THE AFFINENESS CRITERION FOR DOI-KOPPINEN MODULES 223

THEOREM 3.9. *Let (H, A, C) be a Doi-Koppinen datum, with C projective as a k-module. Let $z = \sum_i c_i \otimes a_i \in C \otimes A$ be a grouplike element of (H, A, C). Assume that there exists a total and z-normalized integral $\gamma : C \to \mathrm{Hom}(C, A)$. If the canonical map $\beta : A \otimes_B A \to C \otimes A$, $\beta(a \otimes_B b) = \sum c_i b_{<-1>} \otimes aa_i b_{<0>}$ is surjective, then the induction functor $- \otimes_B A : \mathcal{M}_B \to {}^C\mathcal{M}(H)_A$ is a category equivalence.*

Proof. We have proved in Theorem 3.7 that the unit map $\eta_N : N \to (N \otimes_B A)_z$ is an isomorphism, for all $N \in \mathcal{M}_B$. It remains to be shown that the counit map

$$\beta_M : M_z \otimes_B A \to M, \quad \beta_M(m \otimes_B a) = ma$$

is an isomorphism, for all $M \in {}^C\mathcal{M}(H)_A$. Consider the map

$$\tilde{\beta} : A \otimes A \to C \otimes A, \quad \tilde{\beta}(a \otimes b) = \sum c_i b_{<-1>} \otimes aa_i b_{<0>} = \left(\sum c_i \otimes aa_i\right)b.$$

$\tilde{\beta}$ is surjective since it is the composition of the canonical projection $A \otimes A \to A \otimes_B A$ and the surjective map β. Now consider the map $\zeta : A \otimes A \to C \otimes A$ given by

$$\zeta(a \otimes b) = (\tilde{\beta} \circ \tau)(a \otimes b) = \sum c_i a_{<-1>} \otimes ba_i a_{<0>} = \left(\sum c_i \otimes ba_i\right)a, \qquad (26)$$

where $\tau(a \otimes b) = b \otimes a$ is the flip map. We will now prove that ζ is a morphism in ${}^C\mathcal{M}(H)_A$, where $A \otimes A$ and $C \otimes A$ are Doi-Koppinen modules via (23) and (3). Indeed,

$$\zeta((a \otimes b)a') = \zeta(aa' \otimes b) = \left(\sum c_i \otimes ba_i\right)(aa')$$

$$= \left(\left(\sum c_i \otimes ba_i\right)a\right)a' = \zeta(a \otimes b)a',$$

and

$$\rho_{C \otimes A}(\zeta(a \otimes b)) = \sum \rho_{C \otimes A}(c_i a_{<-1>} \otimes ba_i a_{<0>})$$

$$= \sum c_{i_{(1)}} a_{<-2>} \otimes c_{i_{(2)}} a_{<-1>} \otimes ba_i a_{<0>}$$

$$(12) \quad = \sum c_i a_{<-2>} \otimes c_j a_{i_{<-1>}} a_{<-1>} \otimes ba_j a_{i_{<0>}} a_{<0>}$$

$$= \sum (Id \otimes \zeta)(c_i a_{<-1>} \otimes a_i a_{<0>} \otimes b)$$

$$= (Id \otimes \zeta)\rho_{A \otimes A}(a \otimes b),$$

for all $a, a', b \in A$. Moreover, ζ is surjective as $\tilde{\beta}$ is surjective and τ is bijective. C is projective as a k-module, hence $C \otimes A$ is projective as a right A-module. Using the isomorphism (7), we obtain that $C \otimes A$, with the A-module structure given by (3), is still projective as a right A-module. It follows that the surjective morphism $\zeta : A \otimes A \to C \otimes A$ has a section in the category of right A-modules. In particular, ζ is a k-split epimorphism in ${}^C\mathcal{M}(H)_A$.

Let $M \in {}^C\mathcal{M}(H)_A$. Then $A \otimes A \otimes M \in {}^C\mathcal{M}(H)_A$ via (22), with $V = A \otimes M$. The map

$$\zeta \otimes Id : A \otimes A \otimes M \to C \otimes A \otimes M$$

is a k-split epimorphism in ${}^C\mathcal{M}(H)_A$, where $C \otimes A \otimes M \in {}^C\mathcal{M}(H)_A$ via (4). Let $f : C \otimes A \otimes M \to M$ be the k-split epimorphism in ${}^C\mathcal{M}(H)_A$ constructed in Theorem 2.1. Then the map $g = f \circ (\zeta \otimes Id) : A \otimes A \otimes M \to M$ is given by

$$g(a \otimes b \otimes m) = \sum m_{<0>}\gamma\Big(c_i S(b_{<-1>}a_{i_{<-1>}})\Big)(m_{<-1>})b_{<0>}a_{i_{<0>}}a,$$

and is a k-split epimorphism in $^C\mathcal{M}(H)_A$. Thus we have a k-split epimorphism

$$A \otimes A \otimes M = M_1 \xrightarrow{g} M \longrightarrow 0$$

in $^C\mathcal{M}(H)_A$. It follows from Lemma 3.8 that the counit map β_{M_1} is bijective. Applying Theorem 2.1, and invoking the fact that there exists a total integral $\gamma : C \to \mathrm{Hom}(C, A)$, we obtain that g also splits in $^C\mathcal{M}$. In particular, the sequence

$$(M_1)_z \xrightarrow{g_z} M_z \longrightarrow 0$$

is also exact. Continuing the resolution with $\mathrm{Ker}(g)$ instead of M we obtain an exact sequence in $^C\mathcal{M}(H)_A$

$$M_2 \longrightarrow M_1 \longrightarrow M \longrightarrow 0$$

which splits in $^C\mathcal{M}$ and the adjunction maps for M_1 and M_2 are bijective. Using the Five Lemma we obtain that the adjunction map for M is bijective. $\qquad \square$

4. APPLICATIONS

Relative Hopf modules and Schneider's affineness criterion

Theorem 3.9 can be applied for several Doi-Koppinen data. In particular, for $(H, A, C) = (H, A, H)$ and $z = 1_H \otimes 1_A$ we obtain the right-left version of [12, Theorem 3.5].

COROLLARY 4.1. *Let H be a Hopf algebra with a bijective antipode[1] and projective over k. Let A be a left H-comodule algebra and $B = A^{\mathrm{co}(H)}$. Assume*

1. *A is relative injective as left H-comodule.*
2. *can : $A \otimes_B A \to H \otimes A$, $\mathrm{can}(a \otimes_B b) = \sum b_{<-1>} \otimes ab_{<0>}$ is surjective.*

Then the induction functor $- \otimes_B A : \mathcal{M}_B \to {}^H\mathcal{M}_A$ is an equivalence of categories.

Proof. We consider the Doi-Koppinen datum $(H, A, C) = (H, A, H)$ and $z = 1_H \otimes 1_A$. The assumption 1) is equivalent to the fact that there exists a total integral in the sense of Doi, i.e. a left H-colinear map $\varphi : H \to A$ such that $\varphi(1) = 1$ (see [5, Theorem 1.6]). It follows from [8, Remark 2.3] that the map

$$\gamma = \gamma_\varphi : H \to \mathrm{Hom}(H, A), \quad \gamma(h)(g) = \varphi\Big(S^{-1}(g)h\Big)$$

for all g, $h \in H$ is a total integral for the Doi-Koppinen datum (H, A, H). As $z = 1_H \otimes 1_A$ it follows from Example 3.4 that γ is also a z-normalized integral for (H, A, H). Now we apply Theorem 3.9. $\qquad \square$

Generalized Yetter-Drinfeld modules

Let $G = (H, A, C)$ be a Yetter-Drinfeld datum [3]: H is a Hopf algebra with a bijective antipode, A is an H-bicomodule algebra and C an H-bimodule coalgebra. Then $(H \otimes H^{\mathrm{op}}, A, C)$ is a Doi-Koppinen datum, where A is a left $H \otimes H^{\mathrm{op}}$-comodule algebra and C is a right $H \otimes H^{\mathrm{op}}$-module coalgebra via

$$\rho_A(a) = \sum \Big(a_{<-1>} \otimes S^{-1}(a_{<1>})\Big) \otimes a_{<0>} \text{ and } c \bullet (h \otimes k) = k \cdot c \cdot h,$$

---

[1]This assumption is given by the choice of sides (right-left).

THE AFFINENESS CRITERION FOR DOI-KOPPINEN MODULES

for all $a \in A$, $c \in C$ and $h, k \in H$. Then ${}^C\mathcal{M}(H \otimes H^{\mathrm{op}})_A = {}^C\mathcal{YD}(H)_A$, the category of quantum Yetter-Drinfeld modules: an object of it is a k-module M that is a right A-module and a left C-comodule such that

$$\sum m_{<-1>}a_{<-1>} \otimes m_{<0>} \cdot a_{<0>} = \sum a_{<1>}(m \cdot a_{<0>})_{<-1>} \otimes (m \cdot a_{<0>})_{<0>},$$

for all $m \in M$ and $a \in A$. In this case $\gamma : C \to \mathrm{Hom}(C, A)$ is an integral of $(H \otimes H^{\mathrm{op}}, A, C)$ if and only if

$$\sum c_{(1)} \otimes \gamma(c_{(2)})(d) = \sum S^{-1}\Big(\{\gamma(c)(d_{(1)})\}_{<1>}\Big) d_{(2)} \{\gamma(c)(d_{(1)})\}_{<-1>} \otimes \{\gamma(c)(d_{(1)})\}_{<0>},$$

for all $c, d \in C$. a quantum integral γ will be called a total quantum integral if $\sum \gamma(c_{(1)})(c_{(2)}) = \varepsilon(c)1_A$ for all $c \in C$. Let $g \in C$ be a grouplike element of C and $z = g \otimes 1_A$. In view of Example 3.4, we can apply Theorem 3.9, and obtain Corollary 4.2. In the situation where $C = H$ and $g = 1_H$, we recover [8, Theorem 3.15].

COROLLARY 4.2. *Let (H, A, C) be a a Yetter-Drinfeld datum and assume that C is projective over k. Let $g \in C$ be grouplike, $z = g \otimes 1_A$, and $B = A_z$. Assume that the following conditions hold:*

1. *there exists a total quantum integral $\gamma : C \to \mathrm{Hom}(C, A)$;*
2. *the canonical map*

$$\beta : A \otimes_B A \to C \otimes A, \quad \beta(a \otimes_B b) = \sum S^{-1}(b_{<1>})gb_{<-1>} \otimes ab_{<0>}$$

   is surjective.

Then the induction functor $- \otimes_B A : \mathcal{M}_B \to {}^C\mathcal{YD}(H)_A$ is an equivalence of categories.

Modules graded by G-sets

Let X be a set and $C = kX$ the corresponding grouplike coalgebra, i.e. kX is the free k-module having X as a basis and $\Delta(x) = x \otimes x$, $\varepsilon(x) = 1$, for all $x \in X$. Let $(H, A, C) = (H, A, kX)$ be a Doi-Koppinen datum. The category ${}^{kX}\mathcal{M}(H)_A$ associated to this Doi-Koppinen datum covers a large class of examples of X-graded representations of A, starting from the category of super-graded vector spaces (corresponding to the trivial case $H = A = k$, k a field) to the category of graded modules by G-sets (corresponding to the case $H = kG$, where G is a group acting on the set X). Fix an $x_0 \in X$ and let $z = x_0 \otimes 1_A$.

THEOREM 4.3. *Let X be a set, (H, A, kX) a Doi-Koppinen datum and $x_0 \in X$. Then the induction functor $- \otimes_B A : \mathcal{M}_B \to {}^{kX}\mathcal{M}(H)_A$ is an equivalence of categories if and only if the canonical map*

$$\beta : A \otimes_B A \to kX \otimes A, \quad \beta(a \otimes_B b) = \sum x_0 b_{<-1>} \otimes ab_{<0>}$$

is surjective.

Proof. The map

$$\gamma : kX \to \mathrm{Hom}(kX, A), \quad \gamma(x)(y) = \delta_{xy}1_A,$$

is a total integral of (H, A, kX) ([2, Proposition 3.18]). By Example 3.4, γ is z-normalized and we get the "only if" part from Theorem 3.9. The "if" part follows from the fact that the canonical map $\beta = \beta_{C \otimes A}$ is the counit map of the equivalence $(- \otimes_B A, (-)_z)$. $\square$

In particular, we consider the Doi-Koppinen datum (kG, A, k), where G is a group acting from the right on the set X. Then it is well know that (A, ρ) is a left kG-comodule algebra if and only if $A = \oplus_{g \in G} A_g$ is a G-graded k-algebra: $a_g \in A_g$ if and only if $\rho(a_g) = g \otimes a_g$. Moreover, $^{kX}\mathcal{M}(kG)_A = (G, A, X)$-gr, the category of right A-modules graded by the G-set X, see [6]. Fix $x_0 \in X$ and let $H = \mathrm{Stab}_G(x_0)$. Then it is easy to see that the subalgebra of x_0-coinvariants of A is just $B = A_H = \oplus_{h \in H} A_h$. we then recover a result of del Río [11, Theorem 2.3 and Example 2.6].

COROLLARY 4.4. *Let G be a group, X a right G-set, A a G-graded k-algebra, $x_0 \in X$ and $H = \mathrm{Stab}_G(x_0)$. Then the induction functor $- \otimes_{A_H} A : \mathcal{M}_{A_H} \to (G, A, X)$-gr is a category equivalence if and only if the following two conditions hold:*

1. *the action of G on X is transitive, i.e. $x_0 \cdot G = X$; identify X and $(G/H)_r$, the set of right H-cosets and we fix a system $\Theta \subset G$ of representatives G modulo H such that $1_G \in \Theta$;*
2. *for any $\theta \in \Theta$, there exists a finite set I_θ, a map $\chi = \chi_\theta : I_\theta \to H$ and homogeneous elements $a_{i,\chi(i)\theta} \in A_{(\chi(i)\theta)^{-1}}$, $b_{i,\chi(i)\theta} \in A_{\chi(i)\theta}$ such that*

$$\sum_{i \in I_\theta} a_{i,\chi(i)\theta} \, b_{i,\chi(i)\theta} = 1_A. \tag{27}$$

Proof. We leave it to the reader to verify that the two conditions in the Corollary are equivalent to surjectivity of the canonical map

$$\beta : A \otimes_{A_H} A \to kX \otimes A, \quad \beta(a \otimes_{A_H} b) = \sum_{g \in G} x_0 \cdot g \otimes ab_g.$$

$\square$

In particular, let $X = G$, with the action given by right multiplication, and $x_0 = 1_G$. Then (27) is equivalent to $1_A \in A_{g^{-1}} A_g$, for any $g \in G$, i.e. to the fact that A is a strongly graded algebra. Thus we recover a well-known result of Dade: the induction functor $- \otimes_{A_1} A : \mathcal{M}_{A_1} \to A$-gr is a category equivalence if and only if A is a strongly graded algebra.

REFERENCES

[1] T. Brzeziński, The structure of corings. Induction functors, Maschke-type theorem, and Frobenius and Galois properties, *Algebras Representation Theory* **5** (2002), 389–410.

[2] S. Caenepeel, Bogdan Ion, G. Militaru and Shenglin Zhu, Separable functors for the category of Doi-Hopf modules. Applications, *Adv. Math.* **145** (1999), 239–290 .

[3] S. Caenepeel, G. Militaru and Shenglin Zhu, "Frobenius and separable functors for generalized module categories and nonlinear equations", *Lect. Notes Math.* **1787**, Springer Verlag, Berlin, 2002.

[4] E. Cline, B. Parshall and L. Scott, Induced modules and affine quotient, *Math. Ann.* **230** (1977), 1–14.

[5] Y. Doi, Algebras with total integrals, *Comm. Algebra* **13** (1985), 2137–2159.

[6] Y. Doi, Unifying Hopf modules, *J. Algebra* **153** (1992), 373–385.

[7] M. Koppinen, Variations on the smash product with applications to group-graded rings, *J. Pure Appl. Algebra* **104** (1995), 61–80.

THE AFFINENESS CRITERION FOR DOI-KOPPINEN MODULES

[8] C. Menini and G. Militaru, Integrals, quantum Galois extensions and the affineness criterion for quantum Yetter-Drinfeld modules, *J. Algebra* **247** (2002), 467–508.

[9] C. Menini and M. Zuccoli, Equivalence theorems and Hopf-Galois extensions, *J. Algebra* **194** (1997), 245–247.

[10] U. Oberst, Affine Quotientenschemata nach affinen, algebraischen Gruppen und induzierte Darstellungen, *J. Algebra* **44** (1977), 503–538.

[11] A. del Río, Categorical methods in graded ring theory, *Publ. Mathematiques*, **36** (1992), 489–553.

[12] H.-J. Schneider, Principal homogeneous spaces for arbitrary Hopf algebras, *Israel J. Math.* **72** (1990), 167–195.

Algebra Properties invariant under Twisting

SUSAN MONTGOMERY   University of Southern California
Los Angeles, CA 90089-1113, USA
e-mail address: smontgom@math.usc.edu

> ABSTRACT. For a finite-dimensional Hopf algebra H over a field k and an H-comodule algebra A, we study properties of A which are preserved when A is twisted by a Hopf 2-cocycle σ on H. We prove that if there exists σ such that A_σ is super-commutative, then A being affine imples that A is Noetherian. If also H_σ is commutative, then A is integral over a central subring of A^{coH}. We also consider when A satisfies a polynomial identity.

1. INTRODUCTION

Let H be a Hopf algebra over the field k, let A be a k-algebra, and assume that A is an H-comodule algebra (or an H-module algebra). The object of this paper is to study what properties of the algebra A are preserved under twisting by a 2-cocycle σ on H (or under "cotwisting" by a dual cocycle). We are particularly interested in classical algebraic properties of A, such as finite generation, being Noetherian, satisfying a polynomial identity, or being integral over a central subalgebra of the coinvariants A^{coH} (or invariants A^H).

Such results give a method of proving that a given algebra A has a desired property: namely, by showing that a suitable twist A_σ of A exists such that A_σ has the property. Our best results are obtained when we can find σ such that A_σ is super-commutative.

The easiest example of a non-trivial twisting occurs when A is graded by a group G. In this situation, write $A = \oplus_{g \in G} A_g$ and let $\sigma : G \times G \to k^*$ be a 2-cocycle on G. Then the twisted algebra A_σ has the same underlying k-space as A, with new multiplication

$$a \cdot_\sigma b = \sigma(g, h)ab$$

for all homogeneous elements $a \in A_g, b \in A_h$. Note that A_σ is again a G-graded algebra.

The idea of twisting graded algebras by cocycles is an old one; it appears already in [11, Chap. III, Sec 4.7, Prop 10], where the twisted tensor product of graded algebras is considered. In [35], twisting is used to show that any G-Lie color algebra can be twisted to an ordinary (G-graded) Lie superalgebra. More recent references include [3], where twisting of bi-graded algebras is used; [37, Example 2.9], where equivalence of module categories of graded algebras under twisting is considered;

2000 *Mathematics Subject Classification.* 13A20.
Key words and phrases. twisting.
The author was supported by NSF grant DMS-0100461.

230 S. MONTGOMERY

and [7], where it is shown that when G is finite, satisfying a polynomial identity is preserved under twisting.

These ideas can be extended to arbitrary Hopf algebras, but with a complication. If the algebra A is an H-comodule algebra and $\sigma : H \otimes H \to k$ is a Hopf 2-cocycle, then A can be twisted to a new algebra A_σ; see for example [29, 7.5]. However this twisted algebra is no longer an H-comodule algebra. Rather, it turns out to be a comodule algebra over a twisted version H_σ of H (this complication does not appear for groups since kG is cocommutative). This twisting of H by a cocycle $\sigma : H \otimes H \to k$ was done in [14]. In fact the dual case was actually done first, in [15]: in this case one uses an element $\Omega \in H \otimes H$, which is the formal dual of a cocycle, and twists the comultiplication of H to get a new Hopf algebra H^Ω. In this case, if A is an H-module algebra, then A can be twisted to a new algebra A_Ω which is then an H^Ω-module algebra.

In this paper, we first review the above facts in a precise way in Section 2. In Section 3, we first show that being Noetherian or affine is always preserved by twisting. We then show (Theorem 3.5) that if we can twist the algebra to a super-commutative algebra, then affine will always imply Noetherian, a kind of "Hilbert Basis Theorem". In this situation it then follows that the subalgebra of coinvariants A^{coH} is also affine, provided H is cosemisimple, by applying a result of [29]. We also show that A will be integral over the algebra of central coinvariants if A can be twisted to a PI-ring A_σ in which integrality holds; this will always be true if A_σ is super-commutative and H_σ is commutative.

In Section 4 we discuss results of Etingof and Gelaki for (co)triangular Hopf algebras, and of Bahturin, Fischman, Kochetov, and the present author for cocommutative Hopf algebras, which prove the existence of 2-cocycles with nice properties. These results enable us to twist so-called H-commutative algebras A to super-commutative algebras, and thus to apply the work of the previous section. Finally in Section 5 we consider polynomial identites, and study when the (co)invariants satisfying an identity imply that the algebra must also satisfy one.

One application of our results is to give a new proof of the main results of [13]. They prove there that if H is a semisimple cotriangular Hopf algebra over $k = \mathbb{C}$, and A is an H-comodule algebra which is k-affine and H-commutative, then A^{coH} is also k-affine. In their proof, they first prove that A is integral over A^{coH} (using a generalized determinant argument) and then show that A^{coH} is affine; as a consequence, A is Noetherian. It is unknown in general whether an H-commutative affine algebra is Noetherian.

Our work here gives a different proof of their results, as well as a similar result for cocommutative Hopf algebras in any characteristic; see Theorem 3.5 and Corollaries 3.7, 4.12, and 4.13. That is, we use the results noted above that any affine algebra A which can be twisted to a super-commutative algebra is Noetherian, and that the ring of (co)invariants A^{coH} is affine if H is cosemisimple. The hypothesis that H is semisimple cotriangular Hopf algebra over $\mathbb{C}$ and A is H-commutative is a special case of this situation, since by [16] H can be twisted to $(kG)^*$ for some group G, and since A is H-commutative once can see that the corresponding twist of A is super-commutative.

ALGEBRA PROPERTIES INVARIANT UNDER TWISTING

Some references are [29] for Hopf algebras, [23] for twistings, and [26] for basic ring theory properties. We use here the abbreviated form of the summation notation for $\triangle$; that is, $\Delta(h) = \sum h_1 \otimes h_2$.

The author would like to thank S. Raianu for pointing out Proposition 2.7, Yu. Bahturin for general discussions about twisting algebras, and D. Fischman for discussions about integrality and the paper [13].

2. PRELIMINARIES ON TWISTINGS

We first review the idea of a Hopf algebra twisted by a cocycle or dual cocycle. A reference for these definitions is [23, 10.2.3].

First, recall that for a Hopf algebra H, a *2-cocycle* on H is a convolution-invertible map $\sigma : H \otimes H \longrightarrow k$ satisfying, for all $h, l, m \in H$ the equality

$$\sum \sigma(h_1 l_1, m)\sigma(h_2, l_2) = \sum \sigma(h, l_1 m_1)\sigma(l_2, m_2) \tag{2.1}$$

We assume also that σ is normal, that is,

$$\sigma(h, 1) = \sigma(1, h) = \varepsilon(h)$$

for all $h \in H$. We note that this definition goes back to [36]. The cocycle above is usually called a *right* 2-cocycle; a left 2-cocycle satisfies the condition analogous to (2.1) with all the 1,2 subscripts reversed. Thus when H is cocommutative they are the same. The inverse of a right cocycle is a left cocycle, and conversely.

Similarly, let Ω be an invertible element in $H \otimes H$. Write $\Omega = \sum \Omega^1 \otimes \Omega^2$. Then Ω is a *(dual) 2-cocycle* for H if

$$[(\Delta \otimes id)(\Omega)](\Omega \otimes 1) = [(id \otimes \Delta)(\Omega)](1 \otimes \Omega). \tag{2.2}$$

Moreover we assume also that $(id \otimes \varepsilon)(\Omega) = (\varepsilon \otimes id)(\Omega) = 1$.

These conditions are just the formal duals of (2.1) and normality of an ordinary left 2-cocycle. When there is no confusion, we will also call Ω a cocycle. In [16], Ω is called a "twist". However, we reserve that term for the twisted algebras and Hopf algebras themselves; this seems more consistent with older terminology.

We first deform the multiplication of H, via σ.

DEFINITION 2.3. Let $\sigma : H \otimes H \rightarrow k$ be a 2-cocycle for H. The bialgebra H_σ is called a *(cocycle) twist* of H with respect to σ if:

(i) $H_\sigma = H$ as a coalgebra;

(ii) H_σ has new multiplication given by

$$h \cdot_\sigma l := \sum \sigma^{-1}(h_1, l_1)h_2 l_2 \sigma(h_3, l_3).$$

H_σ becomes a Hopf algebra by defining a new antipode via

$$S_\sigma(h) = \sum \sigma^{-1}(h_1, Sh_2)Sh_3\sigma(Sh_4, h_5).$$

We may also twist the multiplication of an H-comodule algebra.

DEFINITION 2.4. Let A be a right H-comodule algebra and let σ be a 2-cocycle for H. Then the *twisted algebra* A_σ has multiplication

$$a \cdot_\sigma b = \sum \sigma(a_1, b_1)a_0 b_0$$

for all $a, b \in A$.

232   S. MONTGOMERY

Moreover, A_σ is a right H_σ-comodule algebra, using the same coaction of H on the vector space A.

We next deform the comultiplication of H, via Ω.

DEFINITION 2.5. Let Ω be a (dual) cocycle for H. The bialgebra H^Ω is called a *cotwist* of H with respect to Ω if:
  (i) $H^\Omega = H$ as an algebra;
  (ii) H^Ω has comultiplication $\Delta^\Omega(h) := \Omega^{-1}(\Delta h)\Omega$.

H^Ω becomes a Hopf algebra, by defining an antipode which is the dual to that on H_σ.

Analogously to Definiton 2.4, we may twist the multiplication of an H-module algebra.

DEFINITION 2.6. Let A be a left H-module algebra and let Ω be a (dual) cocycle for H. Then the *twisted algebra* A_Ω has multiplication

$$a \cdot_\Omega b := \sum (\Omega^1 \cdot a)(\Omega^2 \cdot b).$$

for all $a, b \in A$.

Moreover, A_Ω is a left H^Ω-module algebra, using the same action of H on the vector space A.

Note that if H is cocommutative, then $H_\sigma = H$, and if H is commutative, then $H^\Omega = H$. If H is finite-dimensional, then $(H^*)^\Omega = (H_\sigma)^*$. For in that case, $(H \otimes H)^* \cong H^* \otimes H^*$, and σ corresponds to an invertible element $\Omega \in H^* \otimes H^*$. The correspondence between σ and Ω is given by

$$\sigma(h, l) = \sum \Omega^1(h)\Omega^2(l).$$

Moreover the two new structures on A coincide. That is, if A is a right H-comodule algebra, then it is a left H^*-module algebra in the usual way, and

$$\begin{aligned} a \cdot_\sigma b &= \sum \sigma(a_1, b_1)a_0 b_0 = \sum \Omega^1(a_1)a_0 \Omega^2(b_1)b_0 \\ &= \sum (\Omega^1 \cdot a)(\Omega^2 \cdot b) = a \cdot_\Omega b. \end{aligned}$$

Thus when H is finite-dimensional, we may consider A as an H-comodule or as an H^*-module, whichever is more convenient.

We will also need the following very useful fact about isomorphisms of smash products, due to Majid.

PROPOSITION 2.7. [25, Proposition 2.9] *Let A be a left H-module algebra and let $\Omega \in H \otimes H$ be a (dual) cocycle. Then $A_\Omega \# H^\Omega \cong A \# H$ as algebras.*

The proof of the proposition follows from checking that the map $\Phi : A \# H \longrightarrow A_\Omega \# H^\Omega$ via $a \# h \mapsto \sum \Omega^1 \cdot a \# \Omega^2 h$ is an algebra isomorphism. In fact Majid shows something stronger, in which A is a Hopf algebra in the Yetter-Drinfel'd category for H, but we only need here the algebra isomorphism.

At this meeting, P. Schauenburg pointed out to us that Proposition 2.7 has a non-computational proof, by interpreting it as a statement about braided categories.

3. FINITENESS PROPERTIES AND SUPER-COMMUTATIVE ALGEBRAS

In this section we are concerned with algebraic properties of A which are invariant under twisting, when H is finite-dimensional. We first consider some properties which are invariant under any twistings, and then specialize to the case when some twist of A is super-commutative.

We usually state our results for an H-comodule algebra A twisted by a cocycle, although as noted in Section 2, this is equivalent to twisting an H^*-module algebra by a dual cocycle.

Let A_σ be any twisting of A. Clearly, for $a \in A, b \in A^{coH}$, $a \cdot_\sigma b = ab$ and $b \cdot_\sigma a = ba$, and so we may identify A^{coH} with $(A_\sigma)^{coH_\sigma}$.

Thus if $f : A \longrightarrow A_\sigma$ is the map which is the identity on the vector space A, then f is a (left and right) A^{coH}-module map and takes subcomodules to subcomodules. In particular if V, W are H-subcomodules of A, then $f(VW) = f(V)f(W)$. Moreover for any $b \in A^{coH}$, b is central in A if and only if $f(b)$ is central in A_σ. Thus, letting $Z(A)$ denote the center of A, we may also identify $Z(A) \cap A^{coH}$ with $Z(A_\sigma) \cap (A_\sigma)^{coH_\sigma}$.

For any $a \in A$, we write $\bar{a} = f(a) \in A_\sigma$; if $b \in A^{coH}$, we will usually just write b for $\bar{b} = f(b)$.

We prove a basic result about twistings. Recall that A is *affine* over k if A is finitely-generated as a k-algebra.

PROPOSITION 3.1. *Let σ be a cocycle on H and let A be an H-comodule algebra. Then*

(1) A is affine $\iff$ A_σ is affine;

(2) if I is a subcomodule of A, then I is an ideal of A $\iff$ $f(I)$ is an ideal of A_σ. Moreover, I is nilpotent $\iff$ $f(I)$ is nilpotent, and thus A is H-semiprime $\iff$ A_σ is H_σ-semiprime.

Assume also that H is finite-dimensional. Then

(3) A is (left) Noetherian $\iff$ A_σ is (left) Noetherian.

(4) [24] A satisfies a polynomial identity (PI) $\iff$ A_σ satisfies a PI.

Proof. (1) We first assume that $A = k[a_1, \ldots, a_m]$. Let V be a finite-dimensional H-subcomodule of A such that $a_i \in V$, for all i. Then $A = k[V] = \sum_{j \geq 0} kV^j$. Since all of the V^j are subcomodules, they are preserved in A_σ; that is, $A_\sigma = k[V_\sigma]$. Thus A_σ is also affine. The converse is similar.

(2) Assume I is an ideal and a subcomodule, and let $\bar{a} \in A_\sigma, \bar{b} \in f(I)$. Then $\bar{a} \cdot \bar{b} = \sum \sigma(a_1, b_1)a_0 b_0 \in f(I)$ since all $b_0 \in I$. Thus $f(I)$ is an ideal of A_σ. Similarly if I is nilpotent of index n, and $\bar{a}_1, \bar{a}_2, \ldots, \bar{a}_n \in f(I)$, then $\bar{a}_1 \bar{a}_2 \cdots \bar{a}_n \in \sum k(a_1)_0 (a_2)_0 \cdots (a_n)_0 = 0$. Thus $f(I)$ is nilpotent. The last fact follows from the definition of H-semiprime; that is, A has no non-zero H-stable nilpotent ideals.

Now assume H is finite-dimensional. Then A is a left H^*-module algebra and $\sigma = \Omega \in (H \otimes H)^* \cong H^* \otimes H^*$ is a dual cocycle for H^*. By Proposition 2.7, we know $A\#H^* \cong A^\Omega \#(H^*)^\Omega$.

Now $A\#H^*$ is a finite free A-module, and so $A\#H^*$ is Noetherian if and only if A is Noetherian. Thus (3) follows. Similarly if A satisfies a PI, then also $A\#H$ satisfies a PI, and so (4) follows. $\square$

234   S. MONTGOMERY

The nilpotent part of (2) was shown in [24], and (4) is [24, Lemma 6.1.2], using [6]. The present proof of (4) was pointed out to us by Yu. Bahturin. We note that (4) greatly improves [7, 2.6], where it was proved for algebras graded by a finite group G. Polynomial identities will be considered in more detail in the last section.

We now consider the following question: if A is an H-comodule algebra, and A is integral over a central subring of A^{coH}, when is this property preserved under twisting?

Although we do not know the answer in the general case, we give one result here. We use a consequence of a theorem of Shirshov [26, 13.8.9, p 476] on rings satisfying a polynomial identity. It says that if R is a PI-ring and K is a central subring, then R is affine and integral over K $\iff$ R is a finite K-module.

PROPOSITION 3.2. *Let H be finite-dimensional, let A be an H-comodule algebra, and let B be a subalgebra of $Z(A) \cap A^{coH}$. Assume that there exists a cocycle σ on H such that A_σ is a PI ring and that A_σ is integral over $B_\sigma = B$. Then A is integral over B.*

Proof. First, B_σ is central in A_σ since B is central in A, from the remarks at the beginning of this section.

Now assume that A_σ is integral over B. Choose any $a \in A$ and let $\bar{a} = f(a)$ denote the corresponding element in A_σ. Now $\bar{a} \in C$, a finite dimensional H-subcomodule of A_σ, so if A'_σ is the B-subalgebra of A_σ generated by C, then A'_σ is B-affine. Also A'_σ is a PI ring and integral over the central subring B. Apply Shirshov to see that A'_σ is a finite B-module. Then $f^{-1}(A'_\sigma) \subset A$ is a finite B-module. Thus $a \in f^{-1}(A'_\sigma)$ is integral over B. $\qquad\Box$

We next show that when A can be twisted to a super-commutative algebra, most of our desired properties hold.

DEFINITION 3.3. The algebra A is *super-commutative* if $A = A_0 \oplus A_1$ is a $\mathbb{Z}_2$-graded algebra and for a, b homogeneous elements in A,

$$ab = (-1)^{|a||b|}ba.$$

A is *H-super-commutative* if A is an H-comodule algebra, A is super-commutative, and each A_i is an H-subcomodule.

First note that if A is super-commutative, then A_0 is always in the center of A and $a_1^2 = 0$ for all $a_1 \in A_1$.

The next lemma is well-known; we include a proof for completeness.

LEMMA 3.4. *Let $A = A_0 \oplus A_1$ be a super-commutative algebra. Then*
(1) A is a PI ring;
(2) if A is affine over the field k, then A is a finite A_0-module, A_0 is affine, and A is Noetherian.
(3) if A_0 is integral over a subalgebra $B \subset A_0$, then A is also integral over B.

Proof. (1) One may check directly that A satisfies the identity

$$[[x, y], z] = 0.$$

(2) First note that A is integral over A_0 of index 2. For, choose any $a = a_0 + a_1 \in A$, where $a_i \in A_i$. Then $a^2 - 2a_0 a + a_0^2 = 0$.

ALGEBRA PROPERTIES INVARIANT UNDER TWISTING

Next, A is a finite A_0-module. For, we may assume the generators of A are homogeneous. Let $\{c_1, \ldots, c_r\}$ be the generators of even degree and $\{d_1, \ldots, d_s\}$ be the generators of odd degree. Then A is generated as an A_0-module by the ordered monomials $\{d_{i_1} \cdots d_{i_k}\}$, where $1 \le i_1 < \ldots < i_k \le s$.

Moreover A_0 is finitely-generated, by all $\{c_i, d_j d_k\}$ with $j < k$. Alternatively one could use the Artin-Tate lemma [26, p 481] to see that A_0 is finitely-generated. Thus by the classical Hilbert Basis theorem, A_0 is Noetherian. But then A is Noetherian as it is a finite A_0-module.

(3) Choose any $a = a_0 + a_1 \in A$, and let $A' := B\langle a_0, a_1 \rangle$ be the subalgebra of A generated by a_0 and a_1 over the (central) subring B. Then A' is B-affine, $\mathbb{Z}_2$-graded, and super-commutative, and moreover $(A')_0$ is integral over B. As in (2), A' is a finite $(A')_0$-module and $(A')_0$ is B-affine. Thus $(A')_0$ is a finite B-module, and so A' is a finite B-module. It follows that a lies in a finite B-module, and so is integral over B. $\qquad\square$

We will apply the above results to show that a kind of "Hilbert Basis Theorem" and "Noether's Theorem" hold when A can be twisted to a super-commutative algebra. We use the generalization of Noether's theorem in [29, 4.3.7]: if A is a left Noetherian H-module algebra A with surjective trace which is k-affine, then A^H is k-affine.

By "surjective trace", we mean that if $\Lambda \ne 0$ is an integral in H, then the map $\tilde{\Lambda} : A \to A^H$ given by $a \mapsto \Lambda \cdot a$ is surjective. This map extends the usual trace map for group actions, and the above theorem extends the arguments for groups in [30]. The trace will always be surjective if H is semisimple.

We note also that if A is a Noetherian H-module algebra with surjective trace, then A^H is always Noetherian [29, 4.3.5].

If A is an H-comodule algebra and H is finite-dimensional, we say that A has a *total integral* for H if, dualizing to the action of H^* on A, the trace map from A to A^{H^*} is surjective.

THEOREM 3.5. *Let H be a finite-dimensional Hopf algebra and let A be an H-comodule algebra. Assume that there exists a cocycle σ on H such that A_σ is H_σ-super-commutative. Then*

(1) If A is k-affine, then A is Noetherian.

(2) If A is k-affine and there is a total integral for the coaction of H on A, then the algebra of coinvariants A^{coH} is affine.

(3) If A_σ is integral over a subalgebra $B \subset Z(A_\sigma) \cap (A_\sigma)^{coH_\sigma}$, then A is also integral over B.

Proof. (1) By Proposition 3.1(3), it suffices to show that for some cocycle σ on H, the twisted algebra A_σ is Noetherian. By assumption we can find a σ such that A_σ is H_σ-super-commutative. Thus A_σ is Noetherian by Lemma 3.4(2).

(2) This follows from (1) and the generalization of Noether's theorem in [29] mentioned above.

(3) This follows from Proposition 3.2, since A_σ is PI by Lemma 3.4(1). $\qquad\square$

We need an extension of a classical result.

LEMMA 3.6. *Let H be finite-dimensional and let A be an H-comodule algebra. Assume that H is commutative and that $A = A_0 \oplus A_1$ is H-super-commutative. Then A is integral over $A_0 \cap A^{coH}$.*

Proof. If R is a commutative H-comodule algebra, then it is known that R is integral over R^{coH}. This follows from an old result of Grothendieck. Another proof has been given by Ferrer-Santos; see [29, 4.2.1]. Thus $R = A_0$ is integral over $R^{coH} = A_0 \cap A^{coH}$. The Lemma follows from this fact and Lemma 3.4(3) applied to $B = A_0 \cap A^{coH}$. $\qquad\square$

COROLLARY 3.7. *Let H be finite-dimensional and let A be an H-comodule algebra. Assume that for some cocycle σ on H, H_σ is commutative and A_σ is H_σ-super-commutative. Then A is integral over a central subring B of A^{coH}.*

Proof. Let $B = (A_\sigma)_0 \cap A_\sigma^{coH}$. Then A_σ is integral over B by Lemma 3.6, and so A is integral over B by Theorem 3.5(3). Also B is central in A since it is central in A_σ. $\qquad\square$

We will see in the next section that there are several natural situations in which the hypotheses in Corollary 3.7 are satisfied.

Note that we only required H to be commutative in Lemma 3.6 and Corollary 3.7 in order to know that the commutative ring R was integral over R^{coH}. The question as to whether a commutative H-comodule algebra R is integral over its coinvariants for an arbitrary finite-dimensional H seems to be open. It is known if $H = kG$, that is R is G-graded, by a special case of a result of G. Bergman (see below), and it is known for any H provided R is Noetherian and algebraically closed, by work of A. Braun [12].

If R is not commutative then it is known to be false for group actions if $|G|R = 0$, even using the more general notion of "fully integral". More positively, Bergman showed that an arbitrary G-graded ring R is fully integral over the identity component (see [32]). Another result in the positive direction is due to D. Quinn, who showed the analogous result if H was commutative and cosemisimple [34]. There may be some hope here if H is cosemisimple.

One might also ask whether we could prove results analogous to Theorem 3.5 and Corollary 3.7 by twisting A to an algebra more general than one which is supercommutative. Certainly Lemma 3.4 is known in much greater generality; for example, (2) is true if A is graded by any finite group G, provided A_1 satisfies a polynomial identity. For (3), some additional assumptions would be needed, as the previous paragraph shows. However as a beginning one might try assuming that A_σ is a PI ring with a large central subring.

4. EXISTENCE OF COCYLES AND H-COMMUTATIVITY

In this section we discuss when we can find cocycles to twist A to a nicer algebra. We will see that this can be done if H is either cocommutative with a skew-symmetric bicharacter or cotriangular and semisimple, and in addition if the original algebra A is H-commutative.

We first review a theorem of Scheunert [35]. Let $H = kG$ and let β be a bicharacter on G; that is, $\beta : G \times G \to k^*$ and β is multiplicative in both entries.

ALGEBRA PROPERTIES INVARIANT UNDER TWISTING

β is called *skew-symmetric* if $\beta(g, h)\beta(h, g) = 1$ for all $g, h \in G$. When β is skew-symmetric, we see that $G = G_+ \cup G_-$, where $G_+ = \{g \in G | \beta(g, g) = 1\}$ and $G_- = \{g \in G | \beta(g, g) = -1\}$. Note that G_+ is a subgroup of G of index at most 2.

As an example, we note that whenever a group G has a subgroup G_+ of index at most 2, we may define the *sign bicharacter* β_0 on G. That is, for any $g, h \in G$,

$$\beta_0(g, h) = \begin{cases} -1 & \text{if } g, h \in G_- \\ +1 & \text{otherwise} \end{cases} \tag{4.1}$$

THEOREM 4.2. [35] *Given a finite abelian group G with a skew-symmetric bicharacter β, there exists a 2-cocycle σ on G such that for all $g, h \in G$,*

$$\beta(g, h) = \sigma(h, g)^{-1}\beta_0(g, h)\sigma(g, h)$$

where β_0 is the sign bicharacter as above.

Although Scheunert only proved this result for finite abelian groups, it is not difficult to see that it extends to all groups [7],[33]. He then used this result to prove that any G-Lie color algebra L can be twisted to an ordinary (G-graded) Lie superalgebra L_σ.

A basic question is to what extent Scheunert's theorem on the existence of cocycles extends to more general Hopf algebras, and consequently to twistings of H-comodule algebras. There has been considerable recent progress in this direction; we next summarize work of [4], [5], and [16], [17].

We first need the definition of a bicharacter.

DEFINITION 4.3. Let H be a Hopf algebra.

(a) A function $\beta \colon H \otimes H \to k$ is called a *bicharacter on H* if β is bilinear, convolution invertible, and for all $h, k, l \in H$,
  (i) $\beta(hk, l) = \sum \beta(h, l_1)\beta(k, l_2)$
  (ii) $\beta(h, kl) = \sum \beta(h_2, k)\beta(h_1, l)$
  (iii) β is normal, i.e. $\beta(h, 1) = \beta(1, h) = \varepsilon(h)$.
(b) β is called *skew-symmetric* if $\beta^{-1} = \beta \circ \tau$.

As examples, first note that for any H, $\beta(h, k) = \varepsilon(h)\varepsilon(k)$ is the *trivial* bicharacter. If $(H, \langle\,,\,\rangle)$ is a coquasitriangular Hopf algebra, then $\beta = \langle\,,\,\rangle$ is a bicharacter, and if H is cotriangular, then β is skew-symmetric. When H is cocommutative, any bicharacter is a cocycle on H, although not conversely.

As another example, we may extend the notion of the sign bicharacter. Consider the case when H is $\mathbb{Z}_2$-graded as a Hopf algebra; that is, $H = H_+ \oplus H_-$ as a $\mathbb{Z}_2$-graded algebra, and H_+ and H_- are subcoalgebras of H. Then we may define the *sign bicharacter* β_0 on H as follows: for any homogeneous $h, k \in H$

$$\beta_0(h, k) = \begin{cases} -\varepsilon(h)\varepsilon(k) & \text{if } h, k \in H_- \\ \varepsilon(h)\varepsilon(k) & \text{otherwise} \end{cases} \tag{4.4}$$

and extend β_0 linearly to H. Note that $\beta_0^{-1} = \beta_0$. One can then verify that β_0 is a skew-symmetric bicharacter on H. See [4, (3.19)].

We may now state more precisely what we mean by extending Scheunert's theorm to Hopf algebras. Letting $\tau : H \otimes H \to H \otimes H$ be the usual flip map and letting $*$

238   S. MONTGOMERY

denote the convolution product, the question becomes for which H, β is there a $\mathbb{Z}_2$ grading on H and a 2-cocycle σ such that

$$\beta = (\sigma \circ \tau)^{-1} * \beta_0 * \sigma. \tag{4.5}$$

This is known in two cases. The first is when H is cocommutative.

THEOREM 4.6. [4][5] *Assume that H is cocommutative and that β is a skew-symmetric bicharacter on H. Then there exists a $\mathbb{Z}_2$-grading of H and a 2-cocycle σ on H such that (4.5) holds.*

The case in which H is generated by group-like and primitive elements is shown in [4], and the general case in [5]. In fact the $\mathbb{Z}_2$-grading on H can be defined directly from β, as follows. Let $u(h) = \sum \beta(h_1, Sh_2)$ and define

$$H_+ = \{h \in H | \sum u(h_2)h_1 = h\} \text{ and } H_- = \{h \in H | \sum u(h_2)h_1 = -h\}.$$

u is the analog of the Drinfel'd element in a cotriangular Hopf algebra.

One can also solve for a cocycle is when H is cotriangular.

THEOREM 4.7. [16] *Let k be algebraically closed of characteristic 0 and assume that H is semisimple and cotriangular via the form $\beta = \langle \, , \, \rangle$. Then there exists a $\mathbb{Z}_2$-grading of H and a 2-cocycle σ on H such that (4.5) holds.*

Moreover there exists a finite group G such that $H \cong ((kG)^*)_\sigma$, the twist of the dual of a group algebra.

The result in [16] is not stated in this form, as it was proved for triangular Hopf algebras. More generally, [18] considered the case of certain infinite-dimensional cotriangular Hopf algebras H. In order to state this result, we first define, for any element $c \in H^*$,

$$R_c := \frac{1}{2}[\varepsilon \otimes \varepsilon + c \otimes \varepsilon + \varepsilon \otimes c - c \otimes c] \in H^* \otimes H^*.$$

R_c may be considered as an element of $(H \otimes H)^*$.

THEOREM 4.8. [18] *Let k be algebraically closed of characteristic 0. Assume that (H, β) is cotriangular and that for every finite-dimensional subcoalgebra C of H, $tr(S^2|_C) = dim(C)$. Then there exists a proalgebraic group G, a central group-like element $c \in H^*$ of order ≤ 2 and a 2-cocycle σ on H such that, as cotriangular Hopf algebras,*

$$(H, \beta) \cong (\mathcal{O}(G)_\sigma, (\sigma \circ \tau)^{-1} * R_c * \sigma)$$

where $\mathcal{O}(G)$ is the function algebra on G.

We check that this specializes to Theorem 4.7. First, the $\mathbb{Z}_2$-grading of H can be recovered explicitly, as follows. Since $c \in G(H^*)$ has order ≤ 2, the map $\Phi : H \to H$ given by $\Phi(h) = c \rightharpoonup h$ is an automorphism of H of order ≤ 2. Thus if we define

$$H_+ = \{h \in H | c \rightharpoonup h = h\} \text{ and } H_- = \{h \in H | c \rightharpoonup h = -h\},$$

we clearly have that $H = H_+ \oplus H_-$ as algebras. Moreover H_+ and H_- are H-subcomodules since c is central. It is easy to see that R_c coincides with the sign bicharacter on H, using the fact that for any H and $f \in H^*$, $f(h) = \varepsilon(f \rightharpoonup h)$, for all $h \in H$. Finally if H is finite-dimensional and semisimple, then $S^2 = id$ by the Larson-Radford theorems and so the condition on S holds.

ALGEBRA PROPERTIES INVARIANT UNDER TWISTING

The proofs in both [16] and [18] use a rather difficult theorem of Deligne on semisimple categories. We note that the proofs in [4], which in fact preceded [18], are relatively elementary.

We now turn to H-commutative algebras.

DEFINITION 4.9. Let A be an H-comodule algebra and let β be a bicharacter on H. Then A is (H, β)-commutative if for all $a, b \in A$, $ab = \sum \beta(a_1, b_1)b_0 a_0$.

In [13], such algebras are called *quantum commutative* when H is cotriangular.

LEMMA 4.10. *Let H have a bicharacter β and let A be an H-comodule algebra. Assume that σ is a cocycle on H and let $\beta' := (\sigma \circ \tau)^{-1} * \beta * \sigma$. Then A is (H, β)-commutative $\iff A_\sigma$ is (H_σ, β')-commutative.*

In particular if A is (H, β)-commutative and $\beta' = \beta_0$, then A_σ is H_σ-super-commutative.

Proof. We write ab for the usual product in A and $a \cdot_\sigma b$ for the product in A_σ, as before. Then

$$
\begin{aligned}
a \cdot_\sigma b &= \sum \sigma(a_1, b_1)a_0 b_0 = \sum \sigma(a_2, b_2)\beta(a_1, b_1)b_0 a_0 \\
&= \sum (\beta * \sigma)(a_1, b_1)b_0 a_0 = \sum [(\sigma \circ \tau) * \beta'](a_1, b_1)b_0 a_0 \\
&= \sum \beta'(a_2, b_2)\sigma(b_1, a_1)b_0 a_0 = \sum \beta'(a_1, b_1)b_0 \cdot_\sigma a_0.
\end{aligned}
$$

For the last statement, assume that $\beta = \beta_0$; then H is $\mathbb{Z}_2$-graded by definition (4.4). Thus if A is an H-comodule algebra via $\delta : A \to A \otimes H$, we may define $A_0 := \delta^{-1}(A \otimes H_+)$ and $A_1 := \delta^{-1}(A \otimes H_-)$ to see that A is $\mathbb{Z}_2$-graded. It follows that if A is (H, β_0)-commutative, then A is H-super-commutative. $\square$

COROLLARY 4.11. *Let H have a skew-symmetric bicharacter β and let A be an H-comodule algebra which is (H, β)-commutative. Assume that either:*

(a) H is cocommutative, or

(b) (H, β) is semisimple, cotriangular, and k is algebraically closed of characteristic 0.

Then there exists a cocycle σ on H such that A_σ is super-commutative.

Proof. This follows from Lemma 4.10 and Theorems 4.6 and 4.7. $\square$

We are now able to give a proof of the main results of [13].

COROLLARY 4.12. [13] *Assume that k is algebraically closed of characteristic 0 and that (H, β) is cotriangular and semisimple. If A is a k-affine H-comodule algebra which is (H, β)-commutative, then A is Noetherian and A^{coH} is k-affine. Moreover, A is integral over A^{coH}.*

Proof. The fact that A is Noetherian and A^{coH} is affine follows from Theorem 3.5 and Corollary 4.11(b); A has a total integral for H since H is cosemisimple. For integrality, one may use Corollary 3.7 since $H_\sigma = (kG)^*$ is commutative. Alternatively a direct proof of integrality for A_σ may be given; see below. $\square$

In the case of a group G acting on a super-commutative algebra A, [17, Remark 1.2] gives a direct proof of Lemma 3.6 which extends the classical result for groups acting on commutative rings. That is, they note that for any $a = a_0 + a_1$, a satisfies the polynomial

$$p(y) = \prod_{g \in G} (y - g \cdot a_0)^2 \in (A_0)^G[y].$$

[17] use this fact and give a proof of the integrality result in [13] similar to that in Corollary 4.12. The present argument seems slightly simpler than theirs since it stays entirely inside the algebras and does not require using any category morphisms. However, although both our proof and the proof in [17] are shorter than the one in [13], they are not more elementary since they both depend on [16], which in turn depends on Deligne's theorem.

The next result seems to be new.

COROLLARY 4.13. *Assume that H is a finite-dimensional cocommutative Hopf algebra with a skew-symmetric bicharacter β. Let A be an H-comodule algebra which is (H, β)-commutative.*
(1) If A is k-affine, then A is Noetherian.
(2) If A is k-affine and A has a total integral for H, then A^{coH} is k-affine.
(3) If H is commutative, then A is integral over A^{coH}.

Proof. The proof of (1) and (2) proceeds as in the previous corollary, using Corollary 4.11(a), since in (2) we assume there is a total integral. For(3), note that since H is cocommutative, $H_\sigma = H$ for any cocycle σ, and so H_σ is both commutative and cocommutative. We now apply Corollary 3.7. $\qquad\square$

If A is not (H, β)-commutative, then some additional hypotheses are needed for a "Noether's theorem" result. Even for actions of $H = kG$, counterexamples have been known for a long time if the characteristic of k divides $|G|$: if A is Noetherian or affine, the fixed algebra A^G need not be Noetherian or affine [30]. For Hopf algebras, there are already difficulties in characteristic 0. A recent example of L. W. Small shows that it is possible for $H = H_4$, Sweedler's 4-dimensional Hopf algebra, to act on an affine Noetherian ring A of characteristic 0 such that the invariant ring A^H is neither affine nor Noetherian. In these examples, H is triangular, although not semisimple.

5. POLYNOMIAL IDENTITIES

Our last main topic is related to the following question: if H is finite-dimensional and A is an H-module algebra such that A^H satisfies a PI, must A also satisfy a PI?

This question was originally raised for group actions by J.-E. Bjork in 1971. Asssuming that $|G|^{-1} \in k$, it was shown to be true for solvable groups by the present author in [27], and was proved for arbitrary finite groups by Kharchenko in [22]; see also [28, p.89]. However if $|G|A = 0$ it is known to be false; in fact by an example of Bergman, there is an action of a finite group G on the algebra $A = M_2(F)$ of matrices over a free algebra F, but $A^G = k$ (see [28, p. 6]). By another theorem of Kharchenko [21], A^G being a PI ring implies A is also a PI ring with no assumption on the order of G, provided the algebra has no nilpotent elements. See also [28], where some shorter proofs are given.

ALGEBRA PROPERTIES INVARIANT UNDER TWISTING

The question is also true for graded rings, that is for actions of $H = (kG)^*$; this was shown in [9],using [31]; a better bound is given in [8].

The case of actions of general Hopf algebras was considered in [6], where some interesting equivalent conditions are given.

THEOREM 5.1. [6] *Let H be a finite-dimensional Hopf algebra. Then the following conditions are equivalent:*

(1) for every H-module algebra A such that A^H is PI, also A is PI;

(2) for every H-module algebra A such that A^H is nilpotent, also A is nilpotent;

(3) for every $k \in \mathbb{N}$, the ideal of the tensor algebra $T(H)$ generated by the k-th power of $T(H)^H$ has finite codimension in $T(H)$.

Moreover any of these three conditions imply that H is semisimple.

Recently Linchenko showed another case of the question, using twisting. Here we give a proof using 3.1(4); Linchenko's original proof used Proposition 3.1(2) and Theorem 5.1(2).

PROPOSITION 5.2. [24, Corollary 6.1.1] *Let k be algebraically closed of characteristic 0, and let H be triangular and semisimple. Assume that A is an H-module algebra and that A^H satisfies a PI. Then A satisfies a PI.*

Proof. By using [17] as before, we see that $H^\Omega \cong kG$ acts on A_Ω with $(A_\Omega)^{H^\Omega} = (A_\Omega)^G \cong A^H$ a PI ring. By Kharchenko's theorem [21], it follows that A_Ω satisfies a PI. By Proposition 3.1(4), A is PI. □

We show here that the question can have a positive answer for certain non-semisimple Hopf algebras, when the algebra A is a prime ring. We use [10, Theorem 7.1], which says that if R is a prime ring with a skew derivation d, and if the invariant ring $R^{\langle d \rangle} = \{r \in R | d(r) = 0\}$ satisfies a PI, then the ring is PI.

LEMMA 5.3. *Let H be a Taft Hopf algebra of dimension n^2 and let A be a prime H-module algebra of characteristic relatively prime to n. Then A^H being PI implies that A is PI.*

Proof. We recall that $H = k\langle g, x | g^n = 1, x^n = 0, xg = \omega gx \rangle$, where ω is a primitive n^{th} root of 1 in k, and where g is group-like and x is a g-skew primitive. Thus g acts as an automorphism of A and x acts as a g-skew derivation of A. In fact, notice that g stabilizes $B = A^{\langle x \rangle}$ and that $B^{\langle g \rangle} := A^H$.

Now if A^H is PI, then B is PI by Kharchenko's result (actually one only needs here the easier result for solvable groups [27]). We are now done by the theorem of [10] stated above. □

The proof of Lemma 5.3 extends to more general Hopf algebras, with an additional restriction on A. We consider *modified supergroup algebras*, as described in [1]. That is, consider a finite-dimensional vector space V and a finite group G acting on V. Let $\wedge V$ denote the exterior algebra and let $\mathcal{H} = \wedge V \# kG$. Then $\mathcal{H}$ becomes a cocommutative Hopf superalgebra by letting V be odd, G be even, and each $\dot{x} \in V$ be (graded) primitive; this is what Kostant calls a supergroup. To describe the modified supergroup algebra, assume that G contains a central element g of order 2 such that $gxg = -x$ for all $x \in V$. Define H by letting $H = \mathcal{H}$ as an algebra, but changing the comultiplication on $\mathcal{H}$ as follows: let $\Delta_H(y) = \Delta_{\mathcal{H}}(y)$ for all $y \in G$ as before, but define $\Delta_H(x) := x \otimes 1 + g \otimes x$ for all $x \in V$. With this definition, (H, Δ_H) becomes an ordinary Hopf algebra, the modified supergroup algebra.

Now let $K := \wedge V \# k\langle g \rangle$; then K is a normal Hopf subalgebra of H, with quotient Hopf algebra $H/HK^+ \cong k(G/\langle g \rangle)$.

The Taft algebra of dimension 4 (that is, Sweedler's example) is the smallest non-trivial example of a such a supergroup algebra.

PROPOSITION 5.4. *Let k be a field of characteristic not 2, let H be a modified supergroup algebra, and let A be an H-module algebra which is a domain. If A^H is PI, then A is PI.*

Proof. Let V have basis $\{x_1, \ldots, x_n\}$. Then $x_i x_j = -x_j x_i$ and $g x_i = -x_i g$ for all i, j. Each of the x_i acts as a skew derivation of A. Thus we may apply the result of [10] to the chain of subrings

$$A \supset A^{\langle x_1 \rangle} \supset \cdots A^{\langle x_1, \ldots, x_n \rangle} = A^{\wedge V}.$$

Note that at each stage, each subring is stable under the action of the next x_i, and all the subrings are domains since by our hypothesis A is a domain. Thus we see that if $A^{\wedge V}$ satisfies a PI, then so does A.

But now for any $\alpha \in G = G(H)$, $\alpha x_i \alpha^{-1}$ is also a skew primitive element of H and thus G stabilizes $A^{\wedge V}$; moreover $A^H = (A^{\wedge V})^G$. Now by the result of Kharchenko for rings with no nilpotent elements, if $(A^{\wedge V})^G$ satisfies a PI, then so does $A^{\wedge V}$. Thus A is PI , by the previous paragraph. $\qquad \square$

There may be some hope of extending Proposition 5.4 to all triangular Hopf algebras. For, recently Etingof and Gelaki have shown that when $k = \mathbb{C}$ and H is triangular, then some twist of H is a modified supergroup algebra [19]. One could try to follow the proof of Proposition 5.4, twisting as in Proposition 5.2. However, the twist of a domain is not a domain, in general, and if A is not a domain, then we cannot apply the inductive argument to the chain of subrings as in Proposition 5.4. Thus we do not know the answer to this question.

It would be of interest to look at the invariant theory of actions of supergroups, or more generally of the braided graded (pointed) Hopf algebras considered in [2]. For (co)actions on a domain A, some of the questions in Section 4 might work as well.

REFERENCES

[1] N. Andruskiewitsch, P. Etingof, and S. Gelaki, Triangular Hopf algebras with the Chevalley property, *Michigan J. Math* 49 (2001), 277-298.

[2] N. Andruskiewitsch and H.-J. Schneider, Pointed Hopf algebras, in *New Directions in Hopf Algebras*, MSRI Publ. Vol 43, Cambridge University Press, 2002.

[3] M. Artin, W. Schelter, and J. Tate, Quantum deformations of GL_n, *Comm. Pure and Applied Math* 44 (1991), 879–895.

[4] Y. Bahturin, D. Fischman, and S. Montgomery, Bicharacters, twistings, and Scheunert's theorem for Hopf algebras, *J. Algebra* 236 (2001), 246-276.

[5] Y. Bahturin, M. Kochetov, and S. Montgomery, Polycharacters of cocommutative Hopf algebras, *Canadian Math. Bull.* 45 (2002), 11–24 .

[6] Y. Bahturin and V. Linchenko, Identities of algebras with actions of Hopf algebras, *J. Algebra* 202 (1998), 634–654.

ALGEBRA PROPERTIES INVARIANT UNDER TWISTING

[7] Y. Bahturin and S. Montgomery, PI-envelopes of Lie superalgebras, *Proc. Amer. Math. Soc.* **127** (1999), 2829–2939 .
[8] Y. Bahturin and M. Zaicev, Identities of graded algebras, *J. Algebra* **205** (1998), 1–12.
[9] J. Bergen and M. Cohen, Actions of commutative Hopf algebras, *Bull. London Math. Soc.* **18** (1986), 159–164.
[10] J. Bergen and P. Grzeszczuk, Skew derivations whose invariants satisfy a polynomial identity, *J. Algebra* **228** (2000), 710–737.
[11] N. Bourbaki, *Algebra*, Hermann, Paris, 1970 (English edition).
[12] A. Braun, Hopf algebra versions of two classical finite group action theorems, unpublished manuscript, 1991.
[13] M. Cohen, S. Westreich, and S. Zhu, Determinants, integrality, and Noether's theorem for Hopf algebras, *Israel J. Math* **96** (1996), 185–222.
[14] Y. Doi, Braided bialgebras and quadratic bialgebras, *Comm. Algebra* **12** (1993), 1731–1749.
[15] V. G. Drinfeld, Almost cocommutative Hopf algebras, *Leningrad Math. J.* **1** (1990), 341–342.
[16] P. Etingof and S. Gelaki, Some properties of finite-dimensional semisimple Hopf algebras, *Math. Research Letters* **5** (1998), 191–197.
[17] P. Etingof and S. Gelaki, A method of construction of finite-dimensional triangular semisimple Hopf algebras, *Math. Research Letters* **5** (1998), 551-561.
[18] P. Etingof and S. Gelaki, On cotriangular Hopf algebras, *Amer. J. Math.* **123** (2001), 699-713.
[19] P. Etingof and S. Gelaki, The classification of finite-dimensional triangular Hopf algebras over an algebraically closed field of characteristic 0, *Moscow Math J.* **3** (2003), 37–43.
[20] C. Kassel, *Quantum Groups*, GTM 155, Springer-Verlag, 1995.
[21] V. K. Kharchenko, Generalized identities with automorphisms, *Algebra i Logika* **14** (1975), 215–237 (in Russian; English translation 1976, 132–148).
[22] V. K. Kharchenko, Fixed elements under a finite group acting on a semiprime ring, *Algebra i Logika* **14** (1975), 328-344 (in Russian; English translation 1976, 409–417).
[23] A. Klimyk and K. Schmüdgen, *Quantum Groups and their Representations*, Springer, Texts and Monographs in Physics, 1997.
[24] V. Linchenko, Some properties of Hopf algebras and Hopf module algebras, Ph D thesis, University of Southern California, 2001.
[25] S. Majid, Quasi-* structure on *q*-Poincare algebras, *J. Geometry and Physics* **22** (1997), 14–58.
[26] J. C. McConnell and J. C. Robson, *Noncommutative Noetherian rings*, 2nd Edition, AMS Graduate Studies in Math, Vol 30, AMS, Providence, RI, 2001.
[27] S. Montgomery, Centralizers satisfying polynomial identities, *Israel J.Math.* **18** (1974), 207–219.
[28] S. Montgomery, *Fixed Rings of Finite Automorphism Groups of Associative Rings*, Lecture Notes in Math, Vol. 818, Springer, Berlin, 1980.
[29] S. Montgomery, *Hopf Algebras and their Actions on Rings*, CBMS Lectures, Vol. 82, AMS, Providence, RI, 1993.
[30] S. Montgomery and L. W. Small, Fixed rings of Noetherian rings, *Bull. Lond. Math. Soc.* **13** (1981), 33-38.
[31] S. Montgomery and M. K. Smith, Algebras with a separable subalgebra whose centralizer satisfies a polynomial identity, *Comm. Algebra* **3** (1975), 151–168.
[32] D. S. Passman, Fixed rings and integrality, *J. Algebra* **68** (1981), 510–519.
[33] H. Pop, A generalization of Scheunert's theorem on cocycle twisting of Lie color algebras, preprint, q-alg 9703002.
[34] D. Quinn, Integral extensions of non-commutative rings, *Israel J. Math.* **73** (1991), 113–121.
[35] M. Scheunert, Generalized Lie algebras, *J. Math. Physics* **20** (1979), 712–720.
[36] M. E. Sweedler, Cohomology of algebras over Hopf algebras, *Trans. Amer. Math. Soc.* **127** (1968), 205–239.
[37] J. J. Zhang, Twisted graded algebras and equivalences of graded categories, *Proc. London Math. Soc.* **72** (1996), 281–311.

Quantum SL(3, ℂ)'s: the missing case

CHRISTIAN OHN   Laboratoire de Mathématiques, I.S.T.V.
Université de Valenciennes, Le Mont Houy, F-59313 Valenciennes Cedex 9, France
e-mail: *christian.ohn@univ-valenciennes.fr*

> ABSTRACT. We study the only missing case in the classification of quantum
> SL(3, ℂ)'s undertaken in our paper [*J. of Algebra* **213** (1999), 721–756], thereby
> completing this classification.

1. INTRODUCTION

The aim of this paper is to complete the classification of quantum SL(3)'s undertaken in [8].

Roughly speaking, we call a quantum SL(3) any Hopf algebra (over ℂ) whose (finite-dimensional) comodules are "similar" to the modules of the (ordinary) group SL(3, ℂ) (see Definition 2.1). Given such a Hopf algebra $\mathcal{A}$, it will have, among other things, two simple (nonisomorphic) comodules V, W of dimension 3, and the usual decomposition rules for tensor products will imply the existence of $\mathcal{A}$-comodule morphisms (1), satisfying certain compatibility conditions (2). The tuple $\mathcal{L}_{\mathcal{A}}$, consisting of those two comodules and eight morphisms, will be called the *basic quantum* SL(3) *datum* (BQD for short) associated to $\mathcal{A}$.

Conversely, starting from an "abstract" BQD $\mathcal{L}$ (i.e. where V, W are just vector spaces and (1) just linear maps satisfying the aforementioned conditions), we may reconstruct a Hopf algebra $\mathcal{A}_{\mathcal{L}}$ via the usual Tannakian procedure.

We may now state our main result.

THEOREM A.

(a) *If $\mathcal{A}$ is a quantum* SL(3)*, then $\mathcal{L}_{\mathcal{A}}$ is a BQD.*
(b) *If $\mathcal{L}$ is a BQD, then $\mathcal{A}_{\mathcal{L}}$ is a quantum* SL(3)*.*
(c) *The correspondences $\mathcal{A} \mapsto \mathcal{L}_{\mathcal{A}}$ and $\mathcal{L} \mapsto \mathcal{A}_{\mathcal{L}}$ are inverse of each other between quantum* SL(3)*'s (up to Hopf algebra isomorphism) and BQDs (up to equivalence; see Definition 2.2).*
(d) *BQDs can be explicitly classified up to equivalence, yielding a classification of quantum* SL(3)*'s up to Hopf algebra isomorphism.*

An almost complete version of this theorem was stated and proved in [8]. More precisely, we found a classification of all BQDs, except for one class related to elliptic curves, and we proved the theorem for all BQDs outside this class. In the

---

2000 *Mathematics Subject Classification.* Primary 20G42; Secondary 14M15, 16S37, 16S38.
Key words and phrases. quantum group, flag variety, noncommutative projective geometry, Sklyanin algebra.

present paper, we settle the study of this last class, thereby proving Theorem A in full.

Moreover, a geometric analysis of this class of BQDs yields the following contribution to a question raised in the Introduction of [2].

THEOREM B. *Let B be the quantum three-space associated to an arbitrary quantum SL(3). Then B cannot be a Sklyanin algebra. In other words, the scheme of point modules [3, 4] of B (which is a cubic divisor in $\mathbb{P}^2$) cannot be an elliptic curve.*

The paper is organized as follows. After some recollections from [8] in Section 2 (definition of a quantum SL(3) and of a BQD, and the correspondences between them), Section 3 recalls and studies the form of the only class of BQDs not covered by the results of [8]; this will finish the classification of BQDs. In Section 4, we introduce the shape algebra [8, Section 5] for this class of BQDs and we determine the associated flag variety (in the sense of [9]). These geometric data suggest to view the shape algebra as a twist (in the sense of [10]) of another algebra, which we show in Section 5 to be isomorphic to the shape algebra of another BQD, already covered by the results of [8]. In Section 6, we finish the proof of Theorem A by carrying over the necessary properties from the untwisted to the twisted shape algebra. Theorem B is proved in Section 7 by picking up some leftovers from Section 3.

In the Appendix, we show a result on twists of Koszul algebras that is needed in Section 6, but may also be of independent interest.

Conventions. We denote by $\mathbb{Z}$ (resp. $\mathbb{N}$, $\mathbb{C}$) the set of integers (resp. nonnegative integers, complex numbers). All vector spaces, algebras and tensor products are over $\mathbb{C}$.

2. RECOLLECTIONS FROM [8]

Recall that the group SL(3) is linearly reductive and that its simple modules are parametrized by their highest weights, which are pairs $(k, l) \in \mathbb{N}^2$. Recall further that the dimension of the simple module of highest weight (k, l) is given by $d_{(k,l)} := (k+1)(l+1)(k+l+2)/2$. For any $\lambda, \mu, \nu \in \mathbb{N}^2$, denote by $m_{\lambda\mu}^\nu$ the multiplicity of the simple module of highest weight ν inside the tensor product of those of (respective) highest weights λ and μ. (Recall also that $m_{\lambda\mu}^\nu$ can, in principle, be determined in a purely combinatorial way.)

Our main objects of interest may now be defined as follows (see [8]).

DEFINITION 2.1. We call a *quantum* SL(3) any (not necessarily commutative) Hopf algebra $\mathcal{A}$ (over $\mathbb{C}$) such that

(a) there is a family $\{V_\lambda \mid \lambda \in \mathbb{N}^2\}$ of simple and pairwise nonisomorphic $\mathcal{A}$-comodules, with $\dim V_\lambda = d_\lambda$,
(b) every $\mathcal{A}$-comodule is isomorphic to a direct sum of these,
(c) for every $\lambda, \mu \in \mathbb{N}^2$, $V_\lambda \otimes V_\mu$ is isomorphic to $\bigoplus_\nu m_{\lambda\mu}^\nu V_\nu$.

In particular, if we write $V := V_{(1,0)}$ and $W := V_{(0,1)}$, then Condition (c) implies the existence of $\mathcal{A}$-comodule morphisms

$$A : V \otimes V \to W, \qquad a : W \to V \otimes V$$
$$B : W \otimes W \to V, \qquad b : V \to W \otimes W$$
$$C : W \otimes V \to \mathbb{C}, \qquad c : \mathbb{C} \to V \otimes W \tag{1}$$
$$D : V \otimes W \to \mathbb{C}, \qquad d : \mathbb{C} \to W \otimes V,$$

each being unique up to a scalar. We showed in [8, Propositions 3.1 and 3.2] that these maps must, for an appropriate choice of these scalars, satisfy the following compatibility conditions:

$$(1_V \otimes C)(c \otimes 1_V) = 1_V, \qquad (D \otimes 1_V)(1_V \otimes d) = 1_V \tag{2a}$$
$$Aa = 1_W, \tag{2b}$$
$$C(A \otimes 1_V) = \omega\, D(1_V \otimes A), \qquad (1_V \otimes a)c = \omega\,(a \otimes 1_V)d \tag{2c}$$
$$\omega\,(C \otimes 1_V)(1_W \otimes a) = B, \qquad (A \otimes 1_W)(1_V \otimes c) = b \tag{2d}$$
$$Dc = \kappa\,1_{\mathbb{C}}, \qquad Cd = \kappa\,1_{\mathbb{C}} \tag{2e}$$

$$(1_V \otimes A)(a \otimes 1_V)(A \otimes 1_V)(1_V \otimes a) = \rho\,(1_{V \otimes W} + cD) \tag{2f}$$
$$(A \otimes 1_V)(1_V \otimes a)(1_V \otimes A)(a_V \otimes 1) = \rho\,(1_{W \otimes V} + dC), \tag{2g}$$

where

- ω is a 3rd root of unity,
- $\kappa = q^{-2} + 1 + q^2$ and $\rho = (q + q^{-1})^{-2}$ for some $q \in \mathbb{C}$, $q \neq 0$, with q^2 either 1 or not a root of unity.

DEFINITION 2.2. A *basic quantum* SL(3) *datum* (BQD for short) is a tuple $\mathcal{L} = (V, W, A, a, B, b, C, c, D, d)$ consisting of two vector spaces V and W of dimension 3, and of eight linear maps (1), satisfying Conditions (2a–g) (with ω, κ, q as indicated). Two BQDs are called *equivalent* if one can be obtained from the other through (any combination of) base change, rescaling of the maps (1), and interchanging $V \leftrightarrow W$, $A \leftrightarrow B$, $a \leftrightarrow b$, $C \leftrightarrow D$, $c \leftrightarrow d$.

Thus, each quantum SL(3), $\mathcal{A}$, gives rise to a BQD $\mathcal{L}_{\mathcal{A}}$ (which is really defined only up to equivalence).

Conversely, start with a BQD $\mathcal{L}$, and define an algebra $\mathcal{A}_{\mathcal{L}}$ with (9+9) generators t^i_j $(i, j = 1, 2, 3)$ and u^α_β $(\alpha, \beta = 1, 2, 3)$, and relations

$$A^\alpha_{ij}\, t^i_k t^j_\ell = u^\alpha_\beta\, A^\beta_{k\ell}, \qquad t^i_k t^j_\ell\, a^{k\ell}_\beta = a^{ij}_\alpha\, u^\alpha_\beta$$
$$B^i_{\alpha\beta}\, u^\alpha_\gamma u^\beta_\delta = t^i_j\, B^j_{\gamma\delta}, \qquad u^\beta_\gamma u^\beta_\delta\, b^{\gamma\delta} = b^{\alpha\beta}_i\, t^i_j$$
$$C_{\alpha i}\, u^\alpha_\beta t^i_j = C_{\beta j}, \qquad t^i_j u^\alpha_\beta\, c^{j\beta} = c^{i\alpha}$$
$$D_{i\alpha}\, t^i_j u^\alpha_\beta = D_{j\beta}, \qquad u^\alpha_\beta t^i_j\, d^{\beta j} = d^{\alpha i}.$$

(Here, we have chosen bases x_1, x_2, x_3 of V and y_1, y_2, y_3 of W, and set $A(x_i \otimes x_j) = A^\alpha_{ij} y_\alpha$, etc.) Then $\mathcal{A}_{\mathcal{L}}$ possesses a Hopf algebra structure given by

$$\Delta(t^i_j) = t^i_k \otimes t^k_j, \qquad \Delta(u^\alpha_\beta) = u^\alpha_\gamma \otimes u^\gamma_\beta$$
$$\varepsilon(t^i_j) = \delta^i_j, \qquad \varepsilon(u^\alpha_\beta) = \delta^\alpha_\beta \tag{3}$$
$$S(t^i_j) = c^{i\beta}\, u^\alpha_\beta\, C_{\alpha j}, \qquad S(u^\alpha_\beta) = d^{\alpha j}\, t^i_j\, D_{i\beta}.$$

248   C. OHN

The main difficulty is to prove that the Hopf algebra $\mathcal{A}_\mathcal{L}$ is indeed a quantum SL(3) in the sense of Definition 2.1. We were able to do this in [8] for all BQDs, except for one class, which we will describe in the next section.

3. CLASSIFICATION OF CASE I.h

If we set $Q^i_j := c^{i\alpha} D_{j\alpha}$, then (2a) implies $(Q^{-1})^i_j = d^{\alpha i} C_{\alpha j}$. Now it follows from (3) that
$$S^2(t^i_j) = Q^i_k t^k_l (Q^{-1})^l_j,$$
so S^2 is "encoded" by the linear map $Q : V \to V$.

In [8, Section 10], we used the possible Jordan normal forms of Q and the value of ω as first criteria for the classification of BQDs. One possibility, called *Type I* in [8], consists in taking $Q = 1_V$ and $\omega = 1$. The condition $Q = 1_V$ amounts to setting $W = V^*$, with C, c, D, d the obvious canonical maps. (In particular, we now have $S^2 = 1_A$, reflecting the fact that W is both the left dual and the right dual of V.) In this case, (2a) and (2e) are automatically satisfied (for $\kappa = 3$, so $q^2 = 1$ and $\rho = \frac{1}{4}$).

Now consider the "quantum determinants," appearing in (2c):
$$e := (1_V \otimes a)c = \omega(a \otimes 1_V)d : \mathbb{C} \to V \otimes V \otimes V$$
$$E := C(A \otimes 1_V) = \omega D(1_V \otimes A) : V \otimes V \otimes V \to \mathbb{C}.$$

If $\mathcal{L}$ is of Type I, then Conditions (2c) imply that
$$e = \lambda + s, \qquad E = \Lambda + S,$$
with λ, Λ totally antisymmetric and s, S totally symmetric. In particular, choosing dual bases x_i in V and y_α in $W(= V^*)$, we may view s and S as homogeneous polynomials of degree 3, i.e. as cubic curves in the projective plane $\mathbb{P}^2$ and in its dual plane $\mathbb{P}^{2*}$, respectively (unless $s = 0$ or $S = 0$).

Using the standard classification of cubic curves in $\mathbb{P}^2$, we were then able in [8] to classify all possible forms of e and E that satisfy Conditions (2bfg), except in the case where s is an elliptic curve.

In this particular case, called *Case I.h* in [8], the bases of V and W may be chosen in such a way that a reads
$$a : y_1 \mapsto \alpha x_2 \otimes x_3 + \beta x_3 \otimes x_2 + \gamma x_1 \otimes x_1$$
$$y_2 \mapsto \alpha x_3 \otimes x_1 + \beta x_1 \otimes x_3 + \gamma x_2 \otimes x_2$$
$$y_3 \mapsto \alpha x_1 \otimes x_2 + \beta x_2 \otimes x_1 + \gamma x_3 \otimes x_3,$$
with $\gamma \neq 0$ and $(\alpha + \beta)^3 + \gamma^3 \neq 0$ (so that s is indeed elliptic). It then follows from (2b) that A must read
$$A : x_1 \otimes x_1 \mapsto \gamma' y_1, \qquad x_1 \otimes x_2 \mapsto \alpha' y_3, \qquad x_1 \otimes x_3 \mapsto \beta' y_2$$
$$x_2 \otimes x_1 \mapsto \beta' y_3, \qquad x_2 \otimes x_2 \mapsto \gamma' y_2, \qquad x_2 \otimes x_3 \mapsto \alpha' y_1$$
$$x_3 \otimes x_1 \mapsto \alpha' y_2, \qquad x_3 \otimes x_2 \mapsto \beta' y_1, \qquad x_3 \otimes x_3 \mapsto \gamma' y_3,$$
with
$$\alpha\alpha' + \beta\beta' + \gamma\gamma' = 1. \tag{4}$$

Moreover, Conditions (2fg) now read

$$P_0 := \alpha^2 \alpha'^2 + \beta^2 \beta'^2 + \gamma^2 \gamma'^2 - 2\alpha\alpha'\beta\beta' - 2\alpha\alpha'\gamma\gamma' - 2\beta\beta'\gamma\gamma' = 0$$
$$P_1 := \alpha^2 \beta'\gamma' + \beta^2 \alpha'\gamma' + \gamma^2 \alpha'\beta' = 0 \tag{5}$$
$$P_2 := \alpha'^2 \beta\gamma + \beta'^2 \alpha\gamma + \gamma'^2 \alpha\beta = 0.$$

PROPOSITION 3.1. *Up to base change, we may assume that $\alpha = \alpha' = 0$.*

Proof. Case 1 : $\alpha = 0$ or $\alpha' = 0$. Substituting into P_2 or P_1 shows that $\alpha = \alpha' = 0$.

Case 2 : $\beta = 0$ or $\beta' = 0$. Permuting two of the three basis vectors takes us back to Case 1.

Case 3 : $\gamma' = 0$ (recall that $\gamma = 0$ has been ruled out by ellipticity of s). Substituting into P_1 shows that $\alpha' = 0$ or $\beta' = 0$, which takes us back to Case 1 or Case 2.

Case 4 : none of $\alpha, \alpha', \beta, \beta', \gamma, \gamma'$ equals zero. Viewing P_0, P_1, P_2 as polynomials in α', their resultants must vanish:

$$0 = \mathrm{Res}_{\alpha'}(P_0, P_1) = \begin{vmatrix} \alpha^2 & -2\alpha(\beta\beta' + \gamma\gamma') & (\beta\beta' - \gamma\gamma')^2 \\ \beta^2\gamma' + \gamma^2\beta' & \alpha^2\beta'\gamma' & 0 \\ 0 & \beta^2\gamma' + \gamma^2\beta' & \alpha^2\beta'\gamma' \end{vmatrix}$$

$$= \beta^2\gamma^4\beta'^4 + 2(\alpha^3 + \beta^3 - \gamma^3)\beta\gamma^2\beta'^3\gamma'$$
$$+ (\alpha^6 + \beta^6 + \gamma^6 + 2\alpha^3\beta^3 + 2\alpha^3\gamma^3 - 4\beta^3\gamma^3)\beta'^2\gamma'^2$$
$$+ 2(\alpha^3 + \gamma^3 - \beta^3)\beta^2\gamma\beta'\gamma'^3 + \beta^4\gamma^2\gamma'^4$$

$$=: Q_1$$

$$0 = \mathrm{Res}_{\alpha'}(P_1, P_2) = \begin{vmatrix} \beta^2\gamma' + \gamma^2\beta' & \alpha^2\beta'\gamma' & 0 \\ 0 & \beta^2\gamma' + \gamma^2\beta' & \alpha^2\beta'\gamma' \\ \beta\gamma & 0 & \alpha(\beta'^2\gamma + \gamma'^2\beta) \end{vmatrix}$$

$$= \alpha\gamma^5\beta'^4 + 2\alpha\beta^2\gamma^3\beta'^3\gamma' + (\alpha^3 + \beta^3 + \gamma^3)\alpha\beta\gamma\beta'^2\gamma'^2 + 2\alpha\beta^3\gamma^2\beta'\gamma'^3 + \alpha\beta^5\gamma'^4$$

$$=: Q_2.$$

(These resultants make sense, because the leading coefficients of P_0, P_1, P_2 are nonzero: in particular, if we had $\beta^2\gamma' + \gamma^2\beta' = 0$, then substituting into P_1 would imply $\alpha = 0$, $\beta' = 0$, or $\gamma' = 0$.)

Viewing Q_1, Q_2 as polynomials in β', their resultant must again vanish:

$$0 = \mathrm{Res}_{\beta'}(Q_1, Q_2) = (\text{a } 6 \times 6 \text{ determinant})$$
$$= \gamma'^{16}(\alpha\beta\gamma)^{10}\left[(\alpha^3 + \beta^3 + \gamma^3)^3 - (3\alpha\beta\gamma)^3\right]. \tag{6}$$

(The author confesses not to have computed this 6×6 determinant by hand!)

Therefore, $\alpha^3 + \beta^3 + \gamma^3 = 3\zeta\alpha\beta\gamma$ for some 3rd root of unity ζ, i.e.

$$(\zeta\gamma + \alpha + \beta)(\zeta\gamma + j\alpha + j^2\beta)(\zeta\gamma + j^2\alpha + j\beta) = 0,$$

where j is a primitive 3rd root of unity. But $\zeta\gamma + \alpha + \beta = 0$ is ruled out by ellipticity of the curve s, so we may assume that $\zeta\gamma + j\alpha + j^2\beta = 0$ (exchanging $j \leftrightarrow j^2$ if necessary).

Now consider the following change of basis in V:

$$\begin{cases} x_1' = \zeta x_1 + x_2 + x_3 \\ x_2' = \zeta x_1 + j x_2 + j^2 x_3 \\ x_3' = \zeta x_1 + j^2 x_2 + j x_3 \end{cases}$$

(together with the dual basis y_1', y_2', y_3' in W). Then the map a reads

$$(3\zeta)\ a : y_1' \mapsto (\zeta\gamma + j^2\alpha + j\beta)\, x_3' \otimes x_2' + (\zeta\gamma + \alpha + \beta)\, x_1' \otimes x_1'$$
$$y_2' \mapsto (\zeta\gamma + j^2\alpha + j\beta)\, x_1' \otimes x_3' + (\zeta\gamma + \alpha + \beta)\, x_2' \otimes x_2'$$
$$y_3' \mapsto (\zeta\gamma + j^2\alpha + j\beta)\, x_2' \otimes x_1' + (\zeta\gamma + \alpha + \beta)\, x_3' \otimes x_3',$$

so we are taken back to Case 1. $\qquad\square$

Substituting $\alpha = \alpha' = 0$ back into (4) and (5) now yields $\beta\beta' = \gamma\gamma' = \frac{1}{2}$. Now we still have one degree of freedom to rescale the maps a and A, so we are left with one essential parameter, say

$$t := -\frac{\gamma}{\beta} = -\frac{\beta'}{\gamma'}.$$

This completes the classification of BQDs, undertaken in [8, Section 10].

4. THE SHAPE ALGEBRA AND ITS FLAG VARIETY

Given a BQD $\mathcal{L}$, recall [8, Section 5] that its *shape algebra* $\mathcal{M}_\mathcal{L}$ is generated by x_i ($i = 1, 2, 3$) and y_α ($\alpha = 1, 2, 3$), with defining relations

$$a_\alpha^{ij} x_i x_j = 0, \qquad\qquad c^{i\alpha} x_i y_\alpha = 0$$
$$b_i^{\alpha\beta} y_\alpha y_\beta = 0, \qquad\qquad y_\alpha x_i = -(q + q^{-1}) a_\alpha^{jk} A_{ki}^\beta x_j y_\beta$$

This algebra has a natural $\mathbb{N}^2$-grading $\mathcal{M}_\mathcal{L} = \bigoplus_{(k,l)\in\mathbb{N}^2} V_{(k,l)}$, with the x_i of degree $(1, 0)$ and the y_α of degree $(0, 1)$. (Moreover, it carries a natural $\mathcal{A}_\mathcal{L}$-comodule algebra structure, with each $V_{(k,l)}$ becoming an $\mathcal{A}_\mathcal{L}$-comodule: these are the natural candidates to show that $\mathcal{A}_\mathcal{L}$ satisfies Definition 2.1.)

When $\mathcal{L}$ is the BQD of Case I.h (with $\alpha = \alpha' = 0$ and $\beta\beta' = \gamma\gamma' = \frac{1}{2}$; cf. Section 3), the defining relations of $\mathcal{M}_\mathcal{L}$ read

$$x_3 x_2 = t x_1^2, \qquad t y_2 y_3 = y_1^2$$
$$x_1 x_3 = t x_2^2, \qquad t y_3 y_1 = y_2^2$$
$$x_2 x_1 = t x_3^2, \qquad t y_1 y_2 = y_3^2$$

$$y_1 x_1 = x_2 y_2, \qquad y_2 x_1 = t x_2 y_3, \qquad t y_3 x_1 = x_2 y_1 \tag{7}$$
$$t y_1 x_2 = x_3 y_2, \qquad y_2 x_2 = x_3 y_3, \qquad y_3 x_2 = t x_3 y_1$$
$$y_1 x_3 = t x_1 y_2, \qquad t y_2 x_3 = x_1 y_3, \qquad y_3 x_3 = x_1 y_1$$

$$x_1 y_1 + x_2 y_2 + x_3 y_3 = 0.$$

Now let us determine the flag variety of $\mathcal{M}_\mathcal{L}$ as defined in [9]. To do this, modify Relations (7) as follows: in each relation, adorn the right factor of each term with

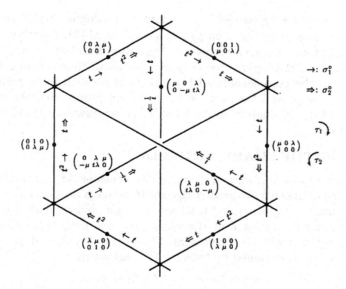

FIGURE 1. Flag variety for Case I.h, with a generic point on each irreducible component (top row in each matrix: coordinates in $\mathbb{P}^2$; bottom row: coordinates in $\mathbb{P}^{2*}$). Single and double arrows represent σ_1° and σ_2°, respectively; each arrow multiplies μ by the coefficient next to it. The automorphisms τ_1, τ_2 are "rotations by 120 degrees" in the directions shown.

a $'$ if the left factor is an x_i, and with a $''$ if the left factor is a y_α (e.g. the first relation on the fourth line now reads $y_1 x_1'' = x_2 y_2'$).

View $(x_1 : x_2 : x_3)$ and $(y_1 : y_2 : y_3)$ as homogeneous coordinates in $\mathbb{P}^2$ and in $\mathbb{P}^{2*}$, respectively. Since Relations (7) are $\mathbb{N}^2$-homogeneous, their modified version may now be seen as defining equations for a subscheme

$$\Gamma \subset (\mathbb{P}^2 \times \mathbb{P}^{2*}) \times (\mathbb{P}^2 \times \mathbb{P}^{2*}) \times (\mathbb{P}^2 \times \mathbb{P}^{2*}).$$

One easily checks that Γ is of the form

$$\Gamma = \{(p, \sigma_1(p), \sigma_2(p)) \mid p \in X\}$$

for some subscheme $X \subset \mathbb{P}^2 \times \mathbb{P}^{2*}$ and some automorphisms σ_1, σ_2 of X: indeed, the scheme X is the variety with nine irreducible components pictured in Figure 1, and the automorphisms σ_1, σ_2 naturally decompose into $\sigma_1 = \tau_1 \sigma_1^\circ$, $\sigma_2 = \tau_2 \sigma_2^\circ$, where τ_1, τ_2 are the automorphisms of $\mathbb{P}^2 \times \mathbb{P}^{2*}$ given by

$$\tau_1 : \begin{pmatrix} x_1 & x_2 & x_3 \\ y_1 & y_2 & y_3 \end{pmatrix} \mapsto \begin{pmatrix} x_3 & x_1 & x_2 \\ y_3 & y_1 & y_2 \end{pmatrix}, \quad \tau_2 : \begin{pmatrix} x_1 & x_2 & x_3 \\ y_1 & y_2 & y_3 \end{pmatrix} \mapsto \begin{pmatrix} x_2 & x_3 & x_1 \\ y_2 & y_3 & y_1 \end{pmatrix},$$

and where $\sigma_1^\circ, \sigma_2^\circ$ are also described in Figure 1, e.g. $\sigma_1^\circ : \begin{pmatrix} 0 & 1 & 0 \\ 0 & \lambda & \mu \end{pmatrix} \mapsto \begin{pmatrix} 0 & 1 & 0 \\ 0 & \lambda & t^2\mu \end{pmatrix}$ and $\sigma_2^\circ : \begin{pmatrix} 0 & 1 & 0 \\ 0 & \lambda & \mu \end{pmatrix} \mapsto \begin{pmatrix} 0 & 1 & 0 \\ 0 & \lambda & t\mu \end{pmatrix}.$

252   C. OHN

INFORMAL REMARK 4.1. Since $\mathcal{M}_\mathcal{L}$ is a quantum analogue of the multihomogeneous coordinate ring of the (ordinary) flag variety of SL(3), nontrivial characters of $\mathcal{M}_\mathcal{L}$ should, in some sense, correspond to "quantum Borel subgroups" of $\mathcal{A}_\mathcal{L}$. On the other hand, such characters obviously correspond to simultaneous fixed points of σ_1, σ_2. But in the case considered here, there is no such fixed point: so once Theorem A will be proved, we will have a quantum group with the same representations as SL(3), although in some sense, it has *no* quantum Borel subgroup to induce them from!

5. TWISTING THE SHAPE ALGEBRA

The natural decompositions $\sigma_i = \tau_i \sigma_i^\circ$ suggest to consider $\mathcal{M}_\mathcal{L}$ as the twist R^τ (in the sense of [10]) of another $\mathbb{N}^2$-graded algebra R, whose associated flag variety will be $(X, \sigma_1^\circ, \sigma_2^\circ)$ instead of (X, σ_1, σ_2) (this works by an obvious multigraded version of [4, Proposition 8.9], noting that the τ_1, τ_2 commute with each other and with $\sigma_1^\circ, \sigma_2^\circ$). The algebra R will also be generated by x_i ($i = 1, 2, 3$) and y_α ($\alpha = 1, 2, 3$), with defining relations obtained by twisting (7) backwards:

$$x_3 x_1 = t\, x_1 x_3, \qquad t\, y_2 y_1 = y_1 y_2$$
$$x_1 x_2 = t\, x_2 x_1, \qquad t\, y_3 y_2 = y_2 y_3$$
$$x_2 x_3 = t\, x_3 x_2, \qquad t\, y_1 y_3 = y_3 y_1$$

$$y_1 x_2 = x_2 y_1, \qquad y_2 x_2 = t\, x_2 y_2, \qquad t\, y_3 x_2 = x_2 y_3 \qquad (8)$$
$$t\, y_1 x_3 = x_3 y_1, \qquad y_2 x_3 = x_3 y_2, \qquad y_3 x_3 = t\, x_3 y_3$$
$$y_1 x_1 = t\, x_1 y_1, \qquad t\, y_2 x_1 = x_1 y_2, \qquad y_3 x_1 = x_1 y_3$$

$$x_1 y_3 + x_2 y_1 + x_3 y_2 = 0.$$

A straightforward computation shows that relations (8), with y_1, y_2, y_3 renamed to y_2, y_3, y_1, turn out to define the shape algebra $\mathcal{M}_{\mathcal{L}^\circ}$, where $\mathcal{L}^\circ$ is the BQD of Type I defined by the maps

$$a : y_1 \mapsto \lambda\, x_2 \otimes x_3 + \mu\, x_3 \otimes x_2$$
$$y_2 \mapsto \lambda\, x_3 \otimes x_1 + \mu\, x_1 \otimes x_3$$
$$y_3 \mapsto \lambda\, x_1 \otimes x_2 + \mu\, x_2 \otimes x_1$$

$$A : \; x_1 \otimes x_1 \mapsto 0, \qquad x_1 \otimes x_2 \mapsto \lambda' y_3, \qquad x_1 \otimes x_3 \mapsto \mu' y_2$$
$$x_2 \otimes x_1 \mapsto \mu' y_3, \qquad x_2 \otimes x_2 \mapsto 0, \qquad x_2 \otimes x_3 \mapsto \lambda' y_1$$
$$x_3 \otimes x_1 \mapsto \lambda' y_2, \qquad x_3 \otimes x_2 \mapsto \mu' y_1, \qquad x_3 \otimes x_3 \mapsto 0$$

(Case I.e in [8, Section 10]), with $\lambda\lambda' = \mu\mu' = \frac{1}{2}$ and $t = -\frac{\mu}{\lambda} = -\frac{\lambda'}{\mu'}$.

6. PROOF OF THEOREM A

In [8], all intermediate results in the (almost complete) proof of Theorem A are valid for an arbitrary BQD $\mathcal{L}$, except for [8, Proposition 5.3] (and the resulting [8, Corollary 5.4]), which relies on a case by case analysis that is not valid in Case I.h. More precisely, the proof of Theorem A will be complete if we show the following facts for this particular case:

(a) the shape algebra $\mathcal{M}_\mathcal{L}$ is a Koszul algebra (when viewed as an N-graded algebra via the total grading),

(b) $\dim V_{(k,l)} = d_{(k,l)}$ for all $(k,l) \in \mathbb{N}^2$.

Since the BQD $\mathcal{L}^\circ$ introduced in Section 5 is already covered by [8], Properties (a) and (b) are true for $\mathcal{M}_{\mathcal{L}^\circ}$.

Therefore, $\mathcal{M}_\mathcal{L}$, being a twist of $\mathcal{M}_{\mathcal{L}^\circ}$, also satisfies those two properties: for Property (b), this is automatic, and for Property (a), it follows from Proposition A.1 in the Appendix. Theorem A follows.

7. PROOF OF THEOREM B

To each BQD $\mathcal{L}$ are associated two "quantum three-spaces," namely the quadratic algebras $\mathcal{B}_\mathcal{L} := T(V)/(\operatorname{Im} a)$ and $\mathcal{C}_\mathcal{L} := T(W)/(\operatorname{Im} b)$. For Case I.h, $\mathcal{B}_\mathcal{L}$ is defined by the following relations:

$$\alpha\, x_2 x_3 + \beta\, x_3 x_2 + \gamma\, x_1^2 = 0$$
$$\alpha\, x_3 x_1 + \beta\, x_1 x_3 + \gamma\, x_2^2 = 0$$
$$\alpha\, x_1 x_2 + \beta\, x_2 x_1 + \gamma\, x_3^2 = 0,$$

(The defining relations of $\mathcal{C}_\mathcal{L}$ are similar.) This algebra is one of the regular algebras of dimension 3 studied in [1], where the matrix Q (see Section 3) also played a role in the classification of such algebras. Algebras for which $Q = 1$ (called of *Type A* in [1]) give rise to the cubic curve s in $\mathbb{P}^2$ as in Section 3, so let us call this curve the *AS-curve*. The AS-curve associated to $\mathcal{B}_\mathcal{L}$ is given by

$$\gamma(x_1^3 + x_2^3 + x_3^3) = 3(\alpha + \beta)x_1 x_2 x_3.$$

Another cubic curve in $\mathbb{P}^2$ has been associated to $\mathcal{B}_\mathcal{L}$ in [3], namely the scheme of its point modules, given by

$$(\alpha\beta\gamma)(x_1^3 + x_2^3 + x_3^3) = (\alpha^3 + \beta^3 + \gamma^3)x_1 x_2 x_3.$$

Call this curve the *ATV-curve* associated to $\mathcal{B}_\mathcal{L}$.

REMARK 7.1. The ATV-curve is defined for an arbitrary regular algebra of dimension 3, whereas the AS-curve is defined only for those of Type A. Note however, as observed in [3], that even when both curves are defined, they *do not* coincide in general!

Recall [3, 7] that $\mathcal{B}_\mathcal{L}$ is called a *Sklyanin algebra* if its ATV-curve is elliptic. But by Proposition 3.1, this is impossible: the ATV-curve of $\mathcal{B}_\mathcal{L}$ must degenerate to a triangle (which may also be viewed as the image of the flag variety X under the projection $\mathbb{P}^2 \times \mathbb{P}^{2^*} \to \mathbb{P}^2$).

On the other hand, a case by case analysis shows that none of the other BQDs, classified in [8, Section 10], can give rise to an elliptic ATV-curve. This proves Theorem B.

REMARK 7.2. The fact that, for the BQD $\mathcal{L}$ of Case I.h, the ATV-curve of $\mathcal{B}_\mathcal{L}$ cannot be elliptic was already visible in (6), which was obtained via an elimination procedure from Conditions (2fg). In turn, the latter were shown in [8, Section 3] to be related to the existence of an endomorphism of $V \otimes V$ satisfying the braid relation.

254 C. OHN

One would of course like to see a more direct and natural link, for an arbitrary BQD $\mathcal{L}$, between the braid relation and the fact that the ATV-curve of $\mathcal{B}_\mathcal{L}$ cannot be elliptic.

APPENDIX: MULTITWISTS PRESERVE THE KOSZUL PROPERTY

Let Γ be a monoid and $A = \bigoplus_{\gamma \in \Gamma} A_\gamma$ a Γ-graded algebra.

Assume that to each $\gamma \in \Gamma$, we associate a graded automorphism τ_γ of A, in such a way that $\tau_{\gamma\gamma'} = \tau_\gamma \tau_{\gamma'}$ for all $\gamma, \gamma' \in \Gamma$. (This is really only a special case of the notion introduced in [10], but it will be sufficient for our purposes: in Section 5, we take $\tau_{(k,l)} := \tau_1^k \tau_2^l$ for each $(k, l) \in \mathbb{N}^2$.)

Recall [10] that the *twisted* algebra A^τ is defined to be the vector space $\bigoplus_{\gamma \in \Gamma} A_\gamma$, endowed with the following new multiplication:

$$x * y := x \tau_\gamma(y) \qquad \text{for all } x \in A_\gamma, \ y \in A.$$

PROPOSITION A.1. *Consider A and A^τ as $\mathbb{N}$-graded algebras via some morphism $h : \Gamma \to \mathbb{N}$, and assume that they are generated by $A_1(= A_1^\tau) = \bigoplus_{h(\gamma)=1} A_\gamma$. Then A^τ is Koszul if and only if A is Koszul.*

REMARK A.2. If $\Gamma = \mathbb{N}$, then the categories of $\mathbb{N}$-graded modules of A and of A^τ are equivalent thanks to [10, Theorem 3.1], and the Koszul property is a homological property in this category, so the result is immediate.

However, for arbitrary Γ, [10, Theorem 3.1] shows that the categories of Γ-graded modules are equivalent, but this does not imply that those of $\mathbb{N}$-graded modules are. Therefore, a proof is still needed.

Proof of Proposition A.1. Of course, since $A = (A^\tau)^{\tau^{-1}}$, we only need to show one way, so assume that A is Koszul. Then A is quadratic, say $A = T(A_1)/(R)$ with $R \subset A_1 \otimes A_1$. Moreover, for each $k \geq 2$, the sublattice (w.r.t. $\cap$ and $+$) of $A_1^{\otimes k}$ generated by R (that is, by all the $A_1^{\otimes(i-1)} \otimes R \otimes A_1^{\otimes(k-i-1)}$) is distributive thanks to Backelin's criterion [5] (see also [6, Lemma 4.5.1]).

Now define $v : A_1^{\otimes k} \to A_1^{\otimes k}$ as follows: view $A_1^{\otimes k}$ as the direct sum of all subspaces $A_{\gamma_1} \otimes \ldots \otimes A_{\gamma_k}$ with $h(\gamma_1) = \cdots = h(\gamma_k) = 1$, and set

$$v_{|A_{\gamma_1} \otimes \ldots \otimes A_{\gamma_k}} := 1_{A_{\gamma_1}} \otimes \tau_{\gamma_1} \otimes (\tau_{\gamma_2} \tau_{\gamma_1}) \otimes \ldots \otimes (\tau_{\gamma_{k-1}} \ldots \tau_{\gamma_1}).$$

By construction, $A^\tau = T(A_1)/(v^{-1}(R))$, so A^τ is again quadratic. Moreover, the sublattice of $A_1^{\otimes k}$ generated by $v^{-1}(R)$ is the image under v^{-1} of that generated by R, so it is still distributive. Applying Backelin's criterion in the reverse direction, we conclude that A^τ is Koszul. $\qquad\square$

REFERENCES

[1] M. Artin and W. F. Schelter, Graded algebras of global dimension 3, *Adv. Math.* **66** (1987), 171–216.

[2] M. Artin, W. F. Schelter, and J. Tate, Quantum deformations of GL_n, *Comm. Pure Appl. Math.* **44** (1991), 879–895.

[3] M. Artin, J. Tate, and M. Van den Bergh, Some algebras associated to automorphisms of elliptic curves, in "The Grothendieck Festschrift," vol. I, Birkhäuser, Basel, 1990.

[4] M. Artin, J. Tate, and M. Van den Bergh, Modules over regular algebras of dimension 3, *Invent. Math.* **106** (1991), 335–388.

[5] J. Backelin, *A distributiveness property of augmented algebras and some related homological results*, Ph. D. Thesis, Stockholm, 1982.

[6] A. A. Beĭlinson, V. A. Ginsburg, and V. V. Schechtman, Koszul duality, *J. Geom. Phys.* **5** (1988), 317–350.

[7] A. V. Odesskiĭ and B. L. Feĭgin, Sklyanin's elliptic algebras, *Funct. Anal. Appl.* **23** (1990), 207–214.

[8] Ch. Ohn, Quantum $SL(3, \mathbb{C})$'s with classical representation theory, *J. Algebra* **213** (1999), 721–756.

[9] Ch. Ohn, "Classical" flag varieties for quantum groups: the standard quantum $SL(n, \mathbb{C})$, *Adv. Math.*, to appear.

[10] J. J. Zhang, Twisted graded algebras and equivalences of graded categories, *Proc. London Math. Soc.* **72** (1996), 281–311.

Cuntz Algebras and Dynamical Quantum Group $SU(2)$

A. PAOLUCCI Dipartimento di Scienza dell'Informazione
Universita' di Genova, Italy
e-mail: *paolucci@disi.unige.it*

> ABSTRACT. In [6, 7] G. Felder introduces a new area in the theory of quantum groups, the so-called theory of dynamical quantum groups. This theory assigns dynamical analogues to various objects related to ordinary Lie algebras and quantum groups, i.e. Hopf algebras, R-matrices, twists, etc. In this paper we study coactions of the dynamical quantum group $SU(2)$ on the $N + 1$-dimensional representation for the trigonometric dynamical R-matrix. A comodule algebra arising from these coactions turns out to be a dynamical analogue of the Cuntz algebra.

1. INTRODUCTION

In [4] and [5] an algebraic framework for studying dynamical R-matrices has been developed. The dynamical quantum groups associated to the latter are called **h**-Hopf algebroids, a notion introduced in [4] and motivated by the work of [8] and [7]. They are constructed from solutions of the dynamical Yang-Baxter equation in a manner analogous to the Faddeev-Reshetikin-Sklyanin-Taktahjan (FRST) construction.

An example of a dynamical quantum group constructed from a trigonometric dynamical R-matrix is studied in [9], thus generalizing the $SU(2)$ quantum group to the dynamical setting.

In this paper we study coactions of such **h**-Hopf algebroid, the $F_R(SU(2))$, on a comodule algebra. This comodule algebra, $\tilde{O}_{N+1}$, turns out to be a dynamical analogue of the Cuntz algebra O_{N+1}. These coactions are the natural generalization to the bialgebroid setting of the coactions of Hopf algebras on Cuntz algebras.

2. NOTATION AND PRELIMINARY RESULTS

Let us start by recalling the fundamental notions of **h**-algebra, **h**-algebroid and **h**-Hopf algebroid as in [9]. These structures are related to the more general Hopf-algebroids introduced by Lu [10].

Let **h** be a finite-dimensional complex vector space, $M_{\mathbf{h}^*}$ denote the field of meromorphic functions on the dual of **h** and $V = \bigoplus_{\alpha \in \mathbf{h}^*} V_\alpha$ a diagonalizable **h**-module.

2000 *Mathematics Subject Classification.* 20G42, 16W30, 81R50.

Key words and phrases. Cuntz algebras, Coaction, Dynamical Yang-Baxter Equation, Dynamical Quantum Group, Hopf-Algebroid.

In the case of dynamical quantum group, $\mathbf{h}$ is a Cartan subalgebra of the corresponding Lie algebra.

The quantum dynamical Yang-Baxter (QDYB) equation is written as

$$R^{12}\left(\lambda - h^{(3)}\right) R^{13}\left(\lambda\right) R^{23}\left(\lambda - h^{(1)}\right) = R^{23}\left(\lambda\right) R^{13}\left(\lambda - h^{(2)}\right) R^{12}\left(\lambda\right). \qquad (1)$$

This is an identity in the algebra of meromorphic functions $\mathbf{h}^* \longrightarrow \mathrm{End}\left(V \otimes V \otimes V\right)$. We denote by $R : \mathbf{h}^* \longrightarrow \mathrm{End}\left(V \otimes V\right)$ a meromorphic function, h denotes the action of $\mathbf{h}$, and the upper indices refer to the position in the tensor product. For instance $R^{12}\left(\lambda - h^{(3)}\right)$ denotes the operator

$$R^{12}\left(\lambda - h^{(3)}\right)\left(u \otimes v \otimes w\right) = \left(R\left(\lambda - \mu\right)\left(u \otimes v\right)\right) \otimes w, \quad w \in V_\mu.$$

We define a dynamical R-matrix to be a solution of the QDYB equation which is $\mathbf{h}$-invariant, i.e.

$$R : \mathbf{h}^* \longrightarrow \mathrm{End}_\mathbf{h}\left(V \otimes V\right).$$

We refer to [3] for the following construction. An $\mathbf{h}$-algebra is a complex associative algebra A with $\mathbf{1}$, bigraded over $\mathbf{h}^*$, $A = \bigoplus_{\alpha, \beta \in \mathbf{h}^*} A_{\alpha, \beta}$ and equipped with two algebra embeddings $\mu_l, \mu_r : M_{\mathbf{h}^*} \longrightarrow A_{00}$, called the left and the right moment maps, such that

$$\mu_l\left(f\right) a = a\mu_l\left(T_\alpha f\right), \quad \mu_r\left(f\right) a = a\mu_r\left(T_\beta f\right), \quad a \in A_{\alpha, \beta}, \quad f \in M_{\mathbf{h}^*},$$
$$\mu_l\left(f\right)\mu_r\left(g\right) = \mu_r\left(g\right)\mu_l\left(f\right), \quad f, g \in M_{\mathbf{h}^*},$$

where T_α denotes the automorphism $T_\alpha f\left(\lambda\right) = f\left(\lambda + \alpha\right)$ of $M_{\mathbf{h}^*}$.

A morphism of $\mathbf{h}$-algebras is an algebra homomorphism preserving the moment maps (and also the bigrading). The matrix tensor product $A \tilde{\otimes} B$ of two $\mathbf{h}$-algebras is the $\mathbf{h}^*$-bigraded vector space with

$$\left(A \tilde{\otimes} B\right)_{\alpha\beta} = \bigoplus_\gamma \left(A_{\alpha\gamma} \otimes_{M_{\mathbf{h}^*}} B_{\gamma\beta}\right),$$

where $\otimes_{M_{\mathbf{h}^*}}$ denotes the usual tensor product modulus the relations

$$\mu_r^A\left(f\right) a \otimes b = a \otimes \mu_l^B\left(f\right) b, \quad a \in A, \quad b \in B, \quad f \in M_{\mathbf{h}^*}.$$

It follows that

$$a\mu_r^A\left(f\right) \otimes b = a \otimes b\mu_l^B\left(f\right)$$

in $A \tilde{\otimes} B$.

The multiplication

$$\left(a \otimes b\right)\left(c \otimes d\right) = ac \otimes bd$$

and the moment maps

$$\mu_l^{A\tilde{\otimes}B}\left(f\right) = \mu_l^A\left(f\right) \otimes \mathbf{1}, \quad \mu_r^{A\tilde{\otimes}B}\left(f\right) = \mathbf{1} \otimes \mu_r^B\left(f\right)$$

make $A \tilde{\otimes} B$ into an $\mathbf{h}$-algebra.

We denote by $D_\mathbf{h}$ the algebra of difference operators on $M_{\mathbf{h}^*}$, consisting of operators

$$\sum_i f_i T_{\beta_i}, \quad f_i \in M_{\mathbf{h}^*}, \quad \beta_i \in \mathbf{h}^*.$$

This is an $\mathbf{h}$-algebra with bigrading defined by $fT_{-\beta} \in \left(D_\mathbf{h}\right)_{\beta\beta}$ and both the moment maps equal the natural embedding.

For any **h**-algebra A, there are canonical **h**-algebra isomorphisms $A \simeq A \tilde{\otimes} D_\mathbf{h} \simeq D_\mathbf{h} \tilde{\otimes} A$, defined by

$$x \simeq x \otimes T_{-\beta} \simeq T_{-\alpha} \otimes x, \quad x \in A_{\alpha\beta}. \tag{2}$$

Thus the algebra $D_\mathbf{h}$ plays the role of unit object in the category of **h**-algebras.

DEFINITION 2.1. An **h**-bialgebroid is an **h**-algebra A equipped with two **h**-algebra homomorphisms, the coproduct $\Delta : A \longrightarrow A \tilde{\otimes} A$ and the counit $\varepsilon : A \longrightarrow D_\mathbf{h}$, such that

$$(\Delta \otimes id) \circ \Delta = (id \otimes \Delta) \circ \Delta, \quad (\varepsilon \otimes id) \circ \Delta = (id \otimes \varepsilon) \circ \Delta = id.$$

DEFINITION 2.2. An **h**-Hopf bialgebroid is an **h**-bialgebroid A equipped with a $\mathbb{C}$-linear map $S : A \longrightarrow A$, called the antipode, satisfying the conditions

$$S\left(\mu_r\left(f\right)a\right) = S\left(a\right)\mu_l\left(f\right)$$
$$S\left(a\mu_l\left(f\right)\right) = \mu_r\left(f\right)S\left(a\right)$$
$$m \circ \left(id \otimes S\right) \circ \Delta\left(a\right) = \mu_l\left(\varepsilon\left(a\right)\mathbf{1}\right)$$
$$m \circ \left(S \otimes id\right) \circ \Delta\left(a\right) = \mu_r\left(T_\alpha\left(\varepsilon\left(a\right)\mathbf{1}\right)\right)$$

for all $a \in A$, $f \in M_{\mathbf{h}^*}$ and $a \in A_{\alpha,\beta}$. Here m denotes multiplication, $\varepsilon\left(a\right)\mathbf{1}$ is obtained by applying the difference operator $\varepsilon\left(a\right)$ to the constant function $\mathbf{1} \in M_{\mathbf{h}^*}$.

We introduce a $*$-structure to an **h**-Hopf bialgebroid by assuming that a conjugation $\lambda \to \overline{\lambda}$ has been chosen on $\mathbf{h}^*$. As in [9] we define a $*$-structure on the **h**-bialgebroid A to be a $\mathbb{C}$-antilinear and antimultiplicative involution $a \to a^*$ on A such that $\mu_l\left(f\right)^* = \mu_l\left(\overline{f}\right)$, $\mu_r\left(f\right)^* = \mu_r\left(\overline{f}\right)$ where $\overline{f}(\lambda) = \overline{f(\overline{\lambda})}$.
It follows that $\left(A_{\alpha\beta}\right)^* = A_{-\overline{\alpha},-\overline{\beta}}$. A $*$-structure on a **h**-bialgebroid is in addition required to satisfy

$$(* \otimes *) \circ \Delta = \Delta \circ *, \quad \varepsilon \circ * = *^{D_\mathbf{h}} \circ \varepsilon$$

where $*^{D_\mathbf{h}}$ si defined by $\left(fT_\alpha\right)^* = \left(T_{-\overline{\alpha}}\overline{f}\right)$. We use complex conjugation on $\mathbf{h} \simeq \mathbb{C}$. Let us now recall the generalized FRST construction. Let $V = \bigoplus_{\alpha \in \mathbf{h}^*} V_\alpha$ be a finite-dimensional diagonalizable **h**-module and $R : \mathbf{h}^* \longrightarrow \mathrm{End}_\mathbf{h}\left(V \otimes V\right)$ be a meromorphic function. To each such R one associates an **h**-bialgebroid A_R. Let $\{e_x\}_{x \in X}$ be a homogeneous basis of V, where X is an index set. Write R_{xy}^{ab} for the matrix elements

$$R\left(\lambda\right)\left(e_a \otimes e_b\right) = \sum_{xy} R_{xy}^{ab}\left(\lambda\right) e_x \otimes e_y$$

of R, and define $w : X \longrightarrow \mathbf{h}^*$ by $e_x \in V_{w(x)}$. The algebra A_R is generated by elements $\{L_{xy}\}_{x,y \in X}$ together with two copies of $M_{\mathbf{h}^*}$ embedded as subalgebras. We write the elements of these two copies as $f\left(\lambda\right)$, $f\left(\mu\right)$ respectively. The defining relations of A_R are:

$$f\left(\lambda\right)L_{xy} = L_{xy}f\left(\lambda + w\left(x\right)\right)$$
$$f\left(\mu\right)L_{xy} = L_{xy}f\left(\mu + w\left(y\right)\right)$$
$$f\left(\lambda\right)g\left(\mu\right) = g\left(\mu\right)f\left(\lambda\right)$$

for $f, g \in M_{\mathbf{h}^*}$ together with the following relations, which we call RLL relations:

$$\sum_{xy} R_{ac}^{xy}\left(\lambda\right)L_{xb}L_{yd} = \sum_{xy} R_{xy}^{bd}\left(\mu\right)L_{cy}L_{ax}.$$

260   A. PAOLUCCI

The bigrading on A_R is defined by $L_{xy} \in A_{w(x),w(y)}$, $f(\lambda)$, $f(\mu) \in A_{00}$ and the moment maps by $\mu_l(f) = f(\lambda)$, $\mu_r(f) = f(\mu)$.

For the RLL -relations to be consistent with the grading R must be **h**-invariant or $R_{xy}^{ab} = 0$ for $w(x) + w(y) \neq w(a) + w(b)$. We can also define a coproduct and a counit on A_R by

$$\Delta(L_{ab}) = \sum_{x \in X} L_{ax} \otimes L_{xb} \; ; \; \Delta(f(\lambda)) = f(\lambda) \otimes \mathbf{1} \; ; \; \Delta(f(\mu)) = \mathbf{1} \otimes f(\mu);$$

$$\varepsilon(L_{ab}) = \delta_{ab} T_{-w(a)} \; ; \; \varepsilon(f(\lambda)) = \varepsilon(f(\mu)) = f.$$

Thus A_R has the structure of an **h**-bialgebroid.

3. COACTIONS ON THE DYNAMICAL GROUP $SU(2)$.

Let us now assume **h** to be one dimensional. It may be viewed as a Cartan subalgebra of $sl(2, \mathbb{C})$. Thus we identify $\mathbf{h} = \mathbf{h}^* = \mathbb{C}$ and take V to be the two dimensional **h**-module: $V = \mathbb{C}e_1 \bigoplus \mathbb{C}e_{-1}$, where e_1, e_{-1} are the unit basis vectors of V. Then in the basis $e_1 \otimes e_1$, $e_1 \otimes e_{-1}$, $e_{-1} \otimes e_1$, $e_{-1} \otimes e_{-1}$ the dynamical R-matrix arising from σ_j-symbols of the quantum algebra $U_q(sl(2))$ (see [3]) has the following form:

$$R(\lambda) = \begin{pmatrix} q & 0 & 0 & 0 \\ 0 & 1 & \frac{q^{-1}-q}{q^{2(\lambda+1)}-1} & 0 \\ 0 & \frac{q^{-1}-q}{q^{-2(\lambda+1)}-1} & \frac{(q^{2(\lambda+1)}-q^2)(q^{2(\lambda+1)}-q^{-2})}{(q^{2(\lambda+1)}-1)^2} & 0 \\ 0 & 0 & 0 & q \end{pmatrix}$$

where q is a fixed number $0 < q < 1$. Denote the corresponding **h**-bialgebroid A_R by $F_R(M(2))$. It is the dynamical analogue of the algebra of polynomials on the space of complex 2×2 matrices.

As in [9] the L-generators are denoted by $\alpha = L_{11}$, $\beta = L_{1,-1}$, $\gamma = L_{-1,1}$, $\delta = L_{-1,-1}$. We introduce the functions

$$F(\lambda) = \frac{q^{2(\lambda+1)} - q^{-2}}{q^{2(\lambda+1)} - 1};$$

$$G(\lambda) = \frac{\left(q^{2(\lambda+1)} - q^2\right)\left(q^{2(\lambda+1)} - q^{-2}\right)}{\left(q^{2(\lambda+1)} - 1\right)^2};$$

$$H(\lambda, \mu) = \frac{\left(q - q^{-1}\right)\left(q^{2(\lambda+\mu+2)} - 1\right)}{\left(q^{2(\lambda+1)} - 1\right)\left(q^{2(\mu+1)} - 1\right)};$$

$$I(\lambda, \mu) = \frac{\left(q - q^{-1}\right)\left(q^{2(\mu+1)} - q^{2(\lambda+1)}\right)}{\left(q^{2(\lambda+1)} - 1\right)\left(q^{2(\mu+1)} - 1\right)}.$$

DEFINITION 3.1. The algebra $F_R(M(2))$ is generated by the four generators α, β, γ, δ together with two copies of $M_{\mathbf{h}^*}$, whose elements we write as $f(\lambda)$, $f(\mu)$. The defining relations are

$$\alpha\beta = qF(\mu - 1)\beta\alpha, \quad \alpha\gamma = qF(\lambda)\gamma\alpha, \quad \beta\delta = qF(\lambda)\delta\beta, \quad \gamma\delta = qF(\mu - 1)\delta\gamma$$

CUNTZ ALGEBRAS AND DYNAMICAL QUANTUM GROUP $SU(2)$

together with any two of the four relations

$$\alpha\delta - \delta\alpha = H(\lambda,\mu)\gamma\beta, \quad G(\mu)\alpha\delta - G(\lambda)\delta\alpha = H(\lambda,\mu)\beta\gamma,$$
$$\beta\gamma - G(\mu)\gamma\beta = I(\lambda,\mu)\delta\alpha, \quad \beta\gamma - G(\lambda)\gamma\beta = I(\lambda,\mu)\alpha\delta,$$

and for arbitrary $f, g \in M_{\mathbf{h}^*}$,

$$f(\lambda)g(\mu) = g(\mu)f(\lambda), \quad f(\lambda)\alpha = \alpha f(\lambda+1), \quad f(\mu)\alpha = \alpha f(\mu+1),$$
$$f(\lambda)\beta = \beta f(\lambda+1), \quad f(\mu)\beta = \beta f(\mu-1),$$
$$f(\lambda)\gamma = \gamma f(\lambda-1), \quad f(\mu)\gamma = \gamma f(\mu+1),$$
$$f(\lambda)\delta = \delta f(\lambda-1), \quad f(\mu)\delta = \delta f(\mu-1).$$

The bigrading $F_R(M(2)) = \bigoplus_{m,n\in\mathbf{Z}, \ m+n\in 2\mathbf{Z}} F_{m,n}$ is defined on the generators by $\alpha \in F_{1,1}$, $\beta \in F_{1,-1}$, $\gamma \in F_{-1,1}$, $\delta \in F_{-1,-1}$, $f(\lambda)$, $f(\mu) \in F_{00}$.
The coproduct

$$\Delta : F_R(M(2)) \longrightarrow F_R(M(2)) \tilde{\otimes} F_R(M(2))$$

and the counit

$$\varepsilon : F_R(M(2)) \longrightarrow D_{\mathbf{h}}$$

are algebra homomorphisms defined on the generators by

$$\Delta(\alpha) = \alpha \otimes \alpha + \beta \otimes \gamma, \quad \Delta(\beta) = \alpha \otimes \beta + \beta \otimes \delta,$$
$$\Delta(\gamma) = \gamma \otimes \alpha + \delta \otimes \gamma, \quad \Delta(\delta) = \gamma \otimes \beta + \delta \otimes \delta,$$
$$\Delta(f(\lambda)) = f(\lambda) \otimes \mathbf{1}, \quad \Delta(f(\mu)) = \mathbf{1} \otimes f(\mu),$$
$$\varepsilon(\alpha) = T_{-1}, \quad \varepsilon(\beta) = \varepsilon(\gamma) = 0, \quad \varepsilon(\delta) = T_1, \quad \varepsilon(f(\lambda)) = \varepsilon(f(\mu)) = f.$$

Any element in $F_R(M(2))$ can be written uniquely as a finite sum

$$\sum_{k,l,m,n} f_{k,l,m,n}(\lambda,\mu)\,\alpha^k\beta^l\gamma^m\delta^n, \quad f_{k,l,m,n} \in M_{\mathbf{h}^*} \otimes M_{\mathbf{h}^*}$$

and in a similar way for any other ordering of the generators. The element

$$c = \frac{F(\lambda)}{F(\mu)}\delta\alpha - \frac{q^{-1}}{F(\mu)}\beta\gamma = \alpha\delta - qF(\lambda)\gamma\beta$$
$$= \frac{F(\lambda-1)}{F(\mu-1)}\alpha\delta - qF(\lambda-1)\beta\gamma = \delta\alpha - \frac{q^{-1}}{F(\mu-1)}\gamma\beta$$

is a central element of $F_R(M(2))$ (see [9, Lemma 2.5]).
A dynamical analogue of the algebra of functions on the group $SL(2,\mathbf{C})$ is the h-Hopf algebroid obtained by adjoining the relation $c = 1$ to the above defining relations. The antipode is defined by

$$S(\alpha) = \frac{F(\lambda)}{F(\mu)}\delta, \ S(\beta) = \frac{-q^{-1}}{F(\mu)}\beta, \ S(\gamma) = -qF(\lambda)\gamma$$
$$S(\delta) = \alpha, \ S(f(\lambda)) = f(\mu), \ S(f(\mu)) = f(\lambda).$$

The elements $\gamma^k\beta^l\alpha^m$, $k,l,m \geq 0$ and $\delta^k\gamma^l\beta^m$, $k > 0$, $l,m \geq 0$ form a basis for $F_R(SL(2))$ as a module over $\mu_l(M_{\mathbf{h}^*})\mu_r(M_{\mathbf{h}^*}) \simeq M_{\mathbf{h}^*} \otimes M_{\mathbf{h}^*}$. The elements $\{\gamma^k\delta^l\alpha^m\beta^n\}_{k+l+m+n=N}$ are linearly independent over $\mu_l(M_{\mathbf{h}^*})\mu_r(M_{\mathbf{h}^*})$, for each N, see [9]. The algebra $F_R(SU(2))$ is the h-Hopf algebroid $F_R(SL(2))$ equipped with the $*$-structure $f(\lambda)^* = \overline{f}(\lambda)$ and $f(\mu)^* = \overline{f}(\mu)$, $\alpha^* = \delta$, $\beta^* = -q\gamma$, $\gamma^* = -q^{-1}\beta$, $\delta^* = \alpha$.

Let O_d be the Cuntz algebra [2] on d generators. This is the universal C^*-algebra generated by $\{S_j : j = 1, 2, ..., d\}$ satisfying $S_i^* S_j = \delta_{i,j}$ and $\sum_j S_j S_j^* = \mathbf{1}$ where $\mathbf{1}$ denotes the identity of O_d.

Let $\mathcal{H}$ denote the vector space spanned by $S_0, ..., S_N$, the generators of the Cuntz algebra O_{N+1}, together with $\mu_l (M_{\mathbf{h}^*})$. Assume the following holds in $\mathcal{H}$:

$$f(\lambda) S_i = S_i T_{2i-N} (f(\lambda))$$

where $T_{2i-N} (f(\lambda)) = f(\lambda + 2i - N)$, $\lambda \in \mathbf{h}^*$, $i = 0, 1, ... N$. As in [12] let $\tilde{O}_{N+1}$ denote the $*$-algebra generated by $M_{\mathbf{h}^*}, S_0, ..., S_N$ satisfying

$$S_j S_i^* = \delta_{ij} c_j(\lambda) \mathbf{1}, \tag{3}$$

$$\sum_{i=0}^{N} S_i^* \frac{1}{c_i(\lambda)} S_i = \mathbf{1}, \tag{4}$$

$$f S_i = S_i (T_{2i-N}(f)), \tag{5}$$

with

$$c_i(\lambda) = \left[\begin{array}{c} N \\ i \end{array} \right]_{q^2} \frac{(q^{-2\lambda}; q^2)_i}{(q^{2(i-N-\lambda-1)}; q^2)_i}, \quad T_\alpha f(\lambda) = f(\lambda + \alpha).$$

Let us note that in the non-dynamical case $\tilde{O}_{N+1}$ reduces to the Cuntz algebra O_{N+1}. Observe that we have reversed the role of S_i and S_i^* in the definition of $\tilde{O}_{N+1}$ in comparison to the usual one of O_{N+1}. This is necessary in order to be consistent with the relations of dynamical $SU(2)$. It should be observed that the Cuntz algebra relations can be written with the roles of S_i and S_i^* interchanged. An $\mathbf{h}$-space V is a vector space over $M_{\mathbf{h}^*}$ such that $V = \oplus_{\alpha \in \mathbf{h}^*} V_\alpha$ with $M_{\mathbf{h}^*} \cdot V_\alpha \subseteq V_\alpha$ for every α. Then a morphism of $\mathbf{h}$-spaces is an $\mathbf{h}$-invariant, i.e. grade preserving, $M_{\mathbf{h}^*}$-linear map. Let A be an $\mathbf{h}$-algebra and V an $\mathbf{h}$-space; we define $A \tilde{\otimes} V = \oplus_{\alpha\beta} A_{\alpha\beta} \otimes_{M_{\mathbf{h}^*}} V_\beta$ where $\otimes_{M_{\mathbf{h}^*}}$ denotes the usual tensor product modulo the relations $\mu_r^A(f) a \otimes v = a \otimes fv$. The grading $A_{\alpha\beta} \otimes_{M_{\mathbf{h}^*}} V_\beta \subseteq (A \tilde{\otimes} V)_\alpha$ and the extension of scalars $f(a \otimes v) = \mu_l^A(f) a \otimes v$ make $A \tilde{\otimes} V$ an $\mathbf{h}$-space. This definition is compatible with the matrix tensor product of $\mathbf{h}$-algebras i.e. $(A \tilde{\otimes} B) \tilde{\otimes} V = A \tilde{\otimes} (B \tilde{\otimes} V)$ when A and B are $\mathbf{h}$-algebras and V an $\mathbf{h}$-space. We can now give the definition of the corepresentation of an $\mathbf{h}$-bialgebroid.

DEFINITION 3.2. We define a (left) corepresentation of an $\mathbf{h}$-bialgebroid A on an $\mathbf{h}$-space V to be an $\mathbf{h}$-space morphism

$$\pi : V \longrightarrow A \tilde{\otimes} V$$

such that $(\Delta \otimes id) \circ \pi = (id \otimes \pi) \circ \pi$, $(\varepsilon \otimes id) \circ \pi = id$.

Pick a homogeneous basis $\{v_k\}_k$ of V (over $M_{\mathbf{h}}^*$), $v_k \in V_{\omega(h)}$, the matrix elements $t_{kj} \in A$ are given by

$$\pi(v_k) = \sum_j t_{kj} \otimes v_j.$$

CUNTZ ALGEBRAS AND DYNAMICAL QUANTUM GROUP $SU(2)$

Recall from [9] that a corepresentation of a *-h-Hopf algebroid A on an h-space V is unitarizable if there exists a basis of V such that the corresponding matrix elements satisfy

$$\Gamma_k(\mu)S(t_{kj})^* = \Gamma_j(\lambda)t_{jk}$$

for some $0 \neq \Gamma_k \in M_{\mathbf{h}^*}$ with $\overline{\Gamma_k} = \Gamma_k$. The functions Γ_k are called normalizing functions for V with respect to the basis.

For the purpose of our study we only consider unitarizable corepresentations.

In the case of the *- h-bialgebroid $F_R(SU(2))$ the Γ_i normalizing functions for the corepresentation are given by $(c_i(\lambda))^{-1}$.

Let M be a right comodule and A an h-bialgebroid. We define a coaction of the Hopf algebroid A on M as follows.

DEFINITION 3.3. A (left) coaction of A on M is a map $\Gamma : M \to A \widetilde{\otimes} M$ satisfying the following coassociativity condition:

$$(id \otimes \Gamma) \circ \Gamma = (\Delta \otimes id) \circ \Gamma$$

and the counity condition:

$$(\varepsilon \otimes id) \circ \Gamma = id.$$

Let $\mathcal{H}$ be the h-vector space defined above such that the following holds:

$$f(\lambda) S_i = S_i T_{2i-N}(f(\lambda))$$

where

$$T_{2i-N}(f(\lambda)) = f(\lambda + 2i - N), \quad \lambda \in \mathbf{h}^*, \quad i = 0, 1, ..., N.$$

For a given unitarizable corepresentation of the *- h-Hopf algebroid $F_R(SU(2))$ there is a mapping Γ on the h-space $\mathcal{H}$ determined by

$$\Gamma(S_j) = \sum_{k=0}^{N} t_{jk}^N \otimes S_k,$$

PROPOSITION 3.4. *The map Γ sending $S_i \longmapsto \sum_{k=0}^{N} t_{ik}^N \otimes S_k$ extends to*

$$\tilde{O}_{N+1} \longrightarrow F_R(SU(2)) \widetilde{\otimes} \tilde{O}_{N+1},$$

*an h-algebra morphism preserving the *. Then Γ is a coaction on $\tilde{O}_{N+1}$ and $\tilde{O}_{N+1}$ is a comodule algebra over the dynamical quantum group $SU(2)$.*

Proof. $\tilde{O}_{N+1}$ is generated by $f \in M_{\mathbf{h}^*}$, $S_0, ..., S_N, S_0^*, ... S_N^*$, satisfying relations (3), (4), (5).

Let $\mathcal{H}$ denote the vector space spanned by $S_0, ..., S_N$, the generators of the Cuntz algebra O_{N+1}, together with $\mu_l(M_{\mathbf{h}^*})$. We define the following map

$$\Gamma : S_i \longmapsto \sum_{k=0}^{N} t_{ik}^N \otimes S_k \qquad (6)$$

We claim that Γ is a coaction on $\tilde{O}_{N+1}$ of the dynamical quantum group $SU(2)$, see [1] for the case of compact quantum groups. Thus we need to prove that (6) is an algebra homomorphism.

First let us observe that

$$S_i^* \longmapsto \sum_{k=0}^N \left(t_{ik}^N\right)^* \otimes S_k^*.$$

Thus the map preserves the * operation. Hence it follows that

$$\Gamma\left(S_j\right)\Gamma\left(S_i^*\right) = \sum_{p,k=0}^N t_{jp}^N \left(t_{ik}^N\right)^* \otimes S_p S_k^*$$

$$= \sum_{k=0}^N t_{jk}^N \left(t_{ik}^N\right)^* \otimes c_k(\lambda)\mathbf{1} = \sum_{k=0}^N \mu_r\left(c_k(\lambda)\right) t_{jk}^N \left(t_{ik}^N\right)^* \otimes \mathbf{1}.$$

Now we use the condition of S being an antipode

$$\delta_{kl} = \sum_j S\left(t_{kj}\right) t_{jl} = \sum_j t_{kj} S\left(t_{jl}\right)$$

and

$$\mu_l\left(c_j(\lambda)\right) S\left(t_{kj}\right)^* = \mu_r\left(c_k(\lambda)\right) t_{jk}.$$

Then since $\bar{c}_k(\lambda) = c_k(\lambda)$, it follows

$$S\left(t_{kj}\right)^* = \frac{\mu_r\left(c_k(\lambda)\right)}{\mu_l\left(c_j(\lambda)\right)} t_{jk}, \quad S\left(t_{kj}\right) = t_{jk}^* \frac{\mu_r\left(c_k(\lambda)\right)}{\mu_l\left(c_j(\lambda)\right)}.$$

Thus

$$\Gamma\left(S_j\right)\Gamma\left(S_i^*\right) = \delta_{ji}\mu_l\left(c_i(\lambda)\right) \otimes \mathbf{1} = \mu_l\left(c_i(\lambda)\right)\delta_{ij}\left(\mathbf{1}\otimes\mathbf{1}\right) = \Gamma\left(S_j S_i^*\right).$$

Γ is then a $*$-homomorphism.

It is easy to see that (3) is preserved. Now let us prove that (4) is satisfied. Since

$$\sum_{i=0}^N \left(\sum_{k=0}^N \left(t_{ik}^N\right)^* \otimes S_k^*\right)\left(\mu_l\left(c_i^{-1}(\lambda)\right)\otimes\mathbf{1}\right)\left(\sum_{p=0}^N t_{ip}^N \otimes S_p\right)$$

$$= \sum_{i,k,p=0}^N \left(t_{ik}^N\right)^* \mu_l\left(c_i^{-1}(\lambda)\right) t_{ip}^N \otimes S_k^* S_p$$

we need to compute

$$\sum_{i=0}^N \left(t_{ik}^N\right)^* \mu_l\left(c_i^{-1}(\lambda)\right) t_{ip}^N.$$

Since $t_{jk}^* \in F_R\left(SU\left(2\right)\right)_{N-2j,N-2k}$, we have

$$\delta_{kl} = \sum_{j=0}^N t_{jk}^* \frac{\mu_r\left(c_k(\lambda)\right)}{\mu_l\left(c_j(\lambda)\right)} t_{jl} = \sum_{j=0}^N \mu_r\left(T_{2k-N}c_k\left(\lambda\right)\right) t_{jk}^*\mu_l\left(c_j^{-1}(\lambda)\right) t_{jl}.$$

Hence

$$\sum_{i=0}^N \left(t_{ik}^N\right)^* \mu_l\left(c_i^{-1}(\lambda)\right) t_{ip}^N = \mu_r\left(T_{2k-N}c_k^{-1}(\lambda)\right)\delta_{kp}\mathbf{1}$$

and

$$\sum_{k,p,i=0}^{N} \left(t_{ik}^N\right)^* \mu_l\left(c_i^{-1}(\lambda)\right) t_{ip}^N \otimes S_k^* S_p$$

$$= \sum_{k=0}^{N} \mu_r\left(T_{2k-N}\, c_k^{-1}(\lambda)\right) \mathbf{1} \otimes S_k^* S_k = \mathbf{1} \otimes \sum_{k=0}^{N} \left(T_{2k-N}\, c_k^{-1}(\lambda)\right) S_k^* S_k.$$

Observe that $\mathcal{H}$ as an **h**-space has a decomposition such that $S_k \in \mathcal{H}_{2k-N}$, $S_k^* \in \mathcal{H}_{N-2k}$. Then

$$\left(T_{2k-N}\, c_k^{-1}(\lambda)\right) S_k^* S_k = S_k^* \left(T_{2k-N+N-2k}\, c_k^{-1}(\lambda)\right) S_k = S_k^* c_k^{-1}(\lambda) S_k$$

and

$$\mathbf{1} \otimes \sum_{k=0}^{N} \left(T_{2k-N}\, c_k^{-1}(\lambda)\right) S_k^* S_k = \mathbf{1} \otimes \sum_{k=0}^{N} S_k^* c_k^{-1}(\lambda) S_k = \mathbf{1} \otimes \mathbf{1}.$$

Hence (4) is satisfied.

Let us now prove that (5) is preserved by the map Γ.

$$\begin{aligned}
\Gamma\left(fS_i\right) &= \left(\mu_l\left(f\right) \otimes \mathbf{1}\right) \left(\sum_{k=0}^{N} t_{ik}^N \otimes S_k\right) \\
&= \sum_{k=0}^{N} \mu_l\left(f\right) t_{ik}^N \otimes S_k = \\
&= \sum_{k=0}^{N} t_{ik}^N \mu_l\left(T_{2i-N}f\right) \otimes S_k \\
&= \left(\sum_{k=0}^{N} t_{ik}^N \otimes S_k\right) \left(\mu_l\left(T_{2i-N}f\right) \otimes \mathbf{1}\right)
\end{aligned}$$

and

$$\Gamma\left(S_i T_{2i-N}f\right) = \left(\sum_{k=0}^{N} t_{ik}^N \otimes S_k\right) \left(\mu_l\left(T_{2i-N}f\right) \otimes \mathbf{1}\right).$$

Thus $\Gamma\left(fS_i\right) = \Gamma\left(S_i T_{2i-N}f\right)$ i.e. (5) holds. Then the mapping Γ preserves the relations (3), (4), (5). It is easy to see that

$$\left(\Delta \otimes id\right) \circ \Gamma = \left(id \otimes \Gamma\right) \circ \Gamma.$$

It follows that Γ is a coaction and preserves the grading. Therefore $\tilde{O}_{N+1}$ is a comodule algebra over the dynamical quantum group $SU(2)$ under the coaction Γ, thus concluding the proof. $\qquad\square$

We have so far assumed that the dynamical quantum group is given and we sought to understand the coaction on a comodule algebra. The comodule algebra is a dynamical analogue of the Cuntz algebra. We will now go in the opposite direction. We assume that the $\widetilde{O}_{N+1}$ is a comodule algebra over a $*$-**h**-Hopf algebroid A. Then we have the following Lemma.

266   A. PAOLUCCI

LEMMA 3.5. *Let A be a $* - $ **h**-Hopf algebroid on an* **h**-*vector space V (of dimension $N + 1$) with generators $t_{j,k}^N$, $\left(t_{j,k}^N\right)^*$, $j, k = 0, ..., N$ defining an unitarizable corepresentation. Assume that $\tilde{O}_{N+1}$ forms a comodule algebra over A where the coaction*

$$\Gamma : \tilde{O}_{N+1} \longrightarrow A \tilde{\otimes} \tilde{O}_{N+1}$$

is given by

$$\Gamma(S_j) = \sum_{k=0}^{N} t_{jk}^N \otimes S_k \text{ and } \Gamma(S_j^*) = \sum_{k=0}^{N} (t_{jk}^N)^* \otimes S_k^*.$$

Then the comultiplication Δ, counit ε, and antipode S of A are given by

$$\Delta\left(t_{ij}^N\right) = \sum_{k=0}^{N} t_{i,k}^N \otimes t_{kj}^N$$

$$\Delta\left((t_{ij}^N)^*\right) = \sum_{k=0}^{N} (t_{i,k}^N)^* \otimes (t_{kj}^N)^*$$

$$\varepsilon\left(t_{ij}^N\right) = \varepsilon\left((t_{ij}^N)^*\right) = \delta_{ij}$$

$$S\left(t_{ij}^N\right)^* = \frac{\mu_r\left(c_i\left(\lambda\right)\right)}{\mu_l\left(c_j\left(\lambda\right)\right)} t_{ji}^N$$

$$S\left(t_{ij}^N\right) = (t_{ji}^N)^* \frac{\mu_r\left(c_i(\lambda)\right)}{\mu_l\left(c_j(\lambda)\right)}$$

and the normalizing functions $\Gamma_i\left(\lambda\right) = (c_i(\lambda))^{-1}$ are provided by the defining relations of $\tilde{O}_{N+1}$.

Proof. The only part to prove is the antipode. Using the defining relations of $\tilde{O}_{N+1}$

$$S_i S_j^* = \delta_{ij} c_j\left(\lambda\right) \mathbf{1}$$

and

$$\sum_{i=0}^{N} S_i^* \frac{1}{c_i\left(\lambda\right)} S_i = 1$$

we have that

$$\Gamma\left(\sum_{k=0}^{N} S_k^* \frac{1}{c_k\left(\lambda\right)} S_k\right) = \sum_{k=0}^{N} \Gamma\left(S_k^* \frac{1}{c_k\left(\lambda\right)} S_k\right)$$

$$= \sum_{k=0}^{N} \Gamma(S_k^*) \frac{1}{c_k\left(\lambda\right)} \Gamma(S_k)$$

$$= \sum_{k=0}^{N} \left(\left(\sum_{i=0}^{N} (t_{ki}^N)^* \otimes S_i^*\right) \frac{1}{c_k\left(\lambda\right)} \left(\sum_j t_{kj}^N \otimes S_j\right)\right)$$

$$= \sum_{k=0}^{N} \left(\sum_{i=0}^{N} (t_{ki}^N)^* \otimes S_i^*\right) \left(\mu_l\left(c_k^{-1}\left(\lambda\right)\right) \otimes 1\right) \left(\sum_{j=0}^{N} (t_{kj}^N) \otimes S_j\right)$$

CUNTZ ALGEBRAS AND DYNAMICAL QUANTUM GROUP $SU(2)$

$$= \sum_{k,i,j=0}^{N} \left(t_{ki}^{N}\right)^{*} \mu_{l}\left(c_{k}^{-1}(\lambda)\right) \left(t_{kj}^{N}\right) \otimes S_{i}^{*}S_{j}.$$

Since Γ is a coaction it preserves the unit, so $\Gamma(1) = 1 \otimes 1$. Hence it follows that

$$\sum_{k=0}^{N} \left(t_{ki}^{N}\right)^{*} \frac{\mu_{r}\left(c_{i}(\lambda)\right)}{\mu_{l}\left(c_{k}(\lambda)\right)} \left(t_{kj}^{N}\right) = \delta_{ij}.$$

On the other hand we have

$$\Gamma\left(S_{i}S_{j}^{*}\right) = \left(\sum_{k=0}^{N} \left(t_{ik}^{N}\right) \otimes S_{k}\right)\left(\sum_{p=0}^{N} \left(t_{jp}^{N}\right)^{*} \otimes S_{p}^{*}\right)$$

$$= \sum_{k,p=0}^{N} \left(t_{ik}^{N}\right)\left(t_{jp}^{N}\right)^{*} \otimes S_{k}S_{p}^{*}$$

Hence it follows

$$\sum_{k=0}^{N} \left(t_{ik}^{N}\right)\left(t_{jk}^{N}\right)^{*} \frac{\mu_{r}\left(c_{k}(\lambda)\right)}{\mu_{l}\left(c_{j}(\lambda)\right)} = \delta_{ij} = \sum_{k=0}^{N} \left(t_{ki}^{N}\right)^{*} \frac{\mu_{r}\left(c_{i}(\lambda)\right)}{\mu_{l}\left(c_{k}(\lambda)\right)} \left(t_{kj}^{N}\right)$$

$$\sum_{i=0}^{N} \left(t_{ik}^{N}\right) S\left(t_{kj}^{N}\right) = \sum_{i=0}^{N} S\left(t_{ik}^{N}\right) t_{kj} = \delta_{ij}.$$

which imply the defining relations of the antipode. Hence

$$S\left(t_{jk}^{N}\right)^{*} = \frac{\mu_{r}\left(c_{j}(\lambda)\right)}{\mu_{l}\left(c_{k}(\lambda)\right)} t_{kj}^{N}.$$

In a similar way we obtain that

$$S\left(t_{jk}^{N}\right) = \left(t_{kj}^{N}\right)^{*} \frac{\mu_{r}\left(c_{j}(\lambda)\right)}{\mu_{l}\left(c_{k}(\lambda)\right)},$$

concluding the proof. $\qquad\square$

Remark 3.6. Lemma 3.5 shows that the h-bialgebroid structure and relations among the generators of A are forced by requiring the particular coaction of the dynamical Cuntz algebra.

Acknowledgement

The author would like to thank H.T. Koelink for useful discussions on the subject.

REFERENCES

[1] A.L. Carey, A. Paolucci, and R.B. Zhang, Quantum group actions on the Cuntz algebra, *Ann. Henri Poincaré* **1** (2000), 1097-1122.

[2] J. Cuntz, Simple C^{*}-algebras generated by isometries, *Comm. Math. Phys.* **57** (1977), 173-185.

[3] P. Etingof and O. Schiffmann, Lectures on the dynamical Yang-Baxter equations, in "Quantum Groups and Lie Theory" (ed. A. Pressley), *London Math. Soc. Lect. Note Ser.* **290**, Cambridge University Press, Cambridge, 2001, 89-129.

[4] P. Etingof and A. Varchenko, Solutions of the quantum dynamical Yang-Baxter equation and dynamical quantum groups; *Comm. Math. Phys.* **196** (1998), 591–640.

[5] P. Etingof and A. Varchenko, Exchange dynamical quantum groups, *Comm. Math. Phys.* **205** (1999), 19-52.

[6] G. Felder, Conformal field theory and integrable systems associated to elliptic curves, in "Proceedings of the International Congress of Mathematicians, Zürich 1994", Birkhaüser, 1994, 1247-1255.

[7] G. Felder, Elliptic quantum groups, Preprint hep-th/9412207, Proceedings of the ICMP, Paris, 1994.

[8] G. Felder and A. Varchenko, On representations of the elliptic quantum group $E_{\tau,\eta}(sl2)$, *Comm. Math. Phys.* **181** (1996) 741-761.

[9] E. Koelink and H. Rosengren, Harmonic analysis on the $SU(2)$ dynamical quantum group, *Acta Appl. Math.* **69** (2001), 163-220.

[10] J.H. Lu, Hopf algebroids and quantum groupoids, *Inter. J. Math.* **7** (1996), 47-70.

[11] S. Montgomery, "Hopf algebras and their actions on rings", American Mathematical Society, Providence, 1993.

[12] A. Paolucci, On a dynamical analogue of the Cuntz algebra, in preparation.

On Symbolic Computations
in Braided Monoidal Categories

BODO PAREIGIS Mathematisches Institut der Universität München
Theresienstr. 39, D-80333 München, Germany
e-mail: *pareigis@lmu.de*

> ABSTRACT. There are some powerful notations and tools to perform computations in with tensors, the Sweedler notation for coalgebras, the Einstein convention to reduce the number of summation signs in computations with tensors, the Penrose notation that has been further developed by Joyal and Street to a graphic calculus in braided monoidal categories. In 1977 I introduced a method of computation that looks very much like computation with ordinary elements or tensors, but can be performed in arbitrary monoidal categories, by using a Yoneda Lemma like technique. In the dual of the category of vector spaces this allows to work with ordinary coalgebras as if they were algebras. I will show how to expand this technique to braided monoidal categories, and develop some of the general rules of computation. As an application I will derive the well known result that the antipode of a Hopf algebra in a braided monoidal category is an algebra antihomomorphism which is expressed by the formulas $S(1) = 1$ and $S(ab) = \langle S(b)S(a), \tau \rangle$.

1. THE BEGINNINGS: THE SWEEDLER-HEYNEMAN NOTATION

To describe the comultiplication of a $\mathbb{K}$-coalgebra in terms of elements we introduce a notation first introduced by Sweedler and Heyneman [3] similar to the notation $\nabla(a \otimes b) = ab$ used for algebras. Instead of $\Delta(c) = \sum c_i \otimes c_i'$ we write

$$\Delta(c) = \sum c_{(1)} \otimes c_{(2)}. \tag{1}$$

Observe that only the complete expression on the right hand side makes sense, not the components $c_{(1)}$ or $c_{(2)}$ which are *not* considered as families of elements of C. This notation alone does not help much in the calculations we have to perform later on. So we introduce a more general notation.

DEFINITION 1.1. (Sweedler notation) Let M be an arbitrary $\mathbb{K}$-module and C be a $\mathbb{K}$-coalgebra. Then there is a bijection between all multilinear maps

$$f : C \times \ldots \times C \to M$$

and all linear maps

$$f^\sharp : C \otimes \ldots \otimes C \to M.$$

These maps are associated to each other by the formula

$$f(c_1, \ldots, c_n) = f^\sharp(c_1 \otimes \ldots \otimes c_n) \tag{2}$$

1991 *Mathematics Subject Classification.* 16W30.
Key words and phrases. braided category, symbolic computation, braided Hopf algebra.

270 B. PAREIGIS

or

$$f = f^{\sharp} \circ \otimes.$$

This follows from the universal property of the tensor product. For $c \in C$ we define

$$\sum f(c_{(1)}, \ldots, c_{(n)}) := f^{\sharp}(\Delta^{n-1}(c)), \tag{3}$$

where Δ^{n-1} denotes the $(n-1)$-fold application of Δ, for example $\Delta^{n-1} = (\Delta \otimes 1 \otimes \ldots \otimes 1) \circ \ldots \circ (\Delta \otimes 1) \circ \Delta$.

In particular we obtain for the bilinear map $\otimes : C \times C \ni (c, d) \mapsto c \otimes d \in C \otimes C$ (with associated identity map)

$$\sum c_{(1)} \otimes c_{(2)} = \Delta(c), \tag{4}$$

and for the multilinear map $\otimes^2 : C \times C \times C \to C \otimes C \otimes C$

$$\sum c_{(1)} \otimes c_{(2)} \otimes c_{(3)} = (\Delta \otimes 1)\Delta(c) = (1 \otimes \Delta)\Delta(c).$$

With this notation one verifies easily

$$\sum c_{(1)} \otimes \ldots \otimes \Delta(c_{(i)}) \otimes \ldots \otimes c_{(n)} = \sum c_{(1)} \otimes \ldots \otimes c_{(n+1)}$$

and

$$\sum c_{(1)} \otimes \ldots \otimes \epsilon(c_{(i)}) \otimes \ldots \otimes c_{(n)} = \sum c_{(1)} \otimes \ldots \otimes 1 \otimes \ldots \otimes c_{(n-1)}$$
$$= \sum c_{(1)} \otimes \ldots \otimes c_{(n-1)}$$

This notation and its application to multilinear maps will also be used in more general contexts like comodules.

2. SYMBOLIC COMPUTATIONS WITH TENSORS

Let C be a monoidal category. For objects $A, X \in C$ define

$$A(X) := \mathrm{Mor}_C(X, A).$$

We consider A as a "graded" or "variable" set with component $A(X)$ of "degree" X. Actually A is a (representable) functor from C into Set.

Let $f : A \to B$ be a morphism in C. Then we get "maps of variable sets" written by abuse of notation as $f : A(X) \to B(X)$ with

$$f(a) := f \circ a. \tag{5}$$

This defines a natural transformation and by the Yoneda Lemma there is a bijection between the morphisms from A to B and the natural transformations from the functor A to the functor B.

In particular two morphisms $f, g : A \to B$ are equal iff

$$\forall X \in C, \forall a \in A(X) : f(a) = g(a).$$

Let $A, B, C \in C$. Then $C(X \otimes Y)$ is a functor in two variables X and Y. Furthermore $A(X) \times B(Y)$ is also a functor in two variables denoted by $A \times B$. A natural transformation of functors in two variables $f : A \times B \to C$ is called a *bimorphism*.

A special example of a bimorphism is

$$\otimes : A(X) \times B(Y) \to A \otimes B(X \otimes Y) \text{ with } \otimes (a, b) := a \otimes b$$

where $a \otimes b : X \otimes Y \to A \otimes B$. An element $a \otimes b \in A \otimes B(X \otimes Y)$ coming from two morphisms a, b is called a *decomposable tensor*.

If $f : A \times B \to C$ is a bimorphism and $g : C \to D$ is a morphism then $gf : A \times B \to D$ is a bimorphism.

If $f : A \times B \to C$ is a bimorphism and $g : U \to A$ and $h : V \to B$ are morphisms then $f(g \times h) : U \times V \to C$ is a bimorphism.

LEMMA 2.1. *For each bimorphism $f : A \times B \to C$ there is exactly one morphism $f^\sharp : A \otimes B \to C$ such that*

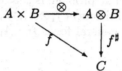

commutes.

Proof. This uses a Yoneda Lemma type argument. For details see [2, Lemma 1.1]. □

Occasionally if $h = f^\sharp$ is given then we write the associated bimorphism as $h^\flat := h \circ \otimes$, so that $(f^\sharp)^\flat = f$ and $(h^\flat)^\sharp = h$.

Given a bimorphism $f = f^\sharp \circ \otimes$ and $a \in A(X), b \in B(Y)$. Let $t = a \otimes b \in A \otimes B(X \otimes Y)$ be a decomposable tensor. Then $f(a,b) = f^\sharp(a \otimes b) = f^\sharp(t)$.

Similar remarks as above hold for *multimorphisms* $f : A_1 \times \ldots \times A_n \to C$ and associated morphisms $f^\sharp : A_1 \otimes \ldots \otimes A_n \to C$. In particular we have for $a_i \in A_i(X_i), i = 1, \ldots, n$ and $t = a_1 \otimes \ldots \otimes a_n$

$$f(a_1, \ldots, a_n) = f^\sharp(t).$$

We introduce a first symbolic expression for all $t \in A_1 \otimes \ldots \otimes A_n(X)$ by

$$\boxed{f(t_1, \ldots, t_n) := f^\sharp(t).} \tag{6}$$

Observe that t is *not* a decomposable tensor in general. We have, however:

For the multimorphism $\otimes^{n-1} : A_1 \times \ldots \times A_n \to A_1 \otimes \ldots \otimes A_n$ and the associated morphism $\otimes^\sharp = \mathrm{id} : A_1 \otimes \ldots \otimes A_n \to A_1 \otimes \ldots \otimes A_n$ we get

$$t_1 \otimes \ldots \otimes t_n = t \tag{7}$$

for all "tensors" $t \in A_1 \otimes \ldots \otimes A_n(X)$. In particular we have then

$$f(t_1, \ldots, t_n) = f^\sharp(t_1 \otimes \ldots \otimes t_n). \tag{8}$$

Given $f^\sharp : A_1 \otimes \ldots \otimes A_n \to B_1 \otimes \ldots \otimes B_m$ and $t \in A_1 \otimes \ldots \otimes A_n(X)$. Then we may consider $f^\sharp(t)$ as an element of $B_1 \otimes \ldots \otimes B_m(X)$ hence

$$\begin{aligned} f^\sharp(t) &= f^\sharp(t)_1 \otimes \ldots \otimes f^\sharp(t)_m = \\ &= f(t_1, \ldots, t_n) = f(t_1, \ldots, t_n)_1 \otimes \ldots \otimes f(t_1, \ldots, t_n)_m. \end{aligned} \tag{9}$$

Since $f^\sharp$ is also an element in $B_1 \otimes \ldots \otimes B_m(A_1 \otimes \ldots \otimes A_n)$ we can write $f^\sharp = f_1^\sharp \otimes \ldots \otimes f_m^\sharp$ and get

$$\begin{aligned} (f_1^\sharp \otimes \ldots \otimes f_m^\sharp)(t) &= f^\sharp(t) = f^\sharp(t)_1 \otimes \ldots \otimes f^\sharp(t)_m \quad \text{or} \\ (f_1^\sharp \otimes \ldots \otimes f_m^\sharp)(t_1 \otimes \ldots \otimes t_n) &= f(t_1, \ldots, t_n)_1 \otimes \ldots \otimes f(t_1, \ldots, t_n)_m. \end{aligned} \tag{10}$$

272   B. PAREIGIS

If in addition $g^\sharp : B_1 \otimes \ldots \otimes B_m \longrightarrow C$ is given then we get

$$g(f^\sharp(t)_1, \ldots, f^\sharp(t)_m) = g^\sharp f(t_1, \ldots, t_n).$$

If $f_i : A_i \longrightarrow B_i, i = 1, \ldots, n$, $f^\sharp := f_1 \otimes \ldots \otimes f_n$, and $t \in A_1 \otimes \ldots \otimes A_n(X)$ are given then we have

$$f_1(t_1) \otimes \ldots \otimes f_n(t_n) = f^\sharp(t) = f^\sharp(t)_1 \otimes \ldots \otimes f^\sharp(t)_n.$$

Observe, we do not admit the same notation for an arbitrary morphism $f^\sharp : A_1 \otimes \ldots \otimes A_n \longrightarrow B_1 \otimes \ldots \otimes B_m$. The problem is that certain natural transformations will commute with morphisms of the form $f_1 \otimes \ldots \otimes f_n : A_1 \otimes \ldots \otimes A_n \longrightarrow B_1 \otimes \ldots \otimes B_n$ but not with morphisms of the general form $f^\sharp : A_1 \otimes \ldots \otimes A_n \longrightarrow B_1 \otimes \ldots \otimes B_m$ even if $m = n$.

LEMMA 2.2. *Given multimorphisms* $f, g : A_1 \times \ldots \times A_n \longrightarrow C$ *with associated morphisms* $f^\sharp, g^\sharp$. *Then the following are equivalent:*

1. $f^\sharp = g^\sharp$,
2. $\forall X_i, \forall a_1, \ldots, a_n \in A_i(X_i) : f(a_1, \ldots, a_n) = g(a_1, \ldots, a_n)$,
3. $\forall X, \forall t \in A_1 \otimes \ldots \otimes A_n(X) : f(t_1, \ldots, t_n) = g(t_1, \ldots, t_n)$.

Proof. $(1) \Longrightarrow f = g \Longrightarrow (3) \Longrightarrow (2) \Longrightarrow f = g \Longrightarrow (1)$. $\qquad\square$

This notation will be used to express and compute certain identities of morphisms. We explain this by the following example. Let $(A, \mu : A \otimes A \longrightarrow A)$ be given. We want to express associativity by elements. Write $ab := \mu(a \otimes b) \in A(X \otimes Y)$. Then $(ab)c = \mu(\mu(a \otimes b) \otimes c) \in A((X \otimes Y) \otimes Z)$. Similarly $a(bc) = \mu(a \otimes \mu(b \otimes c)) \in A(X \otimes (Y \otimes Z))$. In order to compare these two products we apply $A(\alpha) : A(X \otimes (Y \otimes Z)) \longrightarrow A((X \otimes Y) \otimes Z)$ to get $(ab)c = a(bc) \circ \alpha = A(\alpha)(a(bc))$ iff (A, μ) is associative.

Since most such computations can be transferred to a strict monoidal category, we are going to assume from now on that C is a strict monoidal category.

Then $(A, \mu : A \otimes A \longrightarrow A)$ is associative iff $(ab)c = a(bc)$.

3. BRAIDINGS AND TENSORS

Let C be a strict monoidal category that is braided. Let $\rho \in B_n$ be a braid in the braid group with canonical image $\overline{\rho} \in S_n$. Let $\sigma := \overline{\rho}^{-1}$. Let $\rho : A_1 \otimes \ldots \otimes A_n \longrightarrow A_{\sigma(1)} \otimes \ldots \otimes A_{\sigma(n)}$ also denote the associated braid action on the n-fold tensor product. So ρ is a natural transformation of functors in n variables.

Let $f^\sharp : A_{\sigma(1)} \otimes \ldots \otimes A_{\sigma(n)} \longrightarrow B$ be a morphism in C and $f := f^\sharp \circ \otimes^n : A_{\sigma(1)}(X_{\sigma(1)}) \times \ldots \times A_{\sigma(n)}(X_{\sigma(n)}) \longrightarrow B(X_{\sigma(1)} \otimes \ldots \otimes X_{\sigma(n)})$ be the associated multimorphism.

We want to study the application of $f^\sharp$ on a tensor $t \in A_1 \otimes \ldots \otimes A_n(X)$ if ρ is first applied to t. First assume that t is a decomposable tensor of the form $t = a_1 \otimes \ldots \otimes a_n \in A_1 \otimes \ldots \otimes A_n(X_1 \otimes \ldots \otimes X_n)$ with $a_i \in A_i(X_i)$. We get

$$f(\rho(t)_1, \ldots, \rho(t)_n) = f(a_{\sigma(1)}, \ldots, a_{\sigma(n)})\rho \qquad (11)$$

since $f(\rho(t)_1, \ldots, \rho(t)_n) = f^\sharp \rho(t) = f^\sharp \rho(a_1 \otimes \ldots \otimes a_n) = f^\sharp(a_{\sigma(1)} \otimes \ldots \otimes a_{\sigma(n)})\rho = f(a_{\sigma(1)}, \ldots, a_{\sigma(n)})\rho$ where we used that ρ is a natural transformation.

ON SYMBOLIC COMPUTATIONS IN BRAIDED MONOIDAL CATEGORIES   273

Observe that in the symbolic notation ρ is not really applied to $a_1 \otimes \ldots \otimes a_n$ as it is in ordinary computations in braided categories, it changes only the order of the components with $\sigma \in S_n$. We would only be interested in the expression $f(a_{\sigma(1)}, \ldots, a_{\sigma(n)})$ in ordinary computations, but some information about ρ is lost, if we study this term in symbolic calculations. View ρ as an index for this expression and write

$$\langle f(a_{\sigma(1)}, \ldots, a_{\sigma(n)}), \rho \rangle := f(a_{\sigma(1)}, \ldots, a_{\sigma(n)})\rho. \tag{12}$$

In particular we have $\langle f(a_{\sigma(1)}, \ldots, a_{\sigma(n)}), \rho \rangle = f^\sharp(\rho(a_1 \otimes \ldots \otimes a_n))$.

We extend this notation to arbitrary tensors $t = t_1 \otimes \ldots \otimes t_n \in A_1 \otimes \ldots \otimes A_n(X)$ (see equation (7)).

DEFINITION 3.1. We define the map

$$\langle .,.,. \rangle : \mathrm{Nat}(A_{\sigma(1)} \times \ldots \times A_{\sigma(n)}, B) \times (A_1 \otimes \ldots \otimes A_n)(X) \times B_n \longrightarrow B(X)$$

by $\langle f, t, \rho \rangle := f^\sharp(\rho(t)) = f^\sharp \circ \rho \circ t$, where $f : A_{\sigma(1)} \times \ldots \times A_{\sigma(n)} \longrightarrow B$ is a multimorphism, $t = t_1 \otimes \ldots \otimes t_n \in A_1 \otimes \ldots \otimes A_n(X)$ is a variable or argument, and $\rho \in B_n$ is a braid. We write for $\langle f, t, \rho \rangle$ also $\langle f(t_{\sigma(1)}, \ldots, t_{\sigma(n)}), \rho \rangle$ and define

$$\boxed{\langle f(t_{\sigma(1)}, \ldots, t_{\sigma(n)}), \rho \rangle := f^\sharp(\rho(t)).} \tag{13}$$

The expression $f(t_{\sigma(1)}, \ldots, t_{\sigma(n)})$ taken separately is clearly not defined, except in the case where t is a decomposable tensor. Observe that $t : X \longrightarrow A_1 \otimes \ldots \otimes A_n$ and $f^\sharp : A_1 \otimes \ldots \otimes A_n \longrightarrow B$ are morphisms so that ρ can operate on the range of t and on the domain of $f^\sharp$. We tacitly assume in writing down an expression $\langle f(t_{\sigma(1)}, \ldots, t_{\sigma(n)}), \rho \rangle$ that the range and domain of $f^\sharp$ and t are given and fixed. (Otherwise an operation of ρ would not be well defined.) If this separation is not quite clear we also use the notation

$$\langle f[t_{\sigma(1)}, \ldots, t_{\sigma(n)}], \rho \rangle := f^\sharp(\rho(t)).$$

In some cases one has to name the arguments explicitly, which are used in a concrete computation.

THEOREM 3.2. (Comparison Theorem): *Given* $f^\sharp : A_1 \otimes \ldots \otimes A_n \longrightarrow B$ *and* $g^\sharp : A_{\sigma(1)} \otimes \ldots \otimes A_{\sigma(n)} \longrightarrow B$. *Then the following are equivalent:*

1. $g^\sharp \circ \rho = f^\sharp$,
2. $\forall a_1, \ldots, a_n \in A_i(X_i) : f(a_1, \ldots, a_n) = \langle g(a_{\sigma(1)}, \ldots, a_{\sigma(n)}), \rho \rangle$,
3. $\forall t \in A_1 \otimes \ldots \otimes A_n(X) : f(t_1, \ldots, t_n) = \langle g(t_{\sigma(1)}, \ldots, t_{\sigma(n)}), \rho \rangle$.

Proof. $(1) \Longrightarrow (3)$: $f(t_1, \ldots, t_n) = f^\sharp(t) = g^\sharp(\rho(t)) = \langle g(t_{\sigma(1)}, \ldots, t_{\sigma(n)}), \rho \rangle$.

$(3) \Longrightarrow (2)$: Take $t := a_1 \otimes \ldots \otimes a_n$. Then $f(a_1, \ldots, a_n) = f(t_1, \ldots, t_n) = \langle g(t_{\sigma(1)}, \ldots, t_{\sigma(n)}), \rho \rangle = g^\sharp(\rho(t)) = g^\sharp(\rho(a_1 \otimes \ldots \otimes a_n)) = \langle g(a_{\sigma(1)}, \ldots, a_{\sigma(n)}), \rho \rangle$ as in equation (12).

$(2) \Longrightarrow (1)$: Take $X_i = A_i$, $a_i = \mathrm{id}_i$. Then $f(a_1, \ldots, a_n) = \langle g(a_{\sigma(1)}, \ldots, a_{\sigma(n)}), \rho \rangle$ implies $f^\sharp = g^\sharp \circ \rho$. $\square$

4. RULES OF COMPUTATION

4.1. Special cases:

With this symbolic notation we get the following *rules of computation*.

If $\rho = \mathrm{id}$ then

$$\langle f(t_1, \ldots, t_n), \mathrm{id} \rangle = f^\sharp(t) = f(t_1, \ldots, t_n). \tag{14}$$

So the identity braid in the pairing of our notation can be omitted.

If $f^\sharp = \mathrm{id}_{A_{\sigma(1)} \otimes \ldots \otimes A_{\sigma(n)}}$ then we get

$$\langle t_{\sigma(1)} \otimes \ldots \otimes t_{\sigma(n)}, \rho \rangle = \rho(t) = \rho \circ t. \tag{15}$$

4.2. Equality and substitution:

We begin with a warning. Usually certain terms in more complex expressions may be substituted by equal terms. However, separate components of the form $f(t_{\sigma(1)}, \ldots, t_{\sigma(n)})$ in our expression $\langle f(t_{\sigma(1)}, \ldots, t_{\sigma(n)}), \rho \rangle$ may not be replaced, even if it looks as if they could be equal.

For an example let indecomposable tensors $a_1 \otimes a_2, b_1 \otimes b_2 \in A \otimes A(X)$ be given, and let $m^\sharp : A \otimes A \longrightarrow A$ be a multiplication. Assume $m^\sharp(a_1 \otimes a_2) = a_1 a_2 = b_1 b_2 = m^\sharp(b_1 \otimes b_2)$. Then in general

$$\langle a_1 a_2, \tau^2 \rangle \neq \langle b_1 b_2, \tau^2 \rangle.$$

We will find a certain replacement or substitution rule in (17). This expression, however, differs from (17) in that here we have "elements" (or a "function applied to specific elements") whereas we have "functions" in (17). In terms of morphisms we may have

$$(X \xrightarrow{a} A \otimes A \xrightarrow{m^\sharp} A) = (X \xrightarrow{b} A \otimes A \xrightarrow{m^\sharp} A)$$

and at the same time

$$(X \xrightarrow{a} A \otimes A \xrightarrow{\tau^2} A \otimes A \xrightarrow{m^\sharp} A) \neq (X \xrightarrow{b} A \otimes A \xrightarrow{\tau^2} A \otimes A \xrightarrow{m^\sharp} A).$$

If $a = a_1 \otimes a_2$, $b = b_1 \otimes b_2$ are decomposable tensors then we have indeed

$$(X \otimes X \xrightarrow{a} A \otimes A \xrightarrow{\tau^2} A \otimes A \xrightarrow{m^\sharp} A) = (X \otimes X \xrightarrow{b} A \otimes A \xrightarrow{\tau^2} A \otimes A \xrightarrow{m^\sharp} A)$$

since τ^2 is a natural transformation.

By the definition of $\langle ., ., . \rangle$ we may certainly substitute equal expressions for the separate components f, ρ, and t. The following gives a somewhat more general rule for substitutions in case we have decomposable tensors as arguments.

PROPOSITION 4.1. *Given* $f, g : A_{\sigma(1)} \times \ldots \times A_{\sigma(1)} \longrightarrow B$, $\rho \in B$ *and* $a_i, b_i \in A_i(X_i)$ *defining decomposable tensors* $a = a_1 \otimes \ldots \otimes a_n$ *and* $b = b_1 \otimes \ldots \otimes b_n$. *If*

$$f(a_{\sigma(1)}, \ldots, a_{\sigma(n)}) = g(b_{\sigma(1)}, \ldots, b_{\sigma(n)})$$

then

$$\langle f(a_{\sigma(1)}, \ldots, a_{\sigma(n)}), \rho \rangle = \langle g(b_{\sigma(1)}, \ldots, b_{\sigma(n)}), \rho \rangle.$$

Proof. This is a simple computation:

$$\begin{aligned}
\langle f(a_{\sigma(1)}, \ldots, a_{\sigma(n)}), \rho \rangle &= f^{\sharp} \circ \rho \circ a \\
&= f^{\sharp} \circ (a_{\sigma(1)} \otimes \ldots \otimes a_{\sigma(n)}) \circ \rho \\
&= f(a_{\sigma(1)}, \ldots, a_{\sigma(n)}) \circ \rho \\
&= g(b_{\sigma(1)}, \ldots, b_{\sigma(n)}) \circ \rho \\
&= g^{\sharp} \circ (b_{\sigma(1)} \otimes \ldots \otimes b_{\sigma(n)}) \circ \rho \\
&= g^{\sharp} \circ \rho \circ b \\
&= \langle g(b_{\sigma(1)}, \ldots, b_{\sigma(n)}), \rho \rangle.
\end{aligned}$$

$\square$

The proposition shows that $\langle f(a_{\sigma(1)}, \ldots, a_{\sigma(n)}), \rho \rangle$ does indeed only depend on the value of $f(a_{\sigma(1)}, \ldots, a_{\sigma(n)})$ whereas in general it depends separately on f and a. So we may replace the term $f(a_{\sigma(1)}, \ldots, a_{\sigma(n)})$ by its value because the result does not depend on the particular representation.

If $t = \mathrm{id} : A_1 \otimes \ldots \otimes A_n \longrightarrow A_1 \otimes \ldots \otimes A_n$ then $t = \mathrm{id}_{A_1} \otimes \ldots \otimes \mathrm{id}_{A_n}$ is a decomposable tensor. In this case we may apply Proposition 4.1 and have $f(\mathrm{id}_{A_{\sigma(1)}}, \ldots, \mathrm{id}_{A_{\sigma(n)}})$ alone in $\langle f(\mathrm{id}_{A_{\sigma(1)}}, \ldots, \mathrm{id}_{A_{\sigma(n)}}), \rho \rangle$ is defined and we have $f(\mathrm{id}_{A_{\sigma(1)}}, \ldots, \mathrm{id}_{A_{\sigma(n)}}) = f^{\sharp}(\mathrm{id}_{A_{\sigma(1)}} \otimes \ldots \otimes \mathrm{id}_{A_{\sigma(n)}}) = f^{\sharp}$ so that we we can write $\langle f^{\sharp}, \rho \rangle$ and get

$$\langle f^{\sharp}, \rho \rangle = f^{\sharp} \circ \rho. \tag{16}$$

Hence the expression $\langle f^{\sharp}, \rho \rangle$ makes sense without the argument t. The argument can safely be assumed to be $t = \mathrm{id}$.

If $f_1^{\sharp}, f_2^{\sharp} : A_1 \otimes \ldots \otimes A_n \longrightarrow B$ then we have

$$\langle f_1^{\sharp}, \rho \rangle = \langle f_2^{\sharp}, \rho \rangle \iff f_1^{\sharp} = f_2^{\sharp}, \tag{17}$$

since ρ is an isomorphism, and

$$\langle f_1^{\sharp}, \rho_1 \rangle = \langle f_2^{\sharp}, \rho_2 \rangle \iff f_1^{\sharp} = \langle f_1^{\sharp}, \mathrm{id} \rangle = \langle f_2^{\sharp}, \rho_2 \rho_1^{-1} \rangle. \tag{18}$$

4.3. Compatibility with elements of the braid group:

If $t = a_1 \otimes \ldots \otimes a_n$ with $(a_1, \ldots, a_n) \in A_1(X_1) \times \ldots \times A_n(X_n)$ we get

$$a_{\sigma(1)} \otimes \ldots \otimes a_{\sigma(n)} = \rho(a_1 \otimes \ldots \otimes a_n)\rho^{-1} \tag{19}$$

since ρ is a natural transformation where the expression $a_{\sigma(1)} \otimes \ldots \otimes a_{\sigma(n)}$ is the morphism $a_{\sigma(1)} \otimes \ldots \otimes a_{\sigma(n)} : X_{\sigma(1)} \otimes \ldots \otimes X_{\sigma(n)} \longrightarrow A_{\sigma(1)} \otimes \ldots \otimes A_{\sigma(n)}$.

If $t \in A_1 \otimes \ldots \otimes A_n(X_1 \otimes \ldots \otimes X_n)$ then we can use equation (16) to define $\langle \rho t \rho^{-1}, \rho \rangle$ and get

$$\langle \rho t \rho^{-1}, \rho \rangle = \rho t \rho^{-1} \circ \rho = \rho \circ t = \langle t_{\sigma(1)} \otimes \ldots \otimes t_{\sigma(n)}, \rho \rangle$$

In view of equations (17) and (19) we define for $t \in A_1 \otimes \ldots \otimes A_n(X_1 \otimes \ldots \otimes X_n)$

$$t_{\sigma(1)} \otimes \ldots \otimes t_{\sigma(n)} := \rho t \rho^{-1}. \tag{20}$$

We will write $[\rho](t) := \rho t \rho^{-1} = t_{\sigma(1)} \otimes \ldots \otimes t_{\sigma(n)}$ if $t \in A_1 \otimes \ldots \otimes A_n(X_1 \otimes \ldots \otimes X_n)$. Then $[\rho](a_1 \otimes \ldots \otimes a_n) = a_{\sigma(1)} \otimes \ldots \otimes a_{\sigma(n)}$.

4.4. Composition:

Remark: The terms encountered in this section are complicated and usually have no simplification. They are given for completeness and will not be used in the sequel.

Certain of these expressions can be composed or applied to each other. In particular we get the following. Given $\rho_1, \rho_2 \in B_n$. If $f^\sharp : A_{\sigma_2\sigma_1(1)} \otimes \ldots \otimes A_{\sigma_2\sigma_1(n)} \to B$ and $t \in A_1 \otimes \ldots \otimes A_n(X)$ then

$$\langle f^\sharp, \rho_1 \rangle \left(\langle t_{\sigma_2(1)} \otimes \ldots \otimes t_{\sigma_2(n)}, \rho_2 \rangle \right) = \langle f(t_{\sigma_2\sigma_1(1)}, \ldots, t_{\sigma_2\sigma_1(n)}), \rho_1\rho_2 \rangle. \tag{21}$$

The following are immediately clear

$$\langle f^\sharp, \rho_1\rho_2 \rangle = \langle \langle f^\sharp, \rho_1 \rangle, \rho_2 \rangle, \tag{22}$$

$$\langle f_1^\sharp, \rho_1 \rangle \otimes \langle f_2^\sharp, \rho_2 \rangle = \langle f_1^\sharp \otimes f_2^\sharp, \rho_1 \otimes \rho_2 \rangle, \tag{23}$$

If $t \in A_1 \otimes \ldots \otimes A_n(X_1 \otimes \ldots \otimes X_n)$ then

$$\langle f^\sharp, \rho_1 \rangle \circ \langle t, \rho_2 \rangle = \langle f^\sharp \circ [\rho_1](t), \rho_1 \circ \rho_2 \rangle. \tag{24}$$

If $f^\sharp : A_{\sigma(1)} \otimes \ldots \otimes A_{\sigma(n)} \to B$ and $g : B \to C$ are given then

$$g(\langle f(t_{\sigma(1)}, \ldots, t_{\sigma(n)}), \rho \rangle) = \langle gf(t_{\sigma(1)}, \ldots, t_{\sigma(n)}), \rho \rangle. \tag{25}$$

If $t \in A_1 \otimes \ldots \otimes A_n(X)$ then we get from (21)

$$f^\sharp(\langle t_{\sigma(1)} \otimes \ldots \otimes t_{\sigma(n)}, \rho \rangle) = \langle f(t_{\sigma(1)}, \ldots, t_{\sigma(n)}), \rho \rangle. \tag{26}$$

The naturality of braids leads to a very usefull rule for a *change of variables or of arguments*. Given $f_i : A_i \to B_i, i = 1, \ldots, n$, $g^\sharp : B_1 \otimes \ldots \otimes B_n \to C$, and $t \in A_1 \otimes \ldots \otimes A_n(X)$. Then we get

$$
\begin{aligned}
\langle g^\sharp(f_{\sigma(1)}&(t_{\sigma(1)}) \otimes \ldots \otimes f_{\sigma(n)}(t_{\sigma(n)})), \rho \rangle \\
&= \langle g(f_{\sigma(1)}(t_{\sigma(1)}), \ldots, f_{\sigma(n)}(t_{\sigma(n)})), \rho \rangle \\
&= \langle (g^\sharp(f_{\sigma(1)} \otimes \ldots \otimes f_{\sigma(n)}))^\flat(t_{\sigma(1)}, \ldots, t_{\sigma(n)}), \rho \rangle \\
&= \langle (g(f_{\sigma(1)}, \ldots, f_{\sigma(n)}))(t_{\sigma(n)}, \ldots, t_{\sigma(n)}), \rho \rangle,
\end{aligned} \tag{27}
$$

or

$$
\begin{aligned}
\langle g[f_{\sigma(1)}&(t_{\sigma(1)}), \ldots, f_{\sigma(n)}(t_{\sigma(n)})], \rho \rangle \\
&= \langle (g^\sharp(f_{\sigma(1)} \otimes \ldots \otimes f_{\sigma(n)}) \otimes^n)[t_{\sigma(1)}, \ldots, t_{\sigma(n)}], \rho \rangle,
\end{aligned}
$$

where we change from the "arguments" $f_1(t_1), \ldots, f_n(t_n)$ to the "arguments" $t_1, \ldots, t_n$.

More general terms for composition are obtained from $f^\sharp : A_1 \otimes \ldots \otimes A_n \to B_1 \otimes \ldots \otimes B_m$ as

$$\langle f(t_{\sigma(1)}, \ldots, t_{\sigma(n)})_1 \otimes \ldots \otimes f(t_{\sigma(1)}, \ldots, t_{\sigma(n)})_m, \rho \rangle.$$

We get compositions of such terms which in general cannot be simplified. For $g^\sharp : B_1 \otimes \ldots \otimes B_m \to C$ we get

$$
\begin{aligned}
\langle g(\langle f(t_{\sigma_2(1)}&, \ldots, t_{\sigma_2(n)})_{\sigma_1(1)}, \ldots, f(t_{\sigma_2(1)}, \ldots, t_{\sigma_2(n)})_{\sigma_1(m)}, \rho_2 \rangle), \rho_1 \rangle \\
&= \langle g(f_{\sigma_1(1)}, \ldots, f_{\sigma_1(m)}), \rho_1 \rangle \\
&\quad \circ \langle f^\sharp(t_{\sigma_2(1)}, \ldots, t_{\sigma_2(n)})_1 \otimes \ldots \otimes f^\sharp(t_{\sigma_2(1)}, \ldots, t_{\sigma_2(n)})_m, \rho_2 \rangle. \tag{28}
\end{aligned}
$$

ON SYMBOLIC COMPUTATIONS IN BRAIDED MONOIDAL CATEGORIES 277

Furthermore we may take tensor products of terms as follows:

$$\langle f(t_{\sigma_1(1)}, \ldots, t_{\sigma_1(n)})_1 \otimes \ldots \otimes f(t_{\sigma_1(1)}, \ldots, t_{\sigma_1(n)})_m, \rho_1 \rangle \otimes$$
$$\langle g(s_{\sigma_2(1)}, \ldots, s_{\sigma_2(r)})_1 \otimes \ldots \otimes g(u_{\sigma_2(1)}, \ldots, u_{\sigma_2(r)})_s, \rho_2 \rangle =$$
$$\langle f(t_{\sigma(1)}, \ldots, t_{\sigma(n)}, s_{\sigma(n+1)}, \ldots, s_{\sigma2(n+r)})_1 \otimes \ldots$$
$$\ldots \otimes g(t_{\sigma(1)}, \ldots, t_{\sigma(n)}, s_{\sigma(n+1)}, \ldots, s_{\sigma2(n+r)})_{m+s}, \rho_1 \otimes \rho_2 \rangle. \tag{29}$$

5. COALGEBRAS, HOPF ALGEBRAS, AND BEYOND

5.1. Linear algebra:

First we observe some rules from Linear Algebra. Let $\kappa \in I(X)$, $a_i \in A_i(Y_i)$, and $f : A_1 \times \ldots \times A_n \to B$ be a multimorphism. Let κa_i resp $a_i \kappa$ denote the multiplication given by $\lambda : I \otimes A_i = A_i$ and $\rho : A_i \otimes I = A_i$. Then we have

$$f(a_1, \ldots, a_i \kappa, a_{i+1}, \ldots, a_n) = f(a_1, \ldots, a_i, \kappa a_{i+1}, \ldots, a_n) \tag{30}$$

and for any morphism $g : A \to B$

$$\kappa f(a) = f(\kappa a) \text{ and } f(a)\kappa = f(a\kappa). \tag{31}$$

A more interesting formula for a braiding is obtained as

$$\kappa a = \langle a\kappa, \tau \rangle \text{ and } \kappa a = \langle a\kappa, \tau^{-1} \rangle. \tag{32}$$

Let (A, ∇, η) be an algebra then

$$\eta(\kappa) \cdot a = \kappa a \text{ and } a \cdot \eta(\kappa) = a\kappa. \tag{33}$$

5.2. The Sweedler-Heyneman notation:

Let H be a Hopf algebra in $\mathcal{C}$. For $a \in H(X)$ we want to have $\Delta(a) = a_{(1)} \otimes a_{(2)}$.

Let $f : H \times \ldots \times H \to M$ with associated morphism $f^\sharp : H \otimes \ldots \otimes H \to M$ be given. Let $a \in H(X)$, then $\Delta^{n-1}(a) \in H \otimes \ldots \otimes H(X)$. Using the definition in equation (6) we define

$$\boxed{f(a_{(1)}, \ldots, a_{(n)}) := f^\sharp(\Delta^{n-1}(a)).} \tag{34}$$

As in equation (7) (and also as in (4)) this gives the formula $\Delta(a) = a_{(1)} \otimes a_{(2)}$. Then by equation (13)

$$\langle f(a_{(\sigma(1))}, \ldots, a_{(\sigma(n))}), \rho \rangle = f^\sharp(\rho(a_{(1)} \otimes \ldots \otimes a_{(n)})).$$

Using the coassociativity of Δ we get the following rule for a change of the number of arguments

$$\langle f(a_{(\sigma(1))}, \ldots, \Delta(a_{(\sigma(i))}), \ldots, a_{(\sigma(n))}), \rho \rangle = \langle f(a_{(\sigma(1))}, \ldots, a_{(\sigma(n+1))}), \rho_i \rangle \tag{35}$$

where ρ_i acts like ρ but switches the braids i and $i + 1$ in parallel.

Observe, however, that the braid does not change in

$$\langle f(a_{(\sigma(1))}, \ldots, a_{(\sigma(i))}, \ldots, a_{(\sigma(n+1))}), \rho \rangle$$
$$= \langle f(a_{(\sigma(1))}, \ldots, a_{(\sigma(i))(1)}, \ldots, a_{(\sigma(i))(2)}, \ldots, a_{(\sigma(n))}), \rho \rangle. \tag{36}$$

5.3. Hopf algebras:

As usual one gets

$$\eta\varepsilon(a_{(1)})a_{(2)} = a = a_{(1)}\eta\varepsilon(a_{(2)}) \tag{37}$$

and

$$a_{(1)}S(a_{(2)}) = \eta\varepsilon(a) = S(a_{(1)})a_{(2)}. \tag{38}$$

These last equations are to be considered as functions in one argument a, so they allow substitution at any position where a occurs.

The compatibility of multiplication and comultiplication is expressed by

$$(ab)_{(1)} \otimes (ab)_{(2)} = \langle a_{(1)}b_{(1)} \otimes a_{(2)}b_{(2)}, \tau_{23} \rangle \tag{39}$$

where $a \otimes b \in H \otimes H(X)$ and τ is the basic braid map interchanging two factors. Furthermore we have from (32)

$$\eta\varepsilon(a_{(1)})a_{(2)} = a = \langle a_{(2)}\eta\varepsilon(a_{(1)}), \tau \rangle. \tag{40}$$

THEOREM 5.1. *If H is a braided Hopf algebra then the antipode S of the Hopf algebra H is an algebra τ-antihomomorphism, i.e.*

$$S(ab) = \langle S(b)S(a), \tau \rangle. \tag{41}$$

Proof. We compute

$$
\begin{aligned}
S(ab) \quad &= S((ab)_{(1)}\eta\varepsilon((ab)_{(2)})) \\
&\quad \text{(by (37),} \quad \text{the arguments are } a, b) \\[4pt]
&= S((ab)_{(1)})\, \eta\varepsilon((ab)_{(2)}) \\
&\quad \text{(by (31) and (33))} \\[4pt]
&= \langle S(a_{(1)}b_{(1)})\, \eta\varepsilon(a_{(2)}b_{(2)}), \tau_{23} \rangle \\
&\quad \text{(by (39),} \quad \text{change to 4 arguments } a_{(1)}, a_{(2)}, b_{(1)}, b_{(2)}) \\[4pt]
&= \langle S(a_{(1)}b_{(1)})\, \varepsilon(a_{(2)}b_{(2)}), \tau_{23} \rangle \\
&\quad \text{(by (33))} \\[4pt]
&= \langle S(a_{(1)}b_{(1)})\, \varepsilon(a_{(2)})\, \varepsilon(b_{(2)}), \tau_{23} \rangle \\
&\quad (\varepsilon \text{ is multiplicative for } all \text{ elements}) \\[4pt]
&= \langle S(a_{(1)}b_{(1)})\, \eta\varepsilon(a_{(2)})\, \varepsilon(b_{(2)}), \tau_{23} \rangle \\
&\quad \text{(by (33))} \\[4pt]
&= \langle S(a_{(1)}b_{(1)})\, a_{(2)(1)}S(a_{(2)(2)})\, \varepsilon(b_{(2)}), \tau_{23} \rangle \\
&\quad \text{(by (38),} \quad \text{the arguments are still } a_{(1)}, a_{(2)}, b_{(1)}, b_{(2)}) \\[4pt]
&= \langle S(a_{(1)}b_{(1)})\, a_{(2)}S(a_{(3)})\, \varepsilon(b_{(2)}), \tau_{23}\tau_{34} \rangle \\
&\quad \text{(by (35), change of arguments to } a_{(1)}, a_{(2)}, a_{(3)}, b_{(1)}, b_{(2)}) \\
&\quad \text{(change of arguments by (27) to } a_{(1)}, a_{(2)}, S(a_{(3)}), b_{(1)}, \varepsilon(b_{(2)})) \\[4pt]
&= \langle S(a_{(1)}b_{(1)})\, a_{(2)}\, \varepsilon(b_{(2)})\, S(a_{(3)}), \tau_{45}\tau_{23}\tau_{34} \rangle \\
&\quad \text{(by (32), change arguments back to } a_{(1)}, a_{(2)}, a_{(3)}, b_{(1)}, b_{(2)} \text{ by (27))} \\[4pt]
&= \langle S(a_{(1)}b_{(1)})\, a_{(2)}\, \eta\varepsilon(b_{(2)})\, S(a_{(3)}), \tau_{45}\tau_{23}\tau_{34} \rangle \\
&\quad \text{(by (33))}
\end{aligned}
$$

$$= \langle S(a_{(1)}b_{(1)}) \, a_{(2)} \, b_{(2)} \, S(b_{(3)}) \, S(a_{(3)}), \tau_{56}\tau_{45}\tau_{23}\tau_{34} \rangle$$
(as above by (38), (35))

$$= \langle S(a_{(1)}b_{(1)}) \, a_{(2)} \, b_{(2)} \, S(b_{(3)}) \, S(a_{(3)}), \tau_{23}\tau_{56}\tau_{45}\tau_{34} \rangle$$
(change of braid map)

$$= \langle S(a_{(1)(1)}b_{(1)(1)}) \, a_{(1)(2)} \, b_{(1)(2)} \, S(b_{(2)}) \, S(a_{(2)}), \tau_{23}\tau_{56}\tau_{45}\tau_{34} \rangle$$
(by (36), the arguments are $a_{(1)(1)}, a_{(1)(2)}, a_{(2)}, b_{(1)(1)}, b_{(1)(2)}, b_{(2)}$)

$$= \langle S((a_{(1)}b_{(1)})_{(1)}) \, (a_{(1)}b_{(1)})_{(2)} \, S(b_{(2)}) \, S(a_{(2)}), \tau_{34}\tau_{23} \rangle \quad (39)$$
(change to 4 arguments $a_{(1)}, a_{(2)}, b_{(1)}, b_{(2)}$, apply (35) twice)
(read from lower line to upper line)

$$= \langle \eta\varepsilon(a_{(1)}b_{(1)}) \, S(b_{(2)}) \, S(a_{(2)}), \tau_{34}\tau_{23} \rangle$$
(by (38))

$$= \langle \varepsilon(a_{(1)}) \, \varepsilon(b_{(1)}) \, S(b_{(2)}) \, S(a_{(2)}), \tau_{34}\tau_{23} \rangle$$
(by (33) and multiplicativity of ε)

$$= \langle \varepsilon(a_{(1)}) \, S(b) \, S(a_{(2)}), \tau_{23} \rangle$$
(by (38) together with change of arguments to $a_{(1)}, a_{(2)}, b$)

$$= \langle S(b) \, \varepsilon(a_{(1)}) \, S(a_{(2)}), \tau_{12}\tau_{23} \rangle$$
(by (32))

$$= \langle S(b) \, S(a), \tau \rangle$$
(by (38) together with change of arguments to a, b). $\qquad\square$

REFERENCES

[1] A. Joyal and R. Street, The geometry of gensor calculus, I. *Adv. Math.* **88** (1991), 55–112.

[2] B. Pareigis, Non-additive ring and module theory I. General theory of monoids, *Publ. Math. Debrecen* **24** (1977), 190-204.

[3] R. G. Heyneman and M. E. Sweedler, Affine Hopf algebras I, *J. Algebra* **13** (1969), 192–241.

[4] R. Penrose, Applications of negative dimensional tensors., in "Combinatorial Mathematics and its Applications", Academic Press, London, 1971, 221–244.

Quotients of Finite Quasi-Hopf Algebras

PETER SCHAUENBURG Mathematisches Institut der Universität München
Theresienstr. 39, D-80333 München, Germany
e-mail: *schauen@rz.mathematik.uni-muenchen.de*

> ABSTRACT. Let H be a finite-dimensional quasi-Hopf algebra. We show for
> each quotient quasibialgebra Q of H that Q is a quasi-Hopf algebra whose
> dimension divides the dimension of H.

1. INTRODUCTION

In [7] Nichols and Zoeller prove what is now known as the Nichols-Zoeller Theorem:
A finite dimensional Hopf algebra H over a field k is a free module over every Hopf
subalgebra $K \subset H$. This answers affirmatively one of Kaplansky's conjectures on
Hopf algebras in the finite-dimensional case. The Nichols-Zoeller Theorem and
some related results are an important tool in the study of finite-dimensional Hopf
algebras.

Quasi-Hopf algebras, introduced by Drinfeld [2], are a generalization of ordinary
Hopf algebras that can be motivated most easily by looking at representation categories: The category of modules over a Hopf algebra is a monoidal category, with
the module structure on the tensor product over the base field of two modules
given by the diagonal action via comultiplication. The same thing is still true for
quasibialgebras; the difference is that now the tensor product of representations is
associative with an associativity isomorphism that differs from the ordinary one for
vector spaces.

In [11] we have proved the direct generalization of the Nichols-Zoeller Theorem
to quasi-Hopf algebras: A finite-dimensional quasi-Hopf algebra H over a field k
is a free module over every quasi-Hopf subalgebra $K \subset H$. The generalization is
made possible by the introduction of Hopf modules over quasibialgebras by Hausser
and Nill [4]. Hopf modules and the structure theorem for Hopf modules are a
key ingredient in the proof of the Nichols-Zoeller Theorem (and the subject of a
generalization of the statement of the theorem). Although quasi-Hopf algebras are
not coassociative coalgebras, and thus comodules over them are not immediately
defined, one can still define Hopf bimodules, and prove a structure theorem for
them.

One of the standard applications of the Nichols-Zoeller theorem is to investigate the
structure of (semisimple) Hopf algebras by dimension counting: The dimension of a
Hopf subalgebra has to divide the dimension of the large Hopf algebra, a version of
the classical Lagrange theorem for finite groups. This can serve to narrow down the

---

1991 *Mathematics Subject Classification.* 16W30.
Key words and phrases. Quasi-Hopf algebra, Nichols-Zoeller theorem, Hopf module.

possible examples in classification attempts. To be yet more specific, one standard argument is that the number of one-dimensional representations of a Hopf algebra H has to divide the dimenision of H. Unfortunately, the version of the Nichols-Zoeller Theorem for quasi-Hopf algebras provided in [11] does not help at all in this situation: The one-dimensional representations of H are not related to a quasi-Hopf subalgebra, but rather to a quotient quasi-Hopf algebra. For a quotient Hopf algebra Q of an ordinary Hopf algebra H, the Nichols-Zoeller Theorem implies by duality that H is a cofree Q-comodule. In the quasi-Hopf case, this does not even make sense to ask, since Q is not a coassociative coalgebra, hence H is not a comodule in the usual sense.

We shall nevertheless prove that the dimension of a finite-dimensional quasi-Hopf algebra H is divisible by the dimension of any quotient quasi-Hopf algebra Q. The key to this is the construction of an inclusion of quasi-Hopf algebras with the same ratio of dimensions as that between H and Q. In the ordinary Hopf case this is found by simply dualizing. In the quasi-Hopf case, we will find that H and Q^* generate a quasi-Hopf subalgebra $D(Q; H)$ in the Drinfeld double $D(H)$ of H, and $\dim D(Q; H) = \dim H \dim Q$, while $\dim D(H) = (\dim H)^2$.

Without doubt, one could verify the claims just made on $D(Q; H)$ by direct calculations with the rather complicated quasi-Hopf algebra structure of the Drinfeld double given by Hausser and Nill [3] (while the earlier description by Majid [5] is perhaps to indirect for this purpose). Instead, we will do a closer analysis of $D(Q; H)$ in Section 3, giving parallel interpretations for its modules to those of the double $D(H)$. This will allow us to show our claims without calculating much. Moreover, we will be able in Section 4 to show more than the "Lagrange" statement that $\dim Q$ divides $\dim H$: We will show that $\dim Q$ divides the dimension of any Hopf module in $^Q_H\mathcal{M}_H$, parallel to the results of Nichols and Zoeller who also show a freeness result for Hopf modules, not only for the Hopf algebras themselves.

In addition to the results summarized so far, we will show in Section 2 that any quotient quasibialgebra of a finite-dimensional quasi-Hopf algebra is a quotient quasi-Hopf algebra itself. In the case of ordinary Hopf algebras, this is due to Nichols [6], whose arguments we will vary in the necessary manner, replacing a canonical map $H \otimes H \to H \otimes H$ by its quasi-Hopf version due to Drinfeld. Nichols' result applies (by duality) to subbialgebras of finite-dimensional Hopf algebras, whereas it is not true for subquasibialgebras of finite-dimensional quasi-Hopf algebras, although we will give a positive result under additional hypotheses.

In an appendix, we will give a categorical proof, rather free of computations, of the canonical isomorphism $H \otimes H \to H \otimes H$ given by Drinfeld in [2] and used crucially in Section 2.

2. ANTIPODES FOR QUOTIENTS AND SUBOBJECTS

Recall that a quasibialgebra $H = (H, \Delta, \varepsilon, \phi)$ consists of an algebra H, algebra maps $\Delta \colon H \to H \otimes H$ and $\varepsilon \colon H \to k$, and an invertible element $\phi \in H^{\otimes 3}$, the associator, such that

$$(\varepsilon \otimes H)\Delta(h) = h = (H \otimes \varepsilon)\Delta(h), \tag{1}$$

$$(H \otimes \Delta)\Delta(h) \cdot \phi = \phi \cdot (\Delta \otimes H)\Delta(h), \tag{2}$$

$$(H \otimes H \otimes \Delta)(\phi) \cdot (\Delta \otimes H \otimes H)(\phi) = (1 \otimes \phi) \cdot (H \otimes \Delta \otimes H)(\phi) \cdot (\phi \otimes 1), \quad (3)$$

$$(H \otimes \varepsilon \otimes H)(\phi) = 1 \quad (4)$$

hold for all $h \in H$. We will write $\Delta(h) =: h_{(1)} \otimes h_{(2)}$, $\phi = \phi^{(1)} \otimes \phi^{(2)} \otimes \phi^{(3)}$, and $\phi^{-1} = \phi^{(-1)} \otimes \phi^{(-2)} \otimes \phi^{(-3)}$.

We define a morphism or map of quasibialgebras from (H, ϕ) to (L, ψ) to be an algebra map $f \colon H \to L$ compatible with comultiplication $\Delta f = (f \otimes f)\Delta$ and with the counit $\varepsilon f = \varepsilon$, and finally satisfying $(f \otimes f \otimes f)(\phi) = \psi$.

A subquasibialgebra is a subalgebra L of H that has a quasibialgebra structure for which the inclusion is a quasibialgebra map. Equivalently L is a subalgebra of H satisfying $\Delta(L) \subset L \otimes L$ and $\phi \in L \otimes L \otimes L$. A quotient quasibialgebra of H is a quasibialgebra Q for which there is a surjective quasibialgebra map $\nu \colon H \to Q$. Equivalently, Q is isomorphic to H/I for an ideal $I \subset H$ satisfying $\Delta(I) \subset I \otimes H + H \otimes I$ and $\varepsilon(I) = 0$; a coassociator for $Q = H/I$ is then the canonical image of ϕ in $Q \otimes Q \otimes Q$.

Recall further that a quasi-antipode for a quasibialgebra (H, ϕ) is a triple (S, α, β) in which $\alpha, \beta \in H$, and S is an anti-algebra endomorphism of H satisfying

$$S(h_{(1)})\alpha h_{(2)} = \varepsilon(h)\alpha, \qquad h_{(1)}\beta S(h_{(2)}) = \varepsilon(h)\beta,$$
$$\phi^{(1)}\beta S(\phi^{(2)})\alpha\phi^{(3)} = 1, \qquad S(\phi^{(-1)})\alpha\phi^{(-2)}\beta\phi^{(-3)} = 1,$$

for $h \in H$. A quasi-Hopf algebra is a quasibialgebra with a quasi-antipode. Note that our definition differs from Drinfeld's in that we do not require the antipode to be bijective. For finite-dimensional quasi-Hopf algebras it was recently shown by Bulacu and Caenepeel [1] that bijectivity of the antipode is automatic.

We define a quasibialgebra map $f \colon H \to H'$ between two quasi-Hopf algebras $(H, \phi, S, \alpha, \beta)$ and $(H', \phi', S', \alpha', \beta')$ to be a quasi-Hopf algebra map if $S'f = fS$, $f(\alpha) = \alpha'$, and $f(\beta) = \beta'$. We define a quasi-Hopf subalgebra to be a subquasibialgebra $L \subset H$ that has a quasi-antipode so that the inclusion is a quasi-Hopf algebra map. Equivalently, L is a subquasibialgebra such that $S(L) \subset L$, and $\alpha, \beta \in L$. Further we define a quotient quasi-Hopf algebra of H to be a quasi-Hopf algebra Q with a surjective quasi-Hopf algebra map $\nu \colon H \to Q$. Equivalently, a quotient quasi-Hopf algebra is a quotient quasibialgebra $Q \cong H/I$ such that $S(I) = I$. Then S induces an antiautomorphism on the quotient, which is a quasiantipode together with the canonical images of α and β in Q.

THEOREM 2.1. *Let H be a quasi-Hopf algebra over a field k, and let Q be a finite dimensional quotient quasibialgebra of H. Then Q is a quotient quasi-Hopf algebra.*

Proof. By [2, Prop. 1.5] we have an isomorphism

$$\varphi \colon H \otimes H \ni g \otimes h \mapsto gS(\phi^{(-1)})\alpha\phi^{(-2)}h_{(1)} \otimes \phi^{(-3)}h_{(2)} \in H \otimes H \quad (5)$$

with inverse given by

$$\varphi^{-1}(g \otimes h) = g\phi^{(1)}\beta S(h_{(1)}\phi^{(2)}) \otimes h_{(2)}\phi^{(3)}$$

Let $\nu \colon H \ni h \mapsto \overline{h} \in Q$ denote the canonical epi. Define

$$\overline{\varphi} \colon Q \otimes Q \ni p \otimes q \mapsto \overline{pS(\phi^{(-1)})\alpha\phi^{(-2)}}q_{(1)} \otimes \phi^{(-3)}q_{(2)} \in Q \otimes Q.$$

Then the diagram

$$
\begin{array}{ccc}
H \otimes H & \xrightarrow{\;\varphi\;} & H \otimes H \\
\downarrow{\scriptstyle \nu \otimes \nu} & & \downarrow{\scriptstyle \nu \otimes \nu} \\
Q \otimes Q & \xrightarrow{\;\overline{\varphi}\;} & Q \otimes Q
\end{array}
$$

commutes, and thus $\overline{\varphi}$ is onto. Since Q is finite dimensional, $\overline{\varphi}$ is an isomorphism. From the diagram we conclude that

$$
\overline{\varphi}^{-1}(\overline{g} \otimes \overline{h}) = (\nu \otimes \nu)\varphi^{-1}(g \otimes h),
$$

hence

$$
\theta \colon Q \ni q \mapsto (Q \otimes \varepsilon)\overline{\varphi}^{-1}(1 \otimes q) \in Q
$$

satisfies $\theta(\overline{h}) = \overline{\beta S(h)}$ for all $h \in H$. Now define

$$
\overline{S} \colon Q \ni q \mapsto \overline{S(\phi^{(-1)})\alpha\phi^{(-2)}}\,\theta(q\overline{\phi^{(-3)}}) \in Q.
$$

Then for $h \in H$ we have

$$
\overline{S}(\overline{h}) = \overline{S(\phi^{(-1)})\alpha\phi^{(-2)}} \cdot \overline{\beta S(h\phi^{(-3)})} = \overline{S(\phi^{(-1)})\alpha\phi^{(-2)}\beta S(\phi^{(-3)})} \cdot \overline{S(h)} = \overline{S(h)},
$$

showing that S maps the kernel of ν into itself. $\qquad\square$

Remark 2.2. The direct analog of Theorem 2.1 for subquasibialgebras instead of quotients is false for quite trivial reasons. To see this consider a quasi-Hopf algebra $(H, \phi, S, \alpha, \beta)$ and a quasi-Hopf subalgebra $K \subset H$. By the remark following the definition of a quasi-Hopf algebra in [2], we can obtain another quasi-Hopf structure $(H, \phi, S', \alpha', \beta')$ for any unit $u \in H$ by setting $S'(h) = uS(h)u^{-1}$, $\alpha' = u\alpha$, $\beta' = \beta u^{-1}$, while leaving ϕ unchanged. Of course, it may happen that K is not a quasi-Hopf subalgebra for this new quasi-Hopf structure (for example, if α is a unit, and $u \notin K$).

However, we can provide quasiantipodes for subquasibialgebras under some additional assumptions:

PROPOSITION 2.3. *Let H be a quasi-Hopf algebra with coassociator ϕ and quasi-antipode (S, α, β).*
Let $K \subset H$ be a finite-dimensional subquasibialgebra such that

$$
S(\phi^{(-1)})\alpha\phi^{(-2)} \otimes \phi^{(-3)} \in K \otimes K.
$$

Then K is a quasi-Hopf subalgebra.

Proof. Again we consider the canonical map φ from (5). By our extra assumptions, we see that $\varphi(K \otimes K) \subset K \otimes K$, and consider the map $\varphi' \colon K \otimes K \to K \otimes K$ given by restricting φ. It is injective since φ is, hence bijective by finite dimensionality. The inverse of φ' is given by the restriction of φ^{-1}, so we see that for $x \in K$

$$
K \ni (K \otimes \varepsilon)\varphi^{-1}(1 \otimes x) = \beta S(x),
$$

hence in particular $\beta \in K$, and for all $x \in K$

$$
K \ni S(\phi^{(-1)})\alpha\phi^{(-2)}\beta S(x\phi^{(-3)}) = S(x).
$$

Finally $\alpha = (K \otimes \varepsilon)\varphi(1 \otimes 1) \in K$, so K is a quasi-Hopf subalgebra. $\qquad\square$

QUOTIENTS OF FINITE QUASI-HOPF ALGEBRAS

Remark 2.4. As a special case of Proposition 2.3, a finite-dimensional subquasibialgebra $K \subset H$ of a quasi-Hopf algebra H is a quasi-Hopf subalgebra provided that it contains a subquasibialgebra $L \subset K$ which is a quasi-Hopf subalgebra of H.

3. THE PARTIAL DOUBLE

Throughout the section, we let H denote a quasi-Hopf algebra. The key property of a quasibialgebra is that its modules form a monoidal category: The tensor product of $V, W \in {}_H\mathcal{M}$ is their tensor product $V \otimes W$ over k, endowed with the diagonal module structure $h(v \otimes w) = h_{(1)}v \otimes h_{(2)}w$; the neutral object is k with the trivial module structure given by ε. The associativity isomorphism in the category is

$$(U \otimes V) \otimes W \ni u \otimes v \otimes w \mapsto \phi^{(1)}u \otimes \phi^{(2)}v \otimes \phi^{(3)}w \in U \otimes (V \otimes W)$$

for $U, V, W \in {}_H\mathcal{M}$.

The opposite of a quasibialgebra and the tensor product of two quasibialgebras are naturally quasibialgebras. Thus ${}_H\mathcal{M}_H$ is also a monoidal category, with associativity isomorphism

$$(U \otimes V) \otimes W \ni u \otimes v \otimes w \mapsto \phi^{(1)}u\phi^{(-1)} \otimes \phi^{(2)}v\phi^{(-2)} \otimes \phi^{(3)}w\phi^{(-3)} \in U \otimes (V \otimes W).$$

We will make free use of the formalism of (co)algebra and (co)module theory within monoidal categories. When C, D are coalgebras in ${}_H\mathcal{M}_H$, we will use the abbreviations ${}^C_H\mathcal{M}_H, {}_H\mathcal{M}^D_H, {}^C_H\mathcal{M}^D_H$ for the categories of left C-comodules, right D-comodules, and C-D-bicomodules within the monoidal category ${}_H\mathcal{M}_H$.

We see that H itself is a coassociative coalgebra within the monoidal category ${}_H\mathcal{M}_H$. Thus we can define a Hopf module $M \in {}_H\mathcal{M}^H_H$ to be a right H-comodule within the category ${}_H\mathcal{M}_H$. Written out explicitly, this definition is the same as that of Hausser and Nill [4, Def.3.1]. Hausser and Nill have also proved a structure theorem for such Hopf modules, which says that the functor

$$\mathcal{R}: {}_H\mathcal{M} \ni V \mapsto .V \otimes .H. \in {}_H\mathcal{M}^H_H$$

is an equivalence of categories. We have used this equivalence as the basis of a description of the Drinfeld double of H in [10]. In [12, Expl.4.10] we have repeated this description with a general $\mathcal{C}$-categorical technique, which we shall now follow once more to obtain a relative double $D(L; H)$ associated to any quasibialgebra map $\nu: H \to L$.

For any right H comodule M in ${}_H\mathcal{M}_H$ and any $P \in {}_H\mathcal{M}_H$ we can form the right H-comodule $P \otimes M$ in ${}_H\mathcal{M}_H$, which gives us a functor ${}_H\mathcal{M}_H \times {}_H\mathcal{M}^H_H \to {}_H\mathcal{M}^H_H$ that makes ${}_H\mathcal{M}^H_H$ into a left ${}_H\mathcal{M}_H$-category in the sense of Pareigis [8]. Being equivalent to ${}_H\mathcal{M}^H_H$, the category ${}_H\mathcal{M}$ is then also a left ${}_H\mathcal{M}_H$-category, which means that we have a functor $\Diamond: {}_H\mathcal{M}_H \times {}_H\mathcal{M} \to {}_H\mathcal{M}$ and a coherent natural isomorphism $\Omega: (P \otimes Q)\Diamond V \to P\Diamond(Q\Diamond V)$ for $P, Q \in {}_H\mathcal{M}_H$ and $V \in {}_H\mathcal{M}$.

Now let C be a coalgebra in ${}_H\mathcal{M}_H$. Since ${}_H\mathcal{M}$ is a left ${}_H\mathcal{M}_H$-category, it makes sense (see [8]) to talk about C-comodules within ${}_H\mathcal{M}$, which form a category $^C({}_H\mathcal{M})$, which in our situation is naturally equivalent to ${}^C_H\mathcal{M}^H_H$, with the equivalence $^C({}_H\mathcal{M}) \cong {}^C_H\mathcal{M}^H_H$ induced by $\mathcal{R}$.

By [12, Thm.3.3] and the remarks preceding it, $\Diamond$ induces a functor ${}_H\mathcal{M}_H \ni P \mapsto P\Diamond H \in {}_H\mathcal{M}_H$, and we have an isomorphism $(P\Diamond H) \otimes_H V \cong P\Diamond V$, natural in $P \in {}_H\mathcal{M}_H$ and $V \in {}_H\mathcal{M}$.

By [12, Cor.3.8], $C \Diamond H$ has an H-coring structure in such a way that one has a category equivalence $^{C \Diamond H} \mathcal{M} \cong {}^{C}({}_{H}\mathcal{M})$ that commutes with the underlying functors to $_{H}\mathcal{M}$. (Here $^{C \Diamond H} \mathcal{M}$ denotes the category of left comodules over the H-coring $C \Diamond H$.) For any coalgebra morphism $f \colon C \to D$ in $_{H}\mathcal{M}_{H}$ we obtain a commutative diagram of functors

$$
\begin{array}{ccccc}
{}^{C \Diamond H}\mathcal{M} & \longrightarrow & {}^{C}({}_{H}\mathcal{M}) & \overset{\mathcal{R}}{\longrightarrow} & {}^{C}_{H}\mathcal{M}^{H}_{H} \\
\downarrow{\scriptstyle f \Diamond H \mathcal{M}} & & \downarrow{\scriptstyle f({}_{H}\mathcal{M})} & & \downarrow{\scriptstyle {}^{f}_{H}\mathcal{M}^{H}_{H}} \\
{}^{D \Diamond H}\mathcal{M} & \longrightarrow & {}^{D}({}_{H}\mathcal{M}) & \overset{\mathcal{R}}{\longrightarrow} & {}^{D}_{H}\mathcal{M}^{H}_{H}
\end{array}
$$

for the H-coring map $f \Diamond H \colon C \Diamond H \to D \Diamond H$.

From [10, Sec.4] we know that $P \Diamond V \cong P \otimes V$ as vector spaces, functorially in $P \in {}_{H}\mathcal{M}_{H}$ and $V \in {}_{H}\mathcal{M}$. In particular we have $C \Diamond H \cong C \otimes H$ as right H-modules, functorially in the coalgebra C in $_{H}\mathcal{M}_{H}$.

Now assume that C is finite dimensional. Then $C \Diamond H$ is a finitely generated free right H-module, so the dual algebra $\mathrm{Hom}_{-H}(C \Diamond H, H)$ of the H-coring $C \Diamond H$ fulfills $_{(C \Diamond H)^{\vee}}\mathcal{M} \cong {}^{C \Diamond H}\mathcal{M}$. Note that $(C \Diamond H)^{\vee} \cong H \otimes C^{*}$ as vector spaces, functorially in C.

Finally, we specialize to a class of examples of coalgebras in $_{H}\mathcal{M}_{H}$. Whenever $\nu \colon H \to L$ is a morphism of quasibialgebras, we can consider L as a coalgebra in $_{H}\mathcal{M}_{H}$ with respect to its usual comultiplication, and the H-bimodule structure induced along ν. We write $D(L; H) := (L \Diamond H)^{\vee}$ for the dual algebra of the coring $L \Diamond H$. Note that $D(L; H) \cong H \otimes L^{*}$ as vector spaces. The isomorphism is natural in L, meaning that for any morphism $f \colon L \to M$ of quasibialgebras, the induced morphism $D(f; H) \colon D(M; H) \to D(L; H)$ corresponds to $H \otimes f^{*} \colon H \otimes M^{*} \to H \otimes L^{*}$. In particular it is surjective (resp. injective) if f is injective (resp. surjective). The same calculations as those made to prove [10, Lem.3.2] prove more generally that $_{L}^{L}\mathcal{M}^{H}_{H}$ is a monoidal category, the tensor product of $M, N \in {}^{L}_{H}\mathcal{M}^{H}_{H}$ being $M \otimes_{H} N$ with the left and right comodule structures

$$
M \underset{H}{\otimes} N \ni m \otimes n \mapsto m_{(-1)}n_{(-1)} \otimes m_{(0)} \otimes n_{(0)} \in L \otimes \left(M \underset{H}{\otimes} N \right)
$$

$$
M \underset{H}{\otimes} N \ni m \otimes n \mapsto m_{(0)} \otimes n_{(0)} \otimes m_{(1)}n_{(1)} \in \left(M \underset{H}{\otimes} N \right) \otimes H.
$$

Note that the underlying functor $_{H}^{L}\mathcal{M}^{H}_{H} \to {}_{H}\mathcal{M}^{H}_{H}$ is monoidal.

The equivalence $\mathcal{R}$ is a monoidal equivalence. We make $_{D(L;H)}\mathcal{M}$ a monoidal category in such a way that the equivalence $_{D(L;H)}\mathcal{M} \cong {}^{L}_{H}\mathcal{M}^{H}_{H}$ is a monoidal functor. Thus, for any quasibialgebra maps $H \overset{\nu}{\to} L \overset{f}{\to} M$ we obtain a commutative diagram of monoidal functors

$$
\begin{array}{ccccc}
(M \Diamond H)^{\vee}\mathcal{M} & \overset{(f \Diamond H)^{\vee}\mathcal{M}}{\longrightarrow} & (L \Diamond H)^{\vee}\mathcal{M} & \overset{(\epsilon \Diamond H)^{\vee}\mathcal{M}}{\longrightarrow} & {}_{H}\mathcal{M} \\
\downarrow & & \downarrow & & \downarrow \\
{}^{M}_{H}\mathcal{M}^{H}_{H} & \overset{{}^{f}_{H}\mathcal{M}^{H}_{H}}{\longrightarrow} & {}^{L}_{H}\mathcal{M}^{H}_{H} & \longrightarrow & {}_{H}\mathcal{M}^{H}_{H}
\end{array}
$$

By a trivial modification of the monoidal category structures (cf. [9, Rem.5.3] for an analogous trick) we can make sure that the functors in the top row are strict monoidal functors. This implies that $D(L; H)$ is a quasibialgebra, and that $D(f; H) \colon D(M; H) \to D(L; H)$ is a quasibialgebra map.

4. LAGRANGE'S THEOREM

Consider a finite-dimensional quasi-Hopf algebra H, and a quotient quasi-Hopf algebra Q. As a particular case of the constructions in the preceding section, we obtain injective quasibialgebra maps $H \to D(Q; H) \to D(H; H)$. Note that $D(H; H) = D(H)$ is the Drinfeld double of H, which is a quasi-Hopf algebra. As noted in Remark 2.4, it follows from Proposition 2.3 that $D(Q; H)$ is a quasi-Hopf algebra as well, and a quasi-Hopf subalgebra of $D(H)$. By [11], $D(H)$ is a free $D(Q; H)$-module, so in particular $\dim D(Q; H) = \dim Q \dim H$ divides $\dim D(H) = (\dim H)^2$. Cancelling $\dim H$ we get:

COROLLARY 4.1. *Let H be a finite-dimensional quasi-Hopf algebra, and Q a quotient quasibialgebra of H. Then $\dim Q$ divides $\dim H$.*

As an immediate application we have:

COROLLARY 4.2. *Let H be a finite-dimensional quasi-Hopf algebra. Then the number of one-dimensional representations of H divides $\dim H$.*

Proof. We pass to the dual picture: Representations of H are comodules over the dual coalgebra H^*. One-dimensional comodules correspond to grouplike elements of H^*. These grouplikes span a sub-coquasibialgebra of H^*, which corresponds to a quotient quasibialgebra of H, whose dimension divides the dimension of H by the preceding corollary. $\qquad\qquad\square$

Nichols and Zoeller [7] do not only prove a freeness theorem for Hopf algebra inclusions $K \subset H$, but also a freeness theorem for Hopf modules in $^H_K\mathcal{M}$. In particular, for any finite-dimensional $M \in {}^H_K\mathcal{M}$, they show that $\dim K \mid \dim M$. Suppose that H is a finite-dimensional quasi-Hopf algebra and Q a quotient quasibialgebra. Then Q is a coalgebra in the monoidal category $_H\mathcal{M}_H$, so that we have a well-defined notion of Hopf module in $^Q_H\mathcal{M}_H$ (while $^Q_H\mathcal{M}$ is not defined).

PROPOSITION 4.3. *Let H be a finite-dimensional quasi-Hopf algebra, Q a quotient quasibialgebra of H, and $M \in {}^Q_H\mathcal{M}_H$. Then $\dim Q$ divides $\dim M$.*

Proof. Consider the commutative diagram of functors (first without the dotted arrows)

$$
\begin{array}{ccc}
_{D(H)}\mathcal{M} & \xrightarrow{\ \mathcal{R}\ } & ^H_H\mathcal{M}^H_H \\
\mathcal{V}\Big\downarrow\Big\uparrow & & \mathcal{U}\Big\downarrow\Big\uparrow \\
_{D(Q;H)}\mathcal{M} & \xrightarrow{\ \mathcal{R}\ } & ^Q_H\mathcal{M}^H_H
\end{array}
$$

in which the functor $\mathcal{V}$ is the underlying functor induced by the inclusion $D(Q; H) \to D(H)$, and $\mathcal{U}$ is the underlying functor induced by the projection $H \to Q$. The horizontal arrows are induced by the category equivalence $\mathcal{R} \colon {}_H\mathcal{M} \to {}_H\mathcal{M}^H_H$. Since they are equivalences, the diagram also commutes for the dotted arrows, if these denote the right adjoint functors to $\mathcal{V}$ and $\mathcal{U}$. Now the right adjoint to the underlying functor $\mathcal{U}$ is given by cotensor product with H, taken within the monoidal category $_H\mathcal{M}_H$ (dually to the induction functor for an algebra inclusion), whereas the right adjoint to $\mathcal{V}$ is the usual coinduction functor for the algebra inclusion $D(Q; H) \subset D(H)$. This means that for $W \in {}_{D(Q;H)}\mathcal{M}$ we have

$$H \underset{Q}{\square} \mathcal{R}(W) \cong \mathcal{R}(\mathrm{Hom}_{D(Q;H)}(D(H), W)).$$

In particular, since $D(H)$ is a free $D(Q; H)$-module of rank $\dim H / \dim Q$, and since $\mathcal{R}$ multiplies dimensions by $\dim H$, we have

$$\dim(H \underset{Q}{\square} M) = \frac{\dim H}{\dim Q} \dim M$$

whenever $M \in {}^Q_H\mathcal{M}^H_H$ is finite dimensional.

Now consider a finite-dimensional $V \in {}^Q_H\mathcal{M}_H$. Then we can tensor V with $H \in {}_H\mathcal{M}^H_H$ to obtain $M = V \otimes H \in {}^Q_H\mathcal{M}^H_H$. Calculating within the monoidal category $_H\mathcal{M}_H$ we have

$$(H \underset{Q}{\square} V) \otimes H \cong H \underset{Q}{\square} (V \otimes H) = H \underset{Q}{\square} M.$$

It follows that

$$\dim H \dim(H \underset{Q}{\square} V) = \frac{\dim H}{\dim Q} \dim M$$

or $\dim(H \square_Q V) = \frac{\dim H}{\dim Q} \dim V$.

But on the other hand $H \square_Q V \in {}^H_H\mathcal{M}_H$. By the left-right switched version of the structure theorem of Hausser and Nill, any Hopf module in $^H_H\mathcal{M}_H$ is a free left H-module, so that $\dim H$ divides $\dim(H \square_Q V)$. Thus $\dim Q$ divides $\dim V$. $\qquad \square$

Appendix A. A DOGMATIC PROOF OF A FORMULA OF DRINFELD

In this section we return to the isomorphism φ from equation (5). Its proof in Drinfeld's paper [2] is not particularly hard, but does involve a calculation with the coassociator element of H, the pentagon identity (3) and the identities defining a quasi-antipode. The "dogma" alluded to in this section's title says that such calculations should be banned. After all, the pentagon identity precisely ensures that $_H\mathcal{M}$ is a monoidal category, and the quasi-antipode axioms precisely ensure that the dual vector space of a finite-dimensional module can be made into a dual object inside that monoidal category. Since the axioms are more or less equivalent to the categorical properties, no further reference to the former should be necessary, and using the latter should lead to easier and more conceptual proofs.

So now we let H be a finite-dimensional quasi-Hopf algebra and $V \in {}_H\mathcal{M}$ a finite-dimensional H-module. Then V has the dual object $V^* \in {}_H\mathcal{M}$, with module structure given by the transpose of the right module structure V_S induced via the quasi-antipode, evaluation $\mathrm{ev}\colon V^* \otimes V \ni \varphi \otimes v \mapsto \varphi(\alpha v) \in k$ and dual basis $\mathrm{db}\colon k \to V \otimes V^*$, $\mathrm{db}(1) = \beta v_i \otimes v^i$, where v_i and v^i are a pair of dual bases in V and V^*, and summation is suppressed. It follows that the functor

$$\mathcal{F}\colon {}_H\mathcal{M} \ni W \mapsto .W \otimes .V \in {}_H\mathcal{M}$$

QUOTIENTS OF FINITE QUASI-HOPF ALGEBRAS

has the right adjoint

$$\mathcal{G}: {}_H\mathcal{M} \ni X \mapsto X \otimes V^* \in {}_H\mathcal{M}$$

with unit u and counit c of the adjunction given by

$$u = \left(X \xrightarrow{X \otimes \mathrm{db}} X \otimes (V \otimes V^*) \xrightarrow{\Phi^{-1}} (X \otimes V) \otimes V^* \right)$$

$$c = \left((X \otimes V^*) \otimes V \xrightarrow{\Phi} X \otimes (V^* \otimes V) \xrightarrow{X \otimes \mathrm{ev}} X \right).$$

We have the standard isomorphism

$$\Gamma = \left(X \otimes V^* \xrightarrow{\cong} \mathrm{Hom}(V, X) \xrightarrow{\cong} \mathrm{Hom}_{H-}(.H \otimes V, X) \right)$$

with $\Gamma(x \otimes \varphi)(h \otimes v) = h\varphi(v)x$ and $\Gamma^{-1}(f) = f(1 \otimes v_i) \otimes v^i$. Via the isomorphism Γ we have another right adjoint

$$\mathcal{G}': {}_H\mathcal{M} \ni X \mapsto \mathrm{Hom}_{H-}(.H \otimes V, X) \in {}_H\mathcal{M}.$$

The unit u' and counit c' of the adjunction are given by

$$u' = \left(X \xrightarrow{u} (X \otimes V) \otimes V^* \xrightarrow{\Gamma} \mathrm{Hom}_{H-}(Q, X \otimes V) \right)$$

$$c' = \left(\mathrm{Hom}_{H-}(Q, X) \otimes V \xrightarrow{\Gamma^{-1} \otimes V} (X \otimes V^*) \otimes V \xrightarrow{c} X \right)$$

so that

$$u'(x)(h \otimes v) = \Gamma(u(x))(h \otimes v) = \Gamma(\phi^{(-1)}x \otimes \phi^{(-2)}\beta v_i \otimes \phi^{(-3)}v^i)(h \otimes v)$$
$$= h(\phi^{(-3)}v^i)(v) \cdot (\phi^{(-1)}x \otimes \phi^{(-2)}\beta v_i) = hv^i(v) \cdot (\phi^{(-1)}x \otimes \phi^{(-2)}\beta S(\phi^{(-3)})v_i)$$
$$h_{(1)}\phi^{(-1)}x \otimes h_{(2)}\phi^{(-2)}\beta S(\phi^{(-3)})v$$

and

$$c'(f \otimes v) = c(f(1 \otimes v_i) \otimes v^i \otimes v) = \phi^{(1)}f(1 \otimes v_i)(\phi^{(2)}v^i)(\alpha\phi^{(3)}v)$$
$$= f(\phi^{(1)} \otimes v_i)v^i(S(\phi^{(2)})\alpha\phi^{(3)}v) = f(\phi^{(1)} \otimes S(\phi^{(2)})\alpha\phi^{(3)}v).$$

One checks that the relevant H-module structure on $\mathrm{Hom}_{H-}(H \otimes V, X)$ (making Γ an H-module map) is given by $(hf)(g \otimes v) = f(gh_{(1)} \otimes S(h_{(2)})v)$. In other words, we have $\mathcal{G}'(X) = \mathrm{Hom}_{H-}(Q, X)$ for the H-bimodule $Q = .H.\otimes.(V_S)$. In particular $\mathcal{G}'$ is the right adjoint in the standard hom-tensor adjunction with left adjoint $\mathcal{F}' = Q \otimes_H (-)$. We denote the counit and unit of that standard adjunction by c'', u''. Now left adjoints are unique, so that we get mutually inverse isomorphisms

$$\Lambda = \left(Q \underset{H}{\otimes} V \xrightarrow{Q \otimes_H u'} Q \underset{H}{\otimes} \mathrm{Hom}_{H-}(Q, X \otimes V) \xrightarrow{c''} X \otimes V \right)$$

$$\Lambda^{-1} = \left(X \otimes V \xrightarrow{u'' \otimes V} \mathrm{Hom}_{H-}(Q, Q \underset{H}{\otimes} X) \otimes V \xrightarrow{c'} Q \underset{H}{\otimes} X \right).$$

We compute

$$\Lambda(h \otimes v \otimes x) = u'(x)(h \otimes v) = h_{(1)}\phi^{(-1)}x \otimes h_{(2)}\phi^{(-2)}\beta S(\phi^{(-3)})v$$

and

$$\Lambda^{-1}(x \otimes v) = c'(u''(x) \otimes v) = u''(x)(\phi^{(1)} \otimes S(\phi^{(2)})\alpha\phi^{(3)}v) = \phi^{(1)} \otimes S(\phi^{(2)})\alpha\phi^{(3)}v \otimes x$$

Finally, we specialize $V = X = H$, and identify $Q \otimes_H H \cong Q = H \otimes H$ to find $\Lambda(h \otimes g) = h_{(1)}\phi^{(-1)} \otimes h_{(2)}\phi^{(-2)}\beta S(\phi^{(-3)})g$ and $\Lambda^{-1}(h \otimes g) = \phi^{(1)}h_{(1)} \otimes S(h_{(2)})S(\phi^{(2)})\alpha\phi^{(3)}g$.

We have obtained the version of (5) for the opposite and coopposite quasi-Hopf algebra to H.

REFERENCES

[1] D. Bulacu and S. Caenepeel, Integrals for (dual) quasi-Hopf algebras. applications, *J. Algebra* **266** (2003), 552–583.

[2] V. Drinfeld, Quasi-Hopf algebras, *Leningrad Math. J.* **1** (1990), 1419–1457.

[3] F. Hausser and F. Nill, Doubles of quasi-quantum groups. *Comm. Math. Phys.* **199** (1999), 547–589.

[4] F. Hausser and F. Nill, Integral theory for quasi-Hopf algebras, preprint math. QA/9904164.

[5] S. Majid, Quantum double for quasi-Hopf algebras, *Lett. Math. Phys.* **45** (1998), 1–9.

[6] W. D. Nichols, Quotients of Hopf algebras, *Comm. Algebra* **6** (1978), 1789–1800.

[7] W. D. Nichols and M. B. Zoeller, A Hopf algebra freeness theorem, *Amer. J. Math.* **111** (1989), 381–385.

[8] B. Pareigis, Non-additive ring and module theory II. C-categories, C-functors and C-morphisms, *Publ. Math. Debrecen* **24** (1977), 351–361.

[9] P. Schauenburg, Bialgebras over noncommutative rings and a structure theorem for Hopf bimodules, *Appl. Categorical Structures* **6** (1998), 193–222.

[10] P. Schauenburg, Hopf modules and the double of a quasi-Hopf algebra, *Trans. Amer. Math. Soc.* **354** (2002), 3349–3378.

[11] P. Schauenburg, A quasi-Hopf algebra freeness theorem, preprint math. QA/0204141.

[12] P. Schauenburg, Actions of monoidal categories and generalized Hopf smash products, *J. Algebra*, to appear.

Adjointable Monoidal Functors and Quantum Groupoids

KORNEL SZLACHANYI   Theory Division, Research Institute for Particle and Nuclear Physics, PO Box 49, H-1525 Budapest, Hungary
e-mail: *szlach@rmki.kfki.hu*

> ABSTRACT. Every monoidal functor $G\colon C \to M$ has a canonical factorization through the category $_R M_R$ of bimodules in M over some monoid R in M in which the factor $U\colon C \to {}_R M_R$ is strongly unital. Using this result and the characterization of the forgetful functors $M_A \to {}_R M_R$ of bialgebroids A over R given by Schauenburg [15] together with their bimonad description given by the author in [18] here we characterize the "long" forgetful functors $M_A \to {}_R M_R \to M$ of both bialgebroids and weak bialgebras.

1. INTRODUCTION

Takeuchi's $\times_R$-bialgebras [20] or, what is the same [5], Lu's bialgebroids [9] provide far reaching generalizations of the notion of bialgebra. A bialgebroid A is, roughly speaking, a bialgebra over some non-commutative k-algebra R. With noncommutativity of R, however, a new phenomenon appears: the separation of algebra and coalgebra structures into two different categories. While bialgebras are monoids and comonoids in the same category M_k, bialgebroids are monoids in M_k (or in $_{R^e} M_{R^e}$) but comonoids in $_R M_R$. This makes the compatibility conditions rather difficult to formulate.

Simplification, if at all, is expected on passing to the level of categories and functors. Let us take, for example, a k-algebra A and associate to it the monad $T = \_ \otimes A$ on M_k. The monads obtained that way are precisely the monads which have right adjoints. For this characterization of algebras the closed monoidal structure of M_k is essential. The Eilenberg-Moore category of "T-algebras" [11, 2], or perhaps better to say, "T-modules" is nothing but the category M_A of right A-modules equipped with the forgetful functor $M_A \to M_k$. Similarly, we can consider monads on the closed monoidal category $_R M_R \equiv M_{R^e}$, where $R^e = R^{\mathrm{op}} \otimes R$, with right adjoints. It turns out that these are, up to isomorphisms, precisely the monads $\_ \otimes_{R^e} A$ associated to monoids A in $_{R^e} M_{R^e}$, also called R^e-rings. The forgetful functor $U^A\colon M_A \to {}_R M_R$ is monadic, has a right adjoint but has no monoidal structure. At this point bialgebroids enter naturally via Schauenburg's Theorem

---

1991 *Mathematics Subject Classification.* 16W30, 18D10, 16D90.
Key words and phrases. bialgebroid, quantum groupoid, forgetful functor.
This research was partially supported by the Hungarian Scientific Research Fund, OTKA T 034 512.

[15]: the monoidal structures on $\mathcal{M}_A$ such that U^A is strict monoidal are in one-to-one correspondence with (right) bialgebroid structures on the R^e-ring A.

Monoidal structures on $\mathcal{M}_A$ can also be described by opmonoidal structures on the monad $T = {} _- \otimes_{R^e} A$. In a recent paper [18] bialgebroids have been characterized as the *bimonads* on ${}_R\mathcal{M}_R$ the underlying functors of which have right adjoints. A bimonad, or opmonoidal monad [12, 10], is a monad in the 2-category $\mathsf{Mon}_{op}\mathsf{Cat}$ of monoidal categories, opmonoidal functors and opmonoidal natural transformations. More explicitly, a bimonad $\langle T, \gamma, \pi, \mu, \eta \rangle$ on a monoidal category $\langle \mathcal{M}, \otimes, i \rangle$ consists of

1. an endofunctor $T \colon \mathcal{M} \to \mathcal{M}$
2. a natural transformation $\gamma_{x,y} \colon T(x \otimes y) \to Tx \otimes Ty$
3. an arrow $\pi \colon Ti \to i$
4. a natural transformation $\mu_x \colon T^2 x \to Tx$
5. and a natural transformation $\eta_x \colon x \to Tx$

such that $\langle T, \mu, \eta \rangle$ is a monad, $\langle T, \gamma, \pi \rangle$ is an opmonoidal functor, i.e., a monoidal functor in $\langle \mathcal{M}^{op}, \otimes, i \rangle$, and μ and η are opmonoidal natural transformations in the obvious sense. Bimonads, and therefore bialgebroids, too, form a 2-category Bmd and $\mathsf{Bgd} \subset \mathsf{Bmd}$, respectively [18].

The forgetful functors $U^A \colon \mathcal{M}_A \to {}_R\mathcal{M}_R$ of bialgebroids over R can be characterized as the strong monoidal monadic functors to ${}_R\mathcal{M}_R$ that have right adjoints [18, Corollary 4.16]. In this paper we will study analogous characterizations of the *long forgetful functors* $G^A \colon \mathcal{M}_A \to \mathcal{M}_k$. This is motivated by situations where the base algebra R is not given a priori. It is also closer to the classical Tannaka-Krein situation where one reconstructs the "grouplike" object A as the set of natural transformations $G^A \to G^A$. Apart from set theoretical controversies ($\mathcal{M}_A$ is not small, which will be compensated by assuming the existence of left adjoints for our functors) this reconstruction is possible for the long forgetful functor G^A but not for the short forgetful functor U^A. As a matter of fact, comparing $\operatorname{End} G^A$ with $\operatorname{End} U^A$ on sees that the former reconstructs A but the latter is something smaller, carrying no obvious coalgebra structure. With the long forgetful functor, however, we face with a new difficulty: it is not strong monoidal.

To recover R from G^A is in fact very easy. One takes the image of the unit object (the trivial A-module) under G^A. Since the unit object is always a monoid, it is mapped by the monoidal forgetful functor to a monoid in $\mathcal{M}_k$. This gives us the algebra R. This construction is possible for any monoidal functor and a closer look will show in Section 2 that, under mild assumptions on $\mathcal{M}$, every monoidal functor $G \colon \mathcal{C} \to \mathcal{M}$ can be factorized as $\mathcal{C} \xrightarrow{U} {}_R\mathcal{M}_R \to \mathcal{M}$ with U monoidal but strictly unital.

The monoidal functors G for which U is strong monoidal will be called *essentially strong monoidal*. Clearly, the G^A of a bialgebroid is an example of such functors. Finding the extra conditions on an essentially strong monoidal functor $G \colon \mathcal{C} \to \mathcal{M}$ that makes it (1) either factorize through the long forgetful functor G^A of a unique bialgebroid (2) or become isomorphic to such a G^A is the part of a Tannaka duality program for bialgebroids. This has been carried out for "short" forgetful functors in [18]. This type of duality theory uses monad theory to characterize the large module categories of quantum groupoids together with their forgetful functors. With the

ADJOINTABLE MONOIDAL FUNCTORS AND QUANTUM GROUPOIDS

results of the present paper we make some small steps in the direction of extending Tannaka theory from strong monoidal to monoidal functors. As for the state of the art of the traditional method we have to mention the recent papers by Phùng Hò Hái [14] and another one by Hayashi [6] which prove Tannaka duality theorems for Hopf algebroids and for face algebras, respectively. In their approach, as in that of Saavedra-Rivano, Deligne, Ulbrich and others (see [13, 7]) small categories are equipped with strong monoidal functors to a (sometimes rigid) category of bimodules and the task is to find a universal factorization through the *comodule* category of a quantum groupoid.

The organization of the paper is as follows. In Section 2 we pove the canonical factorization of general monoidal functors through a bimodule category. After touching the general case of long forgetful functors of bimonads in Section 3 we determine a class of essentially strong monoidal functors in Section 4 which factorize through the G^A of a bialgebroid. Although we present the proof over the base category $\mathcal{M}_k$, Ab or Set, it is indicated how it could be extended to a general base category $\mathcal{V}$. Then in Section 5 the long forgetful functors of bialgebroids are characterized up to equivalence. Finally, in Section 6, the special case of weak bialgebras are considered, now over $\mathcal{M}_k$, the long forgetful functors of which can be recognized as those that have both monoidal and opmonoidal structures and these two obey compatibility conditions that can be called a *separable Frobenius structure* on the forgetful functor. This characterization of weak bialgebra forgetful functors was already sketched in [17] calling them "split monoidal" functors.

2. THE CANONICAL FACTORIZATION OF MONOIDAL FUNCTORS

Let $\mathcal{C}$ be a monoidal category with monoidal product $\Box : \mathcal{C} \times \mathcal{C} \to \mathcal{C}$ and unit object $e \in \mathcal{C}$. Then we have coherent natural isomorphisms $\mathbf{a}_{a,b,c} : a \Box (b \Box c) \xrightarrow{\sim} (a \Box b) \Box c$, $\mathbf{l}_c : e \Box c \xrightarrow{\sim} c$ and $\mathbf{r}_c : c \Box e \xrightarrow{\sim} c$ satisfying $\mathbf{l}_e = \mathbf{r}_e$, the triangle and the pentagon identity. A monoid $\langle m, \mu, \eta \rangle$ in $\mathcal{C}$ is an object m together with arrows $\mu : m \Box m \to m$, $\eta : e \to m$ satisfying associativity and unit axioms. There is always a canonical monoid: the unit object e equipped with multiplication $\mathbf{l}_e : e \Box e \to e$ and unit the identity arrow $e : e \to e$. Moreover, every object c of $\mathcal{C}$ is a bimodule over this canonical monoid via the actions $\mathbf{l}_c : e \Box c \to c$ and $\mathbf{r}_c : c \Box e \to c$. The bimodule axioms follow simply from recognizing that the three associativity axioms

$$\mathbf{l}_c \circ (e \Box \mathbf{l}_c) = \mathbf{l}_c \circ (\mathbf{l}_e \Box c) \circ \mathbf{a}_{e,e,c} \tag{1}$$

$$\mathbf{l}_c \circ (e \Box \mathbf{r}_c) = \mathbf{r}_c \circ (\mathbf{l}_c \Box e) \circ \mathbf{a}_{e,c,e} \tag{2}$$

$$\mathbf{r}_c \circ (\mathbf{r}_c \Box e) = \mathbf{r}_c \circ (c \Box \mathbf{l}_e) \circ \mathbf{a}_{c,e,e}^{-1} \tag{3}$$

are consequences of special cases of the triangle diagrams valid in any monoidal category while the unit axioms become identities. In this sense every monoidal category is a category of bimodules. The more precise statement will be clear after applying the Theorem below to the identity functor of $\mathcal{C}$. Let us recall an important property of monoidal functors: They map monoids to monoids and (bi)modules to (bi)modules.

LEMMA 2.1. *Let $\langle G, G_2, G_0 \rangle : \langle \mathcal{C}, \Box, e \rangle \to \langle \mathcal{M}, \otimes, i \rangle$ be a monoidal functor.*

1. If $\langle m, \mu, \eta \rangle$ in C is a monoid in C then $\langle Gm, G\mu \circ G_{m,m}, G\eta \circ G_0 \rangle$ is a monoid in $\mathcal{M}$.
2. Let m and n be monoids in C. If $\langle b, \lambda, \rho \rangle$ is an m-n bimodule in C then the triple $\langle Gb, G\lambda \circ G_{m,b}, G\rho \circ G_{b,n} \rangle$ is a Gm-Gn bimodule in $\mathcal{M}$.

Proof. (1) is well-known and can be found e.g. in [16]. (2) is also known to many authors although an explicit proof is difficult to find. Just to advertise the statement we compute here commutativity of the left and right actions:

$$
\begin{aligned}
\lambda' \circ (Gm \otimes \rho') &= G\lambda \circ G_{m,b} \circ (Gm \otimes G\rho) \circ (Gm \otimes G_{b,n}) \\
&= G\lambda \circ G(m \,\square\, \rho) \circ G_{m,b\,\square\,n} \circ (Gm \otimes G_{b,n}) \\
&= G\rho \circ G(\lambda \,\square\, n) \circ Ga_{m,b,n} \circ G_{m,b\,\square\,n} \circ (Gm \otimes G_{b,n}) \\
&= G\rho \circ G(\lambda \,\square\, n) \circ G_{m\,\square\,b,n} \circ (G_{m,b} \otimes Gn) \circ \mathbf{a}_{Gm,Gb,Gn} \\
&= G\rho \circ G_{b,n} \circ (G\lambda \otimes Gn) \circ (G_{m,b} \otimes Gn) \circ \mathbf{a}_{Gm,Gb,Gn} \\
&= \rho' \circ (\lambda' \otimes Gn) \circ \mathbf{a}_{Gm,Gb,Gn}
\end{aligned}
$$

$\square$

Dually, opmonoidal functors map comonoids to comonoids and (bi)comodules to (bi)comodules.

So far $\mathcal{M}$ was an arbitrary monoidal category. In order for the category $_m\mathcal{M}_m$ of bimodules in $\mathcal{M}$ over a monoid m in $\mathcal{M}$ to have a monoidal structure we need the assumptions that $\mathcal{M}$ has coequalizers and the tensor product $\otimes$ preserves coequalizers in both arguments. This is because the tensor product over m of a right m-module $\rho_x : x \otimes m \to x$ with a left m-module $\lambda_y : m \otimes y \to y$ is a coequalizer

$$
\begin{array}{c}
x \otimes (m \otimes y) \xrightarrow{\ x \otimes \lambda_y\ } \\
\Big\downarrow{\scriptstyle\wr} \qquad\qquad\qquad x \otimes y \longrightarrow x \otimes_m y \\
(x \otimes m) \otimes y \xrightarrow{\ \rho_x \otimes y\ }
\end{array}
\tag{4}
$$

The construction of the monoidal product $\otimes_m$ on $_m\mathcal{M}_m$ together with coherence isomorphisms is a long but standard procedure. At the end one obtains a monoidal category $\langle _m\mathcal{M}_m, \otimes_m, m \rangle$ together with a monoidal forgetful functor $\Gamma^m : {}_m\mathcal{M}_m \to \mathcal{M}$ sending the bimodule $\langle x, \lambda, \rho \rangle$ to its underlying object x. The monoidal structure

$$
\Gamma^m_{x,y} : \Gamma^m(x) \otimes \Gamma^m(y) \to \Gamma^m(x \otimes_m y) \tag{5}
$$
$$
\Gamma^m_0 : i \to \Gamma^m(m) \tag{6}
$$

is provided by the chosen coequalizers and by the unit $i \to m$ of the monoid m, respectively.

If $\sigma : m \to n$ is a monoid morphism then there is a functor $\Gamma^\sigma : {}_n\mathcal{M}_n \to {}_m\mathcal{M}_m$ mapping the bimodule $\langle x, \lambda, \rho \rangle$ to the bimodule $\langle x, \lambda \circ (\sigma \otimes x), \rho \circ (x \otimes \sigma) \rangle$. So, $\Gamma^\sigma \Gamma^m = \Gamma^n$. This defines a functor $\Gamma : \mathrm{Mon}\mathcal{M} \to \mathrm{MonCat}/\mathcal{M}$.

THEOREM 2.2. *Let $\langle \mathcal{M}, \otimes, i \rangle$ be a monoidal category with coqualizers and such that $\otimes$ preserves coequalizers in both arguments. If $\langle C, \square, e \rangle$ is a monoidal category and $G : C \to \mathcal{M}$ is a monoidal functor then there is a monoid $\langle R, \mu^R, \eta^R \rangle$ in $\mathcal{M}$ and a strictly unital monoidal functor $U : C \to {}_R\mathcal{M}_R$ such that*

ADJOINTABLE MONOIDAL FUNCTORS AND QUANTUM GROUPOIDS   295

1. G, as a monoidal functor, can be factorized as $\Gamma^R U$, i.e.,

$$G = \Gamma^R U \tag{7}$$

$$G_{a,b} = \Gamma^R U_{a,b} \circ \Gamma^R_{Ua,Ub} \tag{8}$$

$$G_0 = \Gamma^R U_0 \circ \Gamma^R_0 \tag{9}$$

2. If S is a monoid in $\mathcal{M}$ and $V \colon \mathcal{C} \to {}_S\mathcal{M}_S$ is a monoidal functor such that $G = \Gamma^S V$, as monoidal functors, then there exists a unique monoid morphism $\sigma \colon S \to R$ such that $V = \Gamma^\sigma U$.

Proof. By Lemma 2.1 the image under G of the unit monoid e is a monoid $R = \langle Ge, \mu^R, \eta^R \rangle$ with underlying object Ge. Also by the Lemma, every object c in $\mathcal{C}$, as an e-e-bimodule, is mapped by G to the R-R-bimodule

$$Uc := \langle Gc, \lambda_{Uc}, \rho_{Uc} \rangle \tag{10}$$

where

$$\lambda_{Uc} := \quad Ge \otimes Gc \xrightarrow{\ G_{e,c}\ } G(e \,\square\, c) \xrightarrow{\ Gl_c\ } Gc \tag{11}$$

$$\rho_{Uc} := \quad Gc \otimes Ge \xrightarrow{\ G_{c,e}\ } G(c \,\square\, e) \xrightarrow{\ Gr_c\ } Gc \tag{12}$$

Since these actions are natural in c, every arrow $\tau \colon c \to d$ lifts to a bimodule morphism $U\tau \colon Uc \to Ud$. This defines the functor U which obviously satisfies (7). In order to define a monoidal structure for U notice that

$$G_{a,b} \circ (\rho_{Ua} \otimes Gb) \circ \mathbf{a}_{Ga,Ge,Gb} = G_{a,b} \circ (Ga \otimes \lambda_{Ub}) \tag{13}$$

holds true as a consequence of the hexagon of G_2. Therefore, universality of the coequalizer implies the existence of a unique arrow $Ga \otimes_R Gb \to G(a \,\square\, b)$ such that

$$Ga \otimes Gb \xrightarrow{\ \Gamma^R_{Ua,Ub}\ } Ga \otimes_R Gb \equiv \Gamma^R(Ua \otimes_R Ub)$$

with arrows $G_{a,b}$ and $\Gamma^R U_{a,b}$ to

$$G(a \,\square\, b) \tag{14}$$

commutes. Moreover, this new arrow lifts to a bimodule morphism $U_{a,b}$ since the other two in this diagram also lift to ${}_R\mathcal{M}_R$ and $\otimes$ preserves coequalizers. Setting $U_0 \colon {}_R R_R \to Ue$ to be the identity arrow we obtain a monoidal functor $U \colon \mathcal{C} \to {}_R\mathcal{M}_R$ for which (8) and (9) hold and the unit of which, U_0, is an identity arrow, i.e., it is strictly unital. This proves property (1). To prove the universal property (2) we start with uniqueness. If σ exists such that $\Gamma^\sigma U = V$ is a factorization in MonCat then, in particular, $V_0 = \Gamma^\sigma U_0 \circ \Gamma^\sigma_0$. Since U is strict unital, $\Gamma^S V_0 = \Gamma^S \Gamma^\sigma_0$. This means that the underlying arrow in $\mathcal{M}$ of the bimodule morphism $\Gamma^\sigma_0 \colon S \to {}_{\sigma(S)} R_{\sigma(S)}$ is uniquely determined by that of V_0. But this bimodule morphism is the same as the monoid morphism $\sigma \colon S \to R$ (as arrows of $\mathcal{M}$). To prove existence we therefore define $\sigma := \Gamma^S V_0 \colon \Gamma^S S \to \Gamma^R R$, for the time being as an arrow in $\mathcal{M}$. It preserves the unit due to the monoidal factorization,

$$\sigma \circ \eta^S = \Gamma^S V_0 \circ \Gamma_0^S = G_0 = \Gamma_0^R = \eta^R$$

and it is multiplicative because

$$\begin{aligned}
\mu^R \circ (\sigma \otimes \sigma) &= G1_e \circ G_{e,e} \circ (\Gamma^S V_0 \otimes \Gamma^S V_0) \\
&= \Gamma^S V1_e \circ \Gamma^S V_{e,e} \circ \Gamma_{Ve,Ve}^S \circ (\Gamma^S V_0 \otimes \Gamma^S V_0) \\
&= \Gamma^S V1_e \circ \Gamma^S V_{e,e} \circ \Gamma^S (V_0 \otimes_S V_0) \circ \Gamma_{S,S}^S \\
&= \Gamma^S 1_{Ve} \circ \Gamma^S (S \otimes_S V_0) \circ \Gamma_{S,S}^S = \Gamma^S V_0 \circ \Gamma^S 1_S \circ \Gamma_{S,S}^S \\
&= \sigma \circ \mu^S.
\end{aligned}$$

So σ lifts to a monoid morphism $S \to R$. The proof of $\Gamma^\sigma U = V$ requires to show that $\lambda_{Uc} \circ (\sigma \otimes Gc) = \lambda_{Vc}$ and $\rho_{Uc} \circ (Gc \otimes \sigma) = \rho_{Vc}$ for all object c in $\mathcal{C}$. E.g.,

$$\begin{aligned}
\lambda_{Vc} &= \Gamma^S 1_{Vc} \circ \Gamma_{S,Vc}^S \\
&= \Gamma^S V1_c \circ \Gamma^S V_{e,c} \circ \Gamma^S (V_0 \otimes_S V_0) \circ \Gamma_{S,Vc}^S \\
&= G1_c \circ \Gamma^S V_{e,c} \circ \Gamma_{Ve,Vc}^S \circ (\Gamma^S V_0 \otimes Gc) \\
&= G1_c \circ G_{e,c} \circ (\Gamma^S V_0 \otimes Gc) \\
&= \lambda_{Uc} \circ (\sigma \otimes Gc).
\end{aligned}$$

In order to have $\Gamma^\sigma U = V$ in MonCat we still have to show $V_{a,b} = \Gamma^\sigma U_{a,b} \circ \Gamma_{Ua,Ub}^\sigma$ and $V_0 = \Gamma^\sigma U_0 \circ \Gamma_0^\sigma$. The latter follows directly from the definition of σ while the former can be shown by applying the faithful Γ^S on both hand sides and then composing with the coequalizer $\Gamma_{Va,Vb}^S$ from the right. Since a coequalizer is epi, it is right cancellable and the statement follows from (8). $\qquad\square$

We call $G = \Gamma U$ the canonical factorization of the monoidal functor G.

DEFINITION 2.3. A monoidal functor G is called essentially strong monoidal if the U in its canonical factorization $G = \Gamma U$ is strong monoidal.

Equivalently, G is essentially strong monoidal if for all objects a, b of $\mathcal{C}$ the arrow $G_{a,b}$ is a coequalizer in the diagram

$$
\begin{array}{ccc}
Ga \otimes (Ge \otimes Gb) & \xrightarrow{\quad Ga \otimes \lambda_{Ub}\quad} & \\
\Big\downarrow{\wr} & \searrow & Ga \otimes Gb \xrightarrow{\quad G_{a,b}\quad} G(a \,\square\, b) \\
(Ga \otimes Ge) \otimes Gb & \xrightarrow{\quad \rho_{Ua} \otimes Gb\quad} &
\end{array}
\tag{15}
$$

where λ_{Uc}, ρ_{Uc} denote the left and right R-actions (11) and (12) on Gc.

3. THE RELATION WITH THE EILENBERG-MOORE CONSTRUCTION

Let us recall some basic facts about monad theory [11, 2]. Any functor $G \colon \mathcal{C} \to \mathcal{M}$ with a left adjoint F determines a monad $\mathsf{M} = \langle GF, G\varepsilon F, \eta \rangle$ on $\mathcal{M}$ where $\varepsilon \colon FG \to \mathcal{C}$ is the counit and $\eta \colon \mathcal{M} \to GF$ is the unit of the adjunction. The Eilenberg-Moore

ADJOINTABLE MONOIDAL FUNCTORS AND QUANTUM GROUPOIDS

construction associates to any monad M on $\mathcal{M}$ a category $\mathcal{M}^\mathtt{M}$ the objects of which are the M-algebras $\langle x, \alpha \rangle$ where x is an object of $\mathcal{M}$ and $\alpha \colon \mathtt{M}x \to x$ is an arrow of $\mathcal{M}$ satisfying $\alpha \circ \mathtt{M}\alpha = \alpha \circ \mu_x$ and $\alpha \circ \eta_x = x$. The arrows of $\mathcal{M}^\mathtt{M}$ from $\langle x, \alpha \rangle$ to $\langle y, \beta, \rangle$ are the arrows $\tau \colon x \to y$ in $\mathcal{M}$ for which $\tau \circ \alpha = \beta \circ \mathtt{M}\tau$.

The forgetful functor $U^\mathtt{M} \colon \mathcal{M}^\mathtt{M} \to \mathcal{M}$ sending $\langle x, \alpha \rangle$ to x has a left adjoint such that the monad associated to this adjunction is precisely the original M.

Any functor $G \colon \mathcal{C} \to \mathcal{M}$ with a left adjoint can be factorized as $G = U^\mathtt{M} K$, where M is the monad of the adjunction, with a unique $K \colon \mathcal{C} \to \mathcal{M}^\mathtt{M}$, called the comparison functor. Explicitly,

$$c \xmapsto{K} \langle Gc, G\varepsilon_c \rangle \xmapsto{U^\mathtt{M}} Gc. \tag{16}$$

If K is an equivalence of categories the G is called monadic.

Now assume that we have a monoidal functor $G \colon \mathcal{C} \to \mathcal{M}$ with the underlying ordinary functor having a left adjoint F. We briefly say that G is a right adjoint monoidal functor. Then we have two factorizations of G, the $G = \Gamma U$ provided by Theorem 2.2 and $G = U^\mathtt{M} K$ provided by the Eilenberg-Moore construction. In this situation one expects a relation between M-algebras and R-R-bimodules.

Every monoid in $\mathcal{M}$, therefore R too, provides two monads $\mathtt{L} = R \otimes \_$ and $\mathtt{R} = \_ \otimes R$. We can define two monad morphisms

$$\lambda^\mathtt{M} \colon \mathtt{L} \to \mathtt{M} \qquad \lambda^\mathtt{M}_x = G(1_{Fx}) \circ G_{e,Fx} \circ (R \otimes \eta_x) \tag{17}$$

$$\rho^\mathtt{M} \colon \mathtt{R} \to \mathtt{M} \qquad \rho^\mathtt{M}_x = G(\mathbf{r}_{Fx}) \circ G_{Fx,e} \circ (\eta_x \otimes R) \tag{18}$$

They commute in the sense of the diagram

$$
\begin{array}{ccccc}
\mathtt{LR} & \xrightarrow{\lambda^\mathtt{M}\rho^\mathtt{M}} & \mathtt{M}^2 & \xrightarrow{G\varepsilon F} & \mathtt{M} \\
\downarrow{\scriptstyle\wr} & & & & \| \\
\mathtt{RL} & \xrightarrow{\rho^\mathtt{M}\lambda^\mathtt{M}} & \mathtt{M}^2 & \xrightarrow{G\varepsilon F} & \mathtt{M}
\end{array}
\tag{19}
$$

where the vertical isomorphism can be obtained from the associator as $\mathbf{a}_{R,\_,R}$.

LEMMA 3.1. *If $\langle x, \alpha \rangle$ is a M-algebra then $\langle x, \alpha \circ \lambda^\mathtt{M}_x, \alpha \circ \rho^\mathtt{M}_x \rangle$ is an R-R-bimodule.*

This provides the object map of a functor $\Psi \colon \mathcal{M}^\mathtt{M} \to {}_R\mathcal{M}_R$. It is easy to show that if $\tau \colon \langle x, \alpha \rangle \to \langle y, \beta \rangle$ is a M-algebra morphism then it is also an R-R-bimodule morphism. This defines a forgetful functor $\Psi \colon \mathcal{M}^\mathtt{M} \to {}_R\mathcal{M}_R$. Moreover, the Eilenberg-Moore comparison functor $K \colon \mathcal{C} \to \mathcal{M}^\mathtt{M}$ satisfies $\Psi K = U$. Thus we can factorize the original functor G in 3 steps

$$\mathcal{C} \xrightarrow{K} \mathcal{M}^\mathtt{M} \xrightarrow{\Psi} {}_R\mathcal{M}_R \xrightarrow{\Gamma} \mathcal{M} \tag{20}$$

Unfortunately, only the third functor is monoidal. In order to make it a diagram in MonCat we need at least a monoidal structure on the Eilenberg-Moore category $\mathcal{M}^\mathtt{M}$. This could be achieved under the stronger assumption that U is strong monoidal and has a left adjoint. Namely, [18, Theorem 2.8] yields the following result:

Let $U \colon \mathcal{C} \to \mathcal{T}$ be a strong monoidal right adjoint functor. Then
- its monad T is opmonoidal, i.e., it is a monad on $\mathcal{T}$ in the 2-category $\mathrm{Mon}_\mathrm{op}\mathrm{Cat}$;

- the category T^T of T-algebras has a unique monoidal structure such that the Eilenberg-Moore forgetful functor $U^T \colon T^T \to T$ is strict monoidal;
- $U = U^T K$ with a strong monoidal comparison functor $K \colon C \to T^T$.

THEOREM 3.2. *Assume that*

1. $\langle C, \square, e \rangle$ *is a monoidal category where C is complete, well powered and has a small cogenerating set;*
2. $\langle M, \otimes, i \rangle$ *is a monoidal category where M has coequalizers and $\otimes$ preserves them;*
3. $\langle G, G_2, G_0 \rangle$ *is an essentially strong monoidal functor $C \to M$ with G having a left adjoint.*

Then there is a monoid R in M and an opmonoidal monad T on ${}_R M_R$ such that

$$
\begin{array}{ccc}
C & \xrightarrow{\;\;K\;\;} & ({}_R M_R)^T \\
{\scriptstyle G}\big\downarrow & & \big\downarrow{\scriptstyle U^T} \\
M & \xleftarrow{\;\;\Gamma\;\;} & {}_R M_R
\end{array}
\tag{21}
$$

is commutative in MonCat *where K is the strong monoidal comparison functor, U^T is the strict monoidal forgetful functor of T-algebras and Γ is the monoidal forgetful functor of R-R-bimodules.*

Proof. By assumption (2) the canonical factorization $G = \Gamma U$ through ${}_R M_R$ exists. Since G is right adjoint, it preserves limits. But Γ, being the forgetful functor of the monad $R \otimes \_ \otimes R$ on M, is monadic therefore it creates limits. Therefore U preserves limits, too. By the Special Adjoint Functor Theorem [11, Corollary V. 8] assumption (1) ensures that U has a left adjoint. Now using assumption (3) the functor U is strong monoidal right adjoint, therefore, it factorizes monoidally through the Eilenberg-Moore category of its bimonad T by the above quoted [18, Theorem 2.8]. $\square$

The 3 step factorization found in the above Theorem describes what can be expected in general for continuous monoidal functors. Of course, it would be more interesting to replace the "abstract quantum groupoid" T with a concrete bialgebroid, let us say. If M is the category of k-modules over a commutative ring then a necessary and sufficient condition for $({}_R M_R)^T$ to be the monoidal category ${}_A M$ of modules over a bialgebroid A was given in [18, Theorem 4.5]. The condition is very simple: the underlying functor T of the bimonad should have a right adjoint. Unfortunately, it is difficult to find a condition on G that guarantees a right adjoint for U. In the next Section we will study a special case which allows to do so.

We note that even under the conditions of the Theorem the underlying monad of T may not be M, so the factorizations (21) and (20) may be different.

4. MONOIDAL $\mathcal{V}$-FUNCTORS TO $\mathcal{V}$

Now we turn to replace bimonads with bialgebroids. The key observation is that every monoid A in M determines a monad $A \otimes \_$ on M and - under certain conditions on M - these monads are precisely the monads the underlying endofunctor

ADJOINTABLE MONOIDAL FUNCTORS AND QUANTUM GROUPOIDS 299

of which has a right adjoint. The Eilenberg-Moore categories $\mathcal{M}^{\mathsf{T}}$ of such monads are precisely the categories $_A\mathcal{M}$ of modules over A. This can be generalized to (k-linear) bimonads as follows.

THEOREM 4.1. [18, Thm. 4.5] *Let* $\mathcal{M} = \mathcal{M}_k$ *be the category of modules over a commutative ring* k, R *be a* k-*algebra and* T *be a* k-*linear bimonad on* $_R\mathcal{M}_R$. *Then there exists a bialgebroid* A *in* $\mathcal{M}$ *over* R *and an isomorphism* $\mathsf{T} \cong A \otimes\_$ *if and only if* $T\colon \mathcal{M} \to \mathcal{M}$ *has a right adjoint. Moreover,* T *has a right adjoint if and only if* $U^{\mathsf{T}}\colon \mathcal{M}^{\mathsf{T}} \to \mathcal{M}$ *has a right adjoint.*

We want to give an analogous characterization of the long forgetful functors $G\colon {}_A\mathcal{M} \to \mathcal{M}$ of bialgebroids. Clearly, these functors have both left and right adjoints - the induction and coinduction functors - and the factorization through $_R\mathcal{M}_R$ shows that they are essentially strong monoidal. Thus we expect that these properties of a functor leads to the construction of a bialgebroid with Tannaka duality.

We let $\mathcal{V}$ denote either $_k\mathcal{M}$, Ab or Set and work with $\mathcal{V}$-monoidal $\mathcal{V}$-categories and $\mathcal{V}$-functors between them that have $\mathcal{V}$-adjoints [3]. Due to the fact that $\mathcal{V}$ is not only symmetric monoidal closed but its monoidal unit is a generator, we can work with the underlying ordinary categories and consider $\mathcal{V}$-functors as a special class of ordinary functors that preserve some extra structure that is encoded in the choice of $\mathcal{V}$. The set of $\mathcal{V}$-natural transformations between $\mathcal{V}$-functors is the same (under the 2-functor $\mathrm{Hom}_\mathcal{V}(i, \_)$) as the set of ordinary natural transformations. So the complicated formalism of enriched categories can be avoided and proofs of commutativity of diagrams in $\mathcal{V}$ can be done by *elements*. (An element of an object $v \in \mathcal{V}$ is an arrow $i \to v$ in $\mathcal{V}$.) In a $\mathcal{V}$-category $\mathcal{C}$ we denote by $\mathcal{C}(a, b) \in \mathcal{V}$ its hom-objects and by $\mathrm{Hom}_\mathcal{C}(a, b) := \mathrm{Hom}_\mathcal{V}(i, \mathcal{C}(a, b))$ its hom-sets. Otherwise it should be clear from the context whether we speak about the $\mathcal{V}$-category $\mathcal{C}$ or its underlying ordinary category. $\mathcal{V}$ itself is a $\mathcal{V}$-category with $\mathcal{V}(x, y)$ being the internal hom object $\mathrm{hom}(x, y)$ which is defined by

$$\mathrm{Hom}_\mathcal{V}(v \otimes x, y) \cong \mathrm{Hom}_\mathcal{V}(v, \mathrm{hom}(x, y)).$$

If the base category $\mathcal{V}$ has coequalizers then we can apply Theorem 2.2. If it is also complete then we can define the Takeuchi $\times_R$-product as a pullback of equalizers. Readers interested in more general enrichments than the three cases mentioned above can take for $\mathcal{V}$ any complete category with coequalizers which is endowed with a symmetric closed monoidal structure $\langle \mathcal{V}, \otimes, i \rangle$ such that i a projective generator.

We restrict ourselves to study functors with target category being the base category, i.e., G is a $\mathcal{V}$-functor to $\mathcal{V}$. In this situation the existence of a left $\mathcal{V}$-adjoint $F \dashv G$ implies that $G\colon \mathcal{C} \to \mathcal{V}$ is representable:

$$Ga \xrightarrow{\sim} \mathrm{hom}(i, Ga) \xrightarrow{\sim} \mathcal{C}(Fi, a) \tag{22}$$

for $a \in \mathcal{C}$. Thus without loss of generality we may assume that G is a hom-functor $G = \mathcal{C}(g, \_)$. If such a G has a monoidal structure then - by the Yoneda Lemma -

300   K. SZLACHÁNYI

a comonoid structure $\langle g, \gamma, \pi \rangle$ on g arises via

$$G: \mathcal{C} \to \mathcal{V} \qquad\qquad\qquad G = \mathcal{C}(g, \_) \qquad (23)$$

$$G_{a,b}: Ga \otimes Gb \to G(a \,\square\, b) \quad G_{a,b}(\alpha \otimes \beta) = (\alpha \,\square\, \beta) \circ \gamma \qquad (24)$$

$$G_0: i \to Ge \qquad\qquad\qquad G_0 = \pi \qquad (25)$$

where in the last equation one can recognize the usual identification of elements of a hom-object of a $\mathcal{V}$-category with arrows in $\mathcal{V}$.

The monoid R in the canonical factorization $G = \Gamma U$ through ${}_R\mathcal{V}_R$ is nothing but the convolution monoid $\mathcal{C}(g, e)$ with multiplication

$$\rho * \rho' = \mathbf{l}_e \circ (\rho \otimes \rho') \circ \gamma, \qquad \rho, \rho' \in R \qquad (26)$$

and unit π.

We note also that every object $Ga = \mathcal{C}(g, a) \in \mathcal{V}$ has a natural right A-module structure where $A := \mathcal{C}(g, g)$ is the endomorphism monoid in $\mathcal{V}$. So G factorizes through the forgetful functor $\mathcal{V}_A \to \mathcal{V}$. This is compatible with the canonical factorization $G = \Gamma U$ because we have monoid morphisms

$$s: R \to A \qquad s(\rho) := \mathbf{r}_g \circ (g \,\square\, \rho) \circ \gamma \qquad (27)$$

$$t: R^{\mathrm{op}} \to A \qquad t(\rho) := \mathbf{l}_g \circ (\rho \,\square\, g) \circ \gamma \qquad (28)$$

Corresponding to the diagram $i \to R^{\mathrm{op}} \otimes R \xrightarrow{t \otimes s} A$ in $\mathsf{Mon}\mathcal{V}$ there is the diagram $\mathcal{V}_A \to {}_R\mathcal{V}_R \to \mathcal{V}$ in $\mathcal{V}$-Cat.

THEOREM 4.2. *Denoting by $\mathcal{V}$ either $\mathcal{M}_k$, or Ab or Set let $\mathcal{C}$ be a $\mathcal{V}$-monoidal $\mathcal{V}$-category and $G: \mathcal{C} \to \mathcal{V}$ be a representable essentially strong monoidal $\mathcal{V}$-functor. Then there exists a monoid R in $\mathcal{V}$, a right bialgebroid A in $\mathcal{V}$ over R, a strong monoidal $K: \mathcal{C} \to \mathcal{V}_A$ and a monoidal natural isomorphism $G \xrightarrow{\sim} \Gamma^R \Phi^A K$,*

$$
\begin{array}{ccc}
\mathcal{C} & \xrightarrow{\;\; K \;\;} & \mathcal{V}_A \\[4pt]
{\scriptstyle G}\downarrow & \cong & \downarrow{\scriptstyle \Phi^A} \\[4pt]
\mathcal{V} & \xleftarrow{\;\; \Gamma^R \;\;} & {}_R\mathcal{V}_R
\end{array}
\qquad (29)
$$

where Φ^A denotes the strict monoidal forgetful functor of the bialgebroid A.

Proof. As we have explained above, there exists a representing comonoid $\langle g, \gamma, \pi \rangle$ in $\mathcal{C}$ and a monoidal natural isomorphism $G \xrightarrow{\sim} \mathcal{C}(g, \_)$. As a $\mathcal{V}$-functor, $\mathcal{C}(g, \_)$ factorizes as

$$\mathcal{C} \xrightarrow{\;\; K \;\;} \mathcal{V}_A \xrightarrow{\;\; \Phi \;\;} {}_R\mathcal{V}_R \xrightarrow{\;\; \Gamma \;\;} \mathcal{V} \qquad (30)$$

where $A = \mathcal{C}(g, g)$ is the endomorphism monoid, $R = \mathcal{C}(g, e)$ is the convolution monoid and Φ is the $\mathcal{V}$-functor that forgets along $t \otimes s: R^e \to A$ where $R^e = R^{\mathrm{op}} \otimes R$ and we identified ${}_R\mathcal{V}_R$ with $\mathcal{V}_{R^e}$. By the very definition of R we see that the canonical factorization ΓU of $\mathcal{C}(g, \_)$ leads to $U = \Phi K$ which is strong monoidal, hence opmonoidal. Since opmonoidal functors map comonoids to comonoids, the triple

$$C = Ug \equiv \langle A, R \otimes A \xrightarrow{\sim} A \otimes R \xrightarrow{A \otimes t} A \otimes A \xrightarrow{\mu} A, A \otimes R \xrightarrow{A \otimes s} A \otimes A \xrightarrow{\mu} A \rangle \quad (31)$$

$$\delta = \left(C \xrightarrow{U\gamma} U(g \square g) \xrightarrow{U_{g,g}^{-1}} C \otimes_R C \right) \quad (32)$$

$$\varepsilon = \left(C \xrightarrow{U\pi} Ue \xrightarrow{U_0^{-1}} {}_R R_R \right) \quad (33)$$

is a comonoid in ${}_R \mathcal{V}_R$. This comonoid allows to define a monoidal structure on $\mathcal{V}_A$ such that Φ becomes strict monoidal. As a matter of fact, let X and Y be right A-modules in $\mathcal{V}$ and define the A-module $X \square_A Y$ as the object $X \otimes_R Y$ with A-action

$$(x \otimes_R y) \triangleleft \alpha := (x \triangleleft \alpha_{(1)}) \otimes_R (y \triangleleft \alpha_{(2)}), \qquad x \in X, y \in Y, \alpha \in A. \quad (34)$$

But in order for this action to be well-defined we have to show that $\delta(\alpha)$, $\alpha \in A$, belong to the subbimodule $C \times_R C \subset C \otimes_R C$ defined as the intersection of the small set of equalizers $\{e_\rho \,|\, \rho \in R\}$ where

$$C \times_\rho C \xrightarrow{e_\rho} C \otimes_R C \begin{array}{c} Us(\rho) \otimes_R C \\ \Longrightarrow \\ C \otimes_R Ut(\rho) \end{array} C \otimes_R C \quad (35)$$

Here $Us(\rho) = s(\rho) \circ \_$ and $Ut(\rho) = t(\rho) \circ \_$ so they are R-R-endomorphisms of C. Naturality of U_2, the definition of δ and some elementary monoidal calculus yields the following identity in ${}_R \mathcal{V}_R$,

$$\begin{aligned}
U_{g,g} \circ (Us(\rho) \otimes_R Ug) \circ \delta &= U(s(\rho) \square g) \circ U_{g,g} \circ \delta \\
&= U(s(\rho) \square g) \circ U\gamma \\
&= U((\mathbf{r}_g \square g) \circ ((g \square \rho) \square g) \circ (\gamma \square g) \circ \gamma) \\
&= U((g \square \mathbf{l}_g) \circ (g \square (\rho \square g)) \circ (g \square \gamma) \circ \gamma) \\
&= U(g \square t(\rho)) \circ U\gamma \\
&= U_{g,g} \circ (Ug \otimes_R Ut(\rho)) \circ \delta \,.
\end{aligned}$$

Since $U_{g,g}$ is invertible, δ restricts to a map $A \to A \times_R A$. Notice that $A \times_R A$ - as an object in $\mathcal{V}$ - inherits a monoid structure from that of $A \otimes A$ and then δ becomes a morphism of monoids. As a matter of fact,

$$\begin{aligned}
U_{g,g}(\delta(\alpha)\delta(\beta)) &= (\alpha_{(1)} \square \alpha_{(2)}) \circ \gamma \circ \beta \\
&= \gamma \circ \alpha \circ \beta \\
&= U_{g,g}(\delta(\alpha \circ \beta))
\end{aligned}$$

for $\alpha, \beta \in A$. Unitality of δ is obvious. This finishes the definition of $\square_A$ and the associativity coherence isomorphism $\mathbf{a}$ of $\mathcal{V}_A$ as a lift of $\mathbf{a}$ of ${}_R \mathcal{V}_R$. Whether it has a unit object depends on the counit properties.

$$\varepsilon(s(\varepsilon(\alpha)) \circ \beta) = \pi \circ s(\varepsilon(\alpha)) \circ \beta$$
$$= \pi \circ \mathbf{r}_g \circ (g \,\square\, \varepsilon(\alpha)) \circ \gamma \circ \beta$$
$$= \mathbf{r}_e \circ (\pi \,\square\, \varepsilon(\alpha)) \circ \gamma \circ \beta = \varepsilon(\alpha \circ \beta) = \pi \circ \alpha \circ \beta$$
$$= \varepsilon(\alpha \circ \beta)$$

and similarly, $\varepsilon(t(\varepsilon(\alpha)) \circ \beta) = \varepsilon(\alpha \circ \beta)$ holds for all $\alpha, \beta \in A$. Unitality of ε is obvious since $\varepsilon(g) = \pi$ is the unit element of R. Then the monoidal unit E of $\mathcal{V}_A$ becomes $\langle R, R \otimes A \xrightarrow{s \otimes A} A \otimes A \xrightarrow{\mu} A \xrightarrow{\varepsilon} R \rangle$ and the unit coherence isomorphisms $\mathbf{l}$ and $\mathbf{r}$ of $_R\mathcal{V}_R$ lift to become the $\mathbf{l}$ and $\mathbf{r}$ of $\mathcal{V}_A$, respectively.

This finishes the proof of that A is a bialgebroid together with the construction of a monoidal structure on $\mathcal{V}_A$ such that Φ is strict monoidal. Since U is strong and Φ reflects isomorphisms, it follows that K is strong monoidal. $\square$

5. CHARACTERIZING LONG FORGETFUL FUNCTORS OF BIALGEBROIDS

Recall that for every closed monoidal category $\mathcal{V}$ there is a monoidal equivalence (of ordinary categories)

$$\mathcal{V} \simeq \mathsf{L}\text{-}\mathcal{V}\text{-}\mathsf{Fun}(\mathcal{V}, \mathcal{V}), \qquad v \mapsto \_ \otimes v \tag{36}$$

of $\mathcal{V}$ with the strict monoidal category of left adjoint $\mathcal{V}$-functors $\mathcal{V} \to \mathcal{V}$.

Since the $\mathcal{V}$-monads on $\mathcal{V}$ the underlying $\mathcal{V}$-functors of which have right $\mathcal{V}$-adjoints are precisely the monoids in $\mathsf{L}\text{-}\mathcal{V}\text{-}\mathsf{Fun}(\mathcal{V}, \mathcal{V})$, they are mapped by the above equivalence into the monoids in $\mathcal{V}$.

Let $\mathcal{V}\text{-}\mathsf{Cat}/\mathcal{V}$ denote the 2-category with objects the $\mathcal{V}$-functors to $\mathcal{V}$ and with arrows $F \to G$ the $\mathcal{V}$-functors K with an isomorphism $F \cong GK$. One defines $\mathcal{V}\text{-}\mathsf{MonCat}/\mathcal{V}$ similarly.

For a right bialgebroid A in $\mathcal{V}$ we denote by G^A the essentially strong monoidal forgetting $\mathcal{V}$-functor $\mathcal{V}_A \to \mathcal{V}$. G^A is called the long forgetful functor of A.

THEOREM 5.1. *A $\mathcal{V}$-functor $G\colon \mathcal{C} \to \mathcal{V}$ is equivalent in $\mathcal{V}$-$\mathsf{Cat}/\mathcal{V}$ to the long forgetful functor G^A of a bialgebroid in $\mathcal{V}$ iff*

- *G is monadic,*
- *G has a right adjoint and*
- *there is an essentially strong monoidal structure $\langle G, G_2, G_0 \rangle$ on G.*

In this case $G \simeq G^A$ in $\mathcal{V}$-$\mathsf{MonCat}/\mathcal{V}$.

Proof. Necessity: Since the object in $\mathcal{V}$ underlying the monoidal unit of $\mathcal{V}_A$ is just the object underlying the base monoid R, the canonical factorization of G^A is $G^A = \Gamma^R \Phi^A$, in the notation of Theorem 4.2. Thus G^A is essentially strong. As every forgetful functor of a monoid, G^A is monadic. It has a right adjoint, namely, the coinduction functor sending the object x of $\mathcal{V}$ to the object $\mathrm{hom}(A, x)$ endowed with right A-action

$$\varepsilon_x\colon \mathrm{hom}(A, x) \otimes A \to x \tag{37}$$

provided by the x-component of the counit of the adjunction $\_ \otimes A \dashv \mathrm{hom}(A, \_)$.

ADJOINTABLE MONOIDAL FUNCTORS AND QUANTUM GROUPOIDS

Sufficiency: Since G is monadic, it has a left adjoint F and $G = U^{M}J$ where M is the monad with underlying functor GF, U^{M} is its forgetful functor and $J: \mathcal{C} \to \mathcal{V}^{M}$ is an equivalence. Let H be the right adjoint of G. Then GH is a right adjoint of GF, therefore, by the above remark, it is isomorphic to the monad $\_ \otimes B$ associated to a monoid B in $\mathcal{V}$. Therefore $\mathcal{V}^{M}$ is isomorphic to the category $\mathcal{V}_B$ of right B-modules and the decomposition $G = U^{M}J$ can be replaced with $G = G^{B}L$ where $L: \mathcal{C} \to \mathcal{V}_B$ is an equivalence. Using the latter equivalence G^B is given an essentially strong monoidal structure and it has a left adjoint. Therefore Theorem 4.2 provides a monoidal isomorphism $G^B \cong \Gamma\Phi K = G^A K$ where the right bialgebroid A has underlying monoid the endomorphism monoid of the representing object $g = B_B$ of G^B. Clearly, $A = \mathrm{End}\,(B_B) \cong B$. This proves that $K: \mathcal{V}_B \to \mathcal{V}_A$, $X_B \mapsto \mathrm{Hom}\,_B(\,_A B_B, X_B)$, is an isomorphism of (monoidal) categories. Since in the monoidal isomorphism $G \cong G^A KL$ the functor $KL: \mathcal{C} \to \mathcal{V}_A$ is a monoidal equivalence, it defines an equivalence $G \xrightarrow{\sim} G^A$ in $\mathcal{V}$-MonCat/$\mathcal{V}$. $\qquad\square$

6. THE FORGETFUL FUNCTORS OF WEAK BIALGEBRAS

In this section the base category $\mathcal{V}$ is the category $\mathcal{M}_k$ of modules over a commutative ring. So we switch to the convention of writing capital Roman letters for objects and corresponding small case letters for their elements. Weak bialgebras over k are not only special bialgebroids in $\mathcal{M}_k$ but are equipped with some more structure, as well. The extra structure can be recognized in two places: (1) in the difference between a weak bialgebra counit $\epsilon: B \to k$ and a bialgebroid counit $\varepsilon: A \to R$ and (2) in the nontrivial element $\Delta(1) \in A \otimes A$ which is closely related to a separability idempotent of the separable algebra R. In other words, a weak bialgebra A is a bialgebroid over a separable Frobenius algebra R together with a Frobenius structure $\langle \epsilon, \sum_i e_i \otimes f_i \rangle$ in which $\sum_i e_i f_i = 1$ (see [19] for more details). We call $\langle R, \epsilon, \sum_i e_i \otimes f_i \rangle$ a separable Frobenius structure.

Accordingly, the forgetful functor $G^A: \mathcal{M}_A \to \mathcal{M}_k$ of a bialgebroid has more structure than just a monadic essentially strong monoidal functor with right adjoint. It is equipped also with an opmonoidal structure. For concreteness let $\langle A, \Delta, \epsilon \rangle$ be a weak bialgebra [4] and let R be identified with the canonical right subalgebra $A^R = \{1_{(1)}\epsilon(a1_{(2)}) \,|\, a \in A\}$. Then A becomes a right bialgebroid over R with

$$s: R \to A, \qquad\qquad r \mapsto r \tag{38}$$

$$t: R^{\mathrm{op}} \to A, \qquad\qquad r \mapsto \epsilon(r1_{(1)})1_{(2)} \tag{39}$$

$$\delta: A \to A \otimes_R A, \qquad \delta := \tau \circ \Delta \tag{40}$$

$$\varepsilon: A \to R, \qquad\qquad a \mapsto 1_{(1)}\epsilon(a1_{(2)}) \tag{41}$$

where $\tau: A \otimes A \to A \otimes_R A$ is the canonical epimorphism (cf. [19, Lemma 1.1]).

Let G denote the long forgetful functor G^A of the bialgebroid A (see Section 4). Then the monoidal structure

$$G_{X,Y}: GX \otimes GY \to G(X \,\square\,_A Y), \qquad x \otimes y \mapsto x \otimes_R y \tag{42}$$

$$G_0: k \to GE, \qquad\qquad\qquad \kappa \mapsto \kappa \cdot 1_R \tag{43}$$

is built of the canonical epimorphisms $X \otimes Y \to X \otimes_R Y$ for R-R-bimodules and of the unit $k \to R$ of the algebra R. Due to the presence of the separable Frobenius structure it has an opmonoidal counterpart

$$G^{X,Y}: G(X \,\square\, _A Y) \to GX \otimes GY\,, \qquad x \otimes_R y \mapsto \sum_i x \cdot e_i \otimes f_i \cdot y \qquad (44)$$

$$G^0: GE \to k\,, \qquad\qquad\qquad r \mapsto \epsilon(r) \qquad (45)$$

So we have a monoidal $\langle G, G_2, G_0 \rangle$ and an opmonoidal functor $\langle G, G^2, G^0 \rangle$ with the same underlying functor G. This involves that G_2 with G_0 obey one hexagonal and two square diagrams and G^2 together with G^0 obey three analogous, but oppositely oriented, diagrams. In addition, there are compatibility conditions between the monoidal and opmonoidal structures that are not of the bialgebra type but rather of the Frobenius algebra type [1]. Namely,

$$(G_{X,Y} \otimes GZ) \circ \mathbf{a}_{GX,GY,GZ} \circ (GX \otimes G^{Y,Z}) = G^{X \,\square\, Y, Z} \circ G\mathbf{a}_{X,Y,Z} \circ G_{X,Y \,\square\, Z} \qquad (46)$$

$$(GX \otimes G_{Y,Z}) \circ \mathbf{a}_{GX,GY,GZ}^{-1} \circ (G^{X,Y} \otimes GZ) = G^{X,Y \,\square\, Z} \circ G\mathbf{a}_{X,Y,Z}^{-1} \circ G_{X \,\square\, Y,Z} \qquad (47)$$

for all A-modules X, Y, Z and where $\square$ stands for the monoidal product $\square_A$ of A-modules. At last but not least, G_2 is split epi and the splitting map is just G^2, i.e.,

$$G_{X,Y} \circ G^{X,Y} = G(X \,\square\, Y)\,, \qquad X, Y \in \mathcal{M}_A\,. \qquad (48)$$

Now we summarize these experiences in a

DEFINITION 6.1. Let $\langle \mathcal{C}, \square, E \rangle$ and $\langle \mathcal{M}, \otimes, I \rangle$ be monoidal categories. A separable Frobenius structure on a functor $G: \mathcal{C} \to \mathcal{M}$ consists of natural transformations G_2, G^2 and arrows G_0, G^0 such that

1. $\langle G, G_2, G_0 \rangle$ is a monoidal functor,
2. $\langle G, G^2, G^0 \rangle$ is an opmonoidal functor,
3. the Frobenius conditions (46), (47) and the separability condition (48) hold.

Of course, the functor itself may be neither separable nor Frobenius. Only its monoidal-opmonoidal structure is restricted in the above Definition.

LEMMA 6.2. If $\langle G, G_2, G_0, G^2, G^0 \rangle$ is a separable Frobenius structure on the functor $G: \mathcal{C} \to \mathcal{M}$ then (15) is a split coequalizer in $\mathcal{M}$ for all pairs of objects in $\mathcal{C}$. In particular, $\langle G, G_2, G_0 \rangle$ is essentially strong monoidal.

Proof. For each pair X, Y of objects in $\mathcal{C}$ we can define arrows in $\mathcal{M}$ by

$$\partial_0 := (GX \otimes G1_Y) \circ (GX \otimes G_{E,Y})$$

$$\partial_1 := (Gr_X \otimes GY) \circ (G_{X,E} \otimes GY) \circ \mathbf{a}_{GX,GE,GY}$$

$$\gamma := G_{X,Y}$$

$$\sigma := G^{X,Y}$$

$$\tau := (GX \otimes G^{E,Y}) \circ (GX \otimes G1_Y^{-1})$$

Then an elementary calculation gives

$$\begin{array}{lll}
\text{hexagon for } G_2 & \Rightarrow & \gamma \circ \partial_0 = \gamma \circ \partial_1 \\
\text{separability (48)} & \Rightarrow & \gamma \circ \sigma = 1 \\
\text{separability (48)} & \Rightarrow & \partial_0 \circ \tau = 1 \\
\text{Frobenius (46)} & \Rightarrow & \partial_1 \circ \tau = \sigma \circ \gamma
\end{array}$$

ADJOINTABLE MONOIDAL FUNCTORS AND QUANTUM GROUPOIDS

which means precisely that γ is a split coequalizer, hence a coequalizer, of ∂_0 and ∂_1 [11]. Thus G is essentially strong monoidal. $\square$

LEMMA 6.3. *Let $\langle G, G_2, G_0, G^2, G^0 \rangle$ be a separable Frobenius structure on the functor $G \colon C \to M$ between monoidal categories. Then the image GE of the unit object of C is equipped with a separable Frobenius structure in M.*

Proof. GE gets a monoid structure as the image of E by $\langle G, G_2, G_0 \rangle$ as we explained in Lemma 2.1. Dually, the $\langle G, G^2, G^0 \rangle$ maps the comonoid E into a comonoid with underlying object GE. So we obtain

$$R = GE \tag{49}$$

$$\mu = G1_E \circ G_{E,E} \; : \; R \otimes R \to R \tag{50}$$

$$\eta = G_0 \; : \; I \to R \tag{51}$$

$$\sigma = G^{E,E} \circ G1_E^{-1} \; : \; R \to R \otimes R \tag{52}$$

$$\psi = G^0 \; : \; R \to I \tag{53}$$

such that $\langle R, \mu, \eta \rangle$ is a monoid and $\langle R, \sigma, \psi \rangle$ is a comonoid in M. Now (46), (47) and (48) imply that these two structures on GE are compatible in the sense of satisfying

$$(\mu \otimes R) \circ \mathbf{a}_{R,R,R} \circ (R \otimes \sigma) = \sigma \circ \mu \tag{54}$$

$$(R \otimes \mu) \circ \mathbf{a}_{R,R,R}^{-1} \circ (\sigma \otimes R) = \sigma \circ \mu \tag{55}$$

$$\mu \circ \sigma = R \tag{56}$$

which are the defining relations of a separable Frobenius structure on R. $\square$

We note that for the familiar categories, $M = M_k$, the σ is uniquely determined by $\sigma(1_R)$ which in turn is uniquely determined by ψ. If, moreover, k is a field then a separable Frobenius algebra, i.e., an algebra having a separable Frobenius structure (also called an index one Frobenius algebra), is nothing but a separable k-algebra [8].

LEMMA 6.4. *Let $\langle R, \mu, \eta, \sigma, \psi \rangle$ be a separable Fobenius structure in M_k. Then the monoidal forgetful functor of bimodules $\Gamma = \Gamma^R \colon {}_R M_R \to M$ has the following extension to a separable Frobenius structure:*

$$\Gamma^{X,Y} \colon X \otimes_R Y \to X \otimes Y \qquad x \otimes_R y \mapsto \sum_i x \cdot e_i \otimes f_i \cdot y_i \tag{57}$$

$$\Gamma^0 \colon R \to k \qquad r \mapsto \psi(r) \tag{58}$$

where $\sum_i e_i \otimes f_i = \sigma(1_R)$.

Proof. This is left for an exercise. $\square$

After the above preparations Theorem 5.1 has the following

COROLLARY 6.5. *A k-linear functor $G \colon C \to M_k$ - as an object in k-Cat/M_k - is equivalent to the long forgetful functor of a weak bialgebra iff*

- *G is monadic,*
- *G has a right adjoint and*

- *there is a separable Frobenius structure $\langle G, G_2, G_0, G^2, G^0 \rangle$ on G.*

Proof. G is essentially strong monoidal by Lemma 6.2 therefore Theorem 5.1 provides a right bialgebroid $\langle A, R, s, t, \delta, \varepsilon \rangle$ and a monoidal equivalence $G \simeq G^A$. It remains to show that the data on A can be extended to the data of a weak bialgebra. (This extra structure is encoded neither in the monoidal category $\mathcal{M}_A$ nor in the monoidal functor G^A.) A monoidal equivalence is the same thing as an opmonoidal equivalence, so we can use $G \simeq G^A$ to pass the whole separable Frobenius structure of G to G^A. Now Lemma 6.3 implies that $R = G^A E$, the base of A, is given a separable Frobenius algebra structure $\langle R, \mu, \eta, \sigma, \psi \rangle$. Then weak bialgebra comultiplication and counit can be introduced by

$$\Delta := \Gamma^{A,A} \circ \delta \qquad \Delta(a) = \sum_i a_{(1)} s(e_i) \otimes a_{(2)} t(f_i), \qquad (59)$$

$$\epsilon := \Gamma^0 \circ \varepsilon \qquad \epsilon(a) = \psi(\varepsilon(a)). \qquad (60)$$

For the proof of that $\langle A, \Delta, \epsilon \rangle$ is a weak bialgebra we refer to the proof of [8, Proposition 7.4] where this has been done for R a separable k-algebra over a field. However, after providing a separable Frobenius structure on R that proof applies here. $\qquad\square$

REFERENCES

[1] L. Abrams, Modules, comodules and cotensor products over Frobenius algebras, *J. Algebra* **219** (1999), 201–213.

[2] M. Barr and C. Wells, Toposes, Triples and Theories,
http://www.cwru.edu/artsci/math/wells/pub/ttt.html

[3] F. Borceux, "Handbook of Categorical Algebra 2", Cambridge University Press, Cambridge, 1994.

[4] G. Böhm, F. Nill and K. Szlachányi, Weak Hopf algebras, I. Integral theory and C^*-structure, *J. Algebra* **221** (1999), 385–438.

[5] T. Brzeziński and G. Militaru, Bialgebroids, $\times_A$-bialgebras and duality, *J. Algebra* **251** (2002), 279–294.

[6] T. Hayashi, A canonical Tannaka duality for finite semisimple tensor categories, preprint math. QA/9904073.

[7] A. Joyal, R. Street, An introduction to Tannaka duality and quantum groups, in "Proc. of Category Theory, Como 1990", A. Carboni, M.C. Pedicchio and G. Rosolini (eds.), *Lect. Notes Math.* **1488**, Springer Verlag, Berlin, 1991.

[8] L. Kadison, K. Szlachányi, Bialgebroid actions on depth two extensions and duality, *Adv. Math.* **179** (2003), 75–121.

[9] J.-H. Lu, Hopf algebroids and quantum groupoids, *Int. J. Math.* **7** (1996), 47–70.

[10] P. McCrudden, Opmonoidal monads, . *Theory Appl. Categories* **10** (1992), 469–485.

[11] S. Mac Lane, Categories for the working mathematician, second edition, *Graduate Texts in Mathematics* **5**, Springer Verlag, Berlin, 1997.

[12] I. Moerdijk, Monads on tensor categories, *J. Pure Appl. Algebra* **168** (1992), 189–208.

[13] B. Pareigis, *Quantum Groups and Non-commutative geometry*,
http://www.mathematik.uni-muenchen.de/~pareigis/pa_schft.html

[14] Phùng Hồ Hái, Tannaka-Krein duality for bialgebroids, preprint math. QA/0206113.

[15] P. Schauenburg, Bialgebras over noncommutative rings, and a structure Theorem for Hopf bimodules, *Appl. Categorical Structures* **6** (1998), 193–222.

[16] R. Street, "Quantum Groups", available from ftp.mpce.mq.edu.au.

[17] K. Szlachányi, Finite quantum groupoids and inclusions of finite type, *Fields Inst. Comm.* **30** (2001), 393–407.

ADJOINTABLE MONOIDAL FUNCTORS AND QUANTUM GROUPOIDS

[18] K. Szlachányi, The monoidal Eilenberg-Moore construction and bialgebroids, *J. Pure Appl. Algebra* **182** (2003), 287–315.

[19] K. Szlachányi, Galois actions by finite quantum groupoids, in "Locally Compact Quantum Groups and Groupoids", L. Vainerman (ed.), *IRMA Lect. Math. Theoretical Phys.* **2**, De Gruyter, 2003.

[20] M. Takeuchi, Groups of algebras over $A \otimes \overline{A}$, *J. Math. Soc. Japan* **29** (1977), 459–492.

On Galois Corings

ROBERT WISBAUER   University of Düsseldorf, Germany
e-mail address: wisbauer@math.uni-duesseldorf.de

> ABSTRACT. For a long period the theory of modules over rings on the one hand
> and comodules and Hopf modules for coalgebras and bialgebras on the other
> side developed quite independently. In this talk we want to outline how ideas
> from module theory can be applied to enrich the theory of comodules and vice
> versa. For this we consider A-corings $\mathcal{C}$ with grouplike elements over a ring
> A, in particular Galois corings. If A is right self-injective it turns out that $\mathcal{C}$
> is a Galois coring if and only if for any injective comodule N the canonical
> map $\operatorname{Hom}^{\mathcal{C}}(A, N) \otimes_B A \to N$ is an isomorphism, where $B = \operatorname{End}^{\mathcal{C}}(A)$, the
> ring of coinvariants of A. Together with flatness of $_B A$ this characterises A as
> generator in the category of right $\mathcal{C}$-comodules. This is a special case of the
> fact that over any ring A, an A-module M is a generator in the category $\sigma[M]$
> (objects are A-modules subgenerated by M) if and only if M is flat as module
> over its endomorphism ring S and the evaluation map $M \otimes_S \operatorname{Hom}(M, N) \to N$
> is an isomorphism for injective modules N in $\sigma[M]$.

1. INTRODUCTION

Not being born as a member of the Hopf family I lived for many years with modules
and rings without paying attention to the developments in the theory of Hopf algebras. Somehow I had the impression that in the coalgebra world additive categories
are not of central importance and that the inversion of arrows in the definition of
comodules also turned the interest of researchers to different directions. It was only
in recent years that - by comments of colleagues - I became aware of the fact that
the central notion of my own work, the subgenerator of a module category, could
also be of interest to comodule theory. In fact it was known from Sweedler's book
that for coalgebras over fields, every comodule is contained in a direct sum of copies
of C showing that C is a cogenerator as well as a subgenerator for the comodules.
While for coalgebras C over rings in general the cogenerator property of C is lost,
it is easy to see that C is still a subgenerator. This was the motivation for me to
have a closer look at this theory and to investigate how my experience from module
theory could contribute to a better understanding of the coalgebraic world.

Seeing things from a different angle, it was not surprising that I sometimes came up
with interesting answers to questions which native Hopf people had not previously
considered. General (co-)module theory cannot make new contributions to the
classification of finite dimensional (co-) algebras since in this special case the general
notions coincide with more familiar ones. Probably because of this, quite a few

---

2000 *Mathematics Subject Classification.* 16W30.
Key words and phrases. Coring; semisimple coring; module categories; coinvariants functor;
structure theorem; Hopf algebras.

traditionalists doubted if it makes any sense to study Hopf algebras over rings instead of fields. The situation is reminiscent of Jacobson's definition of a radical for any ring, extending the nilpotent radical for finite dimensional algebras. While his radical did not contribute to the classification of simple algebras, it certainly deepened and widened the understanding of ring and module theory.

Familiarity with coalgebras over commutative rings needs only a small step to non-commutative base rings, leading to the notion of corings. The formalism and results from module theory readily apply to this more general situation and in what follows I'll try to give some idea of how they can be used. Many of the observations to be reported result from cooperation and discussions with Tomasz Brzeziński and other colleagues.

2. MODULES AND COMODULES

Let A be any associative ring and denote by $\mathbf{M}_A$ and $_A\mathbf{M}$ the categories of unital right and left A-modules, respectively.

Let C be an A-coring, i.e., an (A, A)-bimodule with coassociative comultiplication $\Delta : C \to C \otimes_A C$ and counit $\varepsilon : C \to A$.

Right C-comodules are right A-modules M with a *right coaction*

$$\varrho^M : M \to M \otimes_A C,$$

which is coassociative and counital. The categories of left and right C-comodules are denoted by $^C\mathbf{M}$ and $\mathbf{M}^C$, respectively.

The investigation of a ring A is strongly influenced by the fact that A is a projective generator for the left and for the right A-modules. An A-coring C need not be a generator or cogenerator for the C-comodules nor is it projective or injective in general. However, every comodule is a subcomodule of a comodule which is generated by C and hence structural properties of C may transfer to comodules.

2.1. C is a subgenerator in $\mathbf{M}^C$. *For $X \in \mathbf{M}_A$, $X \otimes_A C$ is a right C-comodule by*

$$I_X \otimes \Delta : X \otimes_A C \longrightarrow X \otimes_A C \otimes_A C,$$

and for any $M \in \mathbf{M}^C$, the structure map $\varrho^M : M \to M \otimes_A C$ is a comodule morphism.

Moreover any epimorphism $A^{(\Lambda)} \to M$ of A-modules yields a diagram in $\mathbf{M}^C$ with exact bottom row

$$
\begin{array}{c}
M \\
\downarrow{\scriptstyle \varrho^M} \\
A^{(\Lambda)} \otimes_A C \longrightarrow M \otimes_A C \longrightarrow 0,
\end{array}
$$

showing that M is a subcomodule of a C-generated comodule, i.e., C is a subgenerator in $\mathbf{M}^C$.

Let us mention that over a quasi-Frobenius (QF) ring A, any A-coring C is an injective cogenerator in $\mathbf{M}^C$ and in $^C\mathbf{M}$. In fact any comodule is contained in a direct sum of copies of C.

Both the duals of C as left A-module and as right A-module can be defined and are of importance for comodule theory. We concentrate on one side.

ON GALOIS CORINGS

2.2. The dual rings. *Let C be an A-coring. ${}^*C = {}_A \operatorname{Hom}(C, A)$ is a ring with unit ε with respect to the product (for $f, g \in {}^*C$, $c \in C$)*

$$f *^l g \,:\, C \xrightarrow{\Delta} C \otimes_A C \xrightarrow{Ic \otimes g} C \xrightarrow{f} A, \quad f *^l g(c) = \sum f(c_{\underline{1}} g(c_{\underline{2}})),$$

and there is a ring anti-homomorphism $\quad \iota : A \to {}^*C, \ a \mapsto \varepsilon(-a)$.

The bridge from comodules to modules is provided by the following observation.

2.3. C-comodules and ${}^*C$-modules. *Any $M \in \mathbf{M}^C$ is a (unital) left ${}^*C$-module by*

$$\rightharpoonup \,:\, {}^*C \otimes_A M \to M, \quad f \otimes m \mapsto (I_M \otimes f) \circ \varrho^M(m).$$

Any morphism $h : M \to N$ in $\mathbf{M}^C$ is a left ${}^*C$-module morphism, so

$$\operatorname{Hom}^C(M, N) \subset {}_{{}^*C} \operatorname{Hom}(M, N),$$

and there is a faithful functor from $\mathbf{M}^C$ to $\sigma[{}_{{}^*C}C]$, the full subcategory of ${}_{{}^*C}\mathbf{M}$ whose objects are submodules of C-generated ${}^*C$-modules.

Given the basic constructions we pause to think about what we can learn from module theory for comodules.

(1) In case $\mathbf{M}^C = \sigma[{}_{{}^*C}C]$ we can transfer all theorems from module categories of type $\sigma[M]$ to comodules without extra proofs.

(2) More generally we can focus on the situation when C is flat as left A-module, in which case $\mathbf{M}^C$ is a Grothendieck category. Many results and proofs in $\sigma[M]$ can then easily be transferred in this case.

(3) We may study $\mathbf{M}^C$ without any conditions on the A-module structure of C and ask which notions still make sense and which problems can be handled in this general situation. Here the tranfer of results from $\sigma[M]$ needs more caution since monomorphisms in $\mathbf{M}^C$ need no longer be injective maps.

We will take a brief look at the first two situations and then concentrate on certain aspects of the third one in the last section.

To describe the coincidence of $\mathbf{M}^C$ and $\sigma[{}_{{}^*C}C]$ recall that an A-module M is said to be *locally projective* if, for any diagram of left A-modules with exact rows

$$
\begin{array}{ccccc}
0 & \longrightarrow & F & \xrightarrow{\ i\ } & M \\
 & & & & \downarrow{\scriptstyle g} \\
 & & L & \xrightarrow{\ f\ } & N & \longrightarrow & 0,
\end{array}
$$

where F is finitely generated, there exists $h : M \to L$ such that $g \circ i = f \circ h \circ i$.

2.4. $\mathbf{M}^C$ as full subcategory of ${}_{{}^*C}\mathbf{M}$. *The following are equivalent:*

(a) $\mathbf{M}^C = \sigma[{}_{{}^*C}C]$;

(b) *for all $M, N \in \mathbf{M}^C$, $\operatorname{Hom}^C(M, N) = {}_{{}^*C}\operatorname{Hom}(M, N)$;*

(c) *C is locally projective as left A-module;*

(d) *every left ${}^*C$-submodule of C^n, $n \in \mathbb{N}$, is a subcomodule of C^n;*

(e) *the inclusion functor $i : \mathbf{M}^C \to {}_{{}^*C}\mathbf{M}$ has a right adjoint.*

Proof. We refer to [10, 3.5], [2], or [3]. $\qquad\qquad\square$

312   R. WISBAUER

In the situation considered in 2.4 all theorems known for module categories of type $\sigma[M]$ can be formulated for comodules. In particular the decomposition theorems for module categories yield decompositions of comodule categories and coalgebras (e.g., [8]).

In the following case $\mathbf{M}^C$ is a Grothendieck category.

2.5. C as a flat A-module. *The following are equivalent:*

(a) *C is flat as a left A-module;*

(b) *every monomorphism in $\mathbf{M}^C$ is injective;*

(c) *every monomorphism $U \to C$ in $\mathbf{M}^C$ is injective;*

(d) *the forgetful functor $(-)_A : \mathbf{M}^C \to \mathbf{M}_A$ respects monomorphisms.*

If these conditions hold then $\mathbf{M}^C$ is a Grothendieck category.

Proof. See [10, 3.4] or [2]. $\square$

We note that if the category $\mathbf{M}^C$ is Grothendieck then C need not be flat as left A-module (e.g., [3]).

3. GENERATORS IN MODULE CATEGORIES

In any (additive) category a generator P is characterised by the faithfulness of the functor $\mathrm{Hom}(P, -)$. In full module categories the following characterization (due to C. Faith) is well known (e.g., [6, 18.8]).

3.1. Generator in $_A\mathbf{M}$. *For an A-module M with $S = \mathrm{End}(_AM)$, the following are equivalent:*

(a) *M is a generator in $_A\mathbf{M}$;*

(b)   (i) *M_S is finitely generated and S-projective, and*

      (ii) *$A \simeq \mathrm{End}(M_S)$.*

The characterisation of generators in $\sigma[M]$ is more involved.

3.2. Generator in $\sigma[M]$. *For an A-module M with $S = \mathrm{End}(_AM)$, the following are equivalent:*

(a) *M is a generator in $\sigma[M]$;*

(b) *for every $N \in \sigma[M]$, the following evaluation map is an isomorphism:*

$$M \otimes_S \mathrm{Hom}(M, N) \to N, \quad m \otimes f \mapsto f(m);$$

(c)   (i) *M_S is flat,*

      (ii) *for every injective module $V \in \sigma[M]$, the canonical map*

$$M \otimes_S \mathrm{Hom}(M, V) \to V, \quad m \otimes f \mapsto f(m),$$

      is injective (bijective).

If (any of) these conditions are satisfied the canonical map $A \to \mathrm{End}(M_S)$ is dense.

Proof. Most of the implications are well-known (see [6, 15.7,15.9], [2]). Because of its relevance for what follows, we show

ON GALOIS CORINGS 313

$(c) \Rightarrow (a)$ For any $K \in \sigma[M]$, there exists an exact sequence $0 \to K \to Q_1 \to Q_2$, where Q_1, Q_2 are injectives in $\sigma[M]$. We construct an exact commutative diagram (tensoring over S)

$$
\begin{array}{ccccccc}
0 & \longrightarrow & M \otimes_A \mathrm{Hom}(M,K) & \longrightarrow & M \otimes_A \mathrm{Hom}(M,Q_1) & \longrightarrow & M \otimes_A \mathrm{Hom}(M,Q_2) \\
& & \downarrow{\scriptstyle \mu_K} & & \downarrow{\scriptstyle \simeq} & & \downarrow{\scriptstyle \simeq} \\
0 & \longrightarrow & K & \longrightarrow & Q_1 & \overset{g}{\longrightarrow} & Q_2 \ ,
\end{array}
$$

showing that μ_K is an isomorphism and so K is M-generated. $\qquad\square$

For a better understanding of condition (c)(ii) recall the following special case (e.g., [6, 25.5]).

3.3. **Hom-tensor relation.** Given $M, V \in {}_A\mathbf{M}$, $S = \mathrm{End}(M)$, and $L \in \mathbf{M}_S$, consider the map

$$
L \otimes_S \mathrm{Hom}(M,V) \to \mathrm{Hom}_A(\mathrm{Hom}_S(L,M),V), \quad l \otimes f \mapsto [g \mapsto (g(l))f].
$$

This is an isomorphism provided L_S is finitely presented and V is M-injective. If $A \to \mathrm{End}(M_S)$ is dense, setting $M = L$ yields the map

$$
M \otimes_S \mathrm{Hom}(M,V) \to \mathrm{Hom}_A(\mathrm{Hom}_S(M,M),V) \simeq V \ , \quad m \otimes f \mapsto mf \ .
$$

Since every M-injective module $V \in \sigma[M]$ is M-generated, this map is surjective for such modules. To make the map injective, it suffices, for example, to have M_S finitely presented or pure projective, and no flatness condition on M_S is needed. More generally (c)(ii) can be related to descending chain conditions on certain matrix subgroups of M. For details we refer to [7] and [11].

Projectivity of a generator M is also reflected by properties of M as a module over its endomorphism ring.

3.4. **Projective generator in $\sigma[M]$.** *For an A-module M with $S = \mathrm{End}({}_AM)$, the following are equivalent:*

(a) *M is a projective generator in $\sigma[M]$;*

(b)   (i) *M_S is faithfully flat,*

    (ii) *for every injective module $V \in \sigma[M]$, the canonical map*

$$
M \otimes_S \mathrm{Hom}(M,V) \to V \ , \quad m \otimes f \mapsto f(m) \ ,
$$

    is injective (bijective).

Proof. $(a) \Rightarrow (b)$ By the generator property M is a flat module over S (see 3.2). Projectivity of M in $\sigma[M]$ implies $\mathrm{Hom}(M,MI) = I$ (see [6, 18.4]), for every left ideal $I \subset S$, hence $MI \neq M$ if $I \neq S$. This shows that M is faithfully flat (e.g., [6, 12.17]).

$(b) \Rightarrow (a)$ In view of 3.2 it remains to show that M is projective in $\sigma[M]$. For this consider any epimorphism $L \overset{f}{\longrightarrow} N$ in $\sigma[M]$. We obtain the commutative diagram with exact rows

$$M \otimes_S \operatorname{Hom}(M, L) \longrightarrow M \otimes_S \operatorname{Hom}(M, N) \longrightarrow M \otimes_S \operatorname{Coke} \operatorname{Hom}(M, f) \longrightarrow 0$$

$$L \xrightarrow{\quad f \quad} N \longrightarrow 0 ,$$

where the vertical maps are the canonical isomorphisms (see 3.2). From this we conclude $M \otimes_S \operatorname{Coke} \operatorname{Hom}(M, f) = 0$ and faithfulness of M_S implies $\operatorname{Coke} \operatorname{Hom}(M, f) = 0$ which means that $\operatorname{Hom}(M, f)$ is surjective. $\qquad\square$

4. GALOIS CORINGS

Given an A-coring $\mathcal{C}$ we may ask when A is a $\mathcal{C}$-comodule.

4.1. Grouplike elements. A non-zero element g of an A-coring $\mathcal{C}$ is said to be a *grouplike* element if $\Delta(g) = g \otimes g$ and $\varepsilon(g) = 1_A$.
An A-coring $\mathcal{C}$ has a grouplike element g if and only if A is a right or left $\mathcal{C}$-comodule, by the coactions

$$\varrho^A : A \to \mathcal{C}, \ a \mapsto ga, \qquad {}^A\varrho : A \to \mathcal{C}, \ a \mapsto ag.$$

For a proof we refer to [1] or [2]. We write A_g or ${}_gA$, when we consider A with the right or left comodule structure induced by g.

Example. Let $B \to A$ be a ring extension, and let $\mathcal{C} = A \otimes_B A$ be the Sweedler A-coring. Then $g = 1_A \otimes 1_A$ is a grouplike element in $\mathcal{C}$.

4.2. Coinvariants. Given an A-coring $\mathcal{C}$ with a grouplike element g and $M \in \mathbf{M}^{\mathcal{C}}$, the *$g$-coinvariants* of M are defined as the R-module

$$M_g^{co\mathcal{C}} = \{m \in M \mid \varrho^M(m) = m \otimes g\} = \operatorname{Ke}(\varrho^M - (- \otimes g)),$$

and there is an isomorphism

$$\psi_M : \operatorname{Hom}^{\mathcal{C}}(A_g, M) \to M_g^{co\mathcal{C}}, \quad f \mapsto f(1_A).$$

The isomorphism is derived from the fact that any A-linear map with source A is uniquely determined by the image of 1_A.

4.3. Coinvariants of A and $\mathcal{C}$. *Let $\mathcal{C}$ be an A-coring with a grouplike element g. Then:*
  (1) $\operatorname{End}^{\mathcal{C}}(A_g) \simeq A_g^{co\mathcal{C}} = \{a \in A_g \mid ga = ag\}$,
      i.e., subalgebra of A given by the centraliser of g in A.
  (2) *For any $X \in \mathbf{M}_A$, $(X \otimes_A \mathcal{C})^{co\mathcal{C}} \simeq \operatorname{Hom}^{\mathcal{C}}(A_g, X \otimes_A \mathcal{C}) \simeq X$, and for $X = A$,*

$$\mathcal{C}^{co\mathcal{C}} \simeq \operatorname{Hom}^{\mathcal{C}}(A_g, \mathcal{C}) \simeq \operatorname{Hom}_A(A_g, A) \simeq A,$$

  which is a left A- and right $\operatorname{End}^{\mathcal{C}}(A_g)$-morphism.

4.4. The induction functor. *For an A-coring $\mathcal{C}$ with grouplike element g, let $B = A_g^{co\mathcal{C}}$. Given any right B-module M, $M \otimes_B A$ is a right $\mathcal{C}$-comodule via the coaction*

$$\varrho^{M \otimes_B A} : M \otimes_B A \to M \otimes_B A \otimes_A \mathcal{C} \cong M \otimes_B \mathcal{C}, \quad m \otimes a \mapsto m \otimes ga.$$

For any morphism $f: M \to N$ in $\mathbf{M}_B$,
$$f \otimes I_A : M \otimes_A \mathcal{C} \to N \otimes_A \mathcal{C}, \quad m \otimes a \mapsto f(m) \otimes a,$$
is a morphism in $\mathbf{M}^{\mathcal{C}}$ and hence the assignment $M \to M \otimes_B A$ and $f \to f \otimes_B I_A$ defines a functor $- \otimes_B A : \mathbf{M}_B \to \mathbf{M}^{\mathcal{C}}$ known as an *induction functor*.

The g-coinvariants provide a functor in the opposite direction.

4.5. The g-coinvariants functor. *Let $\mathcal{C}$ be an A-coring with a grouplike element g and $B = A^{co\mathcal{C}}$. The functor*
$$\mathrm{Hom}^{\mathcal{C}}(A_g, -) : \mathbf{M}^{\mathcal{C}} \to \mathbf{M}_B,$$
is the right adjoint of the induction functor $- \otimes_B A : \mathbf{M}_B \to \mathbf{M}^{\mathcal{C}}$. *Notice that for $M \in \mathbf{M}^{\mathcal{C}}$, the right B-module structure of* $\mathrm{Hom}^{\mathcal{C}}(A_g, M)$ *is given by* $f \cdot b(a) = f(ba)$.
This functor is isomorphic to the coinvariant functor
$$G_g := (-)_g^{co\mathcal{C}} : \mathbf{M}^{\mathcal{C}} \to \mathbf{M}_B, \quad M \mapsto M^{co\mathcal{C}},$$
which acts on morphisms by restriction of the domain, i.e.,
$$\text{for any } f : M \to N \text{ in } \mathbf{M}^{\mathcal{C}}, \; G_g(f) = f|_{M_g^{co\mathcal{C}}}.$$
For $N \in \mathbf{M}_B$ the unit of the adjunction is given by
$$\eta_N : N \to (N \otimes_B A)^{co\mathcal{C}}, \quad n \mapsto n \otimes 1_A,$$
and for $M \in \mathbf{M}^{\mathcal{C}}$, the counit reads
$$\sigma_M : M^{co\mathcal{C}} \otimes_B A \to M, \quad m \otimes a \mapsto ma.$$
Notice that for any right B-module N, there is a left A-module isomorphism
$$\mathrm{Hom}^{\mathcal{C}}(N \otimes_B A, \mathcal{C}) \cong \mathrm{Hom}_A(N \otimes_B A, A) \cong \mathrm{Hom}_B(N, A).$$

The structure of an A-coring $\mathcal{C}$ with a grouplike element g involves two rings, the algebra A itself and its g-coinvariants algebra B, and a ring map $B \to A$. On the other hand, to any ring extension $B \to A$ one can associate its canonical Sweedler A-coring $A \otimes_B A$ which also has a grouplike element. Thus we have two corings with grouplike elements: the original A-coring $\mathcal{C}$ we started with and the canonical coring associated to the related algebra extension $B \to A$. It is natural to study the relationship between these corings, and, in particular, to analyse corings for which this relationship is given by an isomorphism. This leads to the notion of a *Galois coring* introduced in [1].

Recall that $M \in \mathbf{M}^{\mathcal{C}}$ is said to be $(\mathcal{C}, A)$-*injective* if for every $\mathcal{C}$-comodule map $i : N \to L$ which is a coretraction in $\mathbf{M}_A$, every diagram

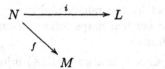

in $\mathbf{M}^{\mathcal{C}}$ can be completed commutatively by some $g : L \to M$ in $\mathbf{M}^{\mathcal{C}}$. This is equivalent to $\varrho^M : M \to M \otimes_A \mathcal{C}$ being a coretraction in $\mathbf{M}^{\mathcal{C}}$.

4.6. Galois corings. *For an A-coring $\mathcal{C}$ with a grouplike element g and $B = A_g^{co\mathcal{C}}$, the following are equivalent:*

(a) *The following evaluation map is an isomorphism:*

$$\varphi_C : \mathrm{Hom}^C(A_g, C) \otimes_B A \to C, \quad f \otimes a \mapsto f(a);$$

(b) *the (A, A)-bimodule map defined by*

$$\chi : A \otimes_B A \to C, \quad 1_A \otimes 1_A \mapsto g,$$

is a (coring) isomorphism;

(c) *for every (C, A)-injective comodule $N \in \mathbf{M}^C$, the evaluation*

$$\varphi_N : \mathrm{Hom}^C(A_g, N) \otimes_B A \to N, \quad f \otimes a \mapsto f(a),$$

is an isomorphism.

(C, g) is called a *Galois coring* if it satisfies the above conditions.

Proof. $(a) \Leftrightarrow (b)$ Observe that the canonical isomorphism

$$h : \mathrm{Hom}^C(A_g, C) \to A, \quad f \mapsto \varepsilon \circ f(1_A),$$

is right B-linear, and we get the commutative diagram

$$
\begin{array}{ccc}
\mathrm{Hom}^C(A_g, C) \otimes_B A & \xrightarrow{\varphi} & C, \\
{\scriptstyle h \otimes I} \downarrow & & \downarrow {\scriptstyle =} \\
A \otimes_B A & \xrightarrow{\chi} & C,
\end{array}
\qquad
\begin{array}{ccc}
f \otimes a & \longmapsto & f(a) \\
\uparrow & & \uparrow {\scriptstyle =} \\
\varepsilon \circ f(1_A) \otimes a & \longmapsto & \varepsilon \circ f(1_A)ga,
\end{array}
$$

where the last equality is obtained by colinearity of f, which implies

$$\varepsilon \circ f(1_A)ga = (\varepsilon \otimes I)(f(1_A) \otimes ga) = (\varepsilon \otimes I) \circ \Delta \, f(a) = f(a).$$

$(b) \Rightarrow (c)$ First observe that for any $X \in \mathbf{M}_A$, χ yields the isomorphisms

$$\mathrm{Hom}^C(A_g, X \otimes_A C) \otimes_B A \simeq X \otimes_B A \simeq X \otimes_A (A \otimes_B A) \simeq X \otimes_A C.$$

Now assume $N \in \mathbf{M}^C$ to be (C, A)-injective and consider the commutative diagram

$$
\begin{array}{ccccccc}
0 & \longrightarrow & \mathrm{Hom}^C(A_g, N) \otimes_B A & \longrightarrow & N \otimes_B A & \longrightarrow & (N \otimes_A C) \otimes_B A \\
& & \downarrow {\scriptstyle \varphi_N} & & \downarrow {\scriptstyle \simeq} & & \downarrow {\scriptstyle \simeq} \\
0 & \longrightarrow & N & \longrightarrow & N \otimes_A C & \longrightarrow & N \otimes_A C \otimes_A C,
\end{array}
$$

where the top row is exact by the purity (splitting) property shown in 4.7 below, and bijectivity of the two vertical maps follows from the preceding remark. From this, bijectivity of φ_N follows.

$(c) \Rightarrow (a)$ This is obvious since C is always (C, A)-injective. $\square$

Let us mention that *weak Galois corings* are considered in [9, 2.4]. For such corings the action of A on C is not required to be unital.

The purity condition needed above arises from the following splitting property (for $L = A$).

ON GALOIS CORINGS

4.7. Splitting induced by (C, A)-injectivity. *For an A-coring C, let $M \in \mathbf{M}^C$ be (C, A)-injective. Then for any $L \in \mathbf{M}^C$, the canonical sequence*

$$0 \longrightarrow \operatorname{Hom}^C(L, M) \xrightarrow{\ i\ } \operatorname{Hom}_A(L, M) \xrightarrow{\ \gamma\ } \operatorname{Hom}_A(L, M \otimes_A C),$$

splits in $\mathbf{M}_B$, where $B = \operatorname{End}^C(L)$ and $\gamma(f) = \varrho^M \circ f - (f \otimes I_C) \circ \varrho^L$. A similar result holds for relative injective left comodules.

Proof. Denote by $h : M \otimes_A C \to M$ the splitting map of ϱ^M in $\mathbf{M}^C$. Then the map

$$\operatorname{Hom}_A(L, M) \simeq \operatorname{Hom}^C(L, M \otimes_A C) \to \operatorname{Hom}^C(L, M), \quad f \mapsto h \circ (f \otimes I_C) \circ \varrho^L,$$

splits the first inclusion in $\mathbf{M}_B$, and the map

$$\operatorname{Hom}_A(L, M \otimes_A C) \to \operatorname{Hom}_A(L, M), \quad g \mapsto h \circ g,$$

yields a splitting map $\operatorname{Hom}_A(L, M \otimes_A C) \to \operatorname{Hom}_A(L, M)/\operatorname{Hom}^C(L, M)$, since for any $f \in \operatorname{Hom}_A(L, M)$,

$$h \circ \gamma(f) = f - h \circ (f \otimes I_C) \circ \varrho^L \in f + \operatorname{Hom}^C(L, M). \qquad \square$$

The next theorem shows which additional condition on A is sufficient to make A_g a comodule generator for a Galois A-coring (C, g). The second part is essentially [1, Theorem 5.6].

4.8. The Galois Coring Structure Theorem. *Let C be an A-coring with grouplike element g and $B = A_g^{coC}$.*

(1) *The following are equivalent:*
    (a) *(C, g) is a Galois coring and $_B A$ is flat;*
    (b) *$_A C$ is flat and A_g is a generator in $\mathbf{M}^C$.*

(2) *The following are equivalent:*
    (a) *(C, g) is a Galois coring and $_B A$ is faithfully flat;*
    (b) *$_A C$ is flat and A_g is a projective generator in $\mathbf{M}^C$;*
    (c) *$_A C$ is flat and $\operatorname{Hom}^C(A_g, -) : \mathbf{M}^C \to \mathbf{M}_B$ is an equivalence with inverse $- \otimes_B A : \mathbf{M}_B \to \mathbf{M}^C$ (cf. 4.5).*

Proof. (1) $(a) \Rightarrow (b)$ Assume (C, g) to be a Galois coring. Then in the diagram of the proof of 4.6, $(c) \Rightarrow (b)$, the top row is exact by flatness of $_B A$ without any condition on $N \in \mathbf{M}^C$. So $\operatorname{Hom}^C(A_g, N) \otimes_B A \to N$ is surjective (bijective) showing that A_g is a generator. Moreover the isomorphism $- \otimes_A C \simeq - \otimes_A (A \otimes_B A)$ implies that $_A C$ is flat.

$(b) \Rightarrow (a)$ If $_A C$ is flat then monomorphisms in $\mathbf{M}^C$ are injective. As for module categories one can show that the generator A_g in the category $\mathbf{M}^C$ is flat over its endomorphism ring B, and $\operatorname{Hom}^C(A_g, M) \otimes_B A \simeq M$, for all $M \in \mathbf{M}^C$.

(2) The proof for 3.4 also works for comodules. $\qquad \square$

If the ring A is right self-injective, then C is injective in $\mathbf{M}^C$ and the reformulation of the characterization of Galois corings and the Structure Theorem is just the description of generators in module categories (compare 3.2, 3.4).

318   R. WISBAUER

4.9. Corollary. *Assume A to be a right self-injective ring and let C be an A-coring with grouplike element g.*

  (1) *The following are equivalent:*

     (a) *(C, g) is a Galois coring;*

     (b) *for every injective comodule $N \in \mathbf{M}^C$, the evaluation*

$$\varphi_N : \mathrm{Hom}^C(A_g, N) \otimes_B A \to N, \quad f \otimes a \mapsto f(a),$$

       is an isomorphism.

  (2) *The following are equivalent:*

     (a) *(C, g) is a Galois coring and $_BA$ is (faithfully) flat;*

     (b) *$_BA$ is (faithfully) flat and for every injective comodule $N \in \mathbf{M}^C$, the following evaluation map is an isomorphism:*

$$\varphi_N : \mathrm{Hom}^C(A_g, N) \otimes_B A \to N, \quad f \otimes a \mapsto f(a).$$

We call a right C-comodule N *semisimple* (in $\mathbf{M}^C$) if every C-monomorphism $U \to N$ is a coretraction, and N is called *simple* if all these monomorphisms are isomorphisms. Semisimplicity of N is equivalent to the fact that every right C-comodule is N-injective. Simple and semisimple left C-comodules and (C, C)-bicomodules are defined similarly.

The coring C is said to be *left (right) semisimple* if it is semisimple as a left (right) comodule. C is called a *simple coalgebra* if it is simple as a (C, C)-bicomodule.

From [2, 3] we recall:

4.10. Semisimple corings. *For an A-coring C the following are equivalent:*

  (a) *C is right semisimple;*

  (b) *$_AC$ is projective and C is a semisimple left $^*C$-module;*

  (c) *C_A is projective and C is a semisimple right C^*-module;*

  (d) *C is left semisimple.*

Note that not every canonical coring associated to an algebra extension $B \to A$ is a Galois coring with respect to a grouplike $1_A \otimes 1_A$. However, if the extension $B \to A$ is faithfully flat than $(A \otimes_B A, 1_A \otimes_B 1_A)$ is a Galois-coring. As a particular example of this one can consider a Galois coring provided by Sweedler's Fundamental Lemma (cf. [5, 2.2 Fundamental Lemma]).

4.11. Fundamental Lemma. *Let A be a division ring. Suppose that C is an A-coring generated by a grouplike element g as an (A, A)-bimodule. Then (C, g) is a Galois coring.*

Proof. Under the given condition A is simple as left C-comodule and it subgenerates C and hence $\mathbf{M}^C$. This implies that C is a simple and right semisimple coring and A is a projective generator in $\mathbf{M}$. So (C, g) is a Galois coring by 4.8. $\square$

More general simple corings with grouplike elements can be characterised (compare also [3]) in the following way.

4.12. Simple corings. *Let C be an A-coring with grouplike element g. Then the following are equivalent:*

ON GALOIS CORINGS

(a) C is a simple and left (or right) semisimple coring;

(b) (C, g) is Galois and $\text{End}^C(A_g)$ is simple and left semisimple;

(c) $\chi : A \otimes_B A \to C$ is an isomorphism and B is a simple left semisimple subring of A;

(d) C_A is flat, $_gA$ is a projective generator in $^C\mathbf{M}$, and $\text{End}^C(_gA)$ is simple and left semisimple.

Proof. Let C be simple and left semisimple. Then there exists only one simple comodule (up to isomorphism) and so every non-zero comodule is a projective generator in $\mathbf{M}^C$. In particular A_g is a finite direct sum of isomorphic simple comodules and hence $\text{End}^C(A_g)$ is simple and left (and right) semisimple. So the assertions follow by 4.8 and 4.6. $\qquad\qquad\qquad\qquad\qquad\qquad\qquad\qquad\square$

As a special case we will consider Hopf algebras. For this we recall the conditions for bialgebras.

4.13. Bialgebras. Let R be a commutative ring. An R-module B which is an algebra and a coalgebra is called a *bialgebra* if $B \otimes_R B$ is a B-coring with bimodule structure

$$a'(a \otimes b)b' = \sum a'a b_{\underline{1}}' \otimes bb_{\underline{2}}', \text{ for } a, a', b, b' \in B,$$

comultiplication

$$\underline{\Delta} : B \otimes_R B \to (B \otimes_R B) \otimes_B (B \otimes_R B) \simeq B \otimes_R B \otimes_R B, \quad a \otimes b \mapsto \sum a \otimes b_{\underline{1}} \otimes b_{\underline{2}},$$

and counit $\underline{\varepsilon} : B \otimes_R B \to B, \ a \otimes b \mapsto a\varepsilon(b)$.

Clearly $1_B \otimes 1_B$ is a grouplike element and the ring of $B \otimes_R B$-covariants of B is isomorphic to R.

$B \otimes_R B$ is a subgenerator in the category $\mathbf{M}^{B \otimes_R B}$ which can be identified with the category of $\mathbf{M}_B^B$ right Hopf modules, the subcategory of $\mathbf{M}^B$ consisting of those comodules M whose structure maps are right B-module morphism, i.e.,

$$\varrho^M(mb) = \varrho^M(m)\Delta(b), \text{ for } m \in M, b \in B.$$

By 4.6 and [9, 5.10] we obtain:

4.14. Hopf algebras. *For a bialgebra B the following are equivalent:*

(a) $B \otimes_R B$ *is a Galois B-coring;*

(b) *the following canonical map is an isomorphism:*

$$\gamma_B : B \otimes_R B \to B \otimes_R B, \quad a \otimes b \mapsto (a \otimes 1)\Delta(b);$$

(c) B *is a Hopf algebra (has an antipode);*

(d) $\text{Hom}_B^B(B, -) : \mathbf{M}_B^B \to \mathbf{M}_R$ *is an equivalence (with inverse $- \otimes_R B$).*

If (any of) these conditions hold, B is a projective generator in $\mathbf{M}_B^B$.

Notice that the coinvariants $B^{B \otimes_R B} = R$ and we get the generator property of B without requiring any flatness condition for B_R. Characterization (d) is essentially the Fundamental Theorem for Hopf algebras (e.g., [2]). Of course there are examples of Hopf algebras which are not flat over the base ring (e.g., [4, Beispiel 1.2.7]).

320 R. WISBAUER

REFERENCES

[1] T. Brzeziński, The structure of corings. Induction functors, Maschke-type theorem, and Frobenius and Galois-type properties, *Algebras and Representation Theory* **5** (2002), 389–410.

[2] T. Brzeziński and R. Wisbauer, "Corings and Comodules", *London Math. Soc. Lect. Note Ser.* **309**, Cambridge University Press, Cambridge, 2003.

[3] L. El Kaoutit, J. Gómez Torrecillas, and F.J. Lobillo, Semisimple corings, preprint 2001.

[4] Ch. Lomp, Primeigenschaften von Algebren in Modulkategorien über Hopf Algebren, Dissertation, Universität Düsseldorf (2002).

[5] M. Sweedler, The predual theorem to the Jacobson-Bourbaki Theorem, *Trans. Amer. Math. Soc.* **213** (1975), 391–406.

[6] R. Wisbauer, "Foundations of Module and Ring Theory", Gordon and Breach, Reading, Paris, 1991.

[7] R. Wisbauer, Static modules and equivalences, in "Interactions Between Ring Theory and Representations of Algebras", F. Van Oystaeyen and M. Saorin (eds), *Lect. Notes Pure Appl. Math.* **210**. Dekker, New York (2000), 423–449.

[8] R. Wisbauer, Decompositions of modules and comodules, *Contemp. Math.* **259** (2000), 547–561.

[9] R. Wisbauer, Weak corings, *J. Algebra* **245** (2001), 123–160.

[10] R. Wisbauer, On the category of comodules over corings, in "Mathematics and Mathematics Education", Proc. 3rd Palestinian conf., Elyadi et al. (eds), World Sci. Publishing, River Edge, NJ, 2002, 325–336.

[11] W. Zimmermann, Modules with chain conditions for finite matrix subgroups, *J. Algebra* **190** (1997), 68–87.